Lecture Notes in Civil Engineering

Volume 48

Lecture Notes in Civil Engineering (LNCE) publishes the latest developments in Civil Engineering - quickly, informally and in top quality. Though original research reported in proceedings and post-proceedings represents the core of LNCE, edited volumes of exceptionally high quality and interest may also be considered for publication. Volumes published in LNCE embrace all aspects and subfields of, as well as new challenges in, Civil Engineering. Topics in the series include:

- Construction and Structural Mechanics
- Building Materials
- Concrete, Steel and Timber Structures
- Geotechnical Engineering
- Earthquake Engineering
- Coastal Engineering
- Hydraulics, Hydrology and Water Resources Engineering
- Environmental Engineering and Sustainability
- Structural Health and Monitoring
- Surveying and Geographical Information Systems
- Heating, Ventilation and Air Conditioning (HVAC)
- Transportation and Traffic
- Risk Analysis
- Safety and Security

To submit a proposal or request further information, please contact the appropriate Springer Editor:

- Mr. Pierpaolo Riva at pierpaolo.riva@springer.com (Europe and Americas);
- Ms. Swati Meherishi at swati.meherishi@springer.com (India);
- Ms. Li Shen at li.shen@springer.com (China);
- Dr. Loyola D'Silva at loyola.dsilva@springer.com (S-E Asia and Australia/NZ).

Indexed by Scopus

More information about this series at http://www.springer.com/series/15087

Marco Pasetto · Manfred N. Partl ·
Gabriele Tebaldi
Editors

Proceedings of the 5th International Symposium on Asphalt Pavements & Environment (APE)

Springer

Editors
Marco Pasetto
DICEA
University of Padua
Padua, Italy

Manfred N. Partl
EMPA
Dübendorf, Switzerland

Gabriele Tebaldi
University of Parma
Parma, Italy

ISSN 2366-2557 ISSN 2366-2565 (electronic)
Lecture Notes in Civil Engineering
ISBN 978-3-030-29778-7 ISBN 978-3-030-29779-4 (eBook)
https://doi.org/10.1007/978-3-030-29779-4

This Springer imprint is published by the registered company Springer Nature Switzerland AG
The registered company address is: Gewerbestrasse 11, 6330 Cham, Switzerland

Preface

International Society for Asphalt Pavements (ISAP) is a volunteer organization of practitioners and experts, academics and industry, set up to share the latest in leading edge asphalt pavement technology worldwide. The Society's distinctly global and inclusive approach is reflected by its international membership, representing all stakeholders in the asphalt industry—users, producers, professionals as well as individuals and organizations. ISAP is one of the most prestigious international associations established to promote asphalt pavement technologies.

The Society developed from a series of widely recognized international technical conferences, starting at the University of Michigan, Ann Arbor, in 1962; since then, the conferences, periodically organized by ISAP, cover a wide spectrum of the current practical topics and research activities on asphalt pavements.

ISAP focuses on improving and presenting engineering concepts in pavement design, construction, maintenance under the aspect of key issues, such as environmental protection, socio-economic impacts and political concerns. In order to support some of these interests, in 2008 ISAP took the initiative to organize a Symposium on *Asphalt Pavements and Environment—APE* (Zurich, Switzerland). This initiative has been repeated, due to the success of the conference. So, the *APE Symposium* was organized again in Fortaleza, Brazil (2012); Sun City, South Africa (2015); Tokyo, Japan (2017).

The fifth symposium is held for the first time in Italy, in Padova, from 11 to 13 September 2019, hosted by ISAP and *Università degli Studi di Padova* (University of Padova). The 2019 Symposium provides an excellent chance to share and disseminate research concerning materials and technologies for asphalt pavements, conceived with the purposes of sustainability and environmental compatibility. The main topics addressed are:

- Sustainable pavement materials
- Marginal materials for asphalt pavements
- Pavement structures
- Test methods and performance
- Hot, warm or cold recycling

- Maintenance and management methods
- Environmentally friendly technologies
- Urban heat island mitigation
- Energy harvesting
- Lifecycle assessment

Some of these topics have become increasingly essential in recent years (e.g. energy harvesting, UHI mitigation, LCA, CO_2 reduction); ISAP has a leading role in enhancing the diffusion of research activities and application techniques suitable for their awareness.

The 5th International ISAP-APE Conference has an important function for this purpose. Forty-eight papers with more than 150 authors, from 20 countries on four continents, have been accepted for publication, after rigorous peer review. They are assigned to the following chapters:

1. Sustainable pavements and environmentally friendly technologies
2. Future trends in asphalt pavements
3. Marginal materials for asphalt pavements
4. Hot, warm and cold recycling
5. Test methods and performance
6. Pavement structures, maintenance and management.

This book aims at presenting the most original and innovative ideas concerning the environment-related achievements in asphalt pavements study and development, in the hope of giving industry, academia and practitioners valid stimuli for better growth of research and sustainable application of asphalt pavements all over the world.

The editors would like to thank all the authors and reviewers from the international scientific committee, for their valuable efforts in producing the highest quality papers for this conference. Many thanks to lecturers and chairmen for their contribution during the symposium. A sincere grateful acknowledgement of the unstinting cooperation of Emiliano Pasquini, Giovanni Giacomello and Andrea Baliello in the organization of the event. Finally, heartfelt thanks to all the supporters who have contributed to the success of the conference.

Marco Pasetto
Manfred N. Partl
Gabriele Tebaldi

Contents

Sustainable Pavements and Environmentally Friendly Technologies

Effectiveness of Rejuvenators for Asphalt Mixtures with High Reclaimed Asphalt Pavement Content in Cold Climates

Marco Pasetto(✉), Giovanni Giacomello, and Emiliano Pasquini

Department of Civil, Environmental and Architectural Engineering (ICEA), University of Padova, Via F. Marzolo 9, 35131 Padua, Italy
marco.pasetto@unipd.it

Abstract. Current needs require the use of low environmental impact technologies in construction works. At the same time, the natural deterioration of roads makes maintenance and rehabilitation operations inevitable, with the consequent production and storage (or disposal) of huge quantities of Reclaimed Asphalt Pavement (RAP). Extensive research has thus been conducted and is still ongoing to maximize RAP recycling, trying to minimize the issues related to production and in-service properties, in particular in cold climates. This paper evaluates the feasibility of using large amounts of RAP in asphalt mixtures subjected to cold temperatures. With this aim, mixes were analyzed with different RAP contents (50% and 70%) and prepared with or without the addition of rejuvenator (two types of rejuvenators were selected). A traditional control mixture was also produced for comparison purposes. Dry and wet indirect tensile strength as well as stiffness and resistance to repeated loading properties at low and mid-service temperatures were assessed. Overall, the results showed a similar efficacy of the two rejuvenators in guaranteeing satisfactory mixture performance at low temperatures.

Keywords: Reclaimed Asphalt Pavement · Hot-recycling · Rejuvenator · Cold climate · Asphalt mixtures · Environmentally-friendly mixtures

1 Introduction

Researchers have dealt with the use of waste materials and industrial by-products (such as steel slag, foundry sand, Reclaimed Asphalt Pavement (RAP), etc.) to replace virgin aggregate in asphalt mixtures (Reyes-Ortiz et al. 2012; Zaumanis and Mallick 2015; Fakhri and Ahmadi 2017; Pasetto and Baldo 2017; Rodríguez-Fernández et al. 2019). The reuse of marginal materials is due to economic (lower cost compared to the virgin material) and environmental reasons (reduction of the consumption of natural and not-renewable resources). The amount of RAP, i.e. the material obtained from the milling of old asphalt layers, is constantly increasing (9 million tons were produced in Italy in 2016) and its storage is becoming a problem. Compared to other aggregates, RAP is a material with a high internal heterogeneity that depends on several factors: the pavement milling method, viscosity and content of bitumen, particle size and distribution of aggregates (Izaks et al. 2015).

M. Pasetto et al. (Eds.): ISAP APE 2019, LNCE 48, pp. 3–13, 2020.
https://doi.org/10.1007/978-3-030-29779-4_1

RAP aggregates can be reused via different techniques: cold mix recycling and hot mix recycling (in-place or in-plant). 100% of the RAP can be reused through cold recycling or hot in-place mix recycling techniques, while a limited percentage can be used (generally no more than 40%, due to the mix-plant facility) with hot in-plant mix recycling method. Many authors studied bituminous mixes with high RAP quantities (between 25% and 100%) and some indicate that RAP replacement at proportions above 50% is feasible (Grilli et al. 2013; Stimilli et al. 2017; Hajj et al. 2013; Zaumanis et al. 2014; Celauro et al. 2010; Tapsoba et al. 2014). In Italy generally no more than 20% of the total RAP production is recycled. In Europe and the United States over 80% of RAP is reused in the construction of roads, but regulations are still strict allowing inclusion of RAP in proportions ranging between 5 and 50% for production of new hot mix asphalt (HMA) mixtures. The major performance limitations for a high RAP content in the mix derive from: the aged RAP binder properties, the degree of blending and diffusion (between the virgin and RAP binder) and RAP aggregate properties (especially the fine content).

The aged RAP bitumen is harder and has less aptitude to cover the aggregates, since it was subjected to an aging process that caused a progressive variation in the chemical-physical properties of the bitumen (Hajj et al. 2013). The inclusion of RAP aggregates in mixtures, only as “black aggregates”, particularly when a high RAP percentage is used, negatively affects the performance of the mixtures (indirect tensile strength and stiffness increase: Celauro et al. 2010; Reyes-Ortiz et al. 2012; Zaumanis and Mallick 2015; Baldo et al. 2016). Therefore, some researches try to restore the workability and mechanical properties of the aged bitumen through different types of additive: softer binder, softening additives, rejuvenators and warm additives (Hugener et al. 2014; Zaumanis et al. 2014; Mazzoni et al. 2018). The most used additives are rejuvenators (vegetable oils, synthetic oils, synthetic waxes, refined tallow and distilled tall oil, etc.) since they help to re-balance the composition of the aged binder (that lost its maltenes during construction and service period). Bitumen-rejuvenator compatibility generally depends on the asphaltene content in the RAP bitumen. There are several studies that evaluate the quantity and dosage procedure of the rejuvenator: a high dosage would lead to an excessive lowering of viscosity (permanent deformation accumulation), while a low quantity would cause less adhesion between the aggregates and RAP bitumen (poor reactivation of aged bitumen) (Shen et al. 2007; Simonen et al. 2013; Zaumanis and Mallick 2015; Hugener et al. 2014).

2 Research Objectives

Since the bitumen in RAP is already very hard and brittle at mid-service temperatures (20–25 °C), low in-service temperatures influence the cracking resistance of asphalt concretes with RAP even more negatively. The main aim of this research is thus to check the feasibility of mixtures with high RAP content, with particular emphasis on their performance for cold regions. The efficacy of using different types of rejuvenators to increase the mixture properties at low and mid-service temperatures is also investigated. This will provide a further contribution to the existing literature addressing such issues (Boz and Solaimanian 2018; Elkashef and Williams 2017; Moon et al. 2017; Stimilli et al. 2017).

3 Materials

3.1 Aggregates, Bitumen and Additives

Virgin limestone (L) as well as RAP aggregates milled from a wearing course were used in this research. Table 1 shows the required granulometric properties of the mixture as well as the particle size distribution of L and aggregates in RAP (i.e. the "white curve"), while Table 2 reports the main physical-mechanical properties of these aggregates. A 70/100 penetration grade bitumen was used as virgin binder to prepare the investigated mixtures. Table 3 shows the main characteristics of the aged RAP bitumen and virgin binder.

Table 1. Reference envelope and aggregates grading curves.

Sieve size [mm]	Reference envelope		Passing [%]			
	Min [%]	Max [%]	L 0/4	L 4/8	L 8/12	RAP
16	100.0	100.0	100.0	100.0	100.0	100.0
12	100.0	100.0	100.0	100.0	97.6	100.0
8	70.0	90.0	100.0	98.5	32.6	99.7
4	40.0	60.0	95.8	19.0	0.2	83.6
2	25.0	38.0	63.5	0.6	0.2	55.8
0.4	11.0	20.0	27.5	0.0	0.2	12.6
0.18	8.0	15.0	19.3	0.0	0.2	2.9
0.075	6.0	10.0	14.2	0.0	0.2	0.6

Table 2. Physical-mechanical properties of aggregates.

Properties	Unit	Standard	Limestone			RAP (before bitumen recovery)
Fractions			L 0/4	L 4/8	L 8/12	
Particle density	Mg/m^3	EN 1097-6	2.76	2.75	2.74	2.35
Shape index	%	EN 933-4	-	12.8	7.5	7.0
Flakiness index	%	EN 933-3	-	10.5	11.8	9.5
Los Angeles coeff.	%	EN 1097-2	-	-	16	-
Sand Equivalent	%	EN 933-8	78	-	-	74

Table 3. Virgin and RAP bitumen: amount and physical characteristics.

Source of material	RAP bitumen				Virgin bitumen (70/100)			
Sample	1	2	3	Mean	1	2	3	Mean
Binder content [%]	4.7	5.3	5.0	5.0	-	-	-	-
Bitumen characteristics								
Penetration [0.1/mm]	8	6	7	7	74	74	78	75
Softening point [°C]	72.7	73.1	86.4	77.4	43.5	43.8	43.9	43.7

The RAP had a very hard bitumen with an average content of 5%. The use of a rejuvenator is thus strongly recommended, in particular for large amounts of RAP in the mixtures. In this study two types of rejuvenating additives were tested: the first, indicated with the letter A, is a chemical rejuvenator and the second, indicated with the letter B, is a natural bio-rejuvenator of vegetable origin. Based on producer recommendations and past researches (Hajj et al. 2013), the dosage of rejuvenators with respect to the weight of the RAP was set to 0.40% and 0.75% for the additive A and B, respectively.

3.2 Mixtures

Three types of bituminous mixture were produced and tested: a reference mixture without RAP (R0), a mixture with 50% of RAP (R50) and a mixture with 70% of RAP (R70). Figure 1 and Table 4 show the particle size distribution and composition of the three types of investigated mixtures, respectively. A mineral limestone filler (particle size < 0.063 mm) was also added to obtain the desired gradation.

The three types of mixtures were then produced with or without the addition of the selected rejuvenators obtaining:

- one reference mixture R0 (without RAP);
- a R50 mixture (with 50% of RAP) without rejuvenators;
- a R70 mixture (with 70% of RAP) without rejuvenators;
- a R50A mixture (with 50% of RAP and rejuvenator A);
- a R70A mixture (with 70% of RAP and rejuvenator A);
- a R50B mixture (with 50% of RAP and rejuvenator B);
- a R70B mixture (with 70% of RAP and rejuvenator B).

Based on a preliminary mix design, an optimum bitumen content of 5% by weight of the aggregates was selected for the reference mixture R0 and the recycled mixtures without rejuvenators R50 and R70, taking into account the bitumen coming from the RAP (Table 5). However, the optimum binder content of the rejuvenated recycled mixtures slightly decreased, regardless of the type of rejuvenator (Table 5).

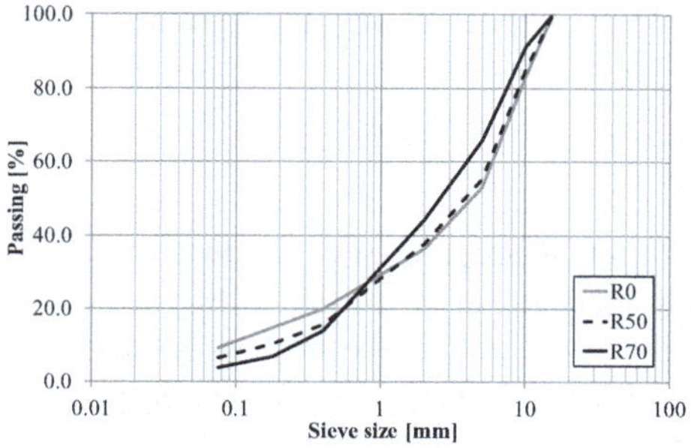

Fig. 1. Particle size distribution of R0, R50 and R70 mixtures

Table 4. Mixtures composition

Fractions	R0 [%]	R50 [%]	R70 [%]
L 0/4 mm	40	1	0
L 4/8 mm	27	19	13
L 8/12 mm	24	21	12
RAP	0	50	70
Filler	9	9	5
Total amount	100	100	100

Table 5. Bitumen content in the investigated mixtures

Mixtures	Bitumen from RAP [%]	Virgin bitumen in the mix [%]	Total bitumen in the mix [%]
R0, R50, R70	0	5.0	5.0
R50A, R50B	2.42	2.45	4.88
R70A, R70B	3.44	1.48	4.92

4 Methods

4.1 Samples Preparation

The laboratory mixing procedure in the case of the mixtures containing RAP tried to simulate what could occur in some plants (hot-recycling) where RAP is blended cold with the overheated virgin aggregates (in this case heated at 170 °C) before the mixing with filler, bitumen and rejuvenator (if applicable).

A shear gyratory compactor (EN 12697-31) was used to prepare cylindrical samples for indirect tensile strength (ITS) and indirect tensile stiffness modulus (ITSM) tests, whereas the prismatic samples for the cyclic four-point bending (4 PB) tests were obtained from slabs compacted by a roller compactor (EN 12697-33). 4% target air voids were selected in both cases. Two cylindrical specimens (150 mm diameter and 60–70 mm height) and five prismatic specimens (50 mm × 50 mm × 400 mm) were obtained from each sample compacted with the gyratory compactor and roller compactor, respectively.

4.2 Testing Methods

A comprehensive experimental program was conducted in the laboratory to evaluate the main volumetric and mechanical properties of the investigated mixtures.

The workability of the investigated mixture was first assessed in terms of air voids content and Compaction Energy Index (CEI) of the specimens (6 replicates) prepared with the gyratory compactor (Bahia and Paye 2004).

The six cylindrical specimens were then subjected to the Indirect Tensile Stiffness Modulus (ITSM) tests at 0 °C and 25 °C, according to EN 12697-26.

Lastly, indirect tensile strength (ITS) in both dry (3 replicates) and wet (3 replicates) conditions was evaluated on the samples at 25 °C, according to EN 1269-23. Dry ITS tests were performed after conditioning the specimens at 25 °C for 4 h whereas wet-conditioned samples were subjected to a freeze-thaw cycle, reproducing cold climate conditions. According to ASTM D 4867, before the final condition at 25 °C for the ITS test, the wet samples were partially saturated, wrapped and placed in a climatic chamber for 15 h at a constant temperature of −18 °C; then they were immersed in a thermostatic bath for 24 h at a constant temperature of 60 °C.

Instead, the prismatic specimens coming from the slabs were tested in 4 PB configuration. First, the LVE properties of the mixtures were evaluated according to EN 12697-26/Annex B at four temperatures (−20 °C, −10 °C, 0 °C and 25 °C) by applying a sinusoidal load at six frequencies (0.1, 0.3, 1, 3, 10, 30 Hz) in controlled strain mode (target strain level equal to 50 $\mu\varepsilon$).

Resistance to repeated loading was then assessed through 4 PB cyclic tests at 0 °C: three specimens for each tested material were subjected to a sinusoidal load, having a 2.5 MPa stress amplitude and 10 Hz test frequency.

5 Test Results and Discussion

5.1 Compactability and Indirect Tensile Stiffness

Figure 2a shows that the addition of large amounts of RAP seems to negatively affect the compactability of the tested mixture since higher average values of both voids and CEI were found. In this sense, it is worth remembering that the higher the CEI, the worse the compactability. However, the use of the selected rejuvenators likely led to some increase in workability since in both cases (additives A and B) the rejuvenated recycled materials achieved a better compaction than the control mixture; with better results observed in the case of the bio-based rejuvenator B.

Figure 2b depicts the average results obtained from ITSM tests: the findings revealed that the stiffness of the rejuvenated mixtures is similar to that of the reference R0, regardless of test temperature and type of rejuvenator, whereas a noticeable increase in stiffness was detected at both 0 °C and 25 °C for the recycled mixtures without rejuvenators (R50 and R70). These results seem to demonstrate that the use of a large amount of RAP markedly stiffened the mixture with possible negative consequences on the fracture resistance at mid and low temperatures (usually, the higher the stiffness, the higher the brittleness); instead, the use of a rejuvenator seemed to guarantee the "restoring" of the stiffness properties while allowing the use of a large amount of RAP. Similar results were found by other researchers (Grilli et al. 2013).

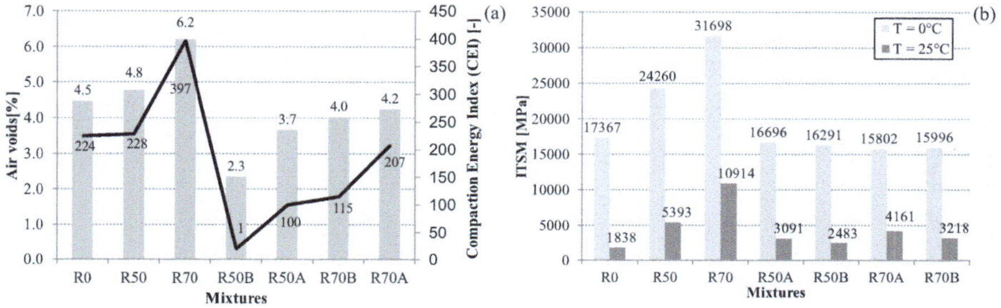

Fig. 2. Compactability (a) and ITSM @ 0 °C and @ 25 °C (b) of tested mixtures

5.2 Flexural Stiffness

Figure 3 shows the overall experimental results of 4 PB flexural tests in terms of average norms of the complex modulus as a function of test frequency. As expected, the higher the temperature and the lower the test frequency, the lower the stiffness.

Generally, according to ITSM results, RAP addition led to an increase in stiffness that was counterbalanced using the selected rejuvenators (with a likely increase in ductility). However, additive A seemed to be generally more effective in "restoring" the flexural stiffness of the investigated mixtures than the bio-based additive B. Such a difference with respect to what emerged from ITSM tests could be attributed to the different compaction procedure, which would have led to inhomogeneous volumetric properties among the different specimens.

5.3 Strength and Moisture Resistance

The strength and resistance to water damage of the investigated recycled materials were evaluated through the ITS tests performed in dry and wet conditions. Figure 4 reports the average ITS values obtained at 25 °C in the two abovementioned conditions: after freeze-thaw (F-T) cycles (ITS wet) and in dry condition (ITS dry). The "dry" values represent the tensile strength of the materials whereas the percentage ratio between "wet" and "dry" values provide information about the moisture resistance of the mixes.

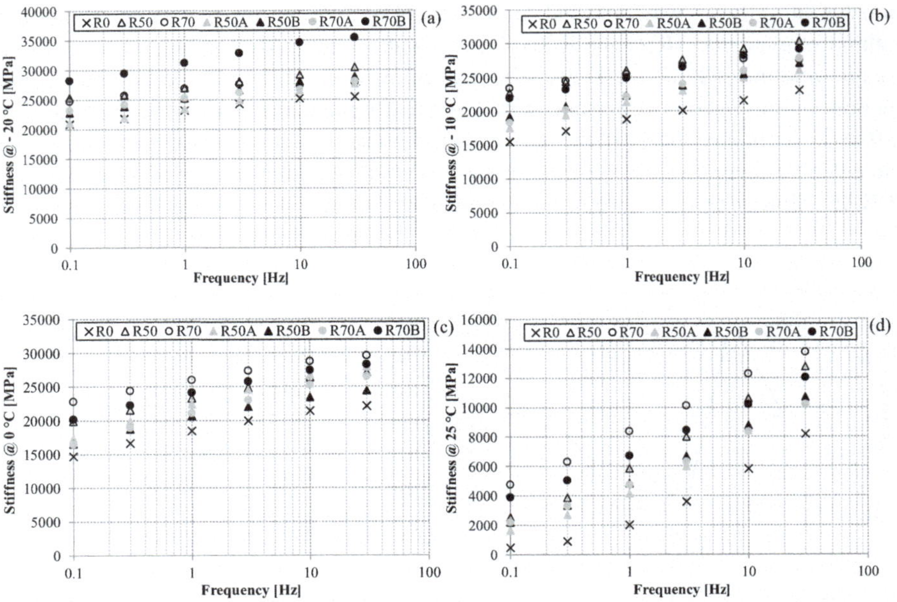

Fig. 3. Flexural stiffness of tested mixtures @ −20 °C (a), −10 °C (b), 0 °C (c) and 25 °C (d)

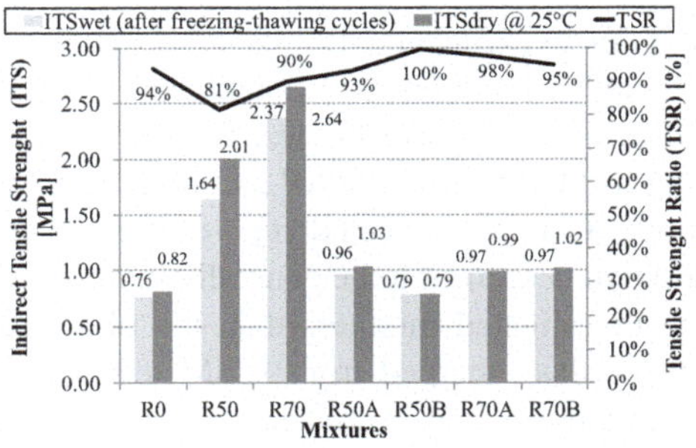

Fig. 4. ITS in dry and wet conditions and Tensile Strength Ratio (TSR) of the tested mixtures

In all cases, the mixtures without additives showed a higher value of ITS than the reference R0: the presence of RAP, as reported in the literature, generally increases the ITS since the aged binder from the RAP stiffens the binder phase of the mixture (Shen et al. 2007). In this sense, the addition of the rejuvenator decreased the ITS compared to the same mixtures without the additive: the rejuvenator softened the binder phase thus leading to a decrease in strength (similar to that of R0) towards a more ductile behavior of the material.

Concerning moisture resistance, it is worth noting that the use of a large amount of RAP led to some decrease of TSR, whereas the addition of rejuvenators guaranteed similar or even higher resistance to water damage with respect to the reference mixture R0.

5.4 Resistance to Repeated Loading

The resistance of the investigated mixtures to repeated loading at 0 °C was preliminarily evaluated by performing 4 PB cyclic stress-controlled tests at one stress amplitude (2.5 MPa) and frequency (10 Hz). Table 6 shows the initial stiffness (i.e. the value measured at the 100th test cycle) and number of cycles at sample failure (assumed as the cycle corresponding to a 50% reduction of the initial stiffness). The mixtures without additives (R50 and R70) had high initial stiffness (due to the non-rejuvenated RAP bitumen) and achieved a larger number of cycles before failure, compared to the other mixtures. Instead, the rejuvenated mixtures R50A, R50B, R70A and R70B showed an initial stiffness very similar to that of the reference mixture R0 but were generally characterized by a higher resistance to repeated loading: this seems to suggest a beneficial effect of the use of rejuvenators on cyclic loading.

Table 6. Resistance to repeated loading: initial stiffness and number of cycles at failure

Mixture	Stiffness @ 100 cycles [MPa]	Number of cycles @ failure [-]
R0	27385	627456
R50	28635	1939719
R70	33632	4610695
R50A	25019	982175
R50B	26437	1061589
R70A	28985	619534
R70B	26173	773164

6 Conclusions

This paper deals with the feasibility of introducing high RAP contents in asphalt mixtures to be used in cold climates. The effectiveness of different rejuvenators (a chemical synthetic additive and a natural bio-rejuvenator of vegetable origin) to help these mixtures in achieving satisfactory workability and mechanical properties was also investigated. With this aim, mixtures with 50% or 70% RAP aggregates were prepared with or without the addition of rejuvenator and were tested in terms of workability, strength, stiffness, moisture susceptibility and resistance to repeated loading. Based on the experimental finding, the following main conclusions can be drawn:

- the addition of a large amount of RAP without the use of rejuvenators can lead to serious issues mainly related to poor workability and excessive embrittlement that could negatively affect the cracking resistance of the mixtures;
- the use of the studied rejuvenators allowed the introduction of a large amount of RAP in asphalt mixtures while guaranteeing satisfactory compactability, mechanical properties and durability (comparable or even better than those of the reference mix), regardless of the type of the rejuvenator.

Overall, this preliminary study seems to demonstrate the feasibility of adopting asphalt mixtures with large amounts of RAP (up to 70% of the aggregates) in cold climates, provided that a proper rejuvenator is also used. This involves strategic environmental benefits such as: less material disposal, reduced need to store the RAP, lower amount of virgin material taken from quarries, less use of virgin bitumen, etc.

Further studies are planned in order to deepen the understanding of thermal and fatigue cracking resistance of the rejuvenated recycled mixtures in cold climates through specific performance-related tests.

References

Bahia HU, Paye BC (2004). Using the gyratory compactor to measure mechanical stability of asphalt mixtures. Wisconsin Highway Research Program 05–02, Department of Civil and Environmental Engineering, University of Wisconsin, Madison, WI, USA

Baldo N, Manthos E, Pasetto M, Nikolaides AF (2016) Comparative analysis of stiffness modulus and fatigue resistance of asphalt concretes containing RAP materials. In: RILEM Bookseries, vol 11, pp 915–926

Boz I, Solaimanian M (2018) Investigating the effect of rejuvenators on low-temperature properties of recycled asphalt using impact resonance test. Int J Pavement Eng 19(11):1007–1016

Celauro C, Celauro B, Boscaion G (2010) Production of innovative, recycled and high-performance asphalt for road pavements. Resour Conserv Recycl 54:337–347

Elkashef M, Williams RC (2017) Improving fatigue and low temperature performance of 100% RAP mixtures using a soybean-derived rejuvenator. Constr Build Mater 151:345–352

Fakhri M, Ahmadi A (2017) Evaluation of fracture resistance of asphalt mixes involving steel slag and RAP: susceptibility to aging level and freeze and thaw cycles. Constr Build Mater 157:748–756

Grilli A, Bocci M, Cardone F, Conti C, Giorgini E (2013) Laboratory and in-plant validation of hot mix recycling using a rejuvenator. Int J Pavement Res Technol 6(4):364–371

Hajj EY, Souliman MI, Alavi MZ, Loria-Salazar GL (2013) Influence of hydrogreen bioasphalt on viscoelastic properties of reclaimed asphalt mixtures. Transp Res Rec 2371:13–22

Hugener M, Partl MN, Morant M (2014) Cold asphalt recycling with 100% reclaimed asphalt pavement and vegetable oil-based rejuvenators. Road Mater Pavement Des 15(2):239–258

Izaks R, Haritonovs V, Klasa I, Zaumanis M (2015) Hot mix asphalt with high RAP content. Procedia Eng 114:676–684

Mazzoni G, Bocci E, Canestrari F (2018) Influence of rejuvenators on bitumen ageing in hot recycled asphalt mixtures. J Traffic Transp Eng (Engl Ed) 5(3):157–168

Moon KH, Falchetto AC, Marasteanu MO, Wistuba MP (2017) Low temperature rheological properties of asphalt mixtures containing different recycled asphalt materials. Int J Pavement Res Technol 10:84–97

Pasetto M, Baldo N (2017) Dissipated energy analysis of four-point bending test on asphalt concretes made with steel slag and RAP. Int J Pavement Res Technol 10:446–453

Reyes-Ortiz O, Berardinelli E, Alvarez AE, Carvajal-Muñoz JS, Fuentes LG (2012) Evaluation of hot mix asphalt mixtures with replacement of aggregates by reclaimed asphalt pavement (RAP) material. Procedia Soc Behav Sci 53:379–388

Rodríguez-Fernández I, Lastra-González P, Indacoechea-Vega I, Castro-Fresno D (2019) Recyclability potential of asphalt mixes containing reclaimed asphalt pavement and industrial by-products. Constr Build Mater 195:148–155

Shen J, Amirkhanian S, Miller JA (2007) Effects of rejuvenating agents on superpave mixtures containing reclaimed asphalt pavement. J Mater Civil Eng 19(5):376–384

Simonen M, Blomberg T, Pellinen T, Valtonen J (2013) Physicochemical properties of bitumens modified with bioflux. Road Mater Pavement Des 14(1):36–48

Stimilli A, Virgili A, Canestrari F, Bahia HU (2017) Estimation of low-temperature performance of recycled asphalt mixtures through relaxation modulus analysis. Cold Reg Sci Technol 133:36–45

Tapsoba N, Sauzéat C, Di Benedetto H, Baaj H, Ech M (2014) Behaviour of asphalt mixtures containing reclaimed asphalt pavement and asphalt shingle. Road Mater Pavement Des 15 (2):330–347

Zaumanis M, Mallick RB, Poulikakos L, Frank R (2014) Influence of six rejuvenators on the performance properties of Reclaimed Asphalt Pavement (RAP) binder and 100% recycled asphalt mixtures. Constr Build Mater 71:538–550

Zaumanis M, Mallick RB (2015) Review of very high-content reclaimed asphalt use in plant-produced pavements: state of the art. Int J Pavement Eng 16(1):39–55

Micromechanical Surface Investigation of Bio-modified RAP Binder

Maria Chiara Cavalli[1(✉)] and Lily D. Poulikakos[2]

[1] École des Ponts ParisTech, 77420 Champs-sur-Marne, France
maria-chiara.cavalli@enpc.fr
[2] Empa, 8600 Dübendorf, Switzerland
lily.poulikakos@empa.ch

Abstract. Pavements are constantly subjected to oxidation and aging during their performance life, causing an increase in stiffness and brittleness and their lower ability to withstand stresses. This is even more pernicious when using reclaimed asphalt pavement (RAP) which cannot be used as it is and needs to be modified with rejuvenators. Up to now, the rejuvenating effect has been evaluated in terms of the improvement of rheological properties whereas the rejuvenation mechanism at micro-scale remains unclear. In this work, RAP with three different bio-based rejuvenators (a natural seed oil, a cashew-nut shell-based oil and a tall based oil) have been analysed by using dynamic shear rheometer (DSR) in combination with atomic force microscopy (AFM). Rheological tests showed how rejuvenators may restore the mechanical properties of RAP binder. On top of, AFM has been used to measure the mechanical properties at the sample's surface such as the Young's modulus. Consequently, quantitative nano-mechanical surface maps have been created and the effect of rejuvenators in restoring the elastic moduli at the RAP's surface has been demonstrated. Furthermore, samples were analyzed after standard aging procedures (RTFOT and PAV) and rejuvenators' effect have been studied. The addition of rejuvenators decreased the elastic moduli of the RAP binder at the sample's surface and the bulk by decreasing the complex moduli. However, aging had a consequence on all binders' surface and this effect was rejuvenators' dependent.

Keywords: RAP binder · Bio-based rejuvenators · DSR · AFM · Morphology · Mechanical properties

1 Introduction

During the last decades, the increasing environmental awareness has brought reclaimed asphalt pavement (RAP) to be used as a component in asphalt pavement mix design. However, RAP binder is stiffer than the virgin binder and it is necessary to recover the properties of the RAP binder by adding a certain amount of specific products called "rejuvenators" (Cavalli et al. 2017). The ability of rejuvenators to counterpoise aged binder rheological properties to virgin-like condition, thus counterpoise the effects of RAP addition to mixture, is a fundamental aspect to be known from rheological measurements. Changes in viscosity has been widely studied also when bitumen have

M. Pasetto et al. (Eds.): ISAP APE 2019, LNCE 48, pp. 14–22, 2020.
https://doi.org/10.1007/978-3-030-29779-4_2

been modified. Understanding the effect of modification regarding rheological properties is fundamental when evaluation the efficiency of certain asphalt additives. The rheological properties of bitumen can normally be improved with the addition of modifiers such as polymers or specific products called rejuvenators. For example, the use of polymer modified bitumens helps improving its rheological properties over a wide range of temperatures and times of loading (Oliviero Rossi et al. 2015). Generally, by adding polymer into the neat bitumen, the stiffness modulus and elasticity values significantly increased (Airey et al. 2004). The effectiveness of rejuvenators is usually measured by comparing the rheological properties of the rejuvenator-aged bitumen blend with reference values of the virgin binder. For instance, the addition of the bio-based oil as recycling agent was observed to generally lower the complex moduli and improve fatigue performances of mixtures with a RAP content equal to or greater than 40% (Mangiafico et al. 2017). In another study, vegetable based oil and maltene based rejuvenators provided similar flow property improvement for highly aged binders (Huang et al. 2015). In (Al-Khateeb and Ramadan 2015), the influence of crumb rubber content on rheological properties of aged bitumen was studied; reporting that the rheological changes in bitumen binder led to improving the mechanical properties of rubberised bitumen binder. Recently, different types of organic oils have been tested as recycling agents to restore the viscosity and elasticity of aged asphalt. For example, recycled cooking oil is a good candidate for improving the low-temperature grade (Raman et al. 2015). However, by correlating rheological and microstructural techniques, the mechanism and performance of rejuvenation can be analysed from different perspectives. Atomic force microscopy (AFM) can give information on the surface morphology of bitumen. Topographic imaging can be conducted by allowing the tip to tap the surface periodically (so called tapping mode or non-contact). The result is that the probe touches the surface only for a short time, thus avoiding the issue of lateral forces and drag across the surface while the cantilever is driven to oscillate up and down at its resonance frequency. The motivation for the use of the tapping mode AFM is to overcome the difficulty of operating the conventional contact mode AFM when studying bituminous materials. The variation of surface topography alters the oscillation amplitude of the cantilever. The topographic image is then generated by the vertical movement of the scanner (García et al. 1999). Jäger et al. (2004) were, to the best of our knowledge, the first ones to investigate the relative stiffness of different phases in asphalt binder. Upon annealing the surface of bitumen films displays three domains: catanphase (the bee structure or undulations), surrounded by the periphase embedded in a matrix, the paraphrase. They identified how the catana phase e and its surrounding matrix showed different relative stiffness. (Allen et al. 2012) evaluated the rheology of the different fractions observed using AFM before and after aging for different asphalt binders. They demonstrated how the difference between the percentage of asphaltenes and saturates was related to the size and the dispersion of the catanaphase and the periphase. (Bhasin and Ganesan 2017) used phase field modelling to hypothesize that the various polar fractions are dispersed in specific domains. In (Pahlavan et al. 2018) the restorative effect of bio-rejuvenator molecules on reducing the asphaltenes interactions through oxidation has been demonstrated by using DFT models in combination with AFM. (Abd et al. 2017) used AFM to show how Sasobit, which is a synthetic wax used in asphalt road, improved the adhesion properties of the bitumen and had an effect on the elastic moduli at the surface of the binder. (Rashid et al. 2019) demonstrated how the elastic modulus at the surface can be correlated with variation in the morphology. They showed how the addition of RAP binder with the base binder increased the stiffness and that the micro-structures as well as the nanomechanistic properties of the blended binders were significantly different from those of the base binder. Aim of the present research is to link the rheological performances with the surface properties of bio-based binders by combining DSR and AFM. The goal is to show how the addition of rejuvenators affected both the surface's topography and the mechanical properties at the bulk of RAP binder before and after aging.

2 Materials and Methods

The bituminous binder from the RAP was extracted using toluene (ASTM 2172). The binder content was 4.60% by weight of the mixture (EN 12697-1). The extracted RAP binder had a penetration of 22×10^{-1} mm at 25 °C (EN 1426) and a softening point equal to 65.7 °C (EN 1427). 50/70 virgin binder used in this study showed a penetration of 62×10^{-1} mm at 25 °C and a softening point of 48.75 °C. Three commercial bio-based rejuvenators were used: rejuvenator "A" from natural seed oil, rejuvenator "B" from cashew nut shell oil and rejuvenator "C" from tall based oil. The rejuvenator dosage was set at 5% by mass of RAP binder following a previous research by the authors (Cavalli et al. 2018b).

2.1 Rejuvenation Procedure

100 grams of RAP binder were left in an oven for 20 min at a temperature equal to the softening point, plus 80 °C (EN 12697-1). Rejuvenators were added to the hot binder and the binder mixture was placed in the Speed Mixer™ and mixed for one minute at 3500 rpm. One minute was visually found suitable for homogenization. Afterwards, the rejuvenated binders were laboratory aged using Rolling Thin Film Oven Test (RTFOT) and subsequently Pressure Aging Vessel (PAV). (Cavalli et al. 2018b). The combined aging procedures are expected to simulate both early stage aging and in situ aging (Mousavi et al. 2016).

2.2 Dynamic Shear Rheometer

The reference temperature Tref in this study was 20 °C. The sigmoidal modal was selected to construct the mastercurves. In order to achieve the most desirable fit between the measured values of the complex modulus and the values described by the sigmoidal model, the shifting algorithm proposed by Gergesova et al. (2011) was used. According to the EN 14770, approximately 1.5 g of each material was placed in a disc shaped silicon mould with diameters of 8 mm and 25 mm and tested with the DSR Physica MCR at temperatures between −10 °C to +80 °C. The 8 mm plate-plate geometry with 2 mm gap was used in strain-controlled mode for the temperature range

−10 °C to +40 °C while the 25 mm plate-plate geometry with 1 mm gap was used for the temperature range +40 °C to +80 °C. Each measurement was repeated four times. Testing frequencies ranged from 0.1 to 20 Hz at each temperature.

2.3 Force Mapping

Following a standardized protocol for AFM measurements (Soenen et al. 2014), around 15 mg of binder were buttered and placed on a glass substrate. The glass slide was covered to prevent dust accumulation with a glass plate and subsequently placed in a ventilated oven at 110 °C for 20 min. Afterwards, the covered sample was left for 24 h prior to testing in a humidity-controlled chamber. Subsequently, AFM measurements were performed at room temperature. Asphalt binders' surface was analysed using an Icon3 AFM device from Bruker. The data were collected using Nanoscope 8.15. (Cavalli et al. 2018a)

The conical tip, consisted of a silicon probe, was RTESPA-150 from Bruker with a resonant frequency of 150 kHz and spring constant of 6 N/m, a cantilever length equal to 125 μm and width of 35 μm. The tip radius was equal to 8 nm. The scan size was 10 μm × 10 μm with a scan rate of 0.5 Hz. The spring constant was determined by the thermal tune method (Hutter and Bechhoefer 1993). To perform quantitative measurements, it was necessary to calibrate the cantilever in order to gather the point of contact and then measure the precise probe geometry as well as the cantilever spring constant. The tip radius was calibrated using Bruker PDMS-12 sample with nominal elastic modulus of 3.5 MPa. QNM mode measurements were performed as follows: the probe was oscillated in the z-direction at 2 kHz at the same time the sample was scanned line by line at a rate of 1 Hz. Every image was built up of 512 per 512 pixels each originating from force curve evaluation. To obtain the Young's moduli at the sample surface, the retraction curve was then fitted with Nanoscope 8.15 software. To obtain the elastic modulus E* at the sample surface, the retraction curve was then fitted by using the Derjaguin–Muller– Toporov (DMT) model (Derjaguin et al. 1975).

3 Results and Discussion

3.1 Rheological Analysis

Figure 1 left, shows the master curves of complex moduli $|G^*|$ as function of reduced frequency at the reference temperature of 20 °C. It follows that the differences between the unmodified RAP binder and the modified RAP binder are significant within the whole frequency spectrum. In particular, RAP + 5% B or RAP + 5% C had lower moduli than the virgin bitumen 50/70. Hence, it was found that at unaged state, the RAP binder modified with 5% rejuvenators B and C were mechanically softer than virgin bitumen 50/70 while RAP binder with rejuvenator A was harder in the lower frequency range. The trend of the phase angle (Fig. 1, right) confirmed what was shown in the rheological measurements: rejuvenator could soften the RAP binder for all frequencies (Cavalli et al. 2018b).

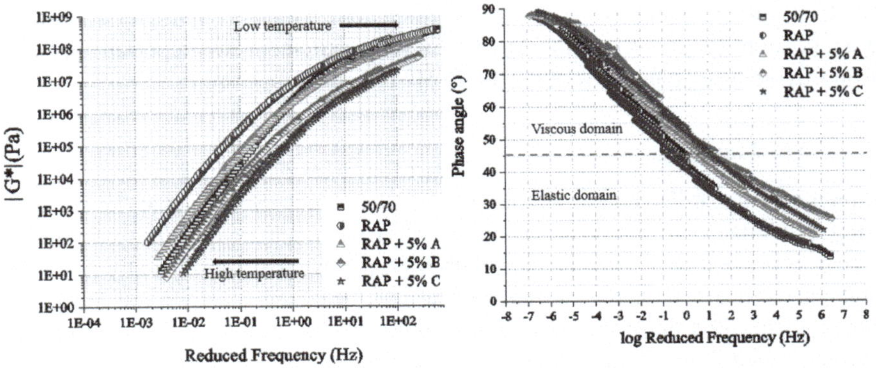

Fig. 1. Master curves of complex shear moduli (left) and phase angle measurements (right) for reference temperature of 20 °C from frequency sweeps (0.1–20 Hz). The presented values are an average of four measured values (Cavalli et al. 2018b).

After ageing, the overall complex moduli of each material tested increased in comparison to the unaged materials as shown in Fig. 2 and in comparison to Fig. 1. It was observed that over most of the frequency range, RAP + 5% A and RAP + 5% C had similar complex moduli after ageing. On the contrary, RAP + 5% B was closer to the complex moduli of the aged RAP binder indicating little improvement in mechanical stiffness after ageing. The aged virgin binder 50/70 had similar values to the unaged RAP binder showing how the standard ageing procedure could simulate closely the ageing caused during its service life. Figure 2 left, shows how, despite the addition of rejuvenators, after ageing there is in general an increase of both complex moduli and phase angle along the whole frequency range. Besides, the difference in the values of complex moduli at aged state is more prominent at low frequencies (higher temperatures, from +40 °C to +80 °C) than at high frequencies (low temperatures, from +30 °C to −10 °C) as the curves converge at high frequencies after ageing showing that rejuvenator's effect is more prominent at high temperatures (Cavalli et al. 2018b).

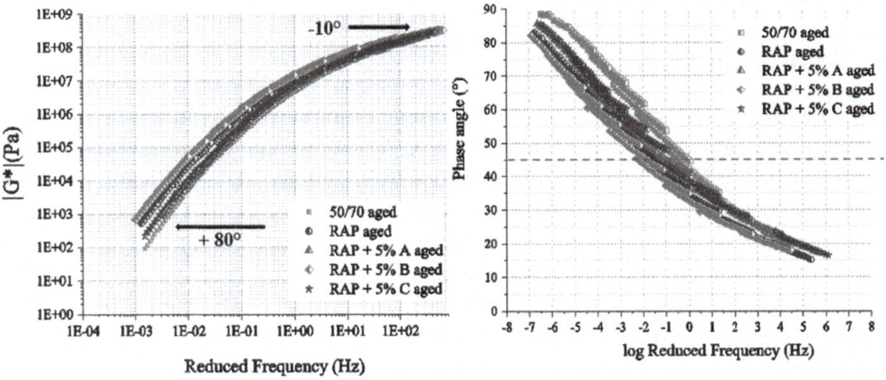

Fig. 2. Master curves of complex shear moduli results of the aged materials measured (left) and phase angle measurements (right) at reference temperature of 20 °C and at frequency sweeps (0.1–20 Hz). The presented values are an average of four measurements (Cavalli et al. 2018b).

3.2 Nano-Mechanical Mapping

Quantitative nano-mechanical (QNM) mapping uses tapping mode to gain information on nano-mechanical properties such as elastic modulus at sample's surface while simultaneously imaging the sample morphology. As can be observed in Fig. 3, the virgin binder images show lower Young's moduli compared to the RAP binder. As can be seen in Fig. 3 on the bottom part, after aging the complex moduli for both RAP binder and virgin binder increase (Cavalli et al. 2019).

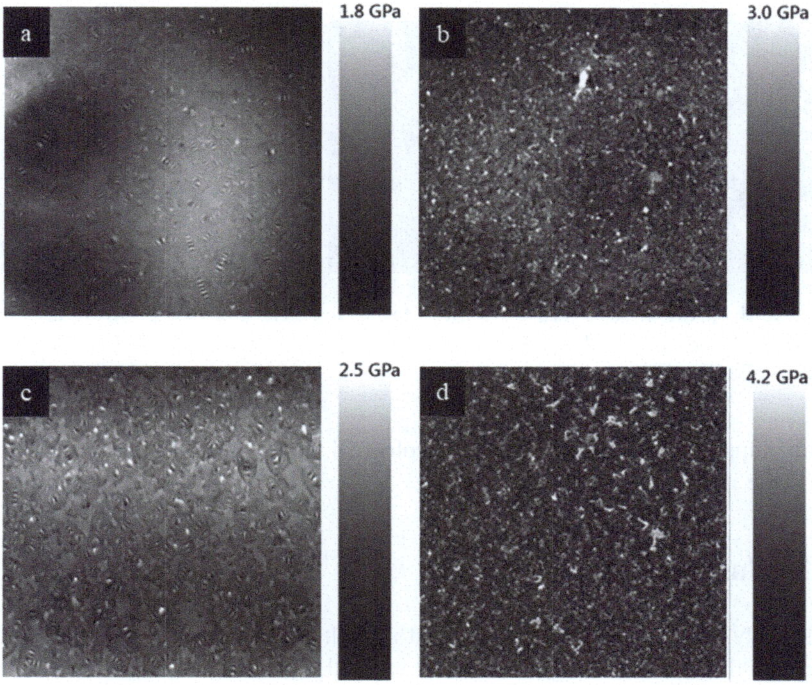

Fig. 3. AFM QNM images (10 × 10 μm). From top left clockwise: (a) Virgin binder 50/70 unaged; (b) RAP binder unaged; (c) Virgin binder 50/70 aged; (d) RAP binder aged.

As can be observed in Fig. 4, RAP + 5% A and RAP + 5% C show the similar branches formations. Additionally, QNM mode indicates how these branches are softer than the matrix in terms of complex moduli. As overall, at unaged stage all the rejuvenated binders, show lower complex moduli than the RAP binder. This can identify the softening potential of rejuvenators in lowering the surface's moduli of the RAP binder. As shown in Fig. 4, RAP + 5% B displays a topography similar to the RAP binder with higher complex moduli than the other two modified binders. As in Fig. 4 from (d) to (f), all binders increased their complex moduli at the surface after laboratory aging. Further, RAP + 5% B was more affected by aging than the other two binders (Cavalli et al. 2019).

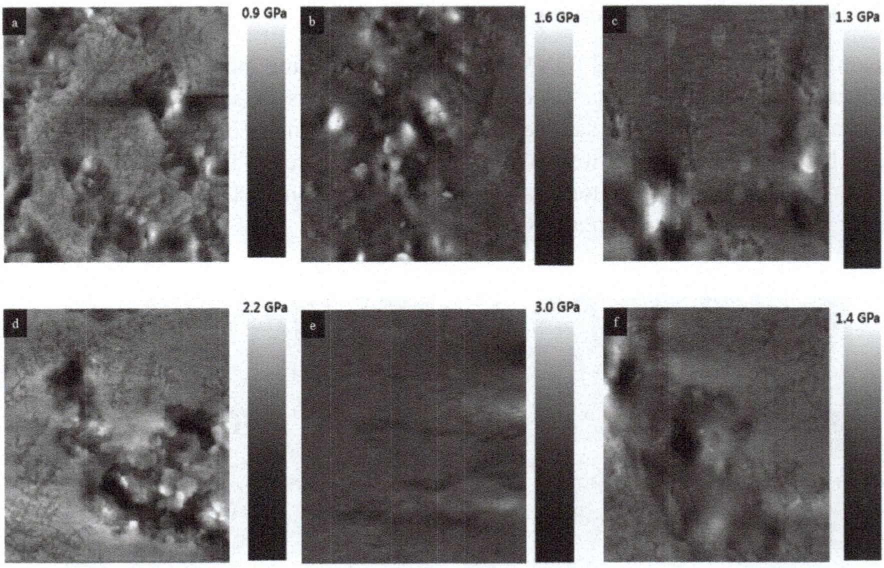

Fig. 4. AFM QNM images (10 × 10 μm). From top left clockwise: (a) RAP + 5% A (b) RAP + 5% B (c) RAP + 5% C (d) RAP + 5% A aged (e) RAP + 5% B aged (f) RAP + 5% C aged.

This supports the findings of the rheological performances were RAP + 5% B resulted the most affected by aging.

4 Conclusions

In this study, a virgin binder, a reclaimed asphalt binder (RAP) and modified RAP binder with three bio-based rejuvenators (a seed oil, a cashew nut shell based oil and a tall based oil) were studied. Dynamic shear rheometer has been used to study the time temperature behaviour while atomic force microscopy (AFM) has been utilized to characterize the elastic moduli at the sample surface. Each material was analysed before and after laboratory aging. The rheological analysis evidenced how the addition of rejuvenators can cause a change in the mechanical performance of RAP. It was demonstrated how the addition of 5%, by mass of RAP binder, of rejuvenator B and 5% of rejuvenator C could reduce the complex moduli at values lower than the 50/70 virgin binder both at low and high frequencies. In general, the complex moduli master curves for all the rejuvenated binders were found to be lower than those for the RAP binder. This suggests that using asphalt binder rejuvenators when producing mixtures containing RAP would result in improving mechanical performance of RAP binder to levels like the virgin binder. However, from the rheological measurements it can be stated that aging plays a significant role in materials rheological properties and should be considered when long term mechanical performance is sought. Regarding the surface properties, undulated domains forming the so called "bee" structures have been seen appearing in the virgin binder only. The addition of rejuvenators could create the formation of new branches at the surface of the RAP binder. Quantitative nano-mechanical mapping of elastic moduli along samples' surface showed how RAP binder was stiffer than the other binders although the addition of rejuvenators decreased the elastic moduli of the RAP binder. As a result of aging, the elastic moduli of all binder increased. This work correlates nano-mechanical properties at the surface with the bulk properties of modified asphalt binders. It was found that a common trend between the material's mechanical properties before and after aging exist for both the bulk and the surface. It has been demonstrated how rejuvenators could soften the RAP's binder both at the bulk and at the surface.

Acknowledgements. Authors would like to thank the Swiss federal office for the environment for the financial support grant number 489.19.14/IDM 2006.2423.487.

References

Abd DM, Al-Khalid H, Akhtar R (2017) Nano-scale properties of warm-modified bituminous binders determined with atomic force microscopy. Road Mater Pavement Des. https://doi.org/10.1080/14680629.2017.1304262

Airey GD, Rahimzadeh B, Collop AC (2004) Linear rheological behavior of bituminous paving materials. J Mater Civil Eng 16(3):212–220. https://doi.org/10.1061/(ASCE)0899-1561(2004)16:3(212)

Al-Khateeb GG, Ramadan KZ (2015) Investigation of the effect of rubber on rheological properties of asphalt binders using superpave DSR. KSCE J Civil Eng 19(1):127–135. https://doi.org/10.1007/s12205-012-0629-2

Allen RG, Little DN, Bhasin A (2012) Structural characterization of micromechanical properties in asphalt using atomic force microscopy. J Mater Civ Eng 24(10):1317–1327. https://doi.org/10.1061/(ASCE)MT.1943-5533.0000510

Bhasin A, Ganesan V (2017) Preliminary investigation of using a multi-component phase field model to evaluate microstructure of asphalt binders. Int J Pavement Eng 18(9):775–782. https://doi.org/10.1080/10298436.2015.1065998

Cavalli MC, Mazza E, Zaumanis M, Partl MN, Poulikakos LD (2018a) Surface nano-mechanical properties of bio-modified reclaimed asphalt binder. Manuscript in Preparation

Cavalli MC, Partl MN, Poulikakos LD (2017) Measuring the binder film residues on black rock in mixtures with high amounts of reclaimed asphalt. J Clean Prod 149:665–672. https://doi.org/10.1016/j.jclepro.2017.02.055

Cavalli MC, Partl MN, Poulikakos LD (2019) Effect of ageing on the microstructure of reclaimed asphalt binder with bio-based rejuvenators. Road Mater Pavement Des 1–12. https://doi.org/10.1080/14680629.2019.1594049

Cavalli MC, Zaumanis M, Mazza E, Partl MN, Poulikakos LD (2018b) Effect of ageing on the mechanical and chemical properties of binder from RAP treated with bio-based rejuvenators. Compos B Eng 141:174–181. https://doi.org/10.1016/j.compositesb.2017.12.060

Derjaguin B, Muller V, Toporov Y (1975) Effect of contact deformations on the adhesion of particles. J Colloid Interface Sci 53(2):314–326. https://doi.org/10.1016/0021-9797(75)90018-1

EUROPEAN STANDARD 12697-1 Bituminous mixtures - Test methods for hot mix asphalt - Part 1: Soluble binder content (2012)

EUROPEAN STANDARD EN 1426: Determination of the needle penetration (2012)

EUROPEAN STANDARD EN 1427: Determination of the softening point-ring and ball method (2012)

García R, Tamayo J, Paulo AS (1999) Phase contrast and surface energy hysteresis in tapping mode scanning force microsopy. Surf Interface Anal 27(5–6):312–316. https://doi.org/10.1002/(SICI)1096-9918(199905/06)27:5/6%3c312:AID-SIA496%3e3.0.CO;2-Y

Gergesova M, Zupančič B, Saprunov I, Emri I (2011) The closed form t-T-P shifting (CFS) algorithm. J Rheol 55(2011):1. https://doi.org/10.1122/1.3503529

Huang SC, Qin Q, Grimes RW, Pauli AT, Glaser R (2015) Influence of rejuvenators on the physical properties of RAP binders. J Test Eval 43(3):594–603. https://doi.org/10.1520/JTE20130314

Hutter JL, Bechhoefer J (1993) Calibration of atomic-force microscope tips. Rev Sci Instrum 64 (7):1868–1873. https://doi.org/10.1063/1.1143970

Jäger A, Lackner R, Eisenmenger-Sittner C, Blab R (2004) Identification of four material phases in bitumen by atomic force microscopy. Road Mater Pavement Des 5(Suppl 1):9–24. https://doi.org/10.1080/14680629.2004.9689985

Mangiafico S, Sauzéat C, Di Benedetto H, Pouget S, Olard F, Planque L (2017) Complex modulus and fatigue performances of bituminous mixtures with reclaimed asphalt pavement and a recycling agent of vegetable origin. Road Mater Pavement Des 18(2):315–330. https://doi.org/10.1080/14680629.2016.1213509

Mousavi M, Pahlavan F, Oldham D, Hosseinnezhad S, Fini EH (2016) Multiscale investigation of oxidative aging in biomodified asphalt binder. J Phys Chem. https://doi.org/10.1021/acs.jpcc.6b05004

Oliviero Rossi C, Spadafora A, Teltayev B, Izmailova G, Amerbayev Y, Bortolotti V (2015) Polymer modified bitumen: rheological properties and structural characterization. Colloids Surf A 480:390–397. https://doi.org/10.1016/J.COLSURFA.2015.02.048

Pahlavan F, Mousavi M, Hung AM, Fini EH (2018) Characterization of oxidized asphaltenes and the restorative effect of a bio-modifier. Fuel 212:593–604. https://doi.org/10.1016/J.FUEL.2017.10.090

Raman NAA, Hainin MR, Hassan NA, Ani FN (2015) A review on the application of bio-oil as an additive for asphalt. Jurnal Teknologi **72**(5), 105–110. https://doi.org/10.11113/jt.v72.3948

Rashid F, Hossain Z, Bhasin A (2019) Nanomechanistic properties of reclaimed asphalt pavement modified asphalt binders using an atomic force microscope. Int J Pavement Eng 20 (3):357–365. https://doi.org/10.1080/10298436.2017.1293268

Soenen H, Besamusca J, Fischer HR, Poulikakos LD, Planche J-P, Das PK, Kringos N, Grenfell JR, Lu X, Chailleux E (2014) Laboratory investigation of bitumen based on round robin DSC and AFM tests. Mater Struct 47(7):1205–1220. https://doi.org/10.1617/s11527-013-0123-4

New Fluxing Agent for the Road Industry – An Overview of Technical Performances and HSE Benefits

Thomas Lebarbé[1(✉)], Frédéric Delfosse[1], Hélène Martin[2], and Arnaud Bourdette[2]

[1] Eurovia Management, 22 Rue Thierry Sabine, 33703 Mérignac, France
thomas.lebarbe@eurovia.com
[2] Solvay Novecare, 40 rue de la Haie Coq, 93306 Aubervilliers, France

Abstract. Fluxing agents (or flux oils) are oil products added to bitumen in order to decrease the viscosity of the binder hence allowing spraying, coating and paving applications more efficiently. Following the application on the road substrate, the flux oil evaporates to recover original binder's properties.

In 2015, Solvay and Eurovia have associated their own expertise in chemicals and road construction respectively, to design a new versatile fluxing agent for the road industry. This new flux oil has an outstanding safety profile (no labelling, and high flash point). It is partly bio-sourced (45% of its carbons are bio-sourced), the other part being a side-product of Solvay's plant production, and fully biodegradable. Because of its unique properties this new flux oil needs less concentration (−20 to 45%) than other fluxing agents to decrease the viscosity of bitumen hence minimizing the use of volatile compounds. Moreover, it enables a quicker cohesion build-up of fluxed bituminous products.

The fluxed binders are devoted to various types of application including surface dressing, micro-surfacing, cold mixes and storable mixes used in maintenance.

This paper aims to describe the numerous technical and other advantages of this new product based on laboratory and jobsite results.

Keywords: Flux oil · Bitumen emulsion · Fluxed bitumen · Surface dressing · Sustainable pavement

1 Introduction

Our society is in logic of major changes where, progressively, alternative solutions are sought, tested and proposed, to answer the challenge of sustainability. The world of road construction is not left behind, since it has undertaken this evolution for some time now with the subjects of recycling, limitation of asphalt manufacturing temperatures and selection of more sustainable additives.

Most of bitumen production ($\sim$85%) is dedicated to paving applications. However, performances of the finished products and bituminous products application constraints make paving grade bitumen (covered by EN 12591 standard) insufficient to answers all road contractors' needs. Three bituminous binders families can be cited to overcome

M. Pasetto et al. (Eds.): ISAP APE 2019, LNCE 48, pp. 23–32, 2020.
https://doi.org/10.1007/978-3-030-29779-4_3

these issues: polymer modified bitumen (EN 14023), fluxed bitumen (EN 15322) and bitumen emulsions (EN 13808). The bituminous binders covered by EN 15322, which includes un-modified and polymer modified bituminous cut-back and fluxed materials, are used in numerous paving techniques where viscosity of the binders must be lowered temporarily. While cut-back bitumen is now rarely used by road contractors due to health, safety and environmental considerations, fluxed bitumen is still widely used because some paving techniques and products such as surface dressing, cold mixes and storable mixes for road maintenance need low viscosity binders to ease application. Bitumen emulsions can also contain significant amounts of flux oil that help their manufacture and curing once applied on the road pavement.

2 New Generation of Fluxing Agents with Advantageous HSE Profile

In France, thousands tons of flux oils are daily used for road maintenance, which justify the necessity to address this problematic from an environmental and sanitary point of view. Indeed, all the manufacturer and laying contractor are concerned. In the past 3 decades, new fluxing agents have been proposed for the road market industries, while others were put aside because of health and environmental issues. A significant step was done in 2005, when recommendations (Usirf 2005) on the use of fluxing agents were given by the Setra (Study department for the transport, road and infrastructures). This organization was appointed by the Administration (direction of Road) and Route de France ex-Usirf (Union of French Road Industries), to give an overview of all the flux oils, used to lower the viscosity of hydrocarbon binder, available on the market. It appeared that some of the flux oils used by the road industry were labeled toxic (R45 – R49, may cause cancer) or harmful (R40, suspected of causing cancer). Today, the regulation (EC) No 1272/2008 on Classification, Labelling and Packaging of Substances and Mixtures, commonly known as CLP Regulation (Regulation 2008) (entered into force on 20 January 2009) would classify those additives as H350 or H351 i.e., with a carcinogen ranking of 1A /1B.

Thus, the main prescriptions of the report were to stop using CMR (carcinogenic mutagenic toxic reproduction) products, to allow the use of labeled fluxing agents, as long as they are not CMR and, to favor the use of non-classified products.

Since the development of biosourced fluxing agent beginning of 2000 (Marcilloux 2003, 2005), only 2 main families are available on the market:

- Fluxing agents from distillation of petroleum and from petro chemistry, called mineral fluxing agents (MF).
- Fluxing agents from natural resources, such as methyl esters, called biosourced fluxing agents (VF).

Beyond the raw material used to produce them, those 2 families distinguish themselves regarding 2 parameters: the technical performances and the health profile.

Regarding the labelling, there is no doubt that biosourced methyl esters are better owing to the fact that they do not show any classification (labelling or hazards sentences): they show an excellent HSE (Health, Safety and Environment) profile. On the opposite, mineral flux oils show a poor HSE profile, with several classifications. They have an impact on both the environment and the human being (Table 1).

Table 1. Characteristics and properties of current fluxing agents

	Mineral fluxing agent	Biosourced fluxing agent
Health profile*	Labelling H304[1] and / or H335[2], H336[3]	Not classified as dangerous
Environnemental profile*	Potentially persistent Labelling H411[4] or H412[5]	Non persistent No labelling
Technical performances / Paving techniques	Good build-up cohesion : - for surface dressing, - for cold mixes - for storable mixes	Slow build-up cohesion: - Mediocre for surface dressing,
Flash point	Low (< 90°C) : could lead to ATEX zone (Explosive atmosphere)	High (> 120°C)
Average annual consumption in France	15 kT	3 kT

The use of mineral fluxing agent is justified only because of their level of performances, better than biosourced flux oils. These differences are mainly explained by their different way of action. Indeed, once the bituminous product, containing the fluxing agent, is applied on the road, the objective is that the bituminous binder gets back its original properties as quick as possible. For the mineral fluxing agent, the mechanism is quite simple as it is the evaporation: the narrower the distillation curve, the quicker the evaporation in the air and thus, the quicker the bitumen gets back to its original properties. Concerning the biosourced flux oil, the mechanism is more complex: it is the additives itself, thanks to its specific structure, which reacts with the oxygen in the air, and then, enables to harden the bituminous product. This oxidation, called "siccativation" is very slow, and the cohesion build-up takes time which is not relevant compared to the short time required to reopen a road. Siccativant (oxidation) agents can be used to accelerate slightly the reaction but, it is not possible nowadays to control and stop this reaction: the hardening on the binder and thus, stiffness is observed on bitumen after a while, causing disorders in some applications. Because of this mechanism, technical performances are limited for the biosourced fluxing agents.

Concerned about the working conditions, the well-being of the residents and the natural habitat, Solvay and Eurovia decided to join their research and innovation strengths by starting a partnership 4 years ago, to propose an alternative to the current fluxing agents. The objective was to design a new fluxing agent for the road industry, efficient and versatile for all the paving applications with a specific focus on the health, safety and environment. Indeed, safety and sustainable development are the priority for these 2 key players in their own industry.

InnRoad Protect (Innovations for Road), also called "IRP" is the answer for the road pavement market, which is looking for greener, safer and more efficient technology. This new generation of flux oils, designed by Solvay, is different from the current technologies.

It combines the best of the two worlds: high level of technical performances, and absence of classification.

The 1st Reach registration tests allowed to show the absence of hazards towards the human health, which would enable to decrease the pressure undergone by the road industry regarding the use of volatile fluxing agents. Another angle to consider is the safety. Indeed, IRP shows a higher flash point (>110 °C) than conventional mineral flux oils allowing limiting the ATEX (Explosive Atmosphere) risks in some industrial configurations.

Finally, the environmental dimension is also advantageous as the IRP is obtained from the reaction of a by-product (and thus avoid the incineration of waste) with a plant-derived compound. The ecological evaluation enabled to consider IRP as easily biodegradable (quality of a substance to be quickly decomposed by microorganisms) and not bioaccumulative (bioaccumulation of products toxic in aquatic species can, for example, affect the biodiversity of species). Thus, its impact on the natural environment is estimated low.

Like VF, IRP fulfills all HSE criteria specified by Eurovia. It differentiates itself by its technical performances, similar and even higher than MF. IRP shows unique evaporation properties, thanks to a very narrow distillation curve (according to the ASTM D86). InnRoad Protect starts its evaporation after the petroleum fluxing agent (Fig. 1) but the full evaporation is quicker, compared to the MF. Thus, InnRoad Protect, characterized by its intrinsic properties, is more efficient, and added-values can be seen in a broad type of applications.

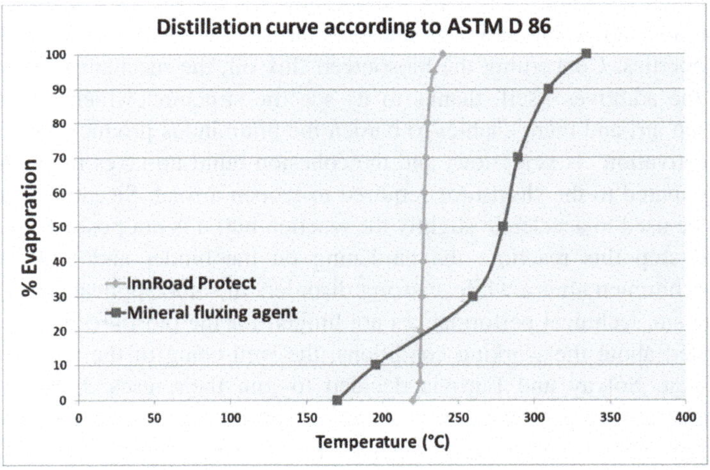

Fig. 1. Distillation curves of InnRoad Protect (IRP) and a mineral fluxing agent according to ASTM D 86

3 Surface Dressing Applications

Surface dressing is a very economical and effective maintenance technique which is capable of greatly extending the life of a road pavement. Flux oils play a key role in the formulation of the binders which are spread. While bitumen emulsions have a predominant part of the market, anhydrous fluxed bitumen is still used in some countries. InnRoad Protect has been designed in order to fulfill the specifications of both kinds of binders.

3.1 Anhydrous Fluxed Bitumen Results

In order to evaluate quantitatively its fluxing properties, IRP has been compared to the benchmark (mineral flux oil) by measuring the efflux time of three types of polymer-modified fluxed bitumen according to EN 12846-2 standard.

As shown in Fig. 2, IRP exhibits good fluxing properties compared to mineral flux oil. The fluxing ability of IRP allows reducing by 20–25% the amount of flux oil in the binder compared to mineral fluxing agent in order to obtain a similar efflux time. As expected, higher amounts of flux oil are needed when the polymer content increases.

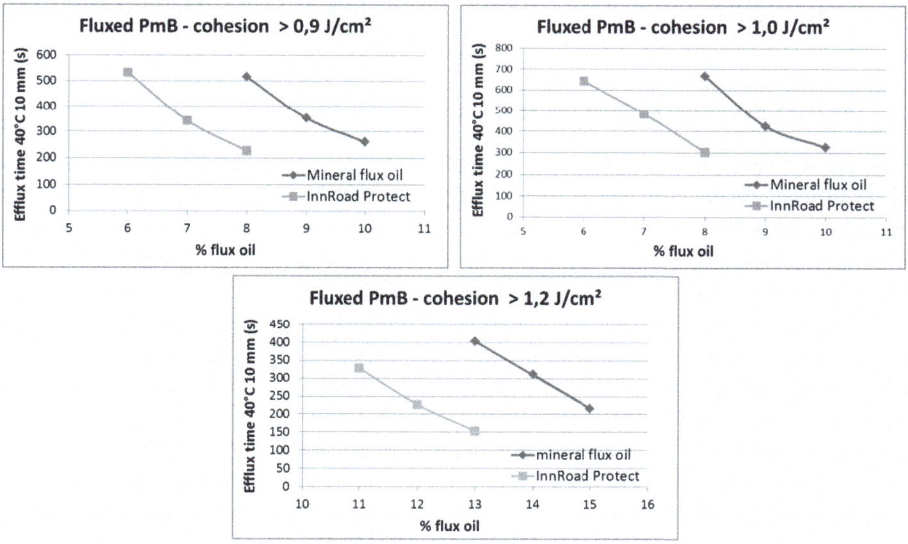

Fig. 2. Efflux time of the binder as a function of flux oil content and type for three levels of cohesion (Polymer content)

Table 2. Properties of the residual binders after stabilization (NF EN 13074-1&2)

Flux oil	Fluxed PmB >0.9 J/cm^2		Fluxed PmB >1.0 J/cm^2		Fluxed PmB >1.2 J/cm^2	
	9% MF	7% IRP	10% MF	8% IRP	14% MF	11% IRP
Flux oil weight loss (%)	63.5	78.6	64.2	81.2	64.4	85.4
Penetration (1/10 mm) NF EN 1426	121	75	115	79	106	56
R&B temperature (°C) NF EN 1427	44.4	49.2	46.4	50.0	52.2	60.2

Viscosity and stiffness recovery of the fluxed binders is a critical step in defining the technical performances of the surface dressing. The stiffness recovery mechanism of the binders fluxed with IRP is similar with the mechanism observed with mineral flux oil. The elimination by evaporation is the main mechanism. Consequently, the different fluxed PmB of this study have been stabilized according to EN 13074-1&2 standard and then analyzed. It is obvious on Table 2 that IRP evaporates much faster than the mineral flux oil as seen from the penetration and R&B temperature values. From these results, a significant improvement in the technical performances of the surface dressing wearing courses can be expected when IRP is used instead of mineral flux oil, specifically at the early stage.

3.2 Fluxed Bitumen Emulsions Results

In bitumen emulsions, flux oils play a significant role in making easier the emulsification process (especially for high viscosity binders). Moreover, after laying on the road, fluxing agents improve the coalescence of the binder's droplets in the emulsion hence accelerating the curing process of the emulsion. Two highly modified (cohesion > 1.2 J/cm^2) bitumen emulsions have been manufactured and characterized (Table 3). In the first emulsion, high amount of mineral flux oil (>3%) has been used in the binder to be in accordance with the EN 13808 specifications. On the contrary, with IRP, the amount of flux oil has been drastically reduced (−25%) with no negative influence on the emulsion properties.

The two emulsions have been stabilized according to EN 13074-1&2 standard and the stabilized binders have been characterized to evaluate the stiffness recovery. Table 4 shows the significant difference in the penetration and R&B temperature values between the two binders. These results confirm that IRP is much more effective in the stiffness recovery of the binders. As expected, the maximum cohesion values are similar between the two binders, because the cohesion is mainly brought by the polymer content in the binder. However the maximum cohesion temperature is slightly shifted to upper values. Thanks to some polar affinities with water, IRP might diffuse to water, explaining good efficiency in the recovery of initial properties of the binders. Some works have been initiated to better understand this potential complementary mechanism.

Table 3. PmB emulsions properties depending on flux oil type

Emulsion type		C69/70 BPF 2	C69/70 BP 2
Flux oil type		Mineral flux oil (MF)	InnRoad protect
Flux oil content in emulsion	%	>3	<3
Binder content: NF EN 16849			
Binder content	%	70,2	69,4
Efflux time: NF EN 12846-1			
4 mm at 40 °C	s	12	17
2 mm at 40 °C	s	140	200
Residue on sieving: NF EN 1429			
0,500 mm sieve	%	0,02	0,05
0,160 mm sieve	%	0,27	0,24
Droplet size distribution - Laser diffraction			
Mean diameter	µm	3,50	2,91
Standard deviation	/	0,41	0,38
Breaking value: NF EN 13075-1			
Breaking value (Forshammer)	/	49	44
Adhesivity: NF EN 13614			
Diorite	%	≥ 75	≥ 75
Quartzite	%	≥ 75	≥ 75

Table 4. Properties of the residual binders after stabilization (NF EN 13074-1&2)

Emulsion type	C69/70 BPF 2	C69/70 BP 2	NF EN 13808 specifications
Flux oil type	MF	IRP	
Flux oil content in emulsion (%)	>3	<3	
Penetration 25 °C (1/10 mm) NF EN 1426	55	40	≤ 100
Ring & Ball temperature (°C) NF EN 1427	58.2	61.0	≥ 50
Cohesion (J/cm^2) NF EN 13588	1.27	1.24	≥ 1.2
Temperature max cohesion (°C)	40	45	/

4 Microsurfacing

4.1 Flux Oil for Early and Late-Season Microsurfacing Applications

Microsurfacing products, which are described and characterized by the EN 12274 European standards, have been widely used over the world for cost-saving pavement preservation. In France, more than 50 million square meters are applied each year. The objective for all microsurfacing formulators is to allow quick cohesion build-up of the microsurfacing in order to get minimal impact on traffic. The microsurfacing curing mechanism is complex and rests on a combination of bitumen emulsion coalescence followed by shape relaxation (Leal-Calderon 2001). The characteristic time for shape relaxation (Tc) is governed by the competition between surface tension and viscous dissipation and is given by the following equation:

$$Tc \propto \frac{R * \eta}{\gamma} \tag{1}$$

Where η is the viscosity of the binder's droplets, R is their characteristic radius and γ is their surface tension.

As seen from Eq. 1, it is then possible to accelerate the shape relaxation and thus the cohesion build-up by decreasing the viscosity of the binder. InnRoad Protect (IRP) flux oil has been evaluated as a mean to lower bitumen's viscosity or increase the penetration grade of the bitumen in order to expect faster shape relaxation of the microsurfacing and thus lower sensitivity of the material to raveling (at the early stage) in harsh curing conditions (early or late-season application periods). Figure 3 illustrates the better fluxing ability of IRP to increase penetration grade of the bitumen compared to mineral flux oil. Only 0.8% of IRP is needed to convert a 50/70 pen grade bitumen to a 70/100 pen grade.

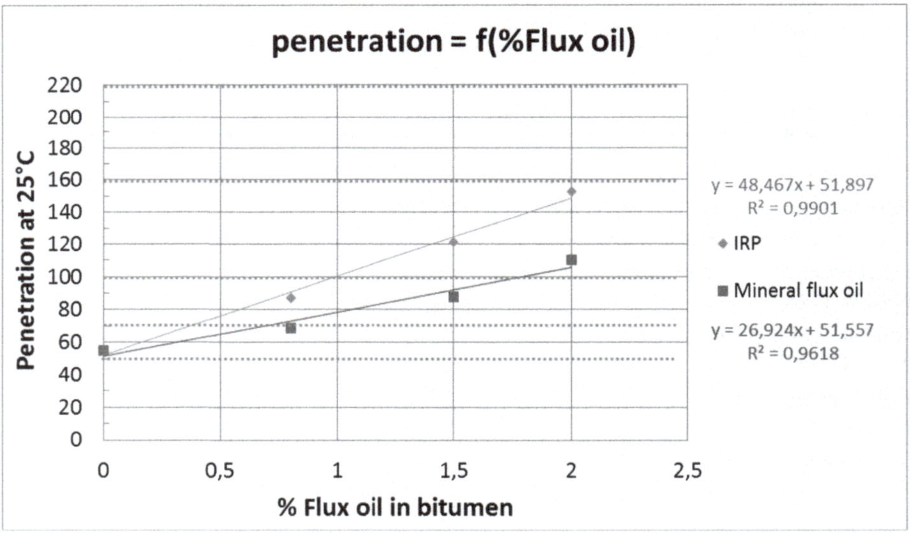

Fig. 3. Penetration as a function of flux oil content for two types of flux oils

4.2 Lab Test Results

The Wet Track Abrasion Test (WTAT) is a laboratory method used to evaluate abrasion resistance of the microsurfacing mix.

However, current standards related to this test fixed curing conditions at 60 °C which are far more comfortable from what is observed on jobsites. Eurovia has fixed his own curing conditions in order to be more predictive from the abrasion issues of the microsurfacing observed on jobsites at the early stage after application. In the case of early and late-season microsurfacing laying, the conditions can reach 10 °C and 100% humidity. These harsh curing conditions have been applied to lab investigations at Eurovia laboratories in order to make sure that the microsurfacing mix design is suitable (weight losses < 25%) for the early and late-season application period.

Three emulsions have been manufactured and investigated in a microsurfacing mix design by ensuring that close workability time (~90–120 s) is obtained for all the mix designs. The first emulsion is the reference with non-fluxed bitumen. The second emulsion has been manufactured with a bitumen modified with 0,8% InnRoad Protect (IRP). Finally, the third emulsion includes 1,5% of mineral oil in the bitumen phase.

The results show that bitumen modification with small amounts of flux oil in the bitumen strongly helps in getting WTAT weight losses within Eurovia's specifications (<25%) for late-season conditions (Table 5). More particularly, besides the fact that IRP amount in bitumen is reduced by 47% compared to the conventional mineral flux oil, IRP flux oil gives a smaller weight losses value. This study shows that IRP is effective in reducing bitumen viscosity to improve bitumen/aggregates active adhesion by shape relaxation. On the top of that, IRP then evaporates over the time in order to recover the binder's stiffness contrary to softer grade bitumen.

Table 5. Microsurfacing mix design details and properties regarding abrasion resistance

Emulsion n°	**1**	**2**	**3**
Binder	No flux oil	**0.8% IRP**	**1.5% mineral flux**
0/6 aggregates	100	100	100
Emulsion (pph)	11.9	11.9	11.9
Breaking agent (pph)	Same amount		
Retarder (pph)	0	0	0
Eau (pph)	Same amount		
Results			
Workability (open) time (s)	120	120	90
WTAT 10°C/100% Humidity (%)	**100**	**9**	**16**

5 Conclusions

The development of a new generation of flux oil took nearly 3 years within the research teams of Eurovia and Solvay to finally identified and commercialized InnRoad Protect as a technical, healthy and environmental solution for road contractors. InnRoad Protect is not classified as dangerous; it is biodegradable, not bio-accumulative and is obtained from the reaction of a by-product with a plant-derived compound. InnRoad Protect shows a better fluxing ability and evaporates faster compared to mineral flux oils. Its versatility has been confirmed in various road techniques including surface dressing (anhydrous and emulsion), microsurfacing, cold mixes, storable mixes, …

Numerous jobsite trials have been carried out with IRP in surface dressing techniques, both with anhydrous binders or bitumen emulsions. In 2018, a total surface of 110 000 m^2 ($\sim$31 km) has been coated with anhydrous binders and 14 000 m^2 ($\sim$4 km) with bitumen emulsions. Microsurfacing have been applied with a surface of 5 000 m^2. Global technical feedbacks for these jobsites are excellent.

References

Usirf (2005) Impacts sanitaires des fluxants, fluidifiants et produits anti-kérosène, groupe de travail Administration/Usirf, mars 2005

Regulation (EC) No 1272/2008 on classification, labelling and packaging of substances and mixtures

Alfos C, David K, Marcilloux J (2003) Formulation of solvents for bitumen based on vegetable oil methyl esters. OCL – Ol Corps Gras Lipides 10:392–399

Antoine J-P, Marcilloux J (2005) Anhydrous binder for mixes or surface dressing road goes to the green. OCL – Ol Corps Gras Lipides 12:304–307

Philip J, Poirier J, Bibette J, Leal-Calderon F (2001) Gelation and coarsening in dispesion of highly viscous droplets. Langmuir 17:3545–3552

Towards a Better Assessment of Recycling Agents Effects on Bitumen During Hot Recycling

Fayçal Lahjiri[1(✉)], Sabine Gazeau[1], Frédéric Delfosse[1], Anne Dony[2], Layella Ziyani[2], Virginie Mouillet[3], and François Henn[4]

[1] Eurovia Management, Centre de Recherche de Mérignac, 22 rue Thierry Sabine, 33703 Mérignac, France
faycal.lahjiri@eurovia.com

[2] Université Paris Est, Institut de Recherche en Constructibilité, Ecole Spéciale des Travaux Publics, 28 avenue du Président Wilson, 94234 Cachan Cedex, France

[3] Cerema, Equipe-projet DIMA, 30 rue Albert Einstein, 13593 Aix-en-Provence, France

[4] Université de Montpellier, Laboratoire Charles Coulomb, UMR 5221, CNRS, 34095 Montpellier, France

Abstract. The current challenge is to produce asphalt mixtures incorporating higher rates (>50%) of Reclaimed Asphalt (RA) while controlling the quality of the final mix. During the service life of a pavement properties change, mainly due to oxidation and need to be restored. The resulting mix must exhibit at least equivalent performances regarding the mechanical and durability properties compared to that from a mix without RA. Hence, recycling agents are sometimes employed to "rejuvenate" aged bitumen. In this study, three recycling agents were used on both RA and neat bitumens, displaying different compositions. The effects on unaged and aged blends were assessed through rheology tests using Dynamic Shear (DSR) and Bending Beam (BBR) Rheometers. Data from DSR obtained with 4 mm and 8 mm parallel plates allowed to calculate the Glover-Rowe criteria and to determine the crossover parameters whilst BBR measurements yield Tm, Ts and ΔTc parameters. The blends resulting from the two bitumens displayed a similar evolution of behavior. This outcome suggests that these tests cannot discriminate the origin of the rejuvenation phenomenon since they only highlight the softening effect of the recycling agent and not its impact at the molecular level, i.e. on asphaltene and maltene compounds.

Keywords: Recycling · Rejuvenation · Reclaimed asphalt · Rheology · Recycling agents · Softening effect

1 Introduction

In France and in 2017, an average of 18.4% of Reclaimed Asphalt Pavements (RA) were reintroduced in new asphalt pavements (Routes de France 2018). Because of environmental and economic benefits, the amount of reused RA has increased in the

M. Pasetto et al. (Eds.): ISAP APE 2019, LNCE 48, pp. 33–43, 2020.
https://doi.org/10.1007/978-3-030-29779-4_4

past few years. Thus, the road industrials are challenged to produce more and more asphalt mixtures incorporating high rates (>50%) of RA. In order to reach that goal, it is crucial to consider and to shed light on the changes in the bitumen properties, as for instance due to oxidation and structural evolution of asphaltenes, and their impact on the asphalt behavior upon ageing (Petersen 2009). It is well known that the material becomes more brittle with time, thus leading to thermal or stress cracking. Nevertheless, an asphalt incorporated RA must exhibit at least equivalent performances regarding the durability compared to that of a mix without RA.

Hence, products known as "rejuvenators" are added to the recycled asphalt in order to restore, if possible, the genuine properties of neat bitumen. "Rejuvenators" can be oils, either extracted from bio or petroleum sources. Besides, it has been proposed by some authors to classify these additives in different categories, i.e. "soluble softeners" or "compatibilizers" (agents targeting the asphaltenes clusters), accordingly to their assumed action on the aged bitumen at the molecular level (Tabatabaee and Kurth 2017). Based on these considerations, the "rejuvenator" is defined as a product associating both the "soluble softener" and the "compatibilizer" effects. However, some shortcuts are sometimes taken to qualify these additives as "rejuvenators" based on some tests that do not fully characterize their interactions with the bitumen at a molecular level. More work is then needed to get to a better qualification of these compounds. We then suggest to keep with the term recycling agent as long as there is no more evidence on their exact role.

This study aims to assess the changes induced by three recycling agents on bituminous binders. To better understand the role of these additives in bitumen, two binders with different physico-chemical compositions, having equivalent penetration grades were used: a RA (recovered binder oxidized on-site for several years) and a neat bitumen (10/20 from a refinery). Since the blends with recycling agents display different origins and complex chemical compositions, our goal is to evaluate the behavior of the mechanical properties, regardless of their chemical features. We thus expect to discriminate softening and compatibilizer effects. Mechanical characteristics are determined using both a dynamic shear rheometer and a bending beam rheometer. Also, since the recycling agents' chemistries are not the same, the dosage in bitumen may vary. Hence, the impact of the dilution rate, or dosage, on the binders' ageing properties was tested. The performances were assessed through oxidative ageing resistance using laboratory ageing procedures.

2 Materials and Methods

2.1 Materials

To evaluate the evolution of the blends' properties, a French RA and a neat penetration graded 10/20 bitumen (EN 13924-1) are used. Some characteristics of these bituminous binders are reported in Table 1.

Table 1. Properties of the base bitumens used in this study

Properties		RA binder	10/20 neat bitumen
Penetration at 25 °C ± 2 [1/10 mm] [EN 1426]		11	12
Ring and Ball softening point (R&B) ± 1 [°C] [EN 1427]		75.2	64.8
Penetration index (PI) [EN 13924-1]		0.5	−0.9
$\|G^*\|_{15\,°C,\ 10\,Hz}$ ± 5% [MPa] [EN 14770]		87.4	117.4
$\delta_{15\,°C,\ 10\,Hz}$ ± 5% [°] [EN 14770]		25.1	26.2
SAR-AD®	Saturates (%)	19	15
	Aromatics (%)	42	58
	Resins (%)	19	15
	Asphaltenes (%)	20	12
Colloidal index (IC)		0.64	0.37

The saturates, aromatics, resins and asphaltenes fractions were determined using the SAR-AD® device from the Western Research Institute (Boysen and Schabron 2013). It can thus be seen that the asphaltene content is almost the double for the RA binder though the penetration values are similar. Hence, properties such as the IC, PI, the R&B or the shear modulus are affected. One can note that the asphaltenes from the RA are a product of native asphaltenes and oxidized components from the bitumen during its lifetime. That is why these two very different bituminous binders, yet similar on the penetration values, are suited for this study.

Three commercial recycling agents were used to blend with both RA and neat binders for the evaluation of the softening effect. The recycling agents A, B and C are a bio-based cashew nut shell oil, a bio-based polyol ester tall oil and a petroleum based aromatic oil respectively.

The dosage was adjusted to obtain a modulus value $|G^*|_{15\,°C,\ 10\,Hz}$ equivalent to a typical French 35/50 neat bitumen (Zaumanis et al. 2014). The recycling agents were added to the bitumen (at 160 °C for the 10/20 and 170 °C for the RA) and then blended using an IKA Eurostar 60® blender at 500 rpm for two min. A 35/50 neat bitumen, from the same refinery as the 10/20, was used as reference.

All the tested binders were assessed at the unaged state and after laboratory ageing procedure. To simulate a very long-term ageing, a pressure ageing vessel (PAV) (EN 14769) was used with a modification of duration: 45 h instead of 20 h of oxidative ageing. As 5 h of PAV at 100 °C has been demonstrated equivalent to RTFOT test, it means that this chosen duration should be equal to RTFOT + 2 PAV according EN 14769 (Migliori and Corté 1998).

2.2 Dynamic Shear Rheometer

The dynamic shear rheometer (DSR) measures the complex shear modulus |G*| of a material through oscillatory type solicitation. The complex shear modulus G* contains several information including the storage modulus, G′, and the loss modulus, G″, related to each other through the phase angle. These values are frequency and temperature dependent.

The tested temperature range has been fixed between −35 °C and 50 °C and the frequencies range is taken between 0.1 and 10 Hz, logarithmically spaced, for 21 frequencies in total. Disc shaped bitumen samples (EN 14770) were tested on a Kinexus Pro + rheometer (Malvern Instruments). 4 mm parallel plates with a 1.75 mm gap were used between −35 °C and 15 °C and 8 mm parallel plates with a 2 mm gap were used between −10 °C and 50 °C (Airey et al. 2002; Laukkanen 2015). The behaviors of the binders were assessed by construction of the master curves. The shift factor a_T was calculated using the Williams-Landel-Ferry – Kaelble equation (Williams et al. 1955; Kaelble 1978).

The reference temperature was set at 15 °C because these master curves are used to determine several indicators including the Glover-Rowe parameter (G-R), which is calculated at 15 °C. The G-R is a value used to evaluate the evolution of the binder mechanical performances and therefore, their durability (Glover et al. 2005; Anderson et al. 2011). Calculated at 15 °C and 0.005 rad/s using the following equation, it is necessary to build a master curve to determine the parameters for these conditions.

$$G - R = \frac{|G^*|.(cos\delta)^2}{sin\delta} \qquad (1)$$

Two thresholds were set to evaluate the durability of a binder. The first limit of 180 kPa corresponds to the onset of fatigue. The second limit of 600 kPa corresponds to a severe cracking limit.

Also, depending on the temperature or the frequency of solicitation, bitumen can exhibit a predominant elastic or viscous behavior. The transition between these two behaviors happens when G′ = G″, i.e. when δ = 45°. This is the viscoelastic transition (VET). At this special point, a crossover temperature and a crossover modulus can be determined.

2.3 Bending Beam Rheometer

The bending beam rheometer (BBR) is a three points bending device that allows us to measure the creep stiffness modulus S(t) of a bitumen, usually at low temperatures. The relaxation rate m(t) translates the time dependency of the stiffness.

We determined $T_{S=300\ MPa}$ and $T_{m=0.3}$ which are respectively the temperature at which S equals 300 MPa for a loading time of 60 s and the temperature at which m equals 0.3 for a loading time of 60 s (EN 14771). For each sample, three temperatures were tested to surround those criteria and then, an interpolation was made to reach them. Several indicators can be calculated including ΔTc, ΔTm or ΔTs, which are $T_{S=300} - T_{m=0.3}$ (for a given ageing state), $T_{m=0.3(\text{after ageing})} - T_{m=0.3(\text{before ageing})}$ and $T_{S=300(\text{after ageing})} - T_{S=300(\text{before ageing})}$, respectively.

3 Results and Discussions

Some classical properties were tested on the blends, before and after ageing and are presented in Table 2. At the unaged state, all the blends exhibit comparable shear moduli because it was our starting point. After ageing, 10/20 blends present higher moduli than RA blends. Nevertheless, the aged blends have lower moduli than their base binders (Table 1). If recycling agents A and B seem to present similar results, C performs better regarding ageing. Also, even if the moduli are restored, RA blends show penetration and R&B values closer to the one from the RA than the reference ones whereas 10/20 blends are closer to the latter. It appears that, for a given recycling agent, not all properties are restored with a single dosage.

Table 2. Properties of the blends before and after ageing

Binders	$\|G^*\|_{15\,°C,\ 10\,Hz}$ ±5% (MPa) [EN 14770]		$\delta_{15\,°C,\ 10\,Hz}$ ±5% (°) [EN 14770]		Penetration ±2 [1/10 mm] [EN 1426]		Ring and Ball softening point ±1 [°C] [EN 1427]		Dosage ±0,2 [%wt.]
	Unaged	45 h PAV	Unaged	45 h PAV	Unaged	45 h PAV	Unaged	45 h PAV	
35/50	42.4	66.4	38.0	28.4	40	15	51.8	67.0	-
10/20 + A	40.5	96.2	38.6	25.6	34	13	56.4	70.4	5
10/20 + B	39.9	91.4	38.8	26.2	30	12	55.2	69.4	5
10/20 + C	39.4	73.4	41.8	30.1	37	13	52.8	65.6	15
RA + A	40.8	60.2	32.6	24.0	19	13	69.6	84.6	5
RA + B	39.2	61.7	33.0	23.7	20	10	68.4	86.6	5
RA + C	38.3	53.7	35.7	27.6	22	13	65.0	81.6	13

Since it is not appropriate to draw conclusions based on a single rheological point, master curves or Black diagrams representations are sometimes used to analyze a bitumen's response through solicitation. To better read the changes that may occur when incorporating a recycling agent in a binder, we follow several rheological indicators covering the high (40–50 °C), intermediate (0–40 °C) and low (<0 °C) temperatures.

Firstly, even if the G-R is calculated at 15 °C, the 0.005 rad/s frequency can give a higher equivalent temperature (Porot 2018). The G-R values have been calculated for all the samples before and after 45 h PAV ageing. Figure 1 shows that, at both ageing states, each tested blend exhibits a lower G-R value than their associated base bitumen. The 10/20 bitumen reaches the first threshold before ageing and after 45 h of PAV ageing, the second threshold. The RA, even at the unaged state, already passes the 600 kPa limit. For the 10/20 blends, at the unaged state, the G-R values are close to the 35/50 one. After ageing, they pass the 180 kPa limit but not the second threshold. Concerning the RA blends, the situation is different with higher G-R values than the reference: the addition of the recycling agents A and B at a dosage of 5% wt. is not enough to stay under the 180 kPa limit. After ageing, all the blends pass the 600 kPa threshold. Blends with recycling agents A and B display the same behavior. One can notice that for both base bitumens, the recycling agent C performs better than the A and B ones. For the 10/20 based blend, the sample prepared with recycling agent C has an even lower G-R value than 35/50 after 45 h of PAV. This can assess a better durability than the reference regarding this criterion.

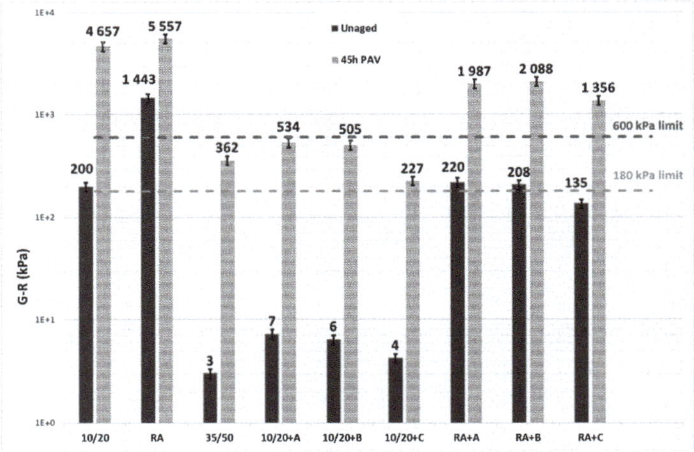

Fig. 1. Glover-Rowe parameters for unaged and aged bitumens

Nevertheless, since the G-R value is calculated through cos δ and sin δ, it is not obvious to make a linear comparison and to assess the softening effect of recycling agents on bitumen. Then, it is interesting to plot the values of |G*| and δ at 15 °C and 0.005 rad/s. Figure 2 illustrates the evolution of the G-R value before and after ageing in the Black space.

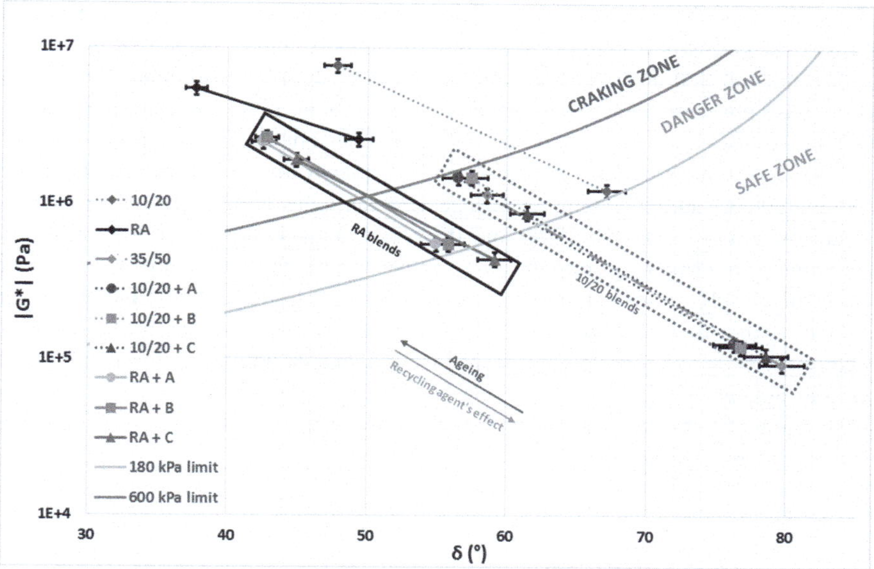

Fig. 2. Glover-Rowe parameter for unaged and aged bitumens represented in the Black space

The same observations can be drawn as previously. However, one can notice that the blends prepared with the same bituminous binder are grouped in the same area, depending on their phase angle. For both bitumens, A and B recycling agents blends have a similar behavior while C provides a |G*| and δ couple less prone to cracking according to the G-R criterion. Two groups, framed in Fig. 2, can be observed highlighting the same phenomena with a shift. For all the blends, the slopes are similar. Then, the G-R parameter may only be used to assess a softening effect for blends including recycling agents.

At intermediates temperatures, the crossover parameters can be obtained. The crossover temperatures (T_{VET}) at 0.1, 1 and 10 Hz were determined, with an error estimated at ± 1 °C, for unaged and aged samples and are displayed on Table 3.

Table 3. T_{VET} (±1 °C) determined for 0.1, 1 and 10 Hz

Binders	Frequency					
	0.1 Hz		1 Hz		10 Hz	
	Unaged	45 h PAV	Unaged	45 h PAV	Unaged	45 h PAV
10/20	16.5	28.5	23.0	36.0	30.0	43.5
RA	27.0	42.0	35.0	51.0	41.5	58.0
35/50	6.0	20.0	13.0	27.0	20.0	34.0
10/20 + A	6.0	21.0	13.5	28.5	21.0	36.0
10/20 + B	5.5	20.0	13.5	27.5	20.5	35.0
10/20 + C	3.5	16.0	11.0	24.0	18.0	31.0
RA + A	18.5	35.0	26.0	43.5	33.0	50.0
RA + B	17.0	35.0	25.5	42.5	32.0	49.0
RA + C	13.0	31.0	22.0	38.5	27.0	44.0

Regardless of the bitumen tested, T_{VET} increases with the frequency. Since this trend was verified with every specimen, only the results at 1 Hz were thus considered. It can be noticed that at the unaged and aged states, the blends exhibit a lower T_{VET} than the base bitumens which were used to prepare them. If we compare the blends to the reference, we can observe that the 10/20 based blends exhibit a similar T_{VET} than the 35/50 one. While recycling agents A and B display a similar behavior in the blends, recycling agent C, added at a higher dosage, decreases the T_{VET} of the base bitumen at an even lower T_{VET} than the 35/50 and the others blends. But, if we look at the RA based blends, we clearly see that the T_{VET} values are higher than that of the reference. These results imply a shift in the transition between the elastic behavior and the viscous one. Once again, the recycling agent C induces the largest decrease of T_{VET} and the related property is not completely restored for RA blends while it is for 10/20 blends, even after ageing.

The graph G′ = G″ versus T_{VET} (see Fig. 3) shows that the behavior strictly follows the employed base bitumen.

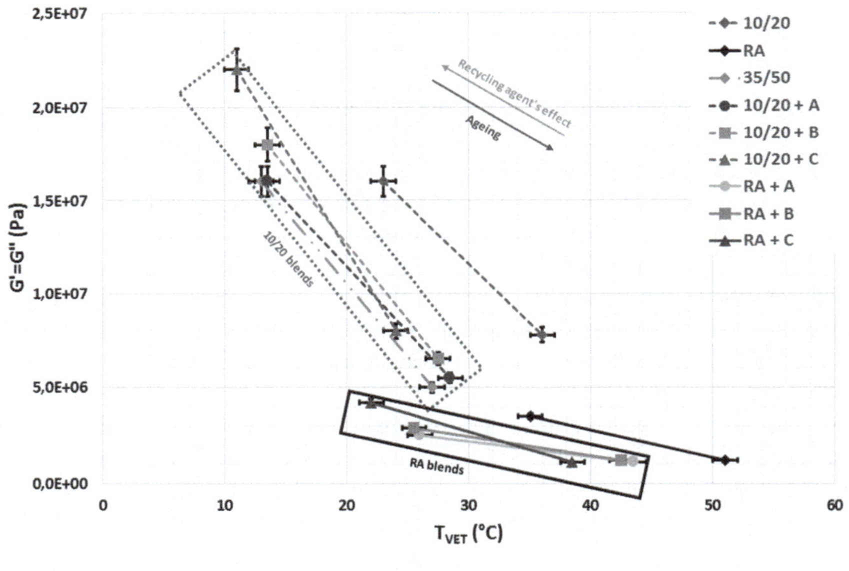

Fig. 3. G′ = G″ versus TVET plot, determined at 1 Hz

The ageing of the blends, considering the errors, are parallel to the ageing of the corresponding base bitumen. In both cases, A and B are almost superimposed whereas C is shifted to lower values of T_{VET} and higher values of G′ = G″. This evidences the impact of the softening effect of the recycling agents. The 10/20 blends are comparable to the 35/50 one, since their initial properties are closer.

Since these results seem to be strongly related to the base used, it is interesting to measure the difference between the T_{VET} of the base bitumen and the T_{VET} of the corresponding blends (Fig. 4). Regardless of the composition of the bitumen, no significative difference for this parameter could be noted for the same recycling agent added. Once again, the biggest softening effect is observed with C while A and B have a similar impact.

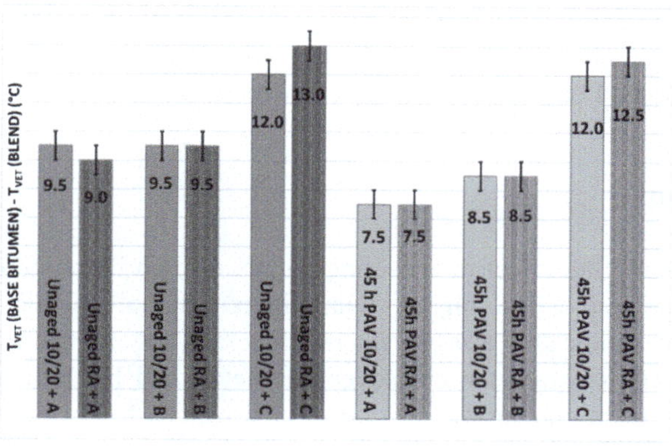

Fig. 4. Difference between T_{VET} from the base binder and T_{VET} from the blend. Same recycling agent at the same aged state are put side to side

Finally, on the colder side of the temperature spectrum, long term ageing susceptibility can be estimated through the BBR. After ageing, ΔTc, which is defined as the difference between $T_{S=300}$ and $T_{m=0.3}$, can help to evaluate a poor bitumen behavior at low temperature when its value is lower than −5 °C. This seems to correspond to a potential severe cracking (Anderson et al. 2011). Table 4 shows ΔTc values after 45 h of PAV for the tested materials.

Table 4. BBR criteria: ΔTc criterion is represented for 45 h PAV aged bitumens. Binders passing the cracking limit are bolded for this indicator.

Binder	10/20	RA	35/50	10/20 + A	10/20 + B	10/20 + C	RA + A	RA + B	RA + C
ΔTc ± 2 °C	−1.7	**−10.3**	−2.6	−0.9	−0.3	2.3	**−12.3**	**−8.6**	**−5.9**
Ts_{45h} ± 1 °C	−2.2	−8.2	−12.1	−10.9	−10.5	−11.7	−15.7	−14.6	−15.0
ΔTs ± 2 °C	4.1	3.5	3.6	4.1	3.7	3.1	2.9	4.9	3.9
Tm_{45h} ± 1° C	−0.5	2.1	−9.5	−10.0	−10.2	−12.2	−3.4	−6.0	−9.1
ΔTm ± 2 °C	7.4	10.4	6.0	7.4	6.9	4.9	12.3	11.0	8.5

After ageing, the behavior changes upon the base of the blend. For the 10/20 blends, ΔTc are above −1 °C. However, for the RA blends, one can note that all the blends after ageing are under the −5 °C limit, which suggests potential severe cracking. Regarding this criterion, only the properties of the 10/20 blends are restored after ageing whereas the properties of the RA blends are not.

To know if the changes in ΔTc mainly come from a gain in stiffness or a loss of relaxation, it is interesting to calculate ΔTs and ΔTm. For all the samples, the loss in relaxation is systematically higher than the gain in stiffness. Even if the trends are the same for all the recycling agents regardless of the bitumen base, the RA blends emphasize the fact that recycling agents A and B yield less relaxation than C. Again, since no significative difference can be found between A and B and since C, dosed at a higher rate, displays a better behavior, only the same conclusion on the softening effect, as the ones mentioned earlier in this paper, can be drawn.

4 Conclusions

This study evaluated the effect of recycling agents on two bituminous binders, i.e. a 10/20 neat binder and a RA binder, with different initial chemical compositions. Several rheological indicators at high, intermediate and low temperatures were calculated, such as the G-R parameter, crossover parameters or BBR indicators, before and after ageing. The main conclusions can be summarized as followed:

- For a given recycling agent, a single dosage did not systematically restore all the properties, as this was observed for the RA tested in this study. It was chosen, as a starting point, to restore $|G^*|_{15\ °C,\ 10\ Hz}$ in order to anticipate future correlations with the asphalt mix. If the 10/20 blends showed equivalent or superior properties than the 35/50 reference, the RA blends could not be considered as fully restored. Even at the unaged state, A and B dosages were not high enough to restore the T_{VET} or the G-R parameter. Note this affirmation is only valid for the 10/20 and RA tested in this study.
- Despite different origins, A and B offered similar results for the whole range of studied temperatures. Since they were mixed at the same dosage, the results obtained could be shifted according to the dosage, for a given bitumen, as the recycling agent C which was added at a higher percentage and which always exhibits better results.
- Long-term ageing is important to check if properties are maintained at an acceptable level. It shows that some blends do not comply with rheological cracking criteria after ageing (600 kPa limit for the G-R parameter, −5 °C for the ΔTc).
- Although the rheological indicators presented in this paper do not explain the accurate action mechanisms of the recycling agents, in particular about their "compatibilizer" effects, it still can make it possible to assess the "soluble softener" impact. Hence, it can be concluded that it is not obvious to qualify a recycling agent as a rejuvenator on the sole basis of rheology tests.

This study opens perspectives as to investigate the effects of recycling agents having similar chemical functionalities but with different viscosities, i.e. blended at different dosages. Additionally, physico-chemical interactions between the binders, the recycling agents and the aggregates should be inspected, since it will be mandatory to evaluate the softening effect at the asphalt mix scale too.

Acknowledgements. The authors would like to thank Jacques-Antoine Decamps for his help during this work.

References

Airey GD, Rahimzadeh B, Collop AC (2002) Linear viscoelastic limits of bituminous binders. Asphalt Paving Technol 71:89–115

Anderson RM, King GN, Hanson DI, Blankenship PB (2011) Evaluation of the relationship between asphalt binder properties and non-load related cracking. J Assoc Asphalt Paving Technol 80:615–663

Boysen RB, Schabron JF (2013) The automated asphaltene determinator coupled with saturates, aromatics, and resins separation for petroleum residua characterization. Energy Fuels 27:4654–4661

Glover CJ, Davison RR, Domke CH, Ruan Y, Juristyarini P, Knorr DB, Jung SH (2005) Development of a new method for assessing asphalt binder durability with field validation. Texas Transportation Institute: College Station, TX

Kaelble DH (1978) Polymer composite reliability. In: Proceedings of the ARPA/AFML Review of Progress in Quantitative NDE, 32

Laukkanen OV (2015) Low-temperature rheology of bitumen and its relationship with chemical and thermal properties. M.Sc. thesis, Aalto University

Migliori F, Corté JF (1998) Comparative study of RTFOT and PAV aging simulation laboratory tests. Transp Res Board 1638:56–63

Petersen JC (2009) A review of the fundamentals of asphalt oxidation. Transp Res Circ E-C140

Porot L (2018) Rheology and bituminous binder, a review of different analyses. In: RILEM 252-CMB Symposium, RILEM Bookseries, 20

Routes de France (2018) Bilan environnemental 2017

Tabatabaee HA, Kurth TL (2017) Analytical investigation of the impact of a novel bio-based recycling agent on the colloidal stability of aged bitumen. Road Mater Pavement Des 18:131–140

Williams ML, Landel RF, Ferry JD (1955) The temperature dependence of relaxation mechanisms in amorphous polymers and other glass-forming liquids. J Am Chem Soc 77:3701–3707

Zaumanis M, Mallick RB, Frank R (2014) Determining the optimum rejuvenator dose for asphalt recycling based on superpave performance grade specifications. Construct Build Mater 69:155–166

Graphene-Enhanced Recycled Asphalt Pavements

Loretta Venturini(✉) and Fabrizio Monti

Iterchimica Srl, Via G. Marconi 21, 24040 Suisio, BG, Italy

Abstract. 100% Italian innovation: a new graphene-enhanced super-modifier (GESM) developed by Iterchimica Srl. In November 2017, the three-year-long research ended with a patent filing. This is the evolution of the traditional Polymer Modified Asphalt (PMA - dry method) technology and it consists of polymeric compound additivated with graphene. An experimental research was carried out with virgin aggregates and it showed greater fatigue resistance (up to +250%), higher resilience and lower thermal susceptibility, throughout a wide range of test temperatures. In September 2018 the first on-site trial section was carried out near Rome. The project consisted in the laying of binder (7 cm) and wearing courses (3 cm) with respectively 40% and 30% of Reclaimed Asphalt Pavement (RAP), and an adequate rejuvenator. A kilometer of road was divided in four segments, each made with a different technology: traditional non-modified asphalt concrete, "soft" PMA with standard polymeric compound, "hard" PMA with SBS (Styrene-Butadiene-Styrene) and PMA with the innovative GESM. During the production/laying phases and the following months, the trial pavement was constantly monitored in collaboration with University La Sapienza of Rome. It will be also monitored for the following 5 years. The paper shows the first comparison results between segments 1 and 4.

Keywords: RAP · Recycle · Performance · Graphene · PMA

1 Introduction

Roads are the first thing we see when we leave the house and the last before returning. They contribute constantly to our life quality, particularly the mobility. Most of the world's paved roads are surfaced with Asphalt Concrete (AC), which has good performance and durability for many years, also under heavy traffic loads. Over the past decade, the mobility increase, loads intensification and climatic change have led to research and to develop new technologies to enhance the service life of highly stressed road pavements. Furthermore, many old pavements have been used for decades and have reached their end of life cycle, so they need frequent maintenance to keep an acceptable level of service. Hence, for such pavements, rutting, fatigue cracking and potholes may occur more quickly and severely. In addition, considering the principles of sustainability, it is also necessary to recycle the materials from the milling of end-of-life (EOL) pavements. So, it is very important that the new technology allows to recycle high percentages of RAP (Reclaimed Asphalt Pavement).

M. Pasetto et al. (Eds.): ISAP APE 2019, LNCE 48, pp. 44–54, 2020.
https://doi.org/10.1007/978-3-030-29779-4_5

The traditional PMA technology has been commonly employed for several years in many Countries worldwide, for example in Romania, for binder layers with Stiffness Modulus (at 15 °C and 2 Hz) higher than 4000 MPa and fatigue resistance (at 15 °C) higher than 400000 cycles, and in Algeria, where it's been used for high modulus asphalt concrete with Stiffness Modulus (at 15 °C and 10 Hz) higher than 12000 MPa and fatigue resistance (at 10 °C and 25 Hz) higher than 1000000 cycles. Typical PMA performance with standard plastomers are reported in Sect. 3.

The new graphene-enhanced super-modifier (GESM) is the result of a three-year-long research, inserted as part of the Ecopave Project funded by the Lombardy Region. The project aimed for developing a super-modifier for AC produced with an innovative system with high efficiency and sustainability. The GESM is manufactured using graphene-based materials and plastics coming from industrial waste and separated waste collection, which are destined to the waste-to-energy plant. The principal research steps are (Fig. 1):

- GESM Chemical Formulation
- Asphalt Concrete Preliminary tests
- Trial Section:

  a. AC Laboratory Mix Design
  b. Production and Laying
  c. Internal Laboratory control tests
  d. Official Laboratory control tests

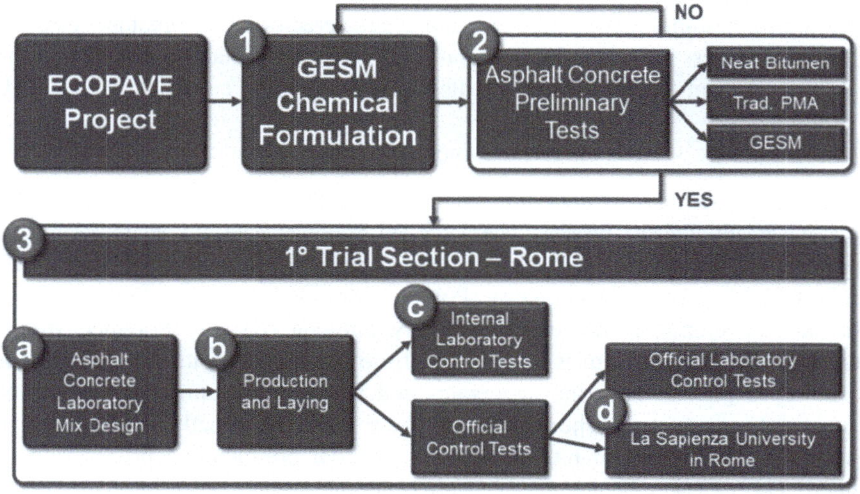

Fig. 1. Research roadmap

2 GESM Chemical Formulation

Throughout the AC production, the ease of use is important both for managing and transportation. The GESM formula is the result of many chemical laboratory tests. Moreover, the compatibility with AC manufacturing process, RAP recycling and production temperature is necessary. The GESM (Fig. 2a) formula is secreted and patented, but is well-known that is composed by:

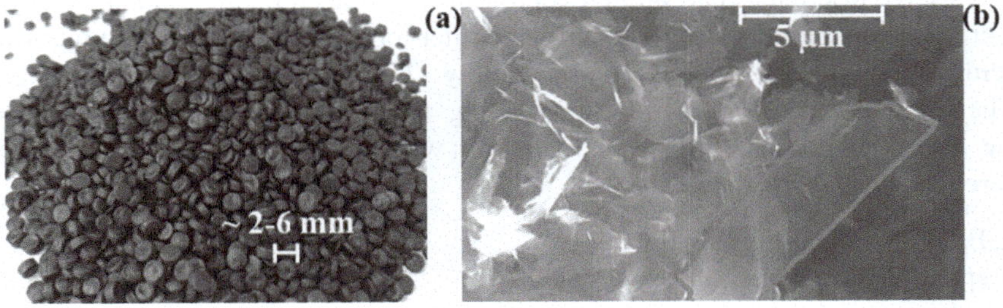

Fig. 2. Graphene nano-platelets (a) and GESM (b)

- recycled plastics: "rigid plastic" materials that cannot be used in the standard recycling chain (materials unrelated to the traditional circuit). These wastes are mainly composed of toys, bins and baskets, plastic tables, etc.;
- graphene-based additive: material composed of pure graphene (single layer of carbon atoms arranged in a honeycomb structure) nano-platelets (Fig. 2b), with optimized morphological and structural characteristics. It is obtained through a patented process based on the physical transformation of natural graphite and it is high-performant, chemical-free, certified as non-toxic and compliant with REACH requirements;
- other additives: various chemical components.

It is very important to underline that GESM granules are produced through an innovative manufacturing process, characterized by lower energy consumption (lower environmental impact) than the polymeric compounds' traditional production method.

The GESM is a PMA technology (Polymer Modified Asphalt – Dry method). In particular, the GESM is not used to modify bitumen (PMB – Polymer Modified Bitumen – Wet method) but is added directly to the mixer (both in continuous and discontinuous plants), after the aggregates and before the bitumen, by means of a pneumatic dosing system. The addition of these additives directly into the pug mill ensures the modification of bituminous mix, without pre-blending process with the bitumen. The new graphene-enhanced super-modifier significantly improves the AC performance, enhancing indirect tensile strength and stiffness modulus, increasing rutting and fatigue resistance. However, the new super-modifier doesn't have rejuvenating properties, therefore it is clear that RAP recycling requires the use of suitable additives.

3 Asphalt Concrete Preliminary Tests

In order to evaluate the behaviour of GESM in the AC, many laboratory tests were performed in comparison with bituminous mixtures produced with traditional technologies. All the mixtures presented in this paper are referred to a binder layer:

- without additives (neat bitumen);
- with standard plastomer – 5% by weight of bitumen;
- with super-modifier (GESM) – 5% by weight of bitumen.

All the mixtures were prepared using the same particle size curve and also the same amount of bitumen (using the Marshall mix design):

- 4.8% by weight of aggregates (Table 1);
- No RAP (only during this phase, to better evaluate the super-modifier behaviour).

Table 1. Specifications of neat bitumen

Parameters	Specification	U.M.	Limits (Class 50/70)	Values
Penetration at 25 °C	UNI EN 1426	0.1 mm	50–70	63
Softening point	UNI EN 1427	°C	46–54	48.5
Breaking point (Fraass)	UNI EN 12593	°C	≤ -8	−12
Values after RTOFT	UNI EN 12607-1			
Residual penetration	UNI EN 1426	%	≥ 50	55
Softening point increase	UNI EN 1427	°C	≤ 9	4.5
Mass variation	UNI EN 12607-1	%	≤ 0.5	0.1

3.1 Physical Analysis – Voids Content

A key feature that highly affects the final performance of asphalt concrete pavements is voids content. For specimen preparation, all the analyzed mixtures were compacted using Gyratory Compactor (UNI EN 12697-31) at 150 °C and 180 cycles (in compliance with ANAS specifications) to achieve the comparable voids content (UNI EN 12697-6) using SSD method (Table 2). It should be noted that the target of this research was to show the different performances and not to verify the absolute values, which changes in function of aggregates and bitumen.

Table 2. Voids content - compaction with gyratory compactor

Mixtures	Voids (%)
AC without additives	2.3
AC with standard plastomer	2.4
AC with super-modifier	2.5

3.2 Mechanical Analysis – Indirect Tensile Strength (ITS)

The ITS (UNI EN 12697-23) increased using super-modifier, this means higher pavement resistance under traffic loading and longer life time (Table 3). In details, for the mixtures containing GESM the ITS values at 25 °C increased by +59.8% compared to the mixtures with neat bitumen.

Table 3. Indirect tensile strength at 25 °C

Mixtures	ITS (MPa)
AC without additives (WA)	1.07
AC with standard plastomer	1.57
AC with super-modifier (GESM)	1.71
Variable percentage GESM-WA	+59.8%

3.3 Dynamic Tests - Stiffness Modulus

The Stiffness Modulus was tested using IT-CY method (EN 12697-26 Annex C). Higher stiffness modulus results in a better vehicle loads distribution to the lower pavement layers, but the behaviour at the different temperatures is conditioned also by the raw materials used. In this case, the stiffness due to the presence of GESM rises until +107.7% at 5 °C and +174.0% at 40 °C, compared to neat bitumen (Table 4). In comparison with these result, the tests performed at 20 °C returned a particular value (+36.7%) that will be kept under control during the Trial Section controls (Sect. 4).

Table 4. Indirect tensile stiffness modulus at different temperatures applying 2 Hz frequency

Mixtures	ITSM (MPa)		
	T = 5 °C	T = 20 °C	T = 40 °C
AC without additives (WA)	10169	5711	1096
AC with standard plastomer	20866	6685	2691
AC with super-modifier (GESM)	21124	7809	3003
Variable percentage GESM-WA	+107.7%	+36.7%	+174.0%

3.4 Dynamic Tests – Rutting

Rutting is caused by the vehicles' pass and it appears as longitudinal pavement deformations. The tests were performed using the Wheel Tracking method (UNI EN 12697-22). The comparison between the different mixtures testes has highlighted the better super-modifier behaviour. In fact, the deformation depths was decreased about −69% (Table 5).

Table 5. Rutting resistance at 60 °C (mm)

Mixtures	Depression depth (mm) at 5000 cycles	Depression depth (mm) at 10000 cycles
AC without additives (WA)	1.39	1.55
AC with standard Plastomer	0.88	0.98
AC with super-modifier (GESM)	0.43	0.48
Variable percentage GESM-WA	−69.1%	−69.0%

3.5 Dynamic Tests – Fatigue Endurance

The pavement's service life is directly linked to the asphalt concrete's fatigue resistance. In laboratory, fatigue test is carried out simulating seasonality and traffic loads.

During this research, the fatigue resistance using ITT method (EN 12697-24 Annex E) was assessed at 40 °C under constant stress-control condition. The results showed that the number of loads to failure increased significantly with GESM (about +570%), in comparison with neat bitumen mixture (Table 6).

Table 6. Fatigue resistance at 20 °C and 2 Hz

Mixtures	Number of loads to failure
AC without additives (WA)	157639
AC with standard Plastomer	473167
AC with super-modifier (GESM)	1056933
Variable percentage GESM-WA	+570.5%

4 Trial Section – Rome

4.1 Description of Trial Section

With the target to verify the new super-modifier on full-scale, the first trial section was laid down in Rome in September 2018, thanks to the collaboration between Iterchimica S.r.l. and Città Metropolitana di Roma Capitale. The reference road section was part of a tender awarded work (SP 3 Ardeatina, on the south lane towards Santa Palomba). The first step was the choice of 1 km path with homogenous physical and mechanical characteristics considering the pavement deterioration (Fig. 3).

Fig. 3. AC condition SP 3 Ardeatina before the trial section

The analysis was done by HWD (Heavy Falling Weight Deflectometer). The trial path was divided in four 250-metres-long segments: one with traditional asphalt concrete, one soft-modified with standard plastomer, one hard-modified with SBS and the last one with the GESM product (Fig. 4).

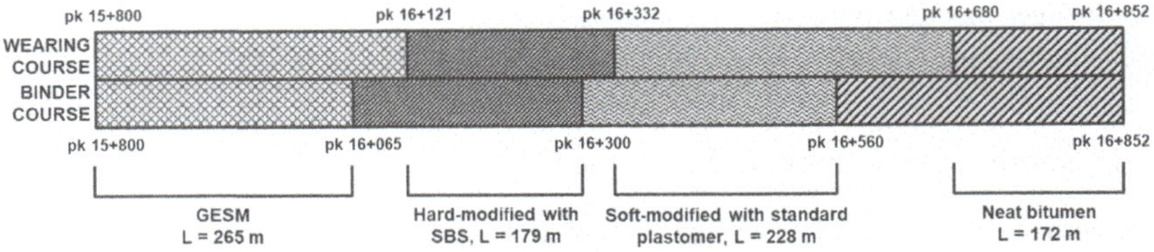

Fig. 4. Trial section path division

Table 7. Blend details of binder course AC

Binder course mixtures	RAP (%)	Rejuvenator on RAP weight (%)	Total bitumen on aggregates weight (%)	Modifier on total bitumen weight (%)
Neat bitumen	40	0.2	4.40	0
GESM	40	0.2	4.40	5
AC hard-modified with SBS	40	0.2	4.40	5
AC soft-modified with standard plastomer	40	0.2	4.40	3

Table 8. Blend details of wearing course AC

Wearing course mixtures	RAP (%)	Rejuvenator on RAP weight (%)	Total bitumen on aggregates weight (%)	Modifier on total bitumen weight (%)
Neat bitumen	30	0.2	5.75	0
GESM	30	0.2	5.75	5
AC hard-modified with SBS	30	0.2	5.75	5
AC soft-modified with standard plastomer	30	0.2	5.75	3

Fig. 5. Trial section

Compared to the preliminary tests, it should be noted that for the trial section (Tables 7, 8 and Fig. 5):

- both binder and wearing course were produced and analyzed;
- the grading curve used for the binder course was different from the one used in the preliminary tests, in order to comply with the local specifications' requirements;
- the bitumen percentage of both mixtures was determined according to the local specifications (the optimum bitumen content was defined by Marshall test);
- the mixtures contained a certain amount of RAP (40% for the binder course and 30% for the wearing course), therefore a suitable rejuvenating agent was also used, adding it into the neat bitumen.

4.2 Trial Section Preliminary in Situ Test Results

HWD (Heavy Falling Weight Deflectometer) tests, that were already used to identify the best road section where to perform the trial, were carried out again one months after the laying. The results regarding neat bitumen and GESM showed that the super-modifier increases the performance from 67% to 76%, compared to the neat bitumen (Fig. 6). In order to monitor the trial section, HWD tests are already planned for the next 5 years.

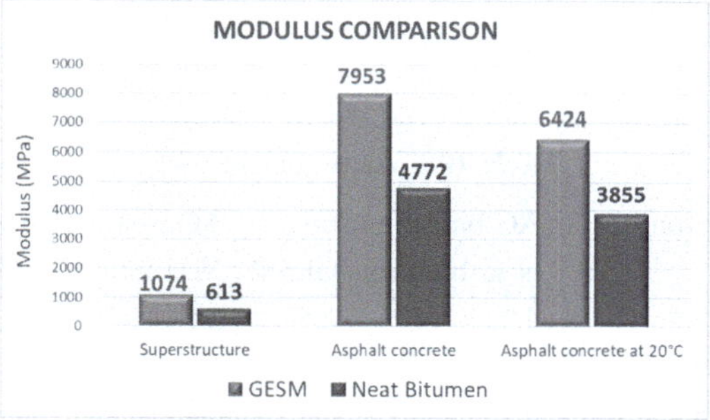

Fig. 6. HWD comparison between neat bitumen and GESM

4.3 Trial Section Preliminary Results in Laboratory

The tests are currently ongoing at an Official Authorized Laboratory, however the first results are already available.

All specimens were compacted with gyratory compactor (180 cycles and 150 °C). Shown in Fig. 7 and Table 9:

- the grading curves are similar to the mix design's ones, so they are acceptable (differences related to the specifications);
- the bitumen contents are slightly different from the mix design.

These differences could lead to other concerns such as residual voids and performances. Tables 9, 10 and 11 make a comparison between the result obtained by

Iterchimica Research and Development department during the different phases of the project: preliminary research tests (R&D), prequalification tests and production controls.

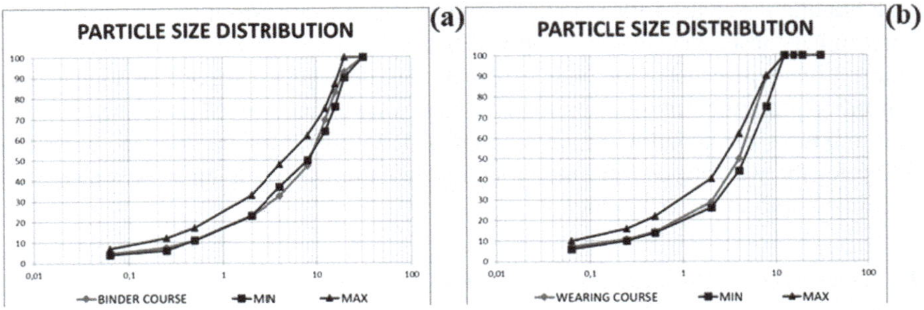

Fig. 7. Particle size distribution production control of binder course (a) and wearing course (b)

Table 9. Bitumen content (% by weight of aggregates)

Comparison	AC binder course		AC wearing course	
	Neat bitumen	GESM	Neat bitumen	GESM
R&D	4.80	4.80	-	-
Prequalification	4.40	4.40	5.75	5.75
Production control	3.59	3.87	5.68	6.02

Table 10. Voids content (%)

Comparison	AC binder course		AC wearing course	
	Neat bitumen	GESM	Neat bitumen	GESM
R&D	2.3	2.5	2.5	2.4
Prequalification	-	2.3	-	2.2
Production control	5.1	5.4	6.2	5.5

Table 11. ITS results (MPa)

Comparison	AC binder course		AC wearing course	
	Neat bitumen	GESM	Neat bitumen	GESM
R&D	1.07	1.71	-	-
Prequalification	-	1.77	-	1.74
Production control	1.22	1.47	1.11	1.50

Despite the quantity of bitumen and the voids content, Indirect Tensile Strength (ITS) values increase from 20% to 35% (Table 11).

Furthermore, after the static tests, dynamic tests were carried out on the AC from the trial section in Rome. Results have shown an improvement in the performances of AC with GESM regarding stiffness, rutting and fatigue resistance (Fig. 8).

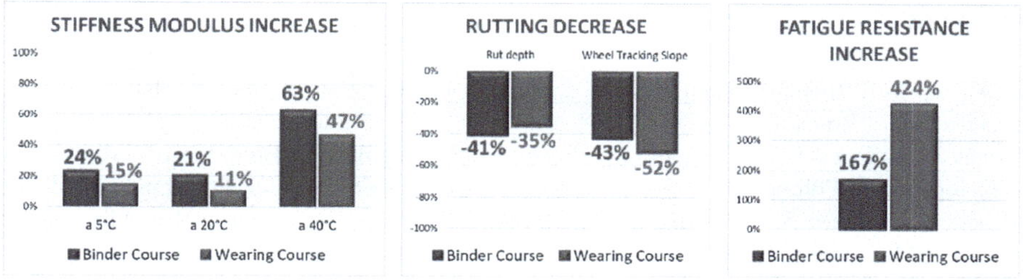

Fig. 8. Mechanical tests results

The mechanical tests results showed an improvement in asphalt concrete performances, especially regarding the fatigue resistance, where it has been improved to 167–424%.

5 Conclusion

Iterchimica took part in Lombardy Region Project in 2017 and the road results showed how the goal has been reached in 2019. Iterchimica R&D developed the new graphene-enhanced super-modifier (GESM) in order to increase the pavement life. The main benefits can be evaluated by modulus, rutting and fatigue. The GESM is manufactured with an innovative system based on high efficiency and sustainability. The raw sources used are graphene-based materials and plastics coming from waste. In September 2018, thanks to the collaboration between Iterchimica S.r.l. and Città Metropolitana di Roma Capitale, Rome has been the first city to pave the first trial section (1 km) using GESM. Iterchimica's laboratory studied the mix design and the job mix formula in order to provide the right recipe containing GESM.

The discrepancies between the results obtained from the laboratory preliminary tests and the trial section can be ascribed to multiple causes, among which:

- different raw materials (aggregates and bitumen);
- different grading curves, that for the trial section had to comply with local specifications;
- the addition of a certain amount of RAP;
- the use of a continuous plant.

Despite these non-optimal conditions, the results provided by Official Authorized Laboratories show that the new technology increases the pavement life. For example, the rut depth decreases by 35–40%, the Wheel Tracking Slope decreases by 40–50% and the fatigue resistance increases by 167–424%.

Acknowledgements. We would like to thank all the Collaborators and Partners who are developing this research: Directa Plus, G.Eco, University of Milan "Bicocca".

References

ANAS S.p.A. (2014) Capitolato speciale di appalto- Parte 2ª Norme tecniche - Pavimentazioni autostradali. Roma

González Ramirez AJ (2016) Studio Sperimentale dei Conglomerati Bituminosi Modificati con Compound di Polimeri e Fibre – Master Thesis in Civil Engineering, Politecnico di Milano

Micaelo R, Gameiro A, Quaresma L, Picado-Santos L (2016) Production of Hot-Mix Asphalt with PMB: compatibility and mechanical behaviour characterization. In: 8th RILEM International Symposium, Ancona

Federal Highway Administration (2011) Reclaimed Asphalt Pavement in Asphalt Mixtures State of the Practice. FHWA-HRT-11-02, Virginia

Abdulshafi A, Kaloush KE (1990) Modifiers for Asphalt Concrete - Resource international, Inc 281 Enterprise Drive, Westerville OH 43081

Properties of Asphalt Binders with Increasing SBS Polymer Modification

Mike Aurilio[1,2(✉)], Peter Mikhailenko[1,3], Hassan Baaj[1], and Lily D. Poulikakos[3]

[1] Centre for Pavement and Transportation Technology (CPATT), Department of Civil and Environmental Engineering, University of Waterloo, Waterloo, Canada
maurilio@yellowline.ca
[2] Yellowline Asphalt Products Ltd., Hamilton, Canada
[3] Empa, Swiss Federal Laboratories for Materials Testing and Research, Laboratory for Road Engineering/Sealing Components, Dübendorf, Switzerland

Abstract. The properties of several asphalt binders with gradual styrene-butane-styrene (SBS) polymer modification were studied. The base binders used were straight-run PG 52-34, 58-28 and 64-28 which were modified with polymers at concentrations of 0, 1, 2, 3 and 4%. Two types of SBS polymers were used with different grain sizes and they were integrated into the binder with the same cross-linking agent. The rheological properties were studied using a Dynamic Shear Rheometer (DSR) with Temperature-Frequency Sweep Master Curves and the Multiple Stress Creep Recovery (MSCR) Test. The Master Curves showed that the polymer modification, up to 4% addition, increases the binder stiffness at high temperatures, which is a good indication of improved rutting resistance. The MSCR test showed improved percent recovery and compliance with increasing concentrations of SBS. To observe the effects of polymer addition on a microscopic level, the unaged binders were analyzed with the Environmental Scanning Electron Microscope (ESEM), finding that there were no significant changes with the level of polymer addition.

Keywords: Polymer modified asphalt binder · Bitumen · PMA · PmB · Rheology · ESEM

1 Introduction

The study of polymer modified asphalts (PMAs) has been crucial in the development of high performance asphalt pavements. Many studies have shown that PMAs have been used successfully to improve the rutting characteristics (Anderson et al. 2011; Bernier et al. 2012; Tayfur et al. 2007; Yildirim 2007) and the fatigue performance of hot mix asphalt (Baaj et al. 2005; Aurilio et al. 2018). Styrene-Butadiene-Styrene (SBS) is one such polymer that has been used successfully to improve the properties of asphalt cement. SBS is an elastomeric co-block polymer which helps to increase the elastomeric properties of asphalt binder (da Silva et al. 2004). The ratio of styrene to butadiene and the asphalt cement chemical composition can impact the material

The original version of this chapter was revised: The first author name has been updated. The correction to this chapter is available at https://doi.org/10.1007/978-3-030-29779-4_49

M. Pasetto et al. (Eds.): ISAP APE 2019, LNCE 48, pp. 55–66, 2020.
https://doi.org/10.1007/978-3-030-29779-4_6

compatibility (Becker et al. 2001). The even dispersion of the polymer in the asphalt binder is critical to the performance of the PMA (Kou et al. 2017) and the compatibility between SBS and asphalt cement are two factors that can affect this (Dong et al. 2014). SBS has been shown to improve the high and low temperature properties of asphalt cement (Zubeck et al. 2003).

In order to evaluate the performance of PMAs, the Multiple Stress Creep Recovery (MSCR) test was developed and has since been adopted in the United States and parts of Canada. This test is performed using the Dynamic Shear Rheometer (DSR) at the local 7-day maximum pavement temperature. The MSCR measures the compliance (Jnr) of the PMA and the ability to recover from an applied load in the form of percent recovery. The compliance has been shown to have a good correlation with rutting performance (Anderson et al. 2011) and shown to be highly reproducible (Hossain et al. 2016). Evaluation of the MSCR also revealed that it is sensitive to optimum blending making it a good tool for the evaluation of PMAs (Hossain et al. 2016).

Polymer modification affects asphalt binder on a microstructural level and Environmental Scanning Electron Microscopy can be used to evaluate this (Mikhailenko et al. 2017a; Rozeveld et al. 1997). The ESEM observation of asphalt binder reveals the development of a fibril structure that evolves with exposure to the electron beam. This fibril formation was shown to be a function of, binder type and aging. This is hypothesized to in some way represent the heavier hydrocarbon fractions of the binder while the lighter hydrocarbons are dispersed (Mikhailenko et al. 2019). It has also been shown that polymer modification affects this structure (Mikhailenko et al. 2017b).

In the following study, the MSCR test was used to evaluate the relationship between increasing concentrations of SBS and different sources of asphalt cement. In addition to this, Temperature-Frequency Sweep Master Curves were used to evaluate how the concentration of SBS changes the rheological properties of asphalt cement. Finally, ESEM analysis was used to evaluate the integration of the SBS in the binder on a microscopic level.

2 Asphalt Binder Preparation and Testing

The asphalt blending process was completed using a Silverson high shear mixer and heating mantle. After preheating the asphalt cement to 170 °C, polymer was added and mixed for one hour. A crosslinking agent was added at the end of the hour and allowed to blend for a further 30 min. The crosslinking agent was added at 10% by weight of the polymer as this was historically a sufficient amount to produce quality PMA in the experience of the authors. After blending the crosslinking agent and polymer, the high shear mixer speed was reduced, and the sample was blended for another hour. The final hour was used as a curing time. Temperatures were monitored and maintained at 180 °C ± 5 °C during the curing process. Three binders with different Superpave performance grades were selected: a PG 64-28, PG 58-28 and PG 52-34, from hardest to softest, respectively. They were blended using the same sources of SBS and at concentrations of 0, 1, 2, 3, and 4%. The two sources of SBS chosen have a linear microstructure. SBS A has a styrene content of 31.1% and SBS B has a styrene content of 31.6%.

AASHTO T350 was followed to test the Jnr 3.2, percent recovery, and Jnr Diff. The climatic pavement high temperature in Southwestern Ontario is 58 °C and this was used for all samples. This test is conducted on Rolling Thin Film Oven (RTFO) aged material. The Jnr is the non-recoverable creep compliance of the binder; Jnr Diff is a measure of the strain sensitivity, while the percent recovery measures the elastic response. The test is run at two different stress levels; the data generated from applying 3.2 kPa of shear stress is used to report the non-recoverable creep compliance and percent recovery, while the 0.1 kPa data is only used to calculate Jnr Diff. Figure 1 shows how the strain changes with time for three cycles. The recovered strain is reported as the percent recovery.

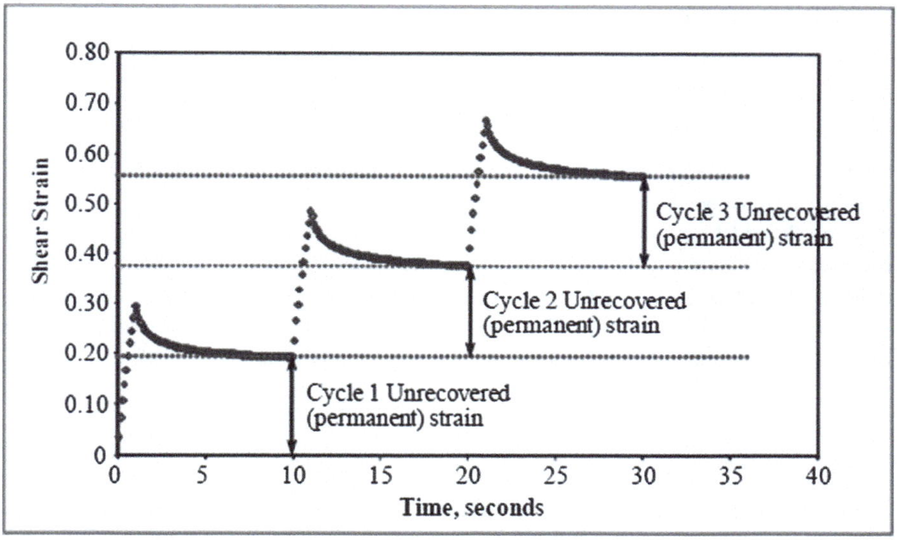

Fig. 1. Example of MSCR data showing shear strain versus time for three loading cycles (Anderson et al. 2011)

Master Curves were generated by performing Temperature-Frequency Sweeps on the PG 58-28 and PG 64-28. At the time of writing this paper, the PG 52-34 had not yet been tested. This was done using the DSR and the standard 25-mm and 8-mm plates to determine the modulus. The Master Curves for each binder were created using RTFOT + Pressure Aging Vessel (PAV) aged material and converting the creep stiffness data from Bending Beam Rheometer (BBR) testing to the modulus. Table 1 shows the temperatures used for DSR testing, based on the PG grading of the source binder. Strain rates of 0.1% were used for intermediate temperature testing and a strain rate of 0.5% was used for high temperature testing. Some intermediate temperatures were omitted if the asphalt cement was too stiff to be tested or had difficulty adhering to the plates.

Table 1. Testing temperatures for development of master curves

Grade	Low temperature (°C)	Intermediate temperature (°C)	High temperature (°C)
PG 58-28	−18	4, 10, 15, 20, 30	52, 58, 64, 70
PG 64-28	−18	10, 15, 20, 30	52, 58, 64, 70

The raw data was fitted into the 2S2P1D model that is used to predict viscoelastic properties at different temperatures and loading frequencies. This model is based on the combination of two springs, two parabolic elements and one dashpot in series. To generate a single curve, the complex modulus is shifted to the reduced frequency using the temperature. The 2S2P1D model is then fitted to the data to produce a master curve. The 2S2P1D model can be defined by Eq. (1):

$$G^*(\omega) = G_0 + \frac{G_g - G_0}{1 + \alpha(i\omega\tau)^{-k} + (i\omega\tau)^{-h} + (i\omega\beta\tau)^{-1}} \tag{1}$$

where G_g is the glassy modulus as the reduced frequency, ω, reaches infinity and is approximated to be $1\text{x}10^9$ Pa. G_0 is assumed to be zero for asphalt binders, the exponents k and h are defined by $0 < k < h < 1$, α is a constant, β is related to the Newtonian viscosity and τ is the characteristic time (Olard and Di Benedetto 2003; Yusoff et al. 2013). BBR data was used in the curves by taking the creep stiffness measurement at different times. The time is converted into a frequency and the creep stiffness is converted into the modulus using Eq. (2) (Booshehrian et al. 2012):

$$G^*(\omega) \approx \frac{S(t)}{3} \tag{2}$$

Where ω is the reduced frequency and S(t) is the creep stiffness measured by the BBR. Master Curves can be produced for asphalt cement and HMA. The 2S2P1D model has been used to characterize asphalt cement and hot mix asphalt master curves (Olard and Di Benedetto 2003). The model parameters were originally taken from work completed by Yusoff et al. (2013) but modified using a combination Microsoft Solver® and trial and error.

AASHTO T313 was completed using the BBR to characterize the low temperature properties. Testing temperatures were chosen using the low temperature performance grade of the base asphalt (PG Low Temperature + 10 °C). The low temperature performance is characterized using the creep stiffness and the m-Value, which is the rate of change of creep stiffness with time (The Asphalt Handbook MS-4).

The binder samples for ESEM were prepared according to a protocol developed previously (Mikhailenko et al. 2017a) and the test was conducted on unaged binders. The observations were conducted with a FEI Quanta 650 FEG ESEM. The observation parameters were an acceleration voltage of 20 keV, a spot size of 3.5, a chamber pressure of 0.8 mbar in low vacuum mode, and a magnification of 1000x in secondary electron (SE) mode.

3 Results and Discussion

The MSCR compliance results can be seen in Fig. 2. The compliance of the asphalt cement decreases as the concentration of polymer increases. This indicates that the material is becoming stiffer and more elastic and therefore more resistant to rutting. As expected based on the PG grading, the PG 52-34 binders consistently showed the highest compliance values and the PG 64-28 consistently shows the lowest compliance values. This is because the modulus of the base asphalt cement also plays a role in the compliance. As the high temperature performance grade increases, so does the modulus. The values converge to very low compliance values at 4% SBS, indicating that the polymer becomes the dominant factor as the concentration increases. This data collected corresponds with previous observations with respect to the ability of SBS to improve the rutting resistance of asphalt cement.

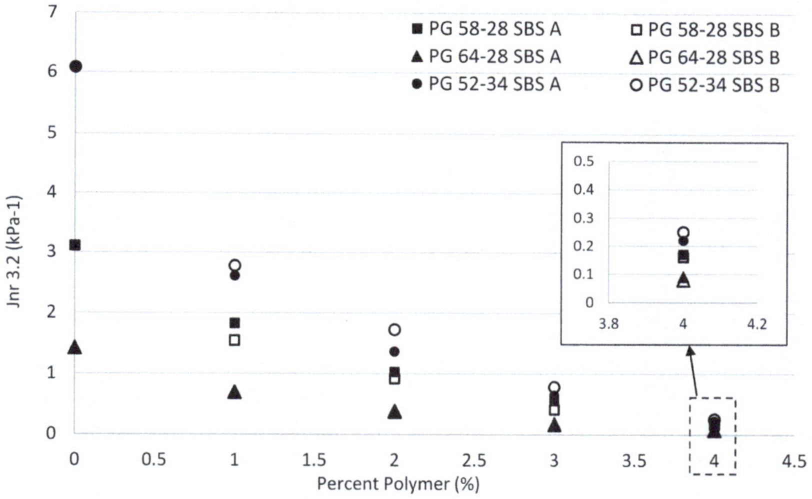

Fig. 2. Relationship between Jnr (3.2 kPa) and concentration of polymer for PG 52-34, PG 58-28 and PG 64-28 with two types of SBS polymer A and B

The percent recovery portion of the testing showed the increasing elastomeric properties of the asphalt cement as the concentration of polymer increased (Fig. 3). The PG 64-28 increases the fastest, but all three asphalt binders have similar percent recovery values at 4%. The convergence at 4% indicates that there is a limitation to improving the elastic performance with SBS. When compared to the MSCR acceptance curve given by AASHTO M332, it can be seen that the PG 52-34 and PG 58-28 require more than 2% SBS in order to meet the limit. PG 64-28 is able to meet the specifications with only 2% SBS. The limit is determined by the compliance value of individual samples. In this case, the softer asphalt binders required more polymer in order to achieve the same elastic recovery. It was expected that the percent recovery would be similar for each SBS at a given concentration of polymer given the similarity in styrene content. This generally holds true with one exception at 3% for the PG 58-28. The exact reason for this is unknown and retesting newly prepared PG 58-28 with 3% SBS A did not change the result.

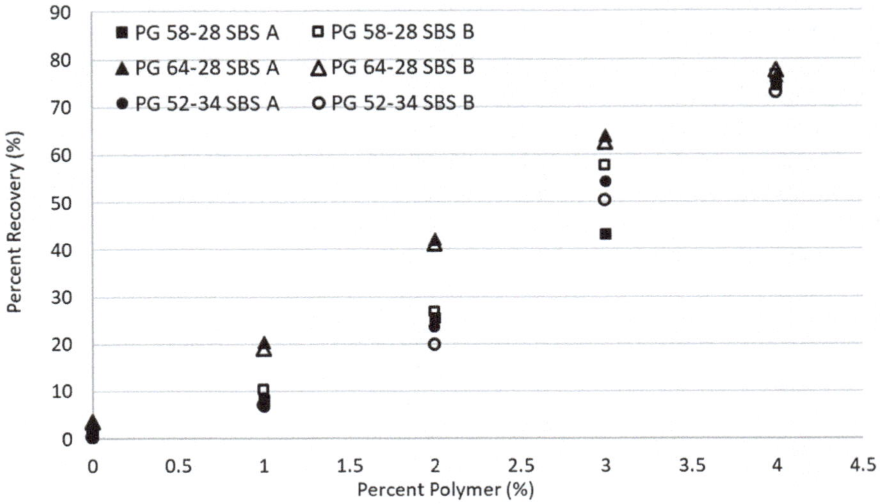

Fig. 3. Relationship between percent recovery and concentration of polymer for PG 52-34, PG 58-28 and PG 64-28 with two types of SBS polymer A and B

Table 2. BBR low temperature m-Value and creep stiffness versus the concentration of polymer

Polymer concentration	Creep stiffness (MPa)					m-Value				
	0%	1%	2%	3%	4%	0%	1%	2%	3%	4%
PG 58-28 SBS A	187	172	167	162	145	0.358	0.359	0.357	0.357	0.373
PG 58-28 SBS B		178	164	150	130		0.354	0.355	0.353	0.354
PG 64-28 SBS A	291	286	270	255	230	0.310	0.313	0.318	0.323	0.332
PG 64-28 SBS B		278	275	256	226		0.312	0.315	0.318	0.328
PG 52-34 SBS A	245	250	244	221	200	0.358	0.328	0.333	0.342	0.354
PG 52-34 SBS B		249	228	216	209		0.343	0.339	0.350	0.340

Table 2 presents the BBR m-Value and creep stiffness data. The m-Value has been said to be analogous to the phase angle (Asphalt Institute MS-25) and higher values of indicate a less elastic and more viscous behaviour (Rowe et al. 2014). However, the results found here varied for different binders. PG 64-28 tends to increase its m-Value at higher concentrations of polymer. The m-Value of the PG 58-28 does not seem to be greatly affected by the concentration of polymer although it appears to increase at 4% SBS A. The addition of polymer to the PG 52-34 appears to decrease the m-Value at lower concentrations of polymer, but overall no consistent relationship is visible. Analysis of low temperature creep stiffness indicates that polymers lower the creep stiffness and provide a benefit at low temperature with increased addition.

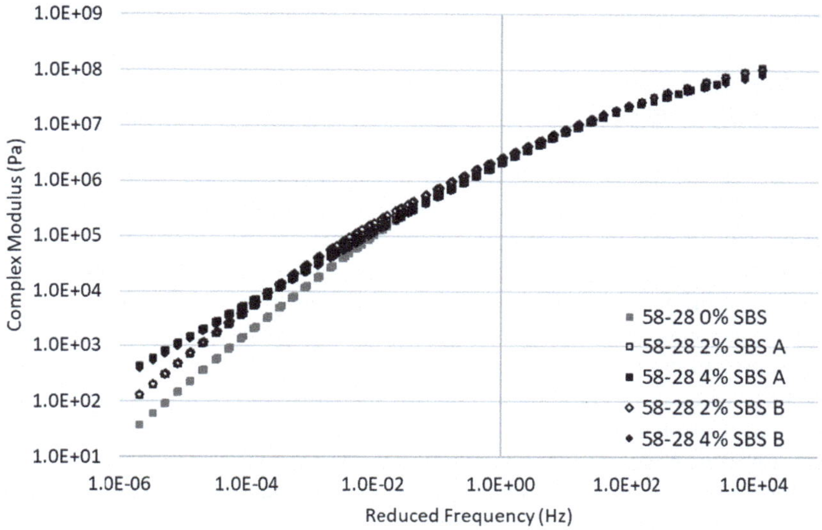

Fig. 4. Master Curve for PG 58-28 SBS A and B: complex modulus versus reduced frequency

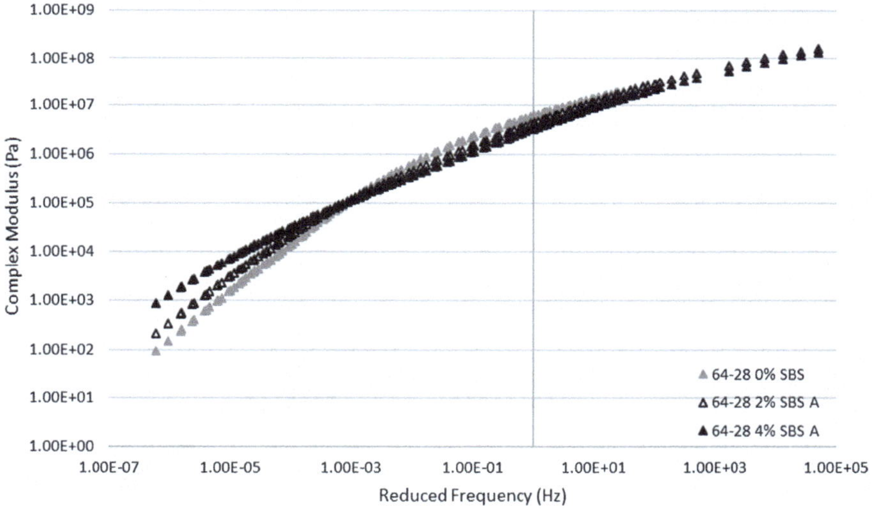

Fig. 5. Master Curve for PG 64-28 SBS A: complex modulus versus reduced frequency

The Master Curves were produced by fitting the raw data for PG 58-28 SBS A and B and PG 64-28 SBS A to the 2S2P1D model. Figure 4 illustrates the relationship between the complex modulus and the reduced frequency for PG 58-28 0, 2 and 4% SBS A and B. The lower frequency portion of the curve, which represents the high temperature performance, indicates that the increasing polymer concentration increases the modulus. There is also a slight decrease in the modulus at high frequency (low temperature) and intermediate frequencies. This same relationship can be seen when looking at the PG 64-28 in Fig. 5. The modulus increases at high temperatures indicating improved resistance to rutting, as found previously with PMA binders (Mazumder et al. 2016). Like the PG 58-28, the PG 64-28 also exhibits a lower modulus at low and intermediate temperatures. The decrease in stiffness at lower temperatures can be an indicator of improved low temperature performance (Pszczola et al. 2018).

Figures 6 and 7 display the black space diagrams for the PG 58-28 and the PG 64-28 produced using the raw DSR data. The low modulus portion of the graph represents the high temperatures and the high modulus portion represents the low temperature. It can be seen from both data sets that the PMAs have much lower phase angles at high temperatures as the concentration of polymer increases, indicating increased elasticity and rutting performance (Olard and Di Benedetto 2003; Said et al. 2011). These regions appear as plateaus in the data indicating that the storage and loss moduli vary proportionally and a shift of the phase angle to lower values indicating the dominance of the elastic part of the complex modulus at higher concentration of polymers. As was also found previously with increased polymer addition to binders (Planche et al. 2008). This may also indicate the PMAs have improved temperature susceptibility. This shape is also characteristic of rubber indicating a more rubber-like behaviour is present (da Silva et al. 2004).

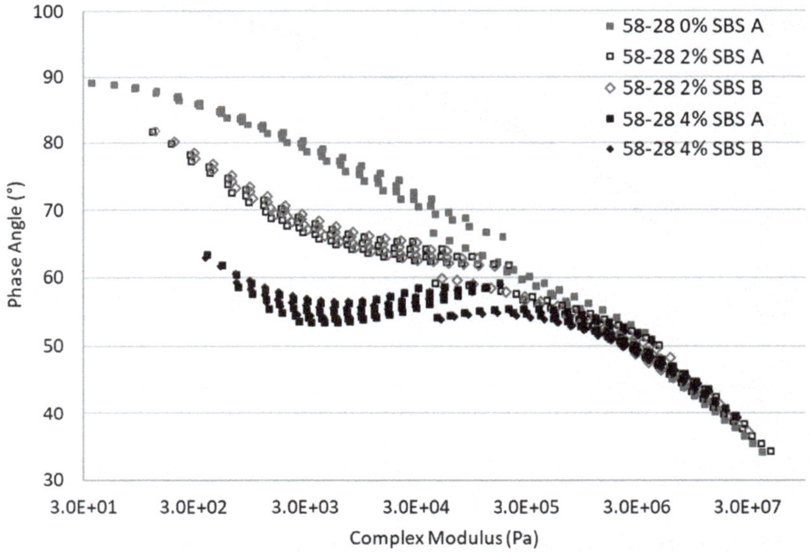

Fig. 6. Black space diagram for PG 58-28 SBS A and B: complex modulus versus phase angle

At 15 °C, the phase angle is highest for both binders at 4% SBS A and B. This can be seen on the black space diagrams at approximately 3×10^7 Pa. The phase angle does not seem to change significantly at 2% SBS. This would suggest a higher concentration of polymer is required to make a significant shift in the viscoelasticity. A higher phase angle at lower temperatures may suggest that the binder is more flexible (Moreno-Navarro et al. 2015), but there may be a limitation to the use of polymers in colder conditions, although this did not appear to be the case with the BBR results.

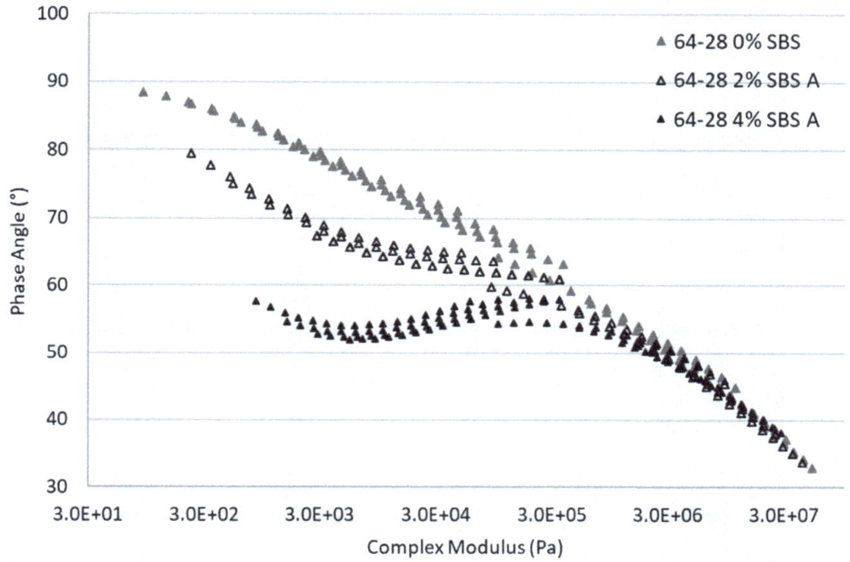

Fig. 7. Black space diagram for PG 64-28 SBS A: complex modulus versus phase angle

The ESEM analysis was conducted on the PG 64-28 binders with SBS A and the PG 58-28 binders with SBS A and B. The ESEM images shown in Fig. 8 indicate that there is some difference between the based binders, with the fibrils for the PG 64-28 binder being somewhat thinner. The difference between the binders plays more of a role and the difference is not that significant between the increments of SBS modification. A previous ESEM analysis of the SBS binder showed a more significant change in the fibril structure with SBS addition (Mikhailenko et al. 2017b), although 10% SBS was used in the referenced work, much higher than the current study. It is also worth considering that the physical tests were conducted on the aged binders which is not the case here. The images also indicate that there was no significant aging during the binder modification process that could manifest into a change in the microstructure. A quantitative analysis of these images will be provided in a journal version of this paper.

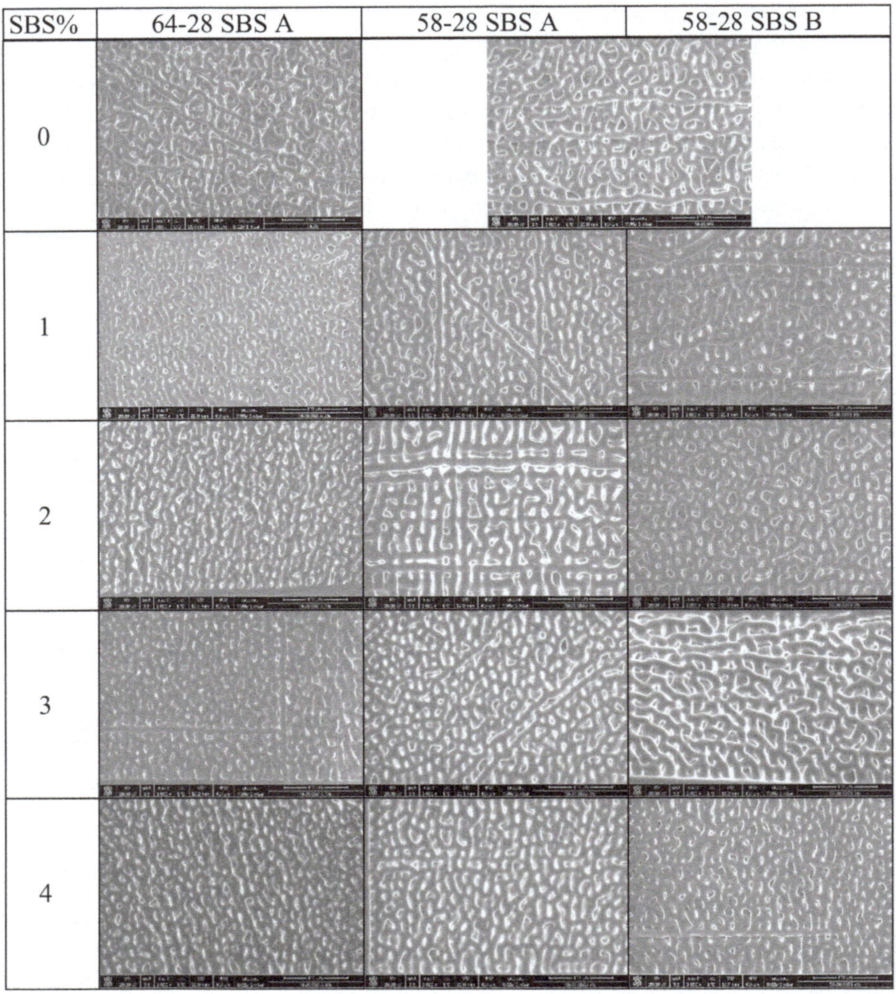

Fig. 8. ESEM Analysis showing the effect of SBS addition (bar on bottom right of images indicates 100 μm)

4 Conclusions

The following conclusions can be drawn from the present study:

- The non-recoverable creep compliance was reduced and the elastic recovery increased significantly with the addition of SBS polymer, up to the maximum addition of 4%. The impact in creep compliance reduction was higher for the softer binders up to where there was little difference between these binders at 4% SBS.
- The BBR creep stiffness was reduced with SBS addition, which is an indicator of improved low-temperature properties.
- The SBS addition also improved the stiffness of the binders at high temperatures.

- The ESEM analysis did not reveal large changes in the fibril structure with SBS modification for the unaged binders, but showed no signs of aging during the modification process.

References

Anderson RM, D'Angelo JA, Bukowski JR (2011) Evaluation of the MSCR test for Canadian asphalt binders. In: Proceedings, Canadian Technical Asphalt Association, vol 56, pp 61–82

Asphalt Binder Testing MS-25 (2014) Asphalt Institute

Aurilio M, Qabur A, Mikhailenko P, Baaj H (2018) Comparing the fatigue performance of HMA samples with PMA to their multiple stress creep recovery and double notched tension test properties. In: Proceedings, Canadian Technical Asphalt Association, vol 63, pp 385–410

Baaj H, Di Benedetto H, Chaverot P (2005) Effect of binder characteristics on fatigue of asphalt pavement using an intrinsic damage approach. Road Mater Pavement Des 6(2):147–174

Becker Y, Mendez M, Rodriguez Y (2001) Polymer modified asphalt. Vis Tecnol 9(1):39–50

Bernier A, Zofka A, Yut I (2012) Laboratory evaluation of rutting susceptibility of polymer-modified asphalt mixtures containing recycled pavements. Constr Build Mater 31:58–66

Booshehrian A, Mogawer WS, Bonaquist R (2012) How to construct an asphalt binder master curve and assess the degree of blending between RAP and virgin binders. J Mater Civ Eng 25 (12):1813–1821

Dong F, Zhao W, Zhang Y, Wei J, Fan W, Yu Y, Wang Z (2014) Influence of SBS and asphalt on SBS dispersion and the performance of modified asphalt. Constr Build Mater 62:1–7

Hossain Z, Ghosh D, Zaman M, Hobson K (2016) Use of the multiple stress creep recovery (MSCR) test method to characterize polymer-modified asphalt binders. J Test Eval 44 (1):507–520

Kou C, Xiao P, Kang A, Mikhailenko P, Baaj H, Wu Z (2017) Protocol for the morphology analysis of SBS polymer modified bitumen images obtained by using fluorescent microscopy. Int J Pavement Eng. https://doi.org/10.1080/10298436.2017.1316647

Mazumder M, Kim H, Lee S-J (2016) Performance properties of polymer modified asphalt binders containing wax additives. Int J Pavement Res Technol 9(2):128–139

Mikhailenko P, Kadhim H, Baaj H, Tighe S (2017a) Observation of asphalt binder microstructure with ESEM. J Microsc 267(3):347–355

Mikhailenko P, Kou C, Baaj H, Tighe S (2017b) Observation of polymer modified asphalt microstructure by ESEM. Presented at the 6th international conference on engineering mechanics and materials CSCE 2017, Vancouver, Canada

Mikhailenko P, Kou C, Baaj H, Poulikakos L, Cannone-Falchetto A, Besamusca J, Hofko B (2019) Comparison of ESEM and physical properties of virgin and laboratory aged asphalt binders. Fuel 235:627–638

Moreno-Navarro F, Sol-Sánchez M, Rubio-Gámez MC (2015) The effect of polymer modified binders on the long-term performance of bituminous mixtures: the influence of temperature. Mater Des 78:5–11

Olard F, Di Benedetto H (2003) General "2S2P1D" model and relation between the linear viscoelastic behaviours of bituminous binders and mixes. Road Mater Pavement Des 4 (2):185–224

Planche JP, Mouillet V, Dumas P, Lapalu L (2008) Aging properties of elastomer modified binders. In: Proceedings of 4th eurasphalt & eurobitume congress. Presented at the 2008 eurasphalt & eurobitume congress, Copenhagen

Pszczola M, Jaczewski M, Rys D, Jaskula P, Szydlowski C (2018) Evaluation of asphalt mixture low-temperature performance in bending beam creep test. Materials 11(1):100

Rowe GM, King G, Anderson M (2014) The influence of binder rheology on the cracking of asphalt mixes in airport and highway projects. J Test Eval 42:1063–1072

Rozeveld SJ, Shin EE, Bhurke A, France L, Drzal LT (1997) Network morphology of straight and polymer modified asphalt cements. Microsc Res Tech 38(5):529–543

Tayfur S, Ozen H, Aksoy A (2007) Investigation of rutting performance of asphalt mixtures containing polymer modifiers. Constr Build Mater 21(2):328–337

Said SF, Hakim H, Oscarsson E, Hjort M (2011) Prediction of flow rutting in asphalt concrete layers. Int J Pavement Eng 12(6):519–532

da Silva LS, de Camargo Forte MM, de Alencastro Vignol L, Cardozo NSM (2004) Study of rheological properties of pure and polymer-modified Brazilian asphalt binders. J Mater Sci 39 (2):539–546

The Asphalt Handbook MS-4 (2007) Asphalt Institute, Lexington, KY

Yildirim Y (2007) Polymer modified asphalt binders. Constr Build Mater 21(1):66–72

Yusoff NIM, Mounier D, Marc-Stéphane G, Rosli Hainin M, Airey GD, Di Benedetto H (2013) Modelling the rheological properties of bituminous binders using the 2S2P1D model. Constr Build Mater 38:395–406

Zubeck HK, Raad L, Saboundjian S, Minassian G, John Ryer PE (2003) Workability and performance of polymer-modified asphalt aggregate mixtures in cold regions. Int J Pavement Eng 4(1):25–36

Non- petroleum- Based Binders for Paving Applications: Rheological and Chemical Investigation on Ageing Effects

Davide Dalmazzo[1(✉)], Ana Jiménez Del Barco Carrión[2], Lucia Tsantilis[1], Davide Lo Presti[2,3], and Ezio Santagata[1]

[1] Department of Environment, Land and Infrastructure Engineering, Politecnico di Torino, Turin, Italy
davide.dalmazzo@polito.it

[2] Nottingham Transportation Engineering Centre, University of Nottingham, Nottingham, UK

[3] Dipartimento di Ingegneria, Universitá degli studi di Palermo, Palermo, Italy

Abstract. The massive exploitation of non-renewable natural resources which has taken place in the last decade has led to significant global environmental concerns. In such a context, the use of non-petroleum-based binders for the construction of bound layers of flexible pavements can represent an effective solution to limit crude oil depletion. The research work presented in this paper focused on the effects of ageing on the rheological and chemical characteristics of a non-bituminous binder, indicated in the study as a "biobinder", and a traditional neat bitumen selected as a reference material. Binders were analyzed in four ageing conditions obtained by making use of the Rolling Thin Film Oven and of the Pressure Ageing Vessel. Rheological behaviour of binders was investigated by means of oscillatory tests carried out in a wide range of temperatures and frequencies with a dynamic shear rheometer. Chemical structure was explored via Thin Layer Chromatographic analyses and Fourier Transform Infrared Spectroscopy. The experimental work demonstrated that mechanisms of ageing which are involved in biobinders completely differ from those experienced by petroleum-based binders. Concerns were expressed with respect to the applicability to non-conventional binders of currently available ageing techniques and of chemical characterization methods.

Keywords: Biobinders · Ageing · Rheology · Chemical characterization

1 Introduction

For more than a century the use of bitumen as the binding agent of mixtures employed for the construction of flexible pavements has been considered as the only available option. However, in the last two decades several researchers have worked on the formulation of non-petroleum-based binders, also known as biobinders, conceived as potential alternatives to bitumen (Su et al. 2018). These efforts have been spurred by the sustainability-driven need of reducing the exploitation of non-renewable natural resources and by the desire of identifying possible reaction strategies to the increase of crude oil and bitumen costs.

M. Pasetto et al. (Eds.): ISAP APE 2019, LNCE 48, pp. 67–76, 2020.
https://doi.org/10.1007/978-3-030-29779-4_7

Although the most attractive application of biobinders is full replacement of bitumen, several studies have addressed alternative strategies of use. In particular, they have focused on the development of products which depending upon the case can be employed as bitumen modifiers (to improve low-temperature performance and workability), extenders (to reduce the demand of bitumen) and rejuvenators (in combination with reclaimed asphalt) (Raouf et al. 2010; Fini et al. 2016; Dashmana et al. 2015; Jiménez del Barco Carrión et al. 2017).

As highlighted by Chailleux et al. (2015), most of the biobinders developed in the course of time for full bitumen replacement are constituted by a combination of a high molecular weight component (usually derived from wood) and a viscous oil. Synthetic polymers and other function-specific materials (such as waxes and fibres) may also be added to the blend. Fabrication conditions and methods vary significantly among the range of products. However, after blending the components at temperatures above their melting point, polymerization reactions usually take place when maintaining the products at high temperatures. In most cases these binders are transparent and can be consequently used in applications in which the natural colour of aggregates is maintained or a specific colour is obtained by means of appropriate pigments.

Despite the remarkable efforts placed in the laboratory characterization of these innovative binders, field experiences have been rather limited, with the placement of few trial sections (Chailleux et al. 2015; Gosselink, 2015). Furthermore, the outcomes of these full-scale trials have been partially disappointing due to the occurrence of failures in a very short time span. It has been reported that mixtures prepared with biobinders tend to excessively stiffen in time, leading to a brittle response under loading. In such a context, it has been postulated that the ageing behaviour of these binders may be quite different from that of bitumen.

As a result of the abovementioned uncertainties, the experimental work described in this paper focused on the assessment of the ageing behaviour of a commercial biobinder which was subjected to several ageing treatments and subsequently characterized from a rheological and chemical point of view. For comparative purposes, a reference standard bitumen was subjected to the same ageing procedures and tests.

2 Materials and Methods

Materials considered in the investigation included a biobinder and a traditional neat bitumen, selected as a reference material. As indicated by its manufacturer, the biobinder is made of pitch (obtained as a by-product of the papermaking industry), rosin and SBS. The employed reference bitumen was a conventional 50/70 penetration grade binder. The two binders were analyzed and compared in four ageing conditions: original state, short-term ageing obtained by means of the Rolling Thin Film Oven (RTFO) test (as per AASHTO T 240), long-term ageing obtained by subjecting the RTFO residue to a Pressure Ageing Vessel (PAV) treatment (at 100 °C for 20 h with an imposed pressure of 2.1 MPa, as per AASHTO R 28), and an extra-long term ageing which involved two PAV cycles after RTFO ageing. As a function of their ageing state, samples of both binders were associated to codes U (unaged), R (RTFO-aged), P (PAV-aged) and PP (aged with double PAV treatment).

Rheological properties of the binders were analyzed by means of a dynamic shear rheometer operated by making use of two different parallel plates measuring systems (8 mm parallel plates with 2 mm gap at test temperatures lower than 34 °C; 25 mm parallel plates with 1 mm gap at test temperatures greater than or equal to 40 °C). Linear viscoelastic properties were investigated by means of strain-controlled frequency sweep tests carried out by using angular frequencies ranging from 1 to 100 rad/s and temperatures comprised between 4 °C and 76 °C with 6 °C increments. In order to evaluate the rheological behaviour within the linear visco-elastic domain, shear strains applied to test specimens were varied depending upon temperature and frequency according to preliminary amplitude sweep tests.

Chemical characterization of the binders was performed by making use of techniques which are typical for standard bitumen. These included Thin Layer Chromatography (TLC) and Fourier Transform Infrared Spectroscopy (FTIR).

TLC tests consist in the separation of binder fractions in specific solvents of increasing polarity. This is done by injecting a binder-dichloromethane solution on silica rods which are then oven-dried and subjected to successive elutions in n-hexane, toluene and a solution of dichloromethane and methanol in a volume ratio of 95:5 (Santagata et al. 2009; Holleran and Holleran 2010). When performed on bitumen, such a test leads to the estimate of the percentages of saturates, aromatics, resins and asphaltenes (of increasing molecular mass, aromatic content and polarity). Since the composition of biobinders is definitely different from that of standard bitumen, in this study the four fractions identified for the considered biobinder were indicated as "p-saturates", "p-aromatics", "p-resins" and "p-asphaltenes" (where "p" stands for "pseudo").

FTIR tests allow the identification of different groups of molecular bonds which are associated to well-defined wavelengths within recorded absorbance spectra. Binder samples are directly spread on the surface of a diamond crystal with high refraction index and FTIR spectra are consequently recorded by using the attenuated total reflectance technology. For both considered binders the FTIR peaks were identified and associated to corresponding bonds by referring to the work performed on bitumen by Petersen (1986) and by Masson et al. (2001).

3 Results and Discussion

3.1 Rheological Tests

As displayed in Fig. 1, values of the norm of the complex modulus ($|G^*|$) and of the phase angle (δ) retrieved from oscillatory tests were plotted, for both binders and for all ageing states, in the so-called Black space. Such a representation allows a global assessment of the rheological response of a given binder and is also functional for the verification of the applicability of the time-temperature superposition principle.

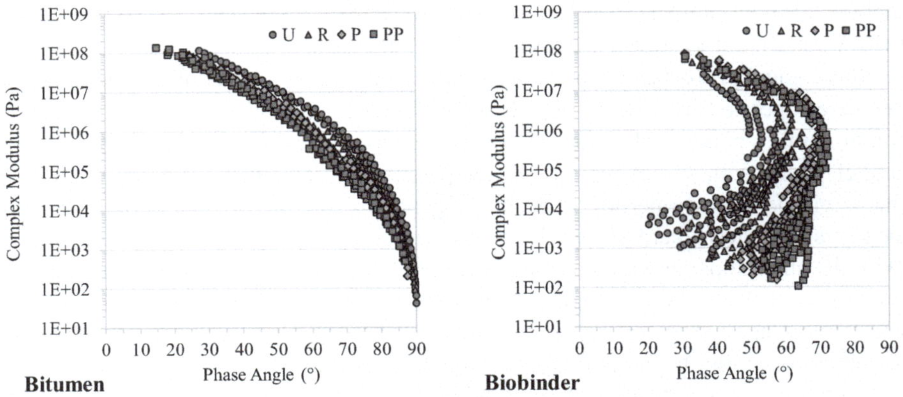

Fig. 1. Black curves of reference bitumen and biobinder

When focusing on the results obtained from tests carried out on materials in their original (unaged) state, a different response was observed for the two binders. The standard bitumen exhibited a single, smooth curve characterized by a gradual transition from the glassy to the viscous state, thus indicating that its behaviour can be modelled according to the time-temperature superposition principle. On the contrary, the biobinder did not show the same type of rheological simplicity, with the presence of several independent curves that did not follow a common trend. In this case it was thus found that the time-temperature superposition principle cannot be applied. Such an outcome can be explained by referring to the composition of the innovative binder, in which the various components may exhibit structural changes due to their different melting points and to the possible formation (and subsequent dismantlement) of semicrystalline domains. These phenomena can significantly affect the relaxation function of the composite material, thus leading to a different sensitivity to temperature and frequency changes.

When considering the effect of ageing, in the case of the reference bitumen it was found that the Black curves were gradually shifted to higher $|G^*|$ and lower δ values. Moreover, it was observed that for the same domain of frequencies and temperatures, the total length of the curves was almost constant, thus indicating that temperature sensitivity was not significantly affected by ageing.

In the case of the biobinder, the effects of ageing on its rheology were totally opposite than those recorded for the reference bitumen. In particular, ageing caused a progressive shift of the Black curves to lower $|G^*|$ and higher δ values. Furthermore, the curves became more consistent and uniform with ageing. These outcomes suggest that the RTFO and PAV treatments caused a progressive loss of internal structure which tended towards a long-term condition closer to rheological simplicity.

For both materials and in all ageing conditions, it can be noticed that at low temperatures the Black curves tend to a zero value of δ and to a constant value of $|G^*|$, of the order of 1 GPa, typical of amorphous materials in their glassy state (Lesueur, 2009). When focusing on the high temperature response, due to the fact that the analysis of Black curves does not easily allow straight comparisons, it was considered beneficial to evaluate the norm of the complex viscosity. In particular, such a parameter was assessed at 76 °C in the investigated range of angular frequency. Corresponding results are displayed in Fig. 2.

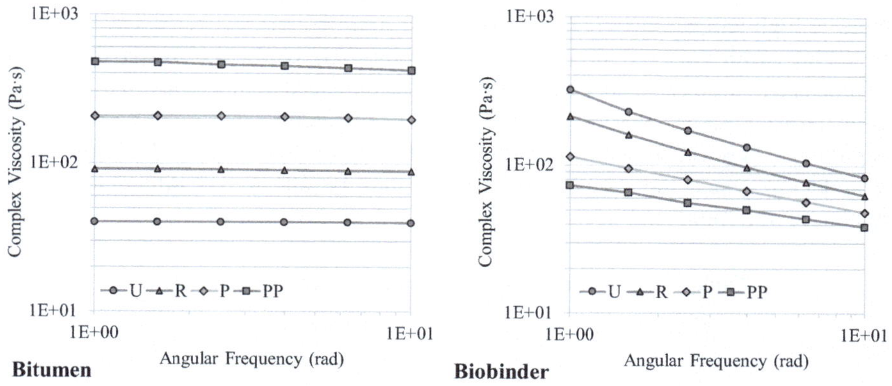

Fig. 2. Complex viscosity at 76 °C of reference bitumen and biobinder

From the plot shown in Fig. 2 it can be observed that the reference bitumen displayed, as expected, an increase of its complex viscosity as a function of progressive ageing. Experimental data highlighted a slight shear-thinning behavior, more evident in long-term aged conditions. When considering the results obtained at low frequency it can be noticed that the complex viscosity tends towards a constant value which can be considered as the zero-shear viscosity.

Due to its unconventional nature, the biobinder exhibited a completely different response. In fact, its viscosity progressively decreased with ageing and in all ageing states the material showed a strong shear-thinning character. Thus, it was clearly observed that the biobinder in the considered conditions behaves like a viscoelastic material and not as a viscous fluid.

In order to quantitatively evaluate ageing effects, the following two indices were calculated after each stage of ageing at several temperatures:

$$AI_{|G^*|} = \frac{|G^*|_i}{|G^*|_O} \qquad AI_\delta = \frac{\delta_i}{\delta_O}$$

where $|G^*|_O$ and δ_O are the complex modulus and the phase angle at 10 rad/s in the unaged condition, while $|G^*|_i$ and δ_i are the complex modulus and the phase angle at 10 rad/s measured in different ageing states. Obtained results are listed in Table 1.

From the analysis of the data provided in Table 1 it can be observed that the trends displayed by the values of the ageing indices of the reference bitumen are in line with what has been widely reported in literature. When considering progressively more severe ageing conditions, $AI_{|G^*|}$ increases and AI_δ decreases. The first parameter displays greater ageing-related variations at higher temperatures, where the response is more distant from the glassy state, whereas the second one shows the greatest changes at lower temperatures, which are distant from the viscous asymptote. The trends observed for the ageing indices of the biobinder are totally different. $AI_{|G^*|}$ decreases with increasing temperatures (with the exception of one data point) and displays a dependency upon ageing which is different at low and high temperatures. At low temperatures (4 °C and 28 °C) $AI_{|G^*|}$ is always greater than 1 and increases with ageing; at high temperatures (52 °C and 76 °C) $AI_{|G^*|}$ is smaller than 1 and decreases with ageing (with the exception of one data point). AI_δ is always greater than 1 and does not show a clear trend as a function of ageing and temperature. However, in general terms it seems that AI_δ increases with temperature and ageing (with some exceptions). These outcomes are consistent with the inherent structural complexity of the binder, in which the various components may be affected in different ways by temperature variations and ageing phenomena.

Table 1. Ageing indices of reference bitumen and biobinder

T (°C)	Reference bitumen						Biobinder					
	$AI_{\|G^*\|}$			AI_δ			$AI_{\|G^*\|}$			AI_δ		
	R	P	PP	R	P	PP	R	P	PP	R	P	PP
4	1.1	1.5	1.7	0.82	0.71	0.60	2.0	3.3	3.3	1.07	1.13	0.99
28	2.4	4.7	7.6	0.81	0.76	0.66	1.0	1.3	1.3	1.08	1.22	1.30
52	2.5	6.0	14.3	0.94	0.86	0.80	0.9	0.4	0.5	1.28	1.57	1.69
76	2.2	4.9	10.6	0.98	0.96	0.92	0.8	0.6	0.5	1.08	1.20	1.33

3.2 Chemical Analyses

Results obtained from TLC tests carried out on the two binders are listed in Table 2, where they are expressed in terms of the percentages of the classical bitumen fractions (saturates, aromatics, resins and asphaltenes) and of the corresponding "pseudo-fractions" obtained in the case of the biobinder.

It can be noticed that the standard reference bitumen displayed an evolution of its composition which was in line with data published in literature (Wang et al. 2019; Siddiqui and Ali 1999). While the percentage of saturates did not change significantly, with the progress of ageing both the asphaltenes and resins showed a remarkable increase. It was also noticed that no relevant additional ageing effects were caused by subjecting the binder to a supplementary PAV treatment.

Table 2. Percentages of fractions and pseudo-fractions of reference bitumen and biobinder

Ageing state	Reference bitumen				Biobinder			
	S (%)	Ar (%)	R (%)	As (%)	p-S (%)	p-Ar (%)	p-R (%)	p-As (%)
U	5.3	42.9	25.2	26.6	0.0	9.7	79.0	11.3
R	5.0	36.9	24.8	33.3	0.0	8.1	68.9	22.9
P	5.6	25.0	35.4	34.0	0.0	7.7	86.4	5.9
PP	5.8	25.8	38.1	30.4	0.0	10.6	82.6	6.8

S: saturates; Ar: aromatics; R: resins; As: asphaltenes
p-S: pseudo-saturates; p-Ar: pseudo-aromatics; p-R: pseudo-resins; p-As: pseudo-asphaltenes

As expected, results obtained for the biobinder were completely different. In particular, it was observed that in all ageing states there was no detectable elution with the first solvent (n-hexane), thus indicating the absence of low polarity molecules in the binder. In practical terms this led to a zero value of the percentage of the pseudo-saturates. It should also be underlined that in all ageing states most of the molecules composing the biobinder were found to belong to the fraction of pseudo-resins. The main chemical changes induced by ageing were found after PAV ageing (either single or double, with no significant differences), which led to an increase of the percentage of the intermediate polarity group (i.e. of the pseudo-resins) that reached 82.6–86.4%. This seems to suggest that the prolonged exposure to high temperatures in the presence of pressure caused a reduction of the degree of internal diversity of the material. Such an outcome is consistent with the rheological results presented in Sect. 3.1 in the form of Black curves, which were found to be progressively more uniform. Finally, it can be noticed that TLC tests carried out on the biobinder after RTFO yielded results which were not consistent with all the others. This may be due to the specific ageing protocol or may have originated by the limitations of the employed characterization technique.

Results obtained from FTIR tests are displayed in the form of absorbance spectra in Fig. 3. From a qualitative point of view, it can be easily recognized that the spectra recorded for the two binders were significantly different and were also sensitive to ageing.

In order to analyze the spectra in quantitative terms, reference was made to previous work performed on bitumen, in which it was shown that the oxidation of hydrocarbons is associated, notably, with the increase of carbonyl groups C = O (corresponding to absorbance signals with wavelengths in the vicinity of 1700 cm^{-1}) and of sulphoxides S = O (corresponding to wavelengths around 1030 cm^{-1}) (Mouillet et al. 2008; Siddiqui and Ali 1999). More specifically, it was found that changes of the sulphoxides are typical of short-term ageing occurring during plant manufacturing of mixtures, whereas changes in carbonyl groups are relevant for long-term ageing which takes place during the service life of a pavement.

Processing of the spectra required the calculation of the sulphoxide index (I_S) and of the carbonyl index (I_C), which were obtained by normalizing the areas underlying the spectra in the relevant wavelength bands to the sum of the areas associated to all peaks. Values of I_S and I_C calculated for the two binders in all ageing states are listed in Table 3. As expected, both indices were found to increase with ageing in the case of the reference bitumen. However, tests carried out on the biobinder led to completely different results. It was observed that in comparison to the reference bitumen the biobinder is characterized by a greater percentage of sulphoxides and carbonyl groups: however, the indices do not change significantly with ageing, showing only a slight decrease. This outcome confirms the conclusion, already drawn from TLC tests, that the mechanisms of ageing which are involved in biobinders may significantly differ from those of standard bitumen.

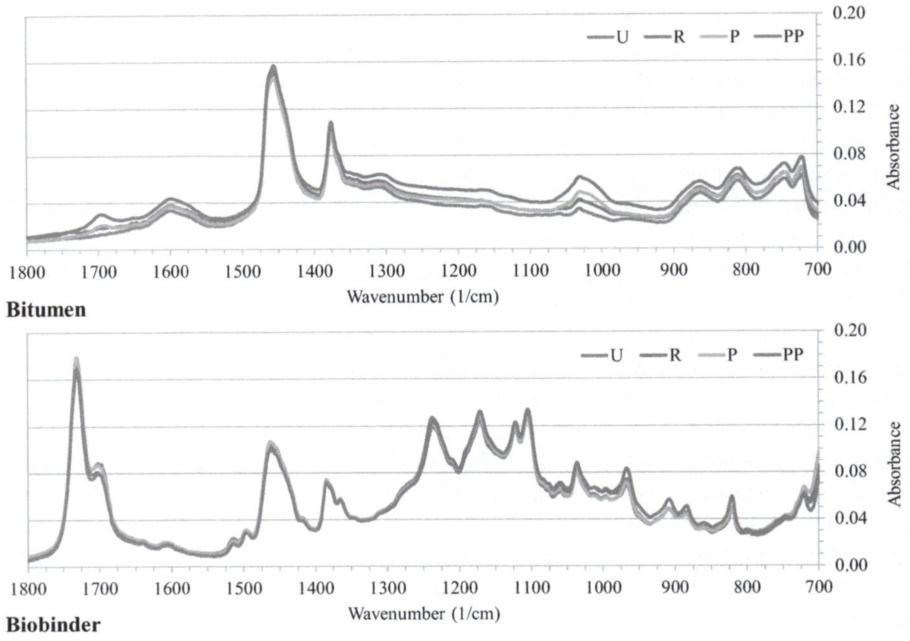

Fig. 3. Absorbance spectra of reference bitumen and biobinder

Table 3. Indices derived from FTIR spectra for reference bitumen and biobinder

Ageing state	Reference bitumen		Biobinder	
	I_S (%)	I_C (%)	I_S (%)	I_C (%)
U	2.28	1.39	4.49	13.72
R	2.25	1.85	4.47	12.61
P	2.60	1.76	4.27	13.66
PP	3.31	2.55	3.83	12.84

4 Conclusions

The experimental investigation described in this paper focused on the comparative evaluation of the ageing effects induced in a commercial biobinder and in a reference standard bitumen. It was found that in comparison to the reference bitumen, which exhibited ageing effects in line with data available in literature, the biobinder showed a counterintuitive behavior. From a rheological viewpoint, as a function of progressive ageing, the biobinder displayed a reduction of stiffness and degree of elasticity. Due to the non-conventional nature of the biobinder, changes in its chemical composition caused by ageing, which were assessed by means of Thin Layer Chromatography (TLC) and Fourier Transform Infrared Spectroscopy (FTIR), were quite difficult to identify. However, in general terms it was found that ageing led to a reduction of the degree of diversity of the binder molecules and that there were no significant variations of sulphoxides and carbonyl groups.

It should be mentioned that the investigation described in this paper focused on a single commercial biobinder and that in literature results reported by other Authors on other products are in some cases completely different from those found in this study. Thus, obtained results cannot be considered of general value.

The Authors believe that the performed investigation highlighted two fundamental issues which should be the subject of further research. First of all, there are serious doubts whether the currently available ageing techniques, originally developed for bituminous binders, can be employed for non-petroleum-based products. In such a context, field validation activities are absolutely necessary. Secondly, there are also concerns with respect to the chemical characterization methods to be employed for the monitoring of ageing effects. In particular, it is questionable whether TLC and FTIR, which are extremely valuable for the analysis of standard bituminous binders, can yield relevant results in the case of non-conventional biobinders.

One of the missions of modern research in the area of pavement engineering is to devise new construction strategies in the context of sustainability. Thus, the Authors believe that investigations in the area of biobinders need to be intensified with the final goal of identifying a true alternative to bitumen. The results obtained in the investigation described in this paper are promising, but they need to be supported and expanded by means of further research.

References

Jiménez del Barco Carrión A, Lo Presti D, Pouget S, Airey G, Chailleux E (2017) Linear viscoelastic properties of high reclaimed asphalt content mixes with biobinders. Road Mater Pavement Des 18:241–251

Chailleux E, Audo M, Goyer S, Queffelec C, Marzouk O (2015) 11 - Advances in the development of alternative binders from biomass for the production of biosourced road binders. Woodhead Publishing, Oxford

Dhasmana H, Ozer H, Al-Qadi IL, Zhang Y, Schideman L, Sharma BK, Chen W-T, Minarick MJ, Zhang P (2015) Rheological and chemical characterization of biobinders from different biomass resources. Transp Res Rec 2505:121–129

Fini EH, Khodaii A, Hajikarimi P (2016) Fractional viscoelastic study of low-temperature characteristics of biomodified asphalt binders. J Mater Civil Eng 28(9):04016078

Gosselink R (2015) Zeeland's Road Trial Section of Bio-Asphalt: A World First. WUR. https://www.wur.nl/en/newsarticle/Zeelands-road-trial-section-of-bioasphalt-a-world-first.htm

Holleran G, Holleran I (2010) Bitumen chemistry using cheaper sources: an improved method of measurement by TLC-FID and the characterisation of bitumen by rheological and compositional means. In: 24th ARRB conference, 12–15 October, Melbourne, Australia

Lesueur D (2009) The colloidal structure of bitumen: consequences on the rheology and on the mechanisms of bitumen modification. Adv Coll Interface Sci 145(1–2):42–82

Masson J-F, Pelletier L, Collins P (2001) Rapid FTIR method for quantification of styrene-butadiene type copolymers in bitumen. J Appl Polym Sci 79:1034–1041

Mouillet V, Lamontagne J, Durrieu F, Planche J-P, Lapalu L (2008) Infrared microscopy investigation of oxidation and phase evolution in bitumen modified with polymers. Fuel 87:1270–1280

Petersen JC (1986) Quantitative functional group analysis of asphalts using differential infrared spectrometry and selective chemical reactions-theory and application. Transp Res Rec 1096:1–11

Raouf MA, Metwally M, Williams RC (2010) Development of non-petroleum based binders for use in flexible pavements. Trans Project Reports 17

Santagata E, Baglieri O, Dalmazzo D, Tsantilis L (2009) Rheological and chemical investigation on the damage and healing properties of bituminous binders. J Assoc Asph Paving Technol 78:567–595

Siddiqui MN, Ali MF (1999) Studies on the aging behavior of the Arabian asphalts. Fuel 78 (9):1005–1015

Su N, Xiao F, Wang J, Cong L, Amirkhanian S (2018) Productions and applications of bio-asphalts – a review. Constr Build Mater 183:578–591

Wang J, Wang T, Hou X, Xiao F (2019) Modelling of rheological and chemical properties of asphalt binder considering SARA fraction. Fuel 238:320–330

Investigation into the Use of Reclaimed Asphalt Pavement in Asphalt Concrete

Olumide Moses Ogundipe(✉)

Department of Civil Engineering, Ekiti State University, Ado-Ekiti, Nigeria
olumide.ogundipe@eksu.edu.ng, momide2002@yaoo.com

Abstract. This paper reviews the recycling of reclaimed asphalt pavement (RAP). The review establishes the various methods of obtaining RAP which include milling, pavement demolition and plant waste. It also looks at the recycling methods which include: hot mix recycling; hot-in-place recycling; cold mix recycling; cold-in-place recycling and full depth reclamation. The benefits and challenges of recycling RAP were also considered. The laboratory study examines the Marshall properties of virgin asphalt concrete and the one containing 20%, 30%, 40%, 50%, 60%, 70% and 100%. The study reveals that the optimum amount of RAP that can be used is 60%. It concludes that recycling of RAP can be used in the asphalt concrete. Therefore, it recommends that for successful recycling of RAP in Nigeria, it is important to: evaluate the properties of RAP; perform enough pavement sampling to estimate variability of material properties; separate and identify by source large quantities of RAP obtained from different sources and commission research into the generation, management and recycling of RAP.

Keywords: Recycling · Reclaimed · Pavement · Marshall · Stability · Flow

1 Introduction

Road transportation is the major mode of transportation in Nigeria. Roads provide increased mobility for people, goods and services. They play key roles in the progress of any nation and drive socio-economic development. Most of our paved roads in Nigeria have asphalt concrete as the surfacing material. This is in part due to the high cost of cement and other advantages associated with asphalt roads. These advantages include its flexibility, cost efficiency, reduction in noise pollution and comfort through the provision of smooth and durable surface and being fast to construct and open to traffic. Asphalt concrete has low initial costs, lasts long, and due to its recyclability, has residual value greater than other pavements.

In Nigeria, when pavement failed, the usual practice is to scarify it and lay new surfacing material. This process generates tonnes of reclaimed asphalt pavement (RAP) that have not been put to good use. Oke *et al.* (2013) reported that Federal Road Maintenance Agency in Nigeria confirmed that most roads in Nigeria were distressed, and as such the entire highway network in the country might face a total collapse, if urgent steps are not taken to rehabilitate, repair or reconstruct them at the appropriate time. Also, they mentioned that it was a known fact that most of the roads managed by

M. Pasetto et al. (Eds.): ISAP APE 2019, LNCE 48, pp. 77–87, 2020.
https://doi.org/10.1007/978-3-030-29779-4_8

the Federal, State and Local governments in Nigeria were not paved, because of the non-availability of fund. Recycling of RAP, if properly implemented may be a sustainable way out. Koch and Ksaibati (2010) defined reclaimed or recycled asphalt pavement (RAP) as the term given to removed and/or reprocessed pavement materials containing asphalt and aggregates. RAP material is generated when damaged pavement is milled, crushed, sometimes fractionated, and stockpiled for use as an additional component in asphalt mixture (Roque *et al.* 2015). The best application of the RAP in Nigeria at present is in unpaved street roads, while a larger percentage is disposed indiscriminately. This is not the case in some developed countries, where recycling of RAP is embraced. Asphalt is one of the most recycled construction products in Europe. It is even possible to create a pavement from almost 100% reclaimed asphalt (EAPA 2015). It makes asphalt pavements sustainable. Less new bitumen is needed to make new asphalt pavements and million tonnes of new aggregate can be saved. Recycling of asphalt pavements dates back to 1915, but it did not become a common practice until the early 1970s when asphalt binder prices skyrocketed as a result of the Arab oil embargo (West 2010). Asphalt paving technologists reacted to this situation by developing recycling methods to reduce the demand for asphalt binder and, thereby, reduce the costs of asphalt paving mixtures (West 2010). According to Al-Qadi *et al.* (2007), the use was favoured over virgin materials in the light of the increasing cost of asphalt, the scarcity of quality aggregates, and the need to preserve the environment. They stated that many State agencies in the United States of America had also reported significant savings when RAP was used. Also, Zaumanis *et al.* (2016) stated that dramatically rising bitumen cost, dwindling budgets, growing traffic loads, and the desire to find more sustainable paving practices were forcing agencies to seek ways for maximizing the re-use of RAP. They observed that while most of the academic and industrial institutions focused on the development of procedures to recycle hot asphalt mixes with up to 40% RAP content, a few industry innovators had refined 100% recycling technologies over the past four decades to a level where routine production of 100% recycled mixes was in clear sight. Griffiths and Krstulovich (2002) stated that the Illinois Department of Transportation (IDOT) used 623,000 tonnes of RAP in highway construction in 2001 and anticipated increasing its use in the near future. In the same vein, Sullivan (1996) reported that about 33% of all asphalt pavement in the United States was recycled into HMA in 1996.

Oliveira *et al.* (2013) in their study on the possibility of using 100% rap in Portugal found that the incorporation of 100% RAP and used motor oil (as a binder rejuvenator) in the production of asphalt mixtures could be a paving solution with a performance as good as conventional asphalt mixtures, as long as adequate storing (moisture) and production (temperature) conditions were assured.

RAP may be obtained from three major sources, which are through milling operations, also known as cold planning, full-depth pavement demolition and asphalt plant waste. Milling involves the removal of the pavement surface using a milling machine, which can remove up to 2 in. (50 mm) thickness in a single pass (Koch and Ksaibati 2010). Milling is a beneficial part of pavement rehabilitation. RAP may also be obtained from complete demolition of an existing pavement using a bulldozer or backhoe. This process is typically limited to small areas of pavement. When pavement rubble is contaminated with underlying layers and soil, it is better for this material to be crushed and used as a shoulder or base material than used in an asphalt mixture (West 2010). All asphalt plant operations generate some waste during plant start-up, transition between mixes, and clean-out. Another form of waste is mix rejected from a project due to incomplete coating or due to the mix temperature being too high or too low for the job. Other situations that may result in wasted mix include trucks loaded with too much mix to finish the job or mix that could not be placed due to inclement weather.

RAP can be used in the following ways for pavement construction and rehabilitation: granular base aggregate; asphalt concrete; stabilized base aggregate; and embankment or fill (Han *et al.* 2011, Koch and Ksaibati 2010). The asphalt recycling methods are as follows (FHWA 1997; Koch and Ksaibati 2010): (i) Hot Mix Recycling; (ii) Hot-in-place Recycling; (iii) Cold Mix Recycling; (iv) Cold-in-place Recycling; (v) Full Depth Reclamation. This study considered the used of RAP as surfacing material and adopted the hot mix recycling method. Recycling the reclaimed asphalt pavement has many benefits which can be grouped as economic, environmental and technical (Han et al. 2011). All these benefits could be harnessed by adopting recycling. It will also ensure mores roads receive attention from government. The recycling of RAP has some challenges which are not insurmountable. Identifying such challenges beforehand will ensure that they are properly addressed. Some of these challenges are as follows:

Non-Uniformity of RAP: RAP derived from different sources can have significantly different gradation, binder type, binder content and binder density (McGarrah 2007).

Rejuvenation of Recycled Binder: The recovered (aged) binder in the RAP must be rejuvenated to achieve good performance. Sondag et al. (2002) stated that the hardened recycled binder must be mixed with a recycling agent or soft asphalt binder to restore its rheological properties. In addition to soft binder, softening agents and rejuvenating agents are commonly used. Roberts et al. (1996) reported that softening agents lower the viscosity of the aged binder while rejuvenating agents restore the physical and chemical properties of the old binder.

Laboratory Characterization of RAP Materials: It is important that laboratory tests are carried out to characterize the RAP. Some of the tests required are as follows (Kuehl et al. 2016): (i) determination of the asphalt Content, (ii) test on extracted asphalt binder (iii) determination of the gradation of the extracted aggregate and (iv) determination of the aggregate properties.

2 Materials and Methods

2.1 Materials

The materials used for the study include RAP that was scarified from the failed Ikere-Ekiti road, Nigeria being rehabilitated, coarse aggregates from crushed granitic rock of single size of 12.7 mm and 9.5 mm and fine aggregates: river sand and stone dust, filler and bitumen. The filler used is non-plastic material. The filler is free from debris and meets the requirement that 75% must pass through sieve No 200. The bitumen is 60/70 penetration grade.

2.2 Methods

The RAP was subjected to extraction test to determine the gradation of the aggregates and the percentage of bitumen in the aggregates. The virgin materials: coarse aggregates, fine aggregates, filler and bitumen were collected. The aggregates were graded. The mix composition for the virgin asphalt concrete was prepared to conform with the General Specifications, Roads and Bridges (FGN 1997). The virgin asphalt mixture was prepared using the mix composition in Table 1 and 5.5%, 6%, 6.5%, 7.0% and 7.5% 60/70 bitumen to determine the optimum bitumen content using Marshall test. Also, the RAP (100%) was heated at 150 °C and compacted in the Marshall mould at 145 °C to determine the Marshall stability, flow and other volumetric properties. The mixture of the RAP and virgin asphalt was prepared by using the optimum bitumen content and replacing the asphalt mixture with 20%, 30%, 40%, 50%, 60% and 70% RAP. The Marshall stability, flow and other volumetric properties of the asphalt concrete containing RAP, RAP only and the virgin asphalt concrete were compared.

Table 1. Mix composition

Aggregates	12.5	9.5	River sand	Quarry dust	Filler
% by mass of total mix	9	18	45	23	5

2.2.1 Experimental Procedures

Particle size distribution: A 1000 g mass of dry sample of aggregate was poured through the nest of sieves stakes in series such that the opening decrease in size from the top sieve downwards with a pan at the bottom of the stack. The test was performed in accordance with BS EN 933-1:1997 (BSI 1997).

Bitumen extraction test for the RAP: A dried sample of RAP material, 1000 g was weighed and soak with petrol and thoroughly washed until the RAP material was free of bitumen. The washed sample was dried for 24 h. The dried sample was weighed to determine the percentage of bitumen in the RAP. Also, the aggregate was graded. The RAP before and after extraction is shown in Fig. 1.

Fig. 1. RAP before and after extraction

Marshall Test: Appropriate mass of aggregate was weighed, bitumen and filler heated in the right proportions according to the mix composition to a temperature of about 150 °C. The heated aggregates and the bitumen were thoroughly mixed together at this temperature. The mix was then placed in the Marshall mould and was well compacted using a rammer with 75 blows on either side. The test specimen in the mould was left to cool for few minutes and was carefully removed from the mould. The asphalt concrete specimens were weighed in air and water and their respective values were recorded before it was loaded into the Marshall Stability machine to determine the Marshall stability and flow and the volumetric properties were calculated.

Preparation of Convectional Asphalt Concrete Mix: 4000 g of aggregate was weighed; bitumen and filler were heated in the right proportions according to the mix composition to a temperature of about 150 °C. The heated aggregates and the bitumen were thoroughly mixed together at this temperature. The mix was separated into 3 through quartering method each with mass of 1200 g. It was then placed in an oiled Marshall mould and was well compacted using a rammer with 75 blows on either side. The test specimen in the mould was left to cool for an hour and was carefully removed from the mould. After 24 h, the prepared mould or test sample was weighed in air and water to determine its bulk density and was later loaded in the Marshall Stability machine in other to get its stability and the volumetric properties.

Preparation of Mix Containing RAP: Samples of mix having different percentages of the virgin and RAP materials were prepared (See Fig. 2). The following percentages of RAP were considered: 20%, 30%, 40%, 50% 60%, 70% and 100% (having 80%, 70%, 60%, 50%, 40%, 30% and 0%, virgin asphalt mixtures, respectively). In each of the samples, 3 specimens were produced having a weight of 1200 g from a total mix weight of 4000 g.

Fig. 2. Conventional asphalt concrete samples and samples containing RAP

3 Results and Discussions

3.1 Particle Size Distribution

The particle size distribution curves for the aggregates, filler and blends of aggregates (Virgin and RAP) are shown in Fig. 3. The blends of aggregates fall within the grading envelope for asphalt concrete wearing course (FGN 1997).

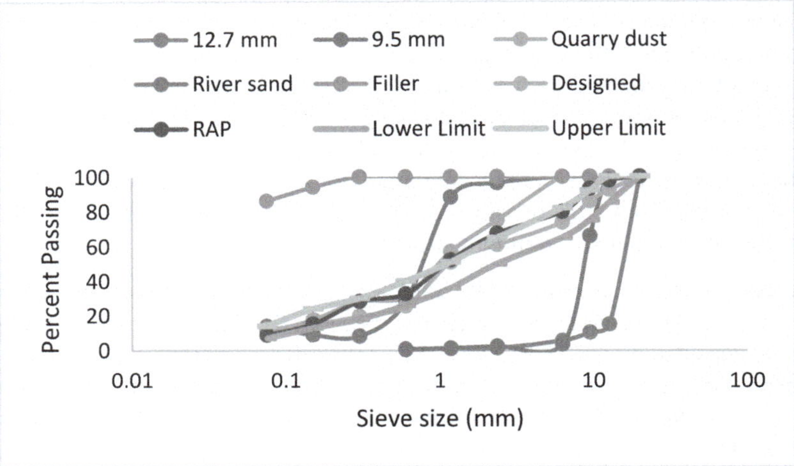

Fig. 3. Particle size distribution for the aggregates, filler and blend of aggregates

3.2 Results of Marshall Test

The results of the Marshall test on the unmodified asphalt concrete indicate that the optimum bitumen content is 6.6%. The virgin mixed with the RAP was produced using 6.6% bitumen content. Figure 4 shows the stability values of the virgin asphalt concrete and asphalt concrete containing RAP. The results show that the stability values increase by 43%, 27.9%, 26.7%, 11.6% and 14% for samples of asphalt concrete containing 20%, 30%, 40%, 50% and 60%, respectively, while those with 70% and 100% RAP reduced by 18.6% and 13.0%, respectively. Both the virgin asphalt concrete and the one containing RAP meet the requirement of not less than 3.5 kN specified for wearing course in the General Specifications (Road and Bridges) of the Federal Republic of Nigeria (FGN 1997). The increase in the stability could be attributed to the fact that the aggregates in the RAP are finer than in the virgin aggregate. Generally, the stability decreases with increasing RAP content.

Figure 5 shows the flow values of the virgin asphalt concrete and asphalt concrete containing RAP. The results show that samples containing 20% to 50% RAP do not meet the requirement of 2 mm to 4 mm specified in the General Specifications (FGN 1997). However, the samples containing 60%, 70% and 100% RAP meet the

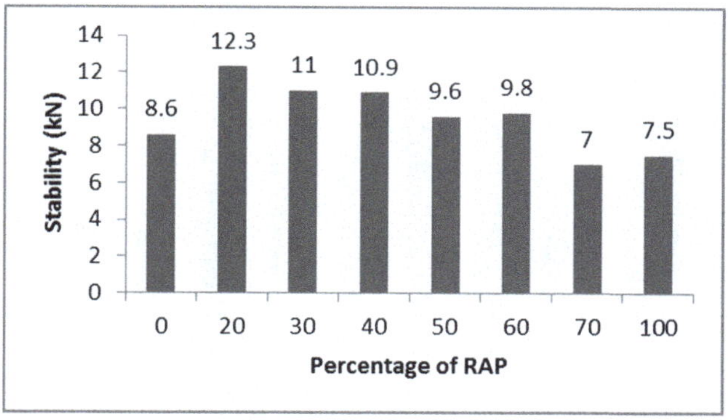

Fig. 4. Stability versus percentage of RAP for asphalt concrete

requirement. It is interesting to note that samples with 70% and 100% RAP has less flow than the virgin asphalt concrete. As explained in the case of stability, because the aggregates in the RAP are finer, having more of it in the mixture of virgin asphalt concrete and RAP results in increased surface area and consequently the ability to accommodate more binder. Figure 6 shows the density of the virgin asphalt concrete and asphalt concrete containing RAP. The results show that the density of the asphalt concrete containing RAP are greater than the virgin asphalt concrete except for those containing 70% and 100% RAP. Also, the density of the asphalt concrete containing RAP decreases with increasing RAP, which implies that an optimum value must be found. The Figure reveals that introducing RAP into the virgin aggregates ensures that it is better compacted, therefore having better stability.

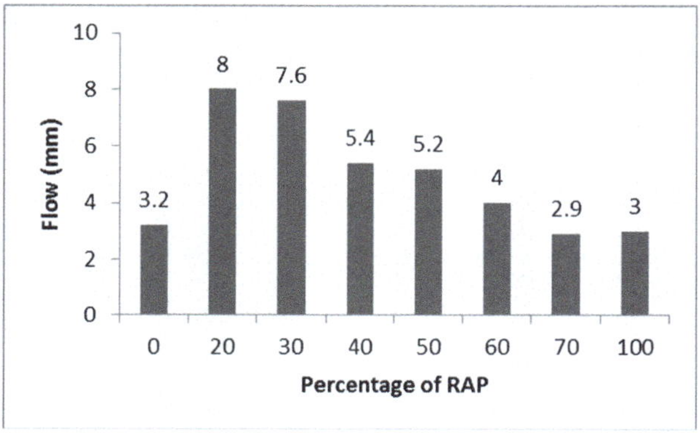

Fig. 5. Flow versus percentage of RAP for asphalt concrete

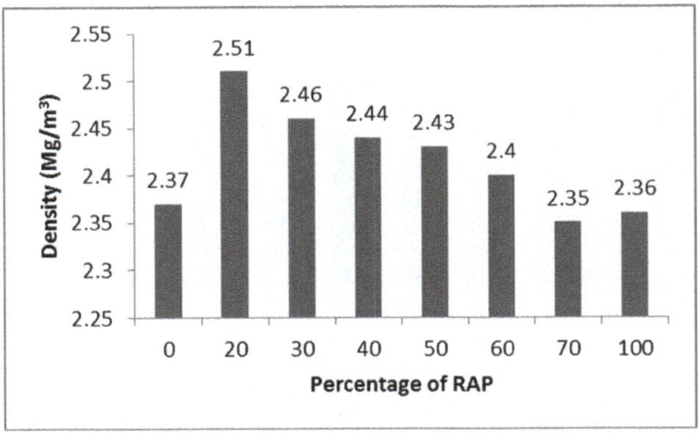

Fig. 6. Density versus percentage of RAP for asphalt concrete

Figure 7 shows the voids in the total mix of the virgin asphalt concrete and asphalt concrete containing RAP. The results show that asphalt concrete with 20%, 30%, 40%, and 50% RAP do not meet the requirement of 3% to 5% specified in the General Specification, Roads and Bridges (FGN 1997), while those with 60%, 70% and 100% meet the requirement. The low value recorded for samples with 20%, 30% and 40% RAP can be attributed to the gradations of the aggregates and the binder content. Also, this accounts for why high flow values were recorded. Figure 8 shows the voids filled with bitumen (VFB) of the virgin asphalt concrete and asphalt concrete containing RAP. The results reveal that only the virgin asphalt concrete and those containing 70% and 100% RAP meet the requirement of 75% to 82% specified in the General Specification, Roads and Bridges (FGN 1997), while others have higher percentage of the voids filled with bitumen. Therefore, accounting for the high flow values. Lastly, Fig. 9 shows the voids in the mineral aggregate (VMA). It can be seen that the asphalt concrete with the virgin aggregates has the highest VMA.

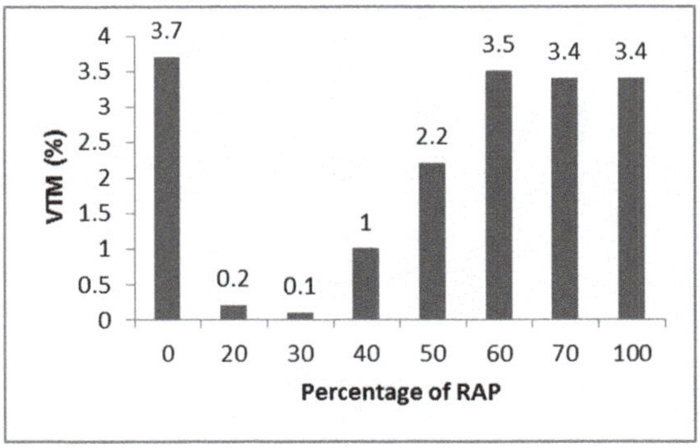

Fig. 7. Voids in the total mix versus percentage of RAP for asphalt concrete

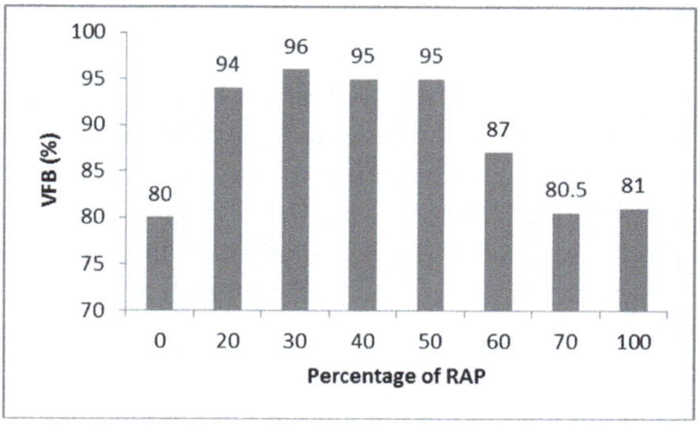

Fig. 8. Voids filled with bitumen (VFB) versus percentage of RAP for asphalt concrete

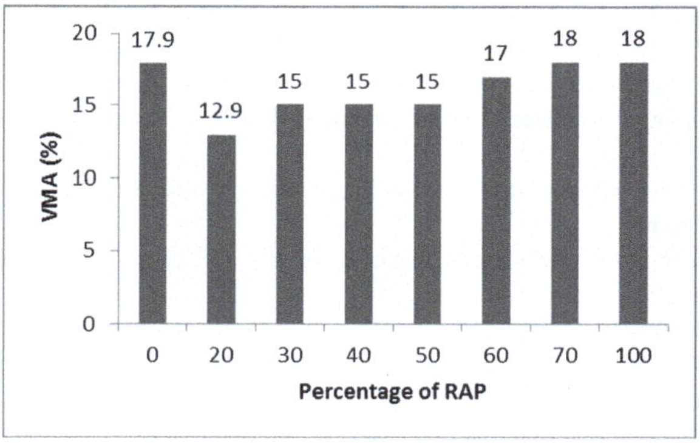

Fig. 9. Voids in mineral aggregate versus percentage of RAP for asphalt concrete

4 Conclusions

Most of our roads in Nigeria are in deplorable state and government resources are not unlimited. Many sectors of the economy like power, water, etc. require government intervention, it is therefore important that we look for a sustainable construction method for our roads. The literature review shows that introduction of RAP into asphalt concrete is practicable and sustainable. It reveals that economic benefits in terms of cost savings on virgin bitumen, which is the most expensive material in asphalt concrete, aggregates and fuels from reduced transportation and processing requirements could be realized. Generally, savings of about 14 to 70% have been documented. Also, it indicates that recycling of reclaimed asphalt pavement reduces the amount of disposable materials and that 35% of CO2 eq reduction per tonne of paved mixture could be achieved for 100% RAP asphalt mixture compared to virgin mix.

It has also been shown that recycled hot mix asphalt concrete is comparable in quality and structural performance to conventional asphalt concrete in terms of rutting, raveling, weathering, and fatigue cracking. The research shows that the optimum amount of RAP that can be used is 60%. This implies that it can be used as surfacing material on Trunks A, B and C roads in Nigeria.

Adopting the recycling of reclaimed asphalt pavement in Nigeria will ensure that amount budgeted by government at the Federal, State and Local levels will be available for more roads that are begging for attention, as less amount will be required for the construction and rehabilitation works, while also reducing the amount of wastes requiring safe disposal. Successful implementation of this sustainable construction method requires that more laboratory and field research works be commissioned to develop guidelines for its use in Nigeria.

References

Al-Qadi IL, Elseifi M, Carpenter SH (2007) Reclaimed asphalt pavement – a literature review. Illinois Center for Transportation (ICT), Department of Civil and Environmental Engineering, University of Illinois at Urbana-Champaign Research. Report FHWA-ICT-07-001

BSI (1997) Tests for geometrical properties of aggregates- determination of particle size distribution (Sieving Method). British Standard Institution, London, UK. (BS EN 933-1:1997)

EAPA (2015) Driving ahead with sustainable asphalt roads. European Asphalt Pavement Association, Belgium

FGN (1997) Government of the Federal Republic of Nigeria, General Specification (Roads and Bridges), Volume II

FHWA (1997) Pavement Recycling Guidelines for State Governments, Federal Highway Agency. FHWA-SA-97

Griffiths CT, Krstulovich JM (2002) Utilization of recycled materials in Illinois highway construction, Illinois Department of Transportation, Springfield, IL. Report No. IL-PRR-142

Han J, Thakur SC, Oswald CO, Parsons RL (2011) Laboratory evaluation of characteristics of recycled asphalt pavement in Kansas. Kansas State University Transportation Center, The University of Kansas. Report No. K-TRAN: KU-09

Koch S, Ksaibati K (2010) Performance of recycled asphalt pavement in gravel roads. Department of Civil and Architectural Engineering, University of Wyoming, Laramie, Wyoming

Kuehl R, Korzilius J, Marti M (2016) Synopsis of recycled asphalt pavement (RAP) material. Minnesota Department of Transportation. MN/RC - 2016RIC08

McGarrah EJ (2007) Evaluation of current practices of reclaimed asphalt pavement/virgin aggregate as base course material. Washington State Department of Transportation. WA-RD 713.1

Oke OL, Aribisala JO, Ogundipe OM, Akinkurolere OO (2013) Recycling of asphalt pavement for accelerated and sustainable road development in Nigeria. Int J Sci Technol Res 2(7):92–98

Oliveira JRM, Hugo MRDS, Jesus CMG, Abreu LPF, Fernandes SRM (2013) Pushing the asphalt recycling technology to the limit. J Pavement Res Technol 6(2):109–116

Roberts FL, Kandhal PS, Brown ER, Lee D, Kennedy TW (1996) Hot mix asphalt materials, mixture design, and construction, 2nd edn. Napa Education Foundation, Lanham

Roque R, Yan Y, Cocconcelli C, Lopp G (2015) Investigation of the effects of increased reclaimed asphalt pavement (rap) levels in dense graded friction courses. Department of Civil and Coastal Engineering. University of Florida. RFP-DOT-11/12-9033-RC

Sondag MS, Chadbourn BA, Drescher A (2002) Investigation of recycled asphalt pavement (rap) mixtures. Minnesota Department of Transportation, St. Paul, MN, Report No. MN/RC – 2002-15

Sullivan J (1996) Pavement recycling executive summary and report, Federal Highway Administration, Washington, D.C. FHWA-SA-95-060

West RC (2010) Reclaimed asphalt pavement management: best practices. National Center for Asphalt Technology, Auburn University, Auburn, Alabama

Zaumanis M, Mallick RB, Frank B (2016) Transp Res Procedia 14:3493–3502

Rheological and Mechanical Properties of HMA Containing Fly Ashes as Alternative Filler

Rosa Veropalumbo(✉), Nunzio Viscione, and Francesca Russo

Department of Civil, Construction and Environmental Engineering (DICEA), University of Naples "Federico II", Via Claudio 21, 80125 Naples, Italy
rosa.veropalumbo@unina.it

Abstract. This study defined the mechanical/volumetric properties and the environmental compatibility of bituminous mixtures containing fly ashes as alternative of traditional limestone filler. The research study is articulated in different steps as follows: (a) characterization of fly ashes according to EN 13043 Standard; (b) environmental compatibility analysis of mastics containing fly ashes by a leaching test; (c) stiffness evaluation of the mastics containing limestone and fly ash filler by static testing, as *Delta Ring and Ball Test,* and dynamic testing, as *Frequency Sweep Test;* (d) mechanical and volumetric characterization of asphalt concrete containing fly ashes and limestone filler starting from a phase of mix design of hot mix asphalt mixtures. The characteristics of stiffness of the bituminous mixtures has been evaluated on the optimum HMA through a dynamic test with a sinusoidal load. Then, the ability of two different HMA to endure a permanent deformation was evaluated by a *Repeated Load Axial Test.* The results has shown that mixtures of Hot Mix Asphalt with flying ashes as filler, compared to Hot Mix Asphalt containing limestone filler, is preferred in terms of mechanical performances and relevant environmental compatibility evaluated with the leaching test.

Keywords: Bituminous mixtures · Fly ashes · Limestone · Alternative fillers

1 Introduction and Literature Review

Every year, in Italy, 1Mt of fly ashes is produced and the possibility to retrieve entirely this by-product, in compliance with strict quality control specifications, caused that the fly ashes are treated more and more like a real product rather than as a by-product. In the road field, there needs to build new pavements and maintain the existing ones even if this requires a large quantity of aggregates and mineral fillers that make up the 95% of the asphalt concrete. The combination of these two problems stems from the possibility of the retrieve of alternative materials, such as fly ashes, in the production of hot mix asphalt for road paving. In the construction and use of roads, it is necessary to consider sustainability criteria (Akbulut et al. (2012); Zanetti et al. (2016); Topini et al. (2018)), for a contribution to the growth and development of nations, which provides access to employment, social, health, and education services.

M. Pasetto et al. (Eds.): ISAP APE 2019, LNCE 48, pp. 88–97, 2020.
https://doi.org/10.1007/978-3-030-29779-4_9

Bocci (2018) evaluated the possibility of the reuse of ladle furnace slag as filler in hot mix asphalt mixtures. From the results of the study it can be affirmed that the reuse of this filler in HMA is actually feasible because the mixture had a higher ITS and an excellent resistance to fatigue when compared with the reference mixture.

Sangiorgi *et al.* (2016) propose a different filler from the limestone one with a study carried out on the evaluation of waste deriving from bleaching of clays for the production of porous asphalts with reference to the bound layers, such as wear, binder and base. Tests results are promising in terms of increasing of Indirect Tensile Strength, stiffness, and resistance to permanent deformations. The authors have shown that a negative effect on the workability of the mixture, depending on the design mixture and the type of binder, can be attributed to the high value of Ridgen voids, the percentage volume of intergranular voids referred to the density of a constipated specimen at dry with a normalized procedure which is able to provide an indirect evaluation of some fundamental properties, such as its mineralogical nature, the size distribution of the particles, their degree of angularity and texture.

Pérez and Pasandín (2017) in order to guarantee sustainable construction, investigated the retrieve of construction and demolition waste (CDW) as recycled concrete aggregate (RCA) for the manufacture of HMA in place of natural aggregates. In according to EN 12697-12:2008 Standard, they evaluated the water resistance of the HMA after a series of eight cylindrical Marshall Samples were compacted with 50 blows for face.

Modarres and Rahmanzadeh (2014) investigated the effect of coal waste powder as filler material in hot mix asphalt. The Marshall method was used to determine the optimum bitumen content that was determined as equal to 5,2%. In comparison to the reference mix containing limestone powder as filler material, the mixture containing coal waste powder has shown an increased for the Marshall Stability, indirect tensile strength and resilient modulus.

The study presented here assessment the recycle of fly ash filler (MFA) in hot mix asphalt design. In the first phase the rheological response of mastics made with fly ashes were evaluated in terms of Frequency Sweep test and Multi Stress Creep Recovery test, compared to mastics containing limestone filler (MFL), at two different F/A ratio (Yan et al. (2013); Cardone et al. (2015); Antunes et al. (2016)). To compare the rheological results to mixtures of asphalt concrete, the optimum bitumen content was evaluated according to Marshall Test and the comparison to traditional limestone mixtures was made in terms of mechanical and volumetric properties with ITS test, water sensitivity, complex modulus and creep test. The results of the study confirm that is possible to reuse fly ashes for construction of HMA, because fly ashes not also improve the mechanical parameters but they confirm the reduction of saving production costs and materials disposal during a reconstruction of the pavement layers.

2 Materials

2.1 Fly Ash Filler

Fly Ashes are the by-product of the coal combustion process in thermoelectric power plants that use coal powder as solid fuel. They are made up of particles of the order of microns, of an illuminated silico nature, of spherical form and substantially amorphous structure. Chemical properties and characteristic of the fly ash filler are given in Table 1a. As shown, the fly ashes have a high content of Silica (18.59%), Aluminium (18.38%) and Carbon (34.02%). To verify that no harmful substances are released by fly ash filler, leaching test was carried out. The results are reported in Table 1.

2.2 Bitumen

The bitumen used in this study was a neat bitumen 50/70 produced in Italy oil refinery. The characteristic of the bitumen are reported in Table 1c.

Table 1. Fillers chemical elements, leaching test and bitumen properties

(a) Chemical properties and characteristic of the fly ash filler											
Fly ash	Content [%]										
	C	O	Ca	Mg	K	Al	Si	Fe	Si	Ti	Ni
	34	19.6	2.68	0.6	0.4	18.4	18.6	0.4	3.4	4.8	0.5

(b) Leaching test results on fly ash fillers		
Parameter	Results (mg/l)	Law Limits (mg/l)
Cobalt	0.06	0.25
Vanadium	0.02	0.25
Zinc	0.33	3
Chloride	35.45	100
Nitrate	0.09	50
Sulphate	4.4	250
DOC	6.1	100
TDS	1.29	100
Ph	9.73	12
COD	5.8	30

(c) Bitumen properties		
	Value	Standards
Penetration at 25 °C	52 dmm	EN 1426
Softening point	49 °C	EN 1427
Dynamic viscosity at 160 °C	0.12 s	EN 13702
Frass	−10 °C	EN 12593

3 Experimental Research

3.1 Mastic Rheological Phase

Filler Classification

The limestone material and fly ashes, in order to be used as filler, were geometrically and volumetrically analysed through a granulometric analysis by sedimentation, performed according to EN 933-10 Standard. The percentage volume of the intergranular voids referred to the density of a dry constipated specimen has been evaluated according to EN 1097-4 Standard. Experimentally it was observed that as the rigden voids increase, the stiffening effect increases. In fact, from the rigden voids values in Fig. 1 it is observed that the fly ashes are stiffer that the limestone filler.

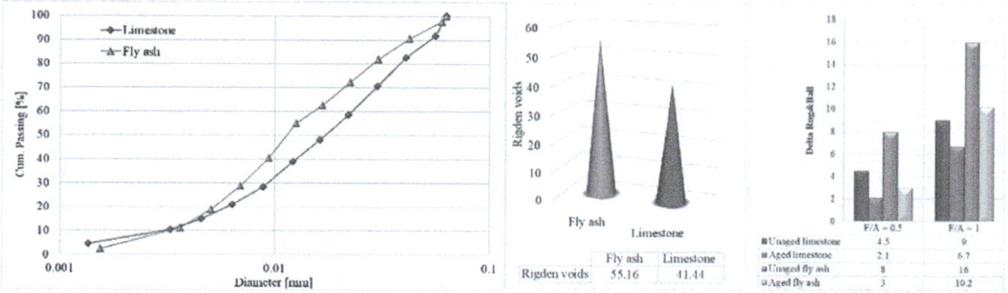

Fig. 1. Limestone and Fly Ash granulometric analysis results

Preparation of Mixtures

The two fillers, limestone and fly ash, were selected to passing at 0.063 mm sieve (Bocci and Giuliani (1998)). The bitumen-filler mastic was produced following a specific protocol and optimized to obtain a homogeneous mixture. The correct mass of the heated bitumen filler was added at the temperature of 170 °C for 2 h. Using a mechanical mixer equipped with an impeller, the mixture was mixed until a homogeneous mastic was obtained. The addition of the filler was completed in about 10 min and the mixing time lasted about 30 min in order to avoid the segregation of the filler. Two fill-to-binder (F/A) ratio for limestone and fly ash filler were considered in this study (0.5 and (1) and for each filler has been evaluated the aged effect. In order to reproduce the real field conditions as accurately as possible and to understand if the effect of time and weather conditions could affect the decay of mechanical characteristics, the bitumen with which the study mastics were packaged were subjected to procedures to age them. For reproduce the aging, RTFOT *(Rolling Thin Film Oven Test)* procedure was implemented for short-term conditions (EN 12607-1 Standard).

Delta Ring and Ball Test

Figure 1 shows that mastics with fly ash filler at both (F/A) ratio show a delta ring and ball twice than the limestone filler at not-aged condition while at aged it continues to be higher than the limestone mastics around 50%.

Frequency Sweep Test

The complex modulus G* was determined with a Frequency Sweep test according to EN 14770 Standard. For each mastics, 5 different temperatures (from 10 °C to 50 °C) and 19 different frequencies (from 0.1 Hz to 10 Hz) have been analysed with a F/A ratio of 0.5 and 1. The results in Fig. 2a show the comparison between master curves obtained for two materials investigated. It is evident that increasing temperatures and decreasing frequency, G* for mastics containing fly ash filler is higher than the limestone mastics at both F/A ratio. The same trend took place after the stage of aging (Fig. 2b).

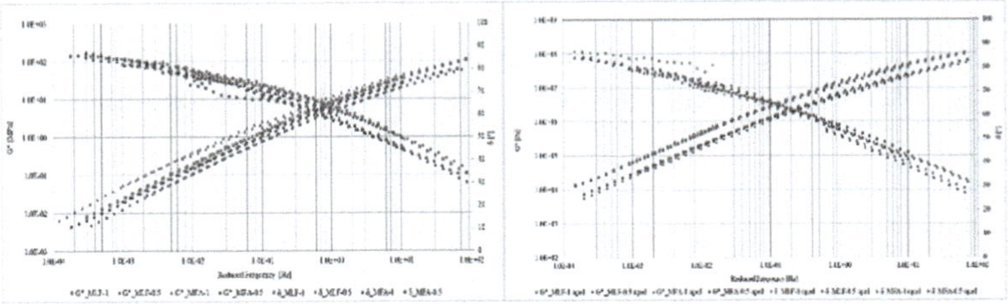

a) Unaged b) Aged

Fig. 2. Fly ashes and limestone master curves

Multi Stress Creep and Recovery Test

To determine the presence of elastic response in mastics containing limestone and fly ash filler, the Multi Stress Creep and Recovery test (EN 16659 Standard) has been used under shear creep and recovery at two stress levels, 0.1 kPa and 3.2 kPa, at a temperature of 40 °C (Simone et al. (2017)). The test consists in loading a sample at constant stress for 1 s, then allowed to recover for 9 s. Ten creep and recovery cycles are run at 0.100 kPa creep stress followed by 10 more cycles at 3.200 kPa creep stress. The shear strain showed in Fig. 3 for mastics containing limestone filler is greater than the fly ash at both F/A ratio. The same is show after aged phase. The stiffening effects of fly ash filler is also confirmed by the non-recoverable compliance values (Jnr). In fact, under 3.2 kPa shear stress the difference of Jnr between fly ash and limestone increase with increase the F/A ratio in both conditions more than 50%.

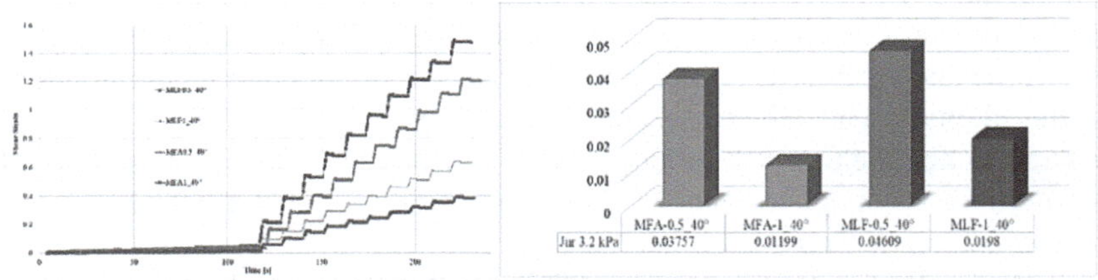

Unaged Shear Strain-Unaged F/A Ratio

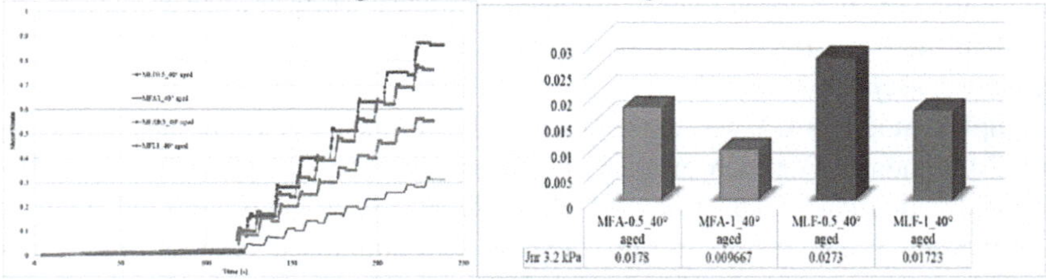

Aged Shear Strain-Aged F/A Ratio

Fig. 3. Multi stress creep and recovery test results

3.2 Asphalt Concrete Phase

Mix Design Phase

In this research study the aggregate mixture has been defined according to dimensional limits expressed by the EN 13108-1 Standard, by managing specification of infrastructure system and by requirement available in the SUPERPAVE system which indicates the grain size through the identification of control points and a sand restriction zone to be avoided. The grading curves obtained in this first phase are been obtained with a variation of filler size percentage by the total weight of aggregates as indicated in Table 2.

Table 2. Overview of the variation of filler size percentage

Aggregate type	Mixture type	
	Limestone (%)	Fly ash (%)
Limestone Nr.2	23	23
Limestone Nr.1	29	29
Limestone 3/6	13	13
Limestone dust	32	32
Filler	3	3

The result obtained, matches to use the same percentage of two different fillers materials (limestone and fly ash) to obtained the same grading curves distribution, as shown in Fig. 4.

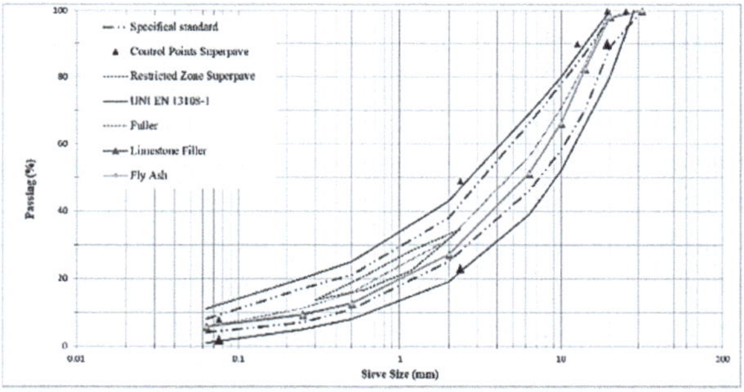

Fig. 4. Mix design grading curve

Once defined the aggregate size distribution, binder content was changed in line with Motorways specification for a binder layer (4–6%), in compliance with a range of 0.25, while Fly Ashes content matches several different iterative percentages such as were 4.00, 4.25% 4.50, 4.75, 5.00, 5.25 and 6%; while for HMA with limestone filler, percentage investigated were 4.75, 5.00, 5.25, 5.50 and 5.75%. To accomplish the mix design process has been used the Marshall method, so has been prepared four bituminous specimens of HMA for each level of bitumen content, according to EN 12697-30 Standard. For each one specimen prepared, have been evaluated the volumetric composition and subsequently they have been submitted to the Marshall test to evaluate the Marshall stability and flow. The Marshall test was carried out according to EN 12697-34 Standard. To determine the best design has also been used the Marshall flow parameter that is a value is an indicator of flexibility and plasticity of bituminous hot mixtures under traffic loads. Another important bituminous mixtures parameter is the "percentage of Voids" (Russo et al. (2018)). This parameter has been calculated, for each specimen, according to EN 12697-08 Standard. The relationships between percentage of bitumen content and Marshall Stability, Marshall Flow and percentage of voids were plotted in Fig. 5.

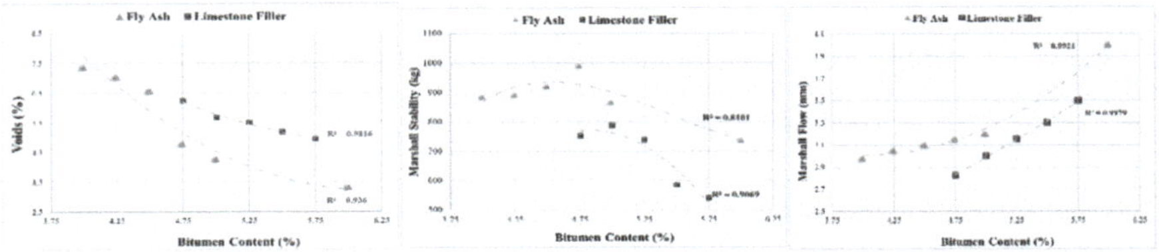

Fig. 5. Relationships between percentage of bitumen content and percentage of voids, Marshall Stability and Marshall Flow.

The mixtures containing fly ash it has a peak of Marshall stability value at 4.75% of bitumen content and it has a stability value of 980.80 kg; at the same time the mixture containing limestone filler has a peak of Marshall stability value at 5.25% of bitumen content and it has a stability value of 738.51 kg. By the analysis of these data, it is possible to see that the optimum values of bitumen content are lower for Fly Ash than Limestone Filler.

Indirect Tensile Strength (ITS)

A mechanical characterization of Hot mix asphalt was carried out by evaluating Indirect Tensile Strength, according to EN 12697-23 and under wet conditions for investigating the water susceptibility of bituminous mixtures according to EN 12697-12 Standard. As it is possible to see in the Fig. 6, is clear the greater ability of resistance to ITS of bituminous mixture containing limestone filler, in both dry and wet condition, compared to hot asphalt mixture containing fly ash. At the same time, it is interesting to underline that the two different mixture analyzed have the same trend in terms of water sensitivity.

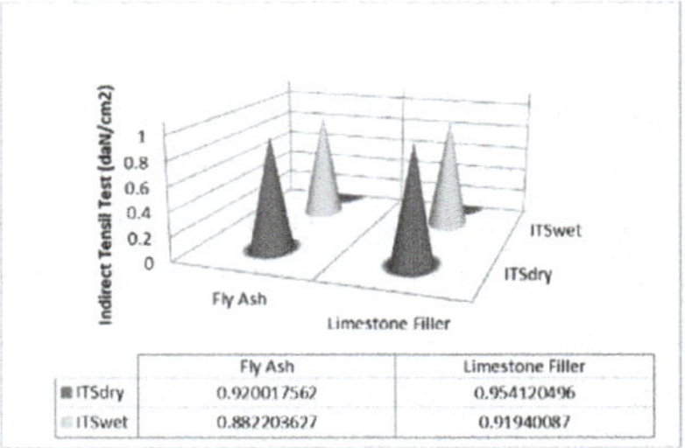

Fig. 6. Indirect tensile test

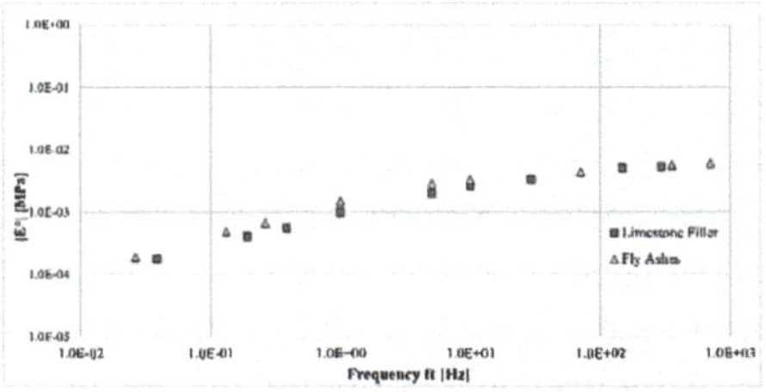

Fig. 7. Direct uniaxial test results

Direct Uniaxial Test

The complex modulus is a parameter used to evaluate the stiffness of bituminous mixture. The evaluation test has been evaluated on 4 cylindrical specimens having an height of 200 mm, for each mixture; the specimens were prepared according to EN 12697-26 Standard, Annex D. Each specimen was submitted to a normal tension according to established amplitude and frequency; the test has been carried out, for each specimen at 4 different temperature value (10 °C, 25 °C, 40 °C) and at 3 different frequencies (1 Hz, 5 Hz, 10 Hz). Starting from the time-temperature superposition principle (25 °C), the master curves (Fig. 7) are obtained using shift factors (a(T)) that are calculated according to WLF equation.

As it is possible to see in the Fig. 7 the complex modulus of HMA containing limestone filler, at high temperature and low frequencies is lower than the complex modulus of HMA containing fly ashes filler. At the same is interesting to see how at higher frequencies and low temperatures the complex modulus of the two mixtures is quiet similar and it tends to overlap. Starting from this results data it is possible to deduce that the HMA containing fly ash filler lend themselves well to elevated temperatures and low frequencies in terms of stiffness.

Repeated Load Axial Test (RLAT)

For further investigation of mechanical characterization of bituminous mixtures has been evaluated the resistance to permanent deformation. The evaluation of this parameter has been used cylindrical specimens that are subjected to a confining stress and a cyclic axial stress. Two specimens for each mixture were tested according to EN 12697-25 Standard. Every cylindrical specimen was placed in a water bath at 40 °C per 4 h and then tested with a cyclic axial pressure obtained as a sinusoidal pressure $\sigma_a(t)$ of 150kN, with amplitude σ_v of 350kN. The test results, in Fig. 8, shows that near the first 500 loading cycles the two mixtures have a similar deformation, after in association with the loading cycles increasing the deformation of the mixture with limestone filler increases dramatically relative to the fly ash.

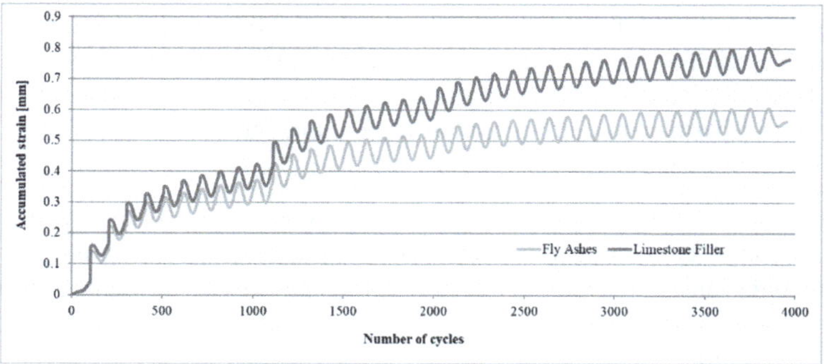

Fig. 8. Accumulated strain of the two mixtures

During the test, the height variation of the specimen is measured at specified numbers of load applications and the cumulative axial strain, εn, has been evaluation.

From these data is it possible to see that the cumulative axial strain for fly ash was 7.5% while the same parameter for limestone filler was 10%. Once again, with a dynamic test as repeated load axial test, the improvement in terms of stiffness of mixtures containing fly ash is confirmed.

4 Results and Conclusions

The aim of this study was to evaluate the possibility of replacing the limestone filler for the design of HMA for the binder layer of with material derived from waste of industrial processes such as fly ashes. From the results of the rheological characterization of mastics at elevated temperatures with Dynamic Shear Rheometer can be drawn the following conclusion:

- The volumetric characterization of fly ashes, based on the analysis of the granulometric curve determined by sedimentation, implemented by the determination of the Ridgen voids values, their smaller size compared to limestone filler increases the percentage of bitumen fixed internally and, consequently the stiffness. Also in terms of delta ring and ball values the mastics made with fly ash have higher values than those made with limestone filler both before and after the aging phase.
- It was observed that, both before and after having subjected the mastics to the aging process by means of RTFOT equipment, the addition of fillers generates an increase in the stiffening effect, above all for low frequencies or for high temperatures, confirming what has been said for the fly ash filler have been recognized as having a G* higher than mastic with limestone filler.
- The Multiple Stress Creep and Recovery Test has shown that the increase in G* modulus leads to a decrease in the accumulation of permanent deformation and a better elastic response of the materials in particular for the mastics containing fly ash filler.

Based on the rheological analysis carried out and the investigated filler and bitumen ratios, it can be said that in the mix design procedure of bituminous mixtures, ordinary fillers can be replaced by innovative fillers made from waste materials (Chen et al. (2011); Ossa et al. (2016)), increasing the performance of the mixture related to the phenomenon of permanent deformation at high temperatures. In fact, from the HMA analysis it can be stated that:

- The amount of bitumen used for the mix design containing fly ash was 0.5% lower than the other containing limestone filler, confirming the assumption made by rigden voids values. In addition, better performance in terms of Marshall Stiffness in the case of mixtures made with fly ash reaching values higher than 25% compared to limestone filler;
- In terms of water sensitivity, both mixtures have the same ITS reduction around 4%.
- Very high complex modulus for mixtures containing fly ash and stiffness confirmed by the creep test where near the end of load cycles examined the mixture was found to settle its deformations unlike the limestone filler where it continued to want to deform.

Moreover, from the results obtained it is possible not only to frame the waste materials analysed from the point of view of mechanical and volumetric performances, achieved to a sufficient degree, but in an environmental sustainability and above all innovative perspective, given the few studies on the use of alternative materials, like fly ashes, in the HMA production.

References

Topini D, Toraldo E, Andena L, Mariani E (2018) Use of recycled fillers in bituminous mixtures for road pavements. Constr Build Mater 159:189–197

Bocci E (2018) Use of ladle furnace slag as filler in hot asphalt mixtures. Constr Build Mater 161:156–164

Sangiorgi C, Tataranni P, Simone A, Vignali V, Lantieri C, Dondi G (2016) Assessment of waste bleaching clay as alternative filler for the production of porous asphalts. Constr Build Mater 109:1–7

Pérez I, Pasandín AR (2017) Moisture damage resistance of hot-mix asphalt made with recycled concrete aggregates and crumb rubber. J Clean Prod 165:405–414

Yan KZ, Xu HB, Zhang HL (2013) Effect of mineral filler on properties of warm asphalt mastic containing Sasobit. Constr Build Mater 48:622–627

Antunes V, Freire AC, Quaresma L, Micaelo R (2016) Effect of the chemical composition of fillers in the filler–bitumen interaction. Constr Build Mater 104:85–91

Modarres A, Rahmanzadeh M (2014) Application of coal waste powder as filler in hot mix asphalt. Constr Build Mater 66:476–483

Cardone F, Frigio F, Ferrotti G, Canestrari F (2015) Influence of mineral fillers on the rheological response of polymer-modified bitumens and mastics. J Traffic Transp Eng (Engl Ed) 2(6):373–381

Simone A, Mazzotta F, Eskandarsefat S, Sangiorgi C, Vignali V, Lantieri C, Dondi G (2017) Experimental application of waste glass powder filler in recycled dense-graded asphalt mixtures. Road Mater Pavement Des 20(3):592–607

Chen M, Lin J, Wu S (2011) Potential of recycled fine aggregates powder as filler in asphalt mixture. Constr Build Mater 25(10):3909–3914

Bocci M, Giuliani F (1998) Caratterizzazione di filler per conglomerati bituminosi. In: The proceedings of the XXIII Convegno Nazionale Stradale AIPCR, Verona, Italy, pp 18–21

Akbulut H, Gürer C, Çetin S, Elmacı A (2012) Investigation of using granite sludge as filler in bituminous hot mixtures. Constr Build Mater 36:430–436

Zanetti MC, Santagata E, Fiore S, Ruffino B, Dalmazzo D, Lanotte M (2016) Evaluation of potential gaseous emissions of asphalt rubber bituminous mixtures. Proposal of a new laboratory test procedure. Constr Build Mater 113:870–879

Russo F, Biancardo SA, Formisano A, Dell'Acqua G (2018) Predicting percent air voids content in compacted bituminous hot mixture specimens by varying the energy laboratory compaction and the bulk density assessment method. Constr Build Mater 164:508–524

Ossa A, García JL, Botero E (2016) Use of recycled construction and demolition waste (CDW) aggregates: a sustainable alternative for the pavement construction industry. J Clean Prod 135:379–386

Future Trends in Asphalt Pavements

Preliminary Study of an Energy Harvesting System for Road Pavements Made with Marginal Aggregate

Marco Pasetto[1(✉)], Andrea Baliello[1], Antonio Galgaro[2], Elisa Mogentale[1], and Anna Sandalo[1]

[1] Department of Civil, Environmental and Architectural Engineering, University of Padova, Via Marzolo 9, 35131 Padua, Italy
marco.pasetto@unipd.it
[2] Department of Geosciences, University of Padova, Via Gradenigo 6, 35131 Padua, Italy

Abstract. The progressive reduction of available energy resources and the continuous increase in demand are providing strong incentives for the use of renewable energies. Asphalt solar collectors are efficient energy harvesting systems for roads, able to extract thermal energy from pavements and convert the heat collected by their surfaces. Indeed, the possible reuse of waste materials in road construction, converted into valuable resources, has a strategic importance and could surely enhance the environmental sustainability of road pavement applications. This paper presents a preliminary experimental study aimed at evaluating the feasibility of a pipe-based energy harvesting system, which allows fluid circulation on a coil embedded in asphalt concrete manufactured with marginal aggregates. For this purpose, two-layer dense-graded asphalt slabs (AC8) were prepared in the laboratory, using different aggregate types (limestone and steel slag). A steel coil positioned at the interface was utilized to establish water circulation below the wearing course. The collected thermal energy was measured varying the water flow characteristics; the system was monitored through thermographic analysis while being subjected to a selected radiative power. Main results indicated that water flow rate was crucial in determining the temperature mitigation effect on asphalt concrete surfaces and the efficiency of the energy harvesting system. Some concerns about the operative approach were evinced (mainly related to the scale of the test); however, steel slag inclusion did not seem to compromise nor enhance the thermal conductivity of mixtures.

Keywords: Asphalt pavement · Energy harvesting · Solar collector · Heat · Conductivity

1 Introduction

Awareness about the topic of sustainability concerns various disciplines and facets of civil engineering (environment, economy, energy, etc.). The progressive reduction of available energy resources and the continuous increase in demand are giving strong

M. Pasetto et al. (Eds.): ISAP APE 2019, LNCE 48, pp. 101–113, 2020.
https://doi.org/10.1007/978-3-030-29779-4_10

incentives for the production and use of renewable energies. Total energy consumption worldwide in 2015 was equal to about 109,136 TWh (International Energy Agency 2017) and is estimated to increase up to 28% by 2040 (U.S. Energy Information Administration 2018). Not surprisingly, growing interest in such topics has also recently been demonstrated in the field of road construction. Within this perspective, the scientific literature recently started to report the design and application of several energy harvesting systems for road pavements, with different targets and operating apparatuses. Depending on the final objectives set for such systems, various working mechanisms and practices applied to flexible and rigid pavements can be cited. Pavement structures equipped with specific devices could be installed to collect vibrational energy created by vehicles on the road (Xu et al. 2018). Special electromagnetic devices attached to moving masses could generate a voltage induction, furnishing energy for storage or re-use (power supply for traffic regulation and monitoring, detection systems, safety and mobility appliances, etc.) (Zuo et al. 2014). Similarly, piezoelectric transducers allow the conversion of mechanical energy to electrical power thanks to the electrical charges generated by the forces due to vehicles passing over the pavement (Williams 1996). Electrostatic technology has also been developed and successfully applied on roads (Boisseau et al. 2010). Instead, geothermal pavement systems are constructed to manage heating, ventilation and air conditioning technologies to reduce primary energy consumption, combining ground-source heat pumps with permeable pavements that collect the water coming from surrounding surfaces (Tota-Maharaj 2010). Additional harvesting technologies concern the collection of thermal energy through pipe systems (installed under the road pavement surfaces). As an example, Chiarelli et al. (2017) proposed a suitable pipe-based structure working with air flows. Other applications considered the circulation of water in coils beneath pavements (Dawson et al. 2012; Guldentops et al. 2016). In general, these systems are reported to be versatile solutions, because they extract or dissipate heat of the pavement by managing the characteristics of the fluid circulating in the pipes. In the typical case of harvesting, a flow of cold fluid (colder than the pavement surface) permits the collection of thermal energy, also decreasing temperatures on top. Thus, concurrently with energy extraction, further benefits due to temperature reductions are connected to the perceived human comfort (and mitigation of the so-called "urban heat island" phenomenon) (Mallick et al. 2009); not least, the abatement of surface heat could enhance the mechanical resistance of pavements (increase of permanent deformation resistance) (Bobes-Jesus et al. 2013). In the case of hot fluid circulation, snow-melting and deicing could be obtained (Chen et al. 2011). This could avoid conventional methods (spreading of chemical compounds or salts), relieving pavements from the well-known stresses due to such practices (Obika et al. 1989), with tangible gains in safety for pedestrians and vehicles. Certainly, the most advanced installations would be designed as integrated systems able to simultaneously fulfil the dual function, i.e. combining thermal energy harvesting and re-use (in summer and winter) for the abovementioned purposes.

In parallel, the possible inclusion of waste/marginal materials in road pavements could assume strategic importance, enhancing the environmental sustainability of the structures. In this perspective, steel slag, i.e. a widely-available by-product derived from the steel-making industry, can be reused as a valuable aggregate to obtain eco-friendly asphalt mixtures, at the same time avoiding its disposal in landfill (and the related environmental concerns) and restricting the provision of precious and limited natural resources. Given all this, the present study aims to investigate the feasibility of utilizing an energy harvesting system under thin-layer pavement wearing courses manufactured with marginal steel slag aggregates. In particular, a small-scale laboratory approach was planned to evaluate the potential of a system composed by a steel coil positioned inside the wearing surface layer (immersed within the asphalt concrete). Significant interest in the project could be justified given the availability of the system also for road maintenance, rather than just in the design of new constructions: it uses resistant, cheap and easily-accessible materials, with low installation costs and negligible need to adapt the paving operations commonly performed. It is also worth considering that the conversion and reuse of steel slags lead to an eco-friendly mixture being obtained that limits any impact on the environment. Indeed, stressing the mandatory need for adequate mechanical and safety standards, it must be specified that this research is only a small part of a wider project addressed to the design of a specific energy harvesting system for road pavements (analysis of overall concerns such as structural performance, durability, efficiency, environmental benefits, economic impacts, optimization of management costs, etc.).

2 Materials and Samples Preparation

In order to reproduce in the laboratory some small-scale segments of road pavement equipped with an energy harvesting system, some asphalt concrete (AC) mixtures were produced in the form of slabs (400 mm long and 300 mm wide). Each slab was composed of a two-layer structure, in which a collecting pipe for fluid circulation was embedded at the interface (between the overlapped layers). The top layer consisted of a dense-graded asphalt mixture AC8 for wearing courses and was chosen to obtain a low-voids structure able to maximize the heat transfer throughout the depth of the pavement (Chen et al. 2015). The lower one was produced in the same manner for the sake of simplicity; it could also simulate an old existing milled wearing course above which the energy harvesting system could be installed. Based on a quite consolidated literature, particular attention was paid to the choice of aggregates since the importance of the lithic matrix and mineralogical compositions in heat transfer phenomena has been confirmed (Highter et al. 1984). Thus, aggregates suitable for road construction such as limestone and steel slag were utilized in different fractions (4/8, 0.075/4 mm and filler); Table 1 shows their main physical characteristics. A 50/70 penetration grade bitumen was used as binder; its basic properties are summarized in Table 2. Different ACs were produced combining bitumen and aggregate fractions. A reference slab (hereafter coded ACL) was manufactured only with limestone; the steel slag-based one (hereafter coded ACS) was prepared replacing 30% (by aggregate weight) of the limestone in the 4/8 and 0.075/4 fractions (filler excluded). Mix-design was performed case-by-case through the Marshall method using a standard grading envelop (prescribed by the Italian technical specifications for wearing courses). Changing the aggregate type, gradations were calibrated in volumetric terms in order to always ensure the same aggregate coverage (in this way, distinct particle densities of the various aggregate types were considered). Table 3 describes the mixtures' properties.

Table 1. Physical properties of aggregates

Property	Standard	Unit	Limestone			Steel slag		
			4/8	0.075/4	Filler	4/8	0.075/4	Filler
Particle density	EN 1097-6	Mg/m^3	2.75	2.76	–	3.89	3.80	–
Shape index	EN 933-4	%	12.8	–	–	7.8	–	–
Flakiness index	EN 933-3	%	10.5	–	–	8.3	–	–
SE	EN 933-8	%	–	78	–	–	92	–
Rigden voids	EN 1097-4	%	–	–	32.7	–	–	60.4

Table 2. Basic properties of bitumen

Property	Standard	Unit	Value
Penetration at 25 °C	EN 1426	0.1 mm	53
Softening point	EN 1427	°C	50.1
Penetration at 25 °C after RTFOT	EN 1426	0.1 mm	37
Softening point after RTFOT	EN 1427	°C	56.8

Table 3. Properties of mixtures

Property	Passing (volumetric) [%]					
Sieve [mm]	8	4	2	0.425	0.18	0.075
Min. prescription	100	45	28	13	8	6
Max. prescription	100	65	45	25	15	10
Gradation target	100	58	43	21	13	7
Binder content (by agg. vol.) [%]	15.2					
Target voids [%]	4.00					

Slabs were prepared with successive operations. First, an AC layer (with height of 20 mm) was compacted at 160 °C (EN 12697-33) to simulate the existing pavement. The energy harvesting system was then placed on the compacted mixture (once cooled). It consisted of a steel coil (outer diameter 12 mm, thickness 1.5 mm). This choice was based on structural and thermal considerations: plastic pipes were not compatible with the high temperatures involved in the AC compaction, whereas copper pipes were considered too expensive. The coil allowed water circulation and was shaped with three bends as schematized in Fig. 1a (the configuration permitted occupation of the slab to be maximized – to enhance the possible thermal exchange – and prevented significant local temperature gradients – dangerous for the mix integrity). Its extremities were equipped with steel fittings located in the proximity of the slab edges (along the longer sides), constituting the input and output sections of the steel circuit (Fig. 1b). A conventional tack coat (C55B2 emulsion) was then sprayed on, obtaining a bituminous film with an approximate dosage of residual binder equal to 0.3 kg/m^2 (according to technical prescriptions). Lastly, an additional bituminous mixture covered the pipe system and was compacted at 160 °C (EN 12697-33), forming the upper layer of the slab (height of 40 mm). The whole configuration (height of 60 mm) was designed in order to guarantee an adequate sealing of the coil within the slab, limiting the border effect next to the slab edges.

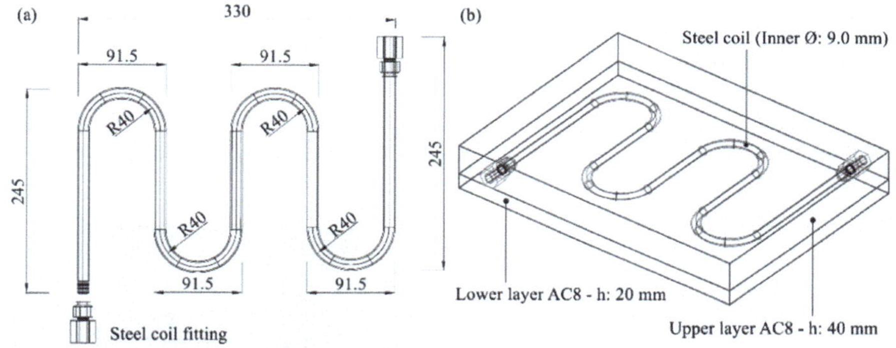

Fig. 1. Steel coil scheme (a) and positioning on asphalt concrete (b)

3 Test Methods

In order to evaluate the potential of the energy harvesting system in collecting heat from the pavement surface, a specific hydraulic circuit was constructed in the laboratory in order to establish a regular water flow and record the thermal responses inside the steel collector immersed in slabs, while exposing the asphalt concrete to a radiation source and monitoring the temperature of the slab. The equipment was arranged according to the scheme represented in Fig. 2. Water was charged through a hydraulic pump (*P*) within the external circuit (*C*) made of plastic pipes (nominal diameter of 9.0 mm). An adjuster of flow A_F was placed just after *P* and governed the flow rate (water at 14 °C from the municipal water supply was used). Two water-resistant thermal sensors (T_I and T_O) were interposed immediately before and after the terminal sections of the steel pipe to measure the input and output temperatures of the fluid circulating in the energy harvesting system (*E*), calculating the heat exchange inside the entire structure. The temperature on the slab was recorded with a thermographic device (*A*), evaluating the heat distribution on the slab surface (see Fig. 3a). Exposure to radiation was simulated with an artificial halogen lamp (*L*) that can emit a maximum power of 382 W and 32,000 lumens; it was lit for 5.5 h (direct irradiation) and then switched off (continuing the recording for 3 h to simulate shading). Figure 3b illustrates one of the tested slabs. In terms of water flow rate, 0.1 or 0.01 dm^3/s were used to simulate a pseudo-laminar or turbulent flow, respectively. A reference monitoring with null water flow was also planned for comparison purposes. All the testing scenarios are summarized in Table 4.

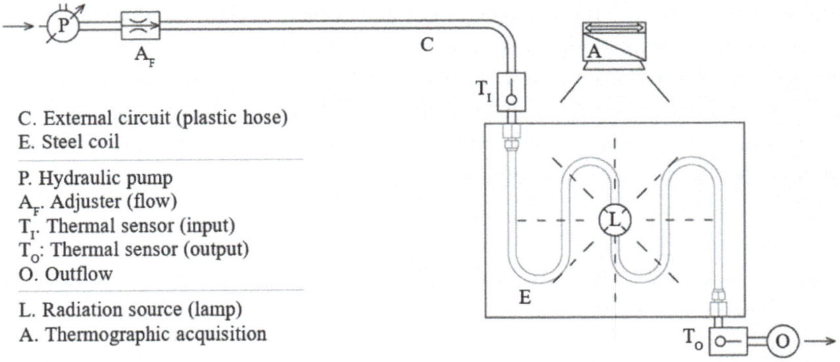

Fig. 2. Hydraulic circuit and equipment used to test the energy harvesting system

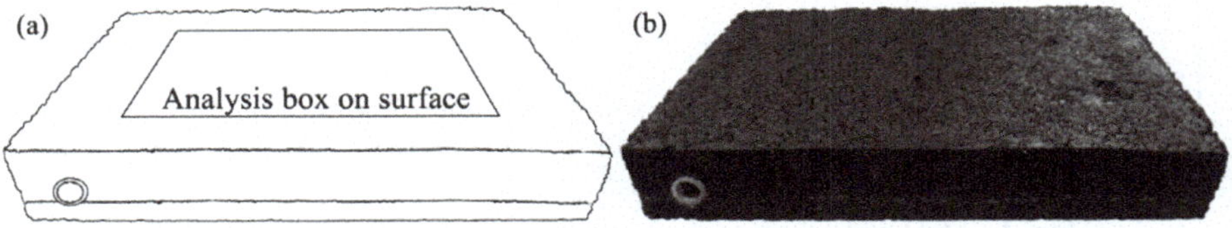

Fig. 3. Asphalt concrete: scheme of thermographic acquisition (a) and example of produced slab (b)

Table 4. Summary of test scenarios

Scenario	L.0	L.1	L.2	S.0	S.1	S.2
Mixture	ACL	ACL	ACL	ACS	ACS	ACS
Water flow rate [dm^3/s]	–	0.1[a]	0.01[b]	–	0.1[a]	0.01[b]

[a] turbulent flow [b] pseudo-laminar flow

Based on the recorded data, the power collected by the fluid Q_w (in W) during the irradiation time (t_i of 5.5 h) was first calculated using Eq. (1), where ΔT derived from the thermal sensors recordings, c_w and ρ_w (4186 J/kg °C and 997 kg/m^3) were the specific heat and density of the water respectively and q was the flow rate. Referring the power to irradiation time, the total collected energy E_w was calculated with Eq. (2) and gave indications about the energy harvesting system efficiency.

$$Q_w = c_w \cdot \rho_w \cdot q \cdot \Delta T \tag{1}$$

$$E_w = \int_0^{t_i} Q_w \cdot dt \tag{2}$$

Further calculations were proposed to assess the heat transfer mechanisms developed within the asphalt concrete mixtures; comparison between the reference solution (ACL) and the steel slag-based one (ACS) was made evaluating the thermal conductivity k of the two mixtures under different test conditions; it was estimated on the basis of the heat balance of the system. The heat balance was computed with a stationary evaluation, considering a small unitary element of the systems (broadly, a 10 × 10 mm square) located in the center of the slab to limit the dissipating effects of the slab edges (see Fig. 4). For this reason, the power collected by the fluid was referred to such a small area ($Q_{w'}$ was calculated using the pipe length). Assuming in such a portion a mean fluid temperature $T_{w,avg}$ (averaged between the input and output sensor readings), Eq. (3) was used to estimate the temperature on the pipe inner wall $T_{p,inn}$.

$$T_{p,inn} = T_{w,avg} + Q_{w'}/(h_c \cdot C \cdot l') \tag{3}$$

where h_c was the convective heat transfer coefficient, dependent on the flow characteristics (Reynolds, Prandtl and Nusselt numbers), C represented the inner circumference of the pipe and l' was the pipe unitary length considered (10 mm). Concerning the slab surface, the light source (382 W, 32,000 lumens) had an emission angle of 135°; according to the literature (Gamito 2016), the assumption led to a solid subtended angle equal to 3.879 steradians, thus a subsequent unitary luminous flux of 8250.16 lm/m^2. Relating this value to the square of the source-object distance (0.3 m) and taking into account the slab dimensions, incident power on the slab surface Q_s was estimated as 131.32 W (corresponding to 1094.29 W/m^2). The corresponding total energy on the slab E_s (calculated under 5.5 h of irradiation) resulted equal to 2600.0 kJ. Then, radiative power Q_s was again referred to the small unitary element and $Q_{s'}$ was obtained dividing Q_s by the unitary element area (element dimensions were chosen equal to 10 and 12 mm – the latter was considered to cover the pipe diameter). In turn, $Q_{s'}$ was split in different portions. The absorbed power $Q_{a'}$ was given by Eq. (4).

$$Q_{a'} = Q_{s'} \cdot (1 - \rho) \tag{4}$$

where ρ was the surface albedo referred to the diffused reflection (it resulted as 0.16). The irradiated power $Q_{irr'}$ was calculated with Eq. (5) (Pasetto et al. 2019).

$$Q_{irr'} = \sigma \cdot A' \cdot \varepsilon \cdot (T_{surf} - T_{air})^4 \tag{5}$$

where σ was the Stefan-Boltzmann constant ($5.67 \cdot 10^{-8}$ W/m^2·°C^4), A' was the unitary element area, ε was the surface emissivity, i.e. the aptitude to emit thermal radiation (considered equal to 0.92), T_{surf} was the temperature recorded on the slab surface (mean temperature on the central unitary element) and T_{air} was the ambient temperature recorded in the proximity of the test environment. The conductive power $Q_{cnd'}$ was then calculated as the difference between $Q_{a'}$ and $Q_{irr'}$ (energy balance on pavement surface). It was used to estimate the heat transfer within the asphalt concrete of the upper layer; based on the theory of the heat conduction in a multi-layer system, the thermal resistance of the mixture (R') was calculated with the Eq. (6).

$$R' = (T_{surf} - T_{p,inn})/Q_{cnd'} - R_p \tag{6}$$

where T_{surf} was the above-mentioned mean surface temperature of the central unitary element and $T_{p,inn}$ was that previously accounted; R_p, i.e. the thermal resistance of the steel coil, depended on the pipe material and shape and was referred to the upper semi-circumference of the pipe, considering a representative length for the R' estimation equal to h^*, i.e. the distance between the slab surface and the half portion of the semi-circumference – see Fig. 4 (constant R_p equal to 0.0197°C/W, h^* equal to 31 mm for geometry). Lastly, the thermal conductivity of the mixture k (referred to the upper layer) was determined – see Eq. (7).

$$k = h^*/(A' \cdot R') \tag{7}$$

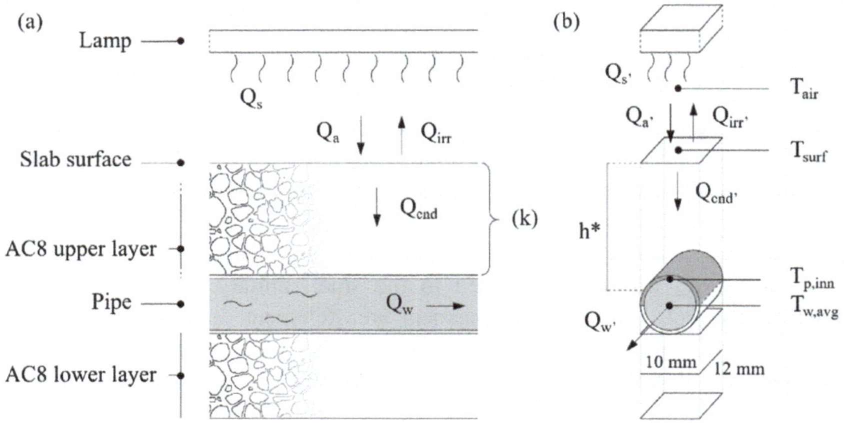

Fig. 4. Energy balance on the system: general scheme (a) and unitary element representation (b)

4 Results and Discussion

The results of the experiment are described in the following Fig. 5a and b (trend of maximum and average temperatures recorded on the central analysis box, respectively). Temperature gradients presented in Fig. 6a illustrate the evolution of ΔT between the maximum slab surface temperature in the scenario and that of the corresponding reference one (with nil water flow). Figure 6b depicts the evolution of the temperature difference recorded between the input and output sections of the steel circuit. Under the proposed test settings, maximum temperature of ACL and ACS (at 5.5 h of irradiation) rose up to 40.8 (in L.2) and 41.1 °C (in S.2), respectively (note that, in the reference scenario, they were 55.6 and 58.5 °C, respectively). At the same time, the maximum gradient recorded on water was equal to 2.6 °C (see Fig. 6b, $L.2.\Delta T_w$ curve). In accordance with previous studies (Bobes-Jesus et al. 2013), a sort of inverse proportionality between the thermal results on the slab surface and temperature of the water within the pipe was detected. Analyzing the temperature gradients (Fig. 6a), ACS always exhibited greater ΔT (against the reference scenario with nil fluid circulation) with respect to the corresponding ACL mix (both in the case of 0.1 and 0.01 dm^3/s).

The higher q value (0.1 dm^3/s) seemed to produce a higher reduction of the slab surface temperatures, but was also responsible for scarcely significant T gradients on water, it could thus partially compromise the possibility of collecting heat and extracting power from the fluid (see square-pointed-curves on Fig. 6a and b). Instead, the lower water flow (0.01 dm^3/s, indicative of a pseudo-laminar flow) resulted in higher ΔT_w values (Fig. 6b) and seemed to enhance the energy harvesting system efficiency (but it also led to less effective mitigation of slab surface temperatures). Some concerns about the system efficacy are then analyzed: Fig. 7 shows the temporal evolutions of power on the slab surface and power collected from water under the different scenarios.

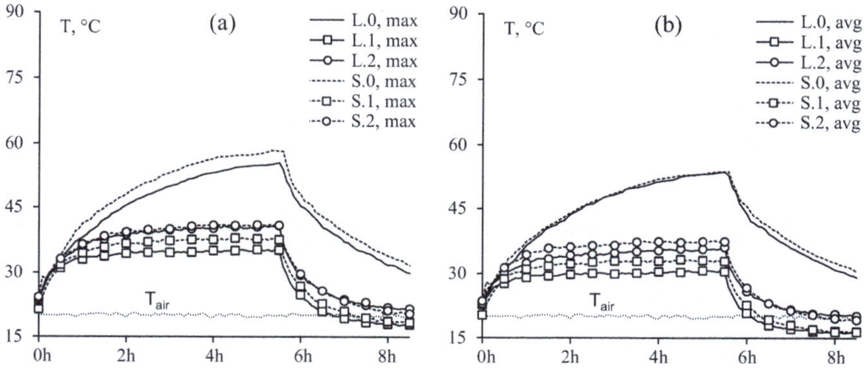

Fig. 5. Evolution of maximum (a) and average (b) T on the central analysis box (slab surface).

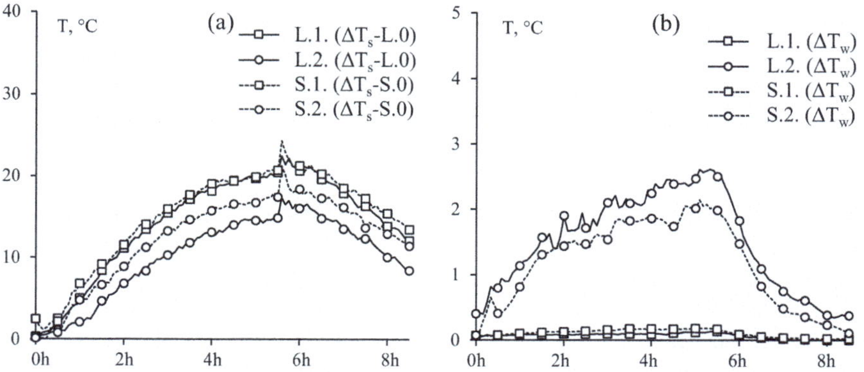

Fig. 6. Temperature gradients: scenarios vs. reference (nill flow rate) slab (a) and input-output water (b)

Based on these data, Table 5 reports the energy calculated on the slab surface (E_s), collected from the water (E_w) and the final efficiencies (η) resulting from the elaborations. η under L.2 scenario (45.2%) tends to maximize the potential of the system, otherwise all the percentages seem to be strongly overestimated; they were thus critically considered also in view of the literature results for similar asphalt collectors, which generally ranged around 20% (Guldentops et al. 2016). This was reasonably ascribed to the small-scale dimension of the system, only partially able to reproduce the real heat transfer mechanism, at least considering the entire slab (strong perturbations were hypothesized). In absolute terms, the higher ΔT_w recorded (for L.2) implied an extracted energy E_w comparable to those spent to light a standard traffic light for about 3 h or a new-generation one (with led bulbs) for about 30 h.

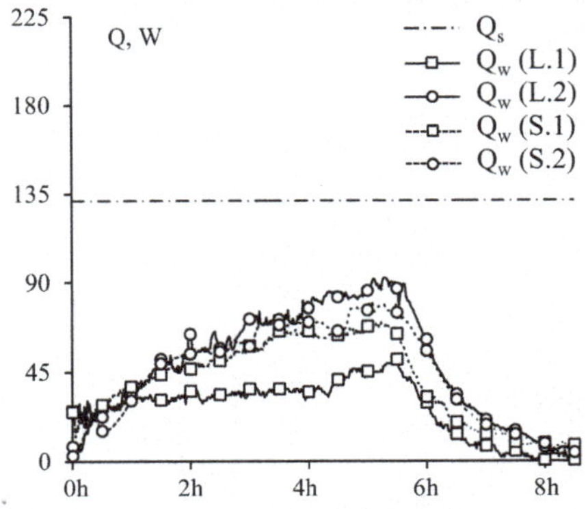

Fig. 7. Temporal evolution of power on slab surface and power collected from water

Table 5. Energy balance referred to slab dimension

Property	Symbol	Unit	Value			
			L.1	L.2	S.1	S.2
Slab total energy	E_s	kJ	2,600.0	2,600.0	2,600.0	2,600.0
Water total energy	E_w	kJ	685.2	1,175.5	1,018.7	1,070.9
System efficiency	η	%	26.4	45.2	39.2	41.2

Dealing with the central unitary element, Table 6 gives some details about the energy balance and heat transfer mechanisms. In addition, Fig. 8 depicts the evolutions of unitary irradiated and conductive power ($Q_{irr'}$ and $Q_{cnd'}$) pertaining to the surface element in the different scenarios. Finally, Fig. 9a shows the temperature trend calculated in the inner pipe wall and that of water (referred to the vertical projection of the unitary element); Fig. 9b points out the computed thermal conductivity (related to the 1.5-5.5 h range to analyze a stable condition). Based on Table 6 results, it is evident that the two selected flow rates corresponded to a different water regime; at 0.1 dm^3/s (turbulent flow – see *Re*), speed *v* induced predominant convective heat transfer (see *Nu*), whereas 0.01 dm^3/s (pseudo-laminar flow – see *Re*) emphasized the conduction: the latter flow rate caused higher water temperatures (L.2 and S.2 scenarios) – see Fig. 9a. As a consequence, $T_{p,inn}$ also increased at the lower flow rate. In general, these behaviors (and the above-described general trends of *T* and ΔT) indicate that the flow rate (related to water speed and occupation-time in the coil) could govern the effectiveness of the energy harvesting or, alternatively, the efficacy in mitigating the pavement surface temperature (at least under the selected test settings). Therefore, studying the unitary element surface balance, the higher surface temperatures detected for ACSs (with respect to ACLs – see Figs. 5 and 6) seemed to cause slightly higher irradiated power (surface dissipation) – see Fig. 8a.

Table 6. Heat transfer on a central unitary surface: calculation details

Parameters	Symbol	Unit	L.1	L.2	S.1	S.2
Prandtl number [a]	*Pr*	–	6.21	6.21	6.21	6.21
Nusselt number [b]	*Nu*	–	99.8	45.9	99.8	45.9
Reynolds number [c]	*Re*	–	14147	1458	14147	1458
Water speed	v	m/s	1.572	0.157	1.572	0.157
Convective coeff.	h_c	W/m^2 °C	6,659	3,061	6,659	3,061

[a] viscous/thermal diffusion ratio [b] convective/conductive transfer ratio
[c] inertia/viscous force ratio

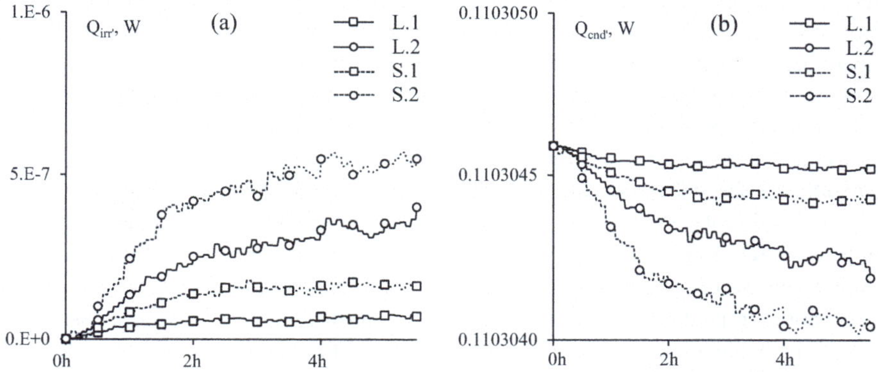

Fig. 8. Energy balance for the central unitary element surface: irradiated (a) and conductive (b) power

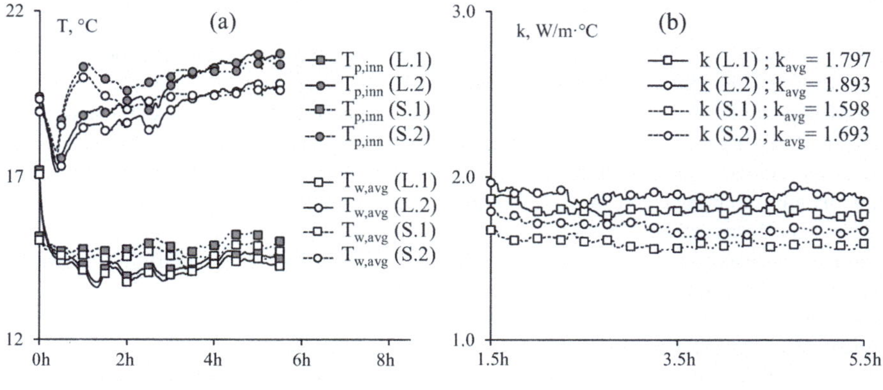

Fig. 9. Unitary element: inner-pipe and water temperatures (a); computed k of AC upper layer (b)

However, these $Q_{irr'}$ resulted always insignificant with respect to the conductive power transferred towards the slab depth: curves of Fig. 8b are almost coincident. The final overall computations gave slightly lower k in the case of ACSs (Fig. 9b). q seemed to assume less influence on conductivity. In general, the identified trends seemed to conform to literature findings, which indicate typical k values for paving materials ranging from 1.2 to 2.2 W/m °C and for steel slag-based mixes of about 1.3–1.7 W/m °C (depending on composition and gradation) (Bai et al. 2015; Mirzanamadi et al. 2018). Slightly lower k values when steel slag is partially included in asphalt concretes (in substitution of natural aggregate) are also reported (Barra et al. 2016; Liu et al. 2017).

5 Conclusions

The preliminary feasibility of an asphalt solar collector composed of a steel coil immersed in a flexible pavement was analyzed in the laboratory. There were some concerns about the operative plan in view of the small-scale approach adopted, otherwise probable findings were evinced analyzing a small unitary element in the center of the system. Mainly, water flow rate resulted crucial to determine an effective temperature mitigation on the concrete surface or the efficiency of the energy harvesting system. Under the testing scenarios, steel slag inclusion in the mixture (suitable for recycling by-products and promote environmental savings) did not seem to compromise nor enhance the thermal conductivity of asphalt concretes. Based on the promising results collected, further studies should be undertaken to propose more realistic small-scale models (as examples avoiding dissipation, increasing the coil length or occupation-time of the fluid) and to enhance the efficiency of the harvesting system by means of the characteristics of the fluid (e.g. speed, temperature) and mixture (composition, gradation). Additionally, field of application, maintenance issues and stress-strain behavior under loads should be studied to adequately determine the suitability of the system.

References

Bai BC, Park DW, Vo HV, Dessouky S, Im JS (2015) Thermal properties of asphalt mixtures modified with conductive fillers. J Nanomater 16:255

Barra M, Aponte D, Vazquez E, Mendez B, Miro R, Valls S (2016) Experimental study of the effect of the thermal conductivity of EAF slag aggregates used in asphaltic concrete of wearing courses on the durability of road pavements. In: Proceedings of the fourth international conference on sustainable construction materials and technologies, 7–11 August 2016, Las Vegas, United States

Bobes-Jesus V, Pascual-Muñoz P, Castro-Fresno D, Rodriguez-Hernandez J (2013) Asphalt solar collectors: a literature review. Appl Energy 102:962–970

Boisseau S, Despesse G, Sylvestre A (2010) Optimization of an electret-based energy harvester. Smart Mater Struct 19:1–10

Chen M, Wu S, Wang H, Zhang J (2011) Study of ice and snow melting process on conductive asphalt solar collector. Solar Energy Mater Solar Cells 95:3241–3250

Chen J, Wang H (2015) Determination of effective thermal conductivity of asphalt concrete with random aggregate microstructure. J Mater Civil Eng 27(12):1–9

Chiarelli A, Dawson AR, García A (2017) Field evaluation of the effects of air convection in energy harvesting asphalt pavements. Sustain Energy Technol Assess 21:50–58

Dawson AR, Dehdezi PK, Hall MR, Wang J, Isola R (2012) Enhancing thermal properties of asphalt materials for heat storage and transfer applications. Road Mater Pavement Des 13 (4):784–803

Gamito NM (2016) Solid angle sampling of disk and cylinder lights. In: Eurographics symposium on rendering, vol 35, no 4, pp 1–12

Guldentops G, Nejad AM, Vuye C, Van den Bergh W, Rahbar N (2016) Performance of a pavement solar energy collector: model development and validation. Appl Energy 163:180–189

Highter WH, Wall DJ (1984) Thermal properties of some asphaltic concrete mixes. Transp Res Rec 968:38–45

International Energy Agency (2017) Key World Energy Statistics. France, Paris

Liu Q, Li B, Schlangen E, Sun Y, Wu S (2017) Research on the mechanical, thermal, induction heating and healing properties of steel slag/steel fibers composite asphalt mixture. Appl Sci 1088:1–13

Mallick RB, Chen BL, Bhowmick C (2009) Harvesting energy from asphalt pavements and reducing the heat island effect. Int J Sustain Eng 2(3):214–228

Mirzanamadi R, Johansson P, Grammatikos SA (2018) Thermal properties of asphalt concrete: a numerical and experimental. Constr Build Mater 158:774–785

Obika B, Freer-Hewish RJ, Fookes PG (1989) Soluble salt damage to thin bituminous road and runway surfaces. Q J Eng Geol 22:59–73

Pasetto M, Pasquini E, Giacomello G, Baliello A (2019) Innovative pavement surfaces as Urban Heat Islands mitigation strategy: chromatic, thermal and mechanical characterization of clear/colored mixtures. Road Mater Pavement Des – Spec Issue. https://doi.org/10.1080/14680629.2019.1593230

Tota-Maharaj K (2010) Geothermal paving systems for urban runoff treatment and renewable energy efficiency. Ph.D. thesis. University of Edinburgh, Scotland, United Kingdom

U.S. Energy Information Administration (2018) International Energy Outlook 2018 Executive Summary. Washington, DC, United States

Williams CB, Yates RB (1996) Analysis of a micro-electric generator for microsystems. Sens Actuators 52(1–3):8–11

Xu X, Cao D, Yang H, He M (2018) Application of piezoelectric transducer in energy harvesting in pavement. Int J Pavement Res Technol 11:388–395

Zuo L, Ban J, Wang L, Park J, Zhou W (2014) On-road energy harvesting from running vehicles. Final Report. University Transportation Research Center, State University of New York, New York, United States

Electric Energy Harvesting Systems from Urban Road Pavements: Analysis and Preliminary Simulation

Sandro Colagrande(✉) and Gino D'Ovidio

Department of Civil, Building-Architectural, Environmental Engineering (DICEAA), University of L'Aquila, Via G. Gronchi, 18, 67100 L'Aquila, Italy
{sandro.colagrande,gino.dovidio}@univaq.it

Abstract. This paper discusses the topic of energy harvesting from urban roads illustrating an overview of new green road pavement integrated technologies to generate electric power. Two kinds of integrated road pavement technologies for clean electric energy harvesting compatible with loads and stresses generated by urban vehicular traffic were examined. The first concerns piezoelectric devices integrated into the road surface for generating electricity from the dynamic actions due to vehicular transit; the latter refers the photovoltaic panels distributed on the road surface for converting solar energy into electricity. Finally, preliminary systems performance simulations over an urban road section located in L'Aquila city (central Italy) with real environment and traffic conditions have been performed. Results are compared in terms of specific energy production. The presented work is a numerical study for understanding the potential of road energy harvesting technologies in real urban environment.

Keywords: Road energy harvesting · Piezoelectric roadways · Solar roadways · Renewable energy

1 Introduction

The transportation energy consumption is growing exponentially every year especially in urban areas, even if compared to the past decade vehicles have become more energy efficient. Currently, EU transport still depends on oil and oil products for 96% of its energy needs.

Low CO_2 alternatives to oil are also indispensable for gradual decarbonization of the transport sector. For smart, sustainable and inclusive growth, the EU strategy is going towards the drastic reduction of world greenhouse gas emissions by 80–95% below 1990 levels by 2050 (European Commission 2013). This is the reason why researchers have been looking for other green energy alternatives. The seek for alternative energy sources has also led to the exploration of road pavement integrated piezoelectric and photovoltaic technologies (Mighall and Phillips 2013), (Beck et al. 2015). In this scenario, innovations in technology on "zero" emission vehicles and on energetically sustainable infrastructures play a key role in emissions reduction.

M. Pasetto et al. (Eds.): ISAP APE 2019, LNCE 48, pp. 114–122, 2020.
https://doi.org/10.1007/978-3-030-29779-4_11

In the last decade a lot of research activities and industrial work concerning alternative vehicle power train innovation were done. In this direction, new technologies have been studied and proposed by Authors: superconducting magnetic levitated train resistance-free ordinary motion - except aerodynamic drag - (Lanzara et al. 2014; D'Ovidio et al. 2008; D'Ovidio et al. 2011), hybrid mini-buses (D'Ovidio et al. 2014; D'Ovidio 2007; D'Ovidio et al. 2016; Ciancetta et al. 2016; Ciancetta et al. 2016) and trains (D'Ovidio et al. 2017) hydrogen-fueled with "zero" emission energy cycle.

For what concerns linear infrastructures (roads and railways), research activities are focused to conceive and transform them from simple structure only dedicated to vehicles transit to smart backbones of services (vehicular traffic and information), able to also generate clean energy (Colagrande and D'Ovidio 2018; Xiong et al. 2012).

In this paper, an analysis of more promising electric energy harvesting technologies is proposed and analyzed for simulation in an urban environment.

Section 2 of this article illustrates the road integrated piezoelectric and photovoltaic technologies and design criteria for electric energy harvesting. Systems performance simulation in an urban environment and key results are illustrated in Sect. 3. The concluding remarks are in Sect. 4.

2 Road Energy Harvesting

In the international arena, different technologies are developed and tested in order to generate electrical or thermal energies from road pavements. These technologies are still under development and the experimental applications are limited to few cases; therefore, limited scientific publications are available in the technical literature.

In this section, piezoelectric and photovoltaic technologies for electricity production from road pavement are reviewed and analyzed. Performances of these technological systems are influenced by the features of materials, the distribution on the road and the characteristics of vehicle flows. Piezoelectric system performance rises with the increase of vehicular traffic, on the contrary, photovoltaic system output decrease with traffic flow increase.

2.1 Piezoelectric Technology

Piezoelectric devices are experimentally used for harvesting energy from external loads for roads and walkways. The applied mechanical stress caused by external actions (wheel loads) produces a storable electric charge. Piezoelectric technology used in pavement generates electricity by the means of piezoelectric crystals, which are implanted few centimeters below the road surface. Their deformation under the load of traffic produces electric current (Shukla and Ansari 2018).

Piezoelectric materials (Barium Titanate, Lead Titanate, Lead Zirconate Titanate, etc.) are crystals that have the property to generate current when compressed or vibrated, vice-versa they generate a stress when a voltage is applied (Hill et al. 2014; Nelson 2010). Therefore, piezoelectric devices, if appropriately integrated into an electromechanical system, can be suitable for road applications in order to convert vehicles motion into electrical power. The working principle is based on the piezoelectricity effect; whereby piezoelectric crystals generate an electric voltage from elastic deformations.

Different products (Innowattech, Cook Chennnault, Virginia Tech, Oregon DOT, Berkeley tech.), (Dikshit et al. 2010), have been designed to be installed under the asphalt pavement or inside of the railway sleepers, in order to generate direct current (DC) from the transit of vehicles/trains (Kour and Charif 2016).

As illustrated in Fig. 1, the DC output of the piezoelectric units electrically connected in series can be converted by DC/AC converter to alternate current (AC) for directly feeding electrical devices and power grid, or can be accumulated in a storage system (e.g. battery).

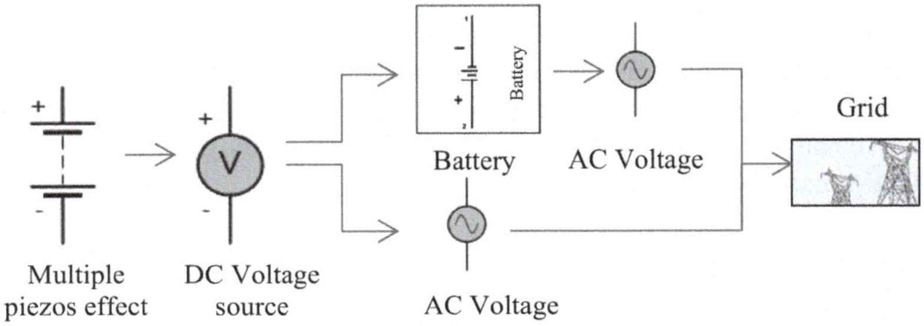

Fig. 1. Conceptual working scheme

The efficiency of a piezoelectric material is greatly influenced by the crystal and its properties (geometry, thickness, fixation, and structure). Thinner and tapered geometry is suitable for better performance (Ibrahim and Ali 2012).

In addition to the properties of materials, speed and weight of the vehicle also affect the energy output (Agarwal 2014; Chari and Kour 2015; Kumar 2013). An interesting application for the production of piezoelectric energy from roadways comes directly from Innowatech, an Israeli company (Edery-Azulay 2010; Songsukthawan and Jettanasen 2015). This solution, when applied to a road, produces electricity as a function of the number of vehicles, their weight, and speed.

Main steps for the installation of piezoelectric generators installation on the road pavement, are: (i) cutting of the pavement surface, (ii) lay down of a quick-setting concrete, (iii) positioning of the piezoelectric generators and drowning in concrete, (iv) connecting the cables, (v) overlaying the generators with an asphalt layer, (vi) laydown of asphalt wearing layer. First tests by Innowatech dealt with the installation of piezoelectric nano-generators along a stretch of 10 m in a road asphalt pavement. In this case, it is reported that generators could potentially produce about 2 kWh. This trial allowed to experimentally verify that the system works better when traffic is at least 600 vehicles/hour with an average speed of about 72 km/h. Currently, the system is under testing, and it is characterized by high implementation costs, that could be reduced if mass-production will be promoted (Kurzweilai 2011).

2.2 Photovoltaic Technology

Photovoltaic (PV) panel is used to convert solar radiation into electric power. The solar panels consist of a P-type semiconductor and N-type semiconductor. When sunlight reaches the semiconductor materials of the PV panel, free electrons are forced to flow in a certain direction. The negatively charged electrons move toward the N-type semiconductor, while the positively charged move toward the P-type semiconductor. The flow of moving electrons creates an electrical current when connected to an electrical load (Wakulat 2016).

The photovoltaic panel placed on the road surface is a pioneering idea by the American engineer Scott Brusaw who, supported by his working team, implemented the "Solar Panel Road" (Stephy and Keerthi 2017).

The solar roadway panel, designed to substitute the asphalt wearing course, is composed of the following three layers (Alark and Kulkarni 2013):

(1) surface layer, made of rough glass, anti-abrasive, self-cleaning and highly resistant, which contains photovoltaic cells and Light Emission Diodes (LED);
(2) intermediate electronic layer, which contains a microprocessor for controlling and monitoring loads and lighting;
(3) bottom layer, which carries the energy collected by the intermediate layer to various storage systems connected to the roadway and transmits the pavement load to sub-grade layer.

The DC is converted to AC energy by a DC/AC converter or solar micro inverter, then is fed to the electrical grid. The energy gain could be significant: it has been estimated that, for an average daily solar irradiation of 4 h, each PV panel should be able to produce around 7.6 kWh per day (PhysOrg.com 2009). However, maintenance procedures for dust accumulation, duration of PV cells and high costs still make the PV panel for road surfaces in need of improvements. The Solar Roadways is currently being tested in a section of a highway (70 km long), located between Coeur D'Alene and Sandpoint in Idaho.

Also in France, a PV pavement has been realized and named with the explicit term of Wattway (Wattway in France 2016).

3 Simulation

An urban road, SS 17 named (Fig. 2) and located in the city of L'Aquila (central Italy), was selected in order to simulate, in real traffic and environmental conditions, the performances of the piezoelectric and PV energy harvesting technology applications described in the previous sections.

The road section (Fig. 3) is 6.4 km long, with an average width of 8.0 m. It has been divided into three stretches (1-2, 2-3, 3-4) according to the occurring different vehicular traffic flows (Tables 1 and 2).

Authors detected traffic flows of the three stretches of the road section and elaborated them in order to compute the daily average traffic flows.

Measurements were made in the absence of vehicle speed detectors.

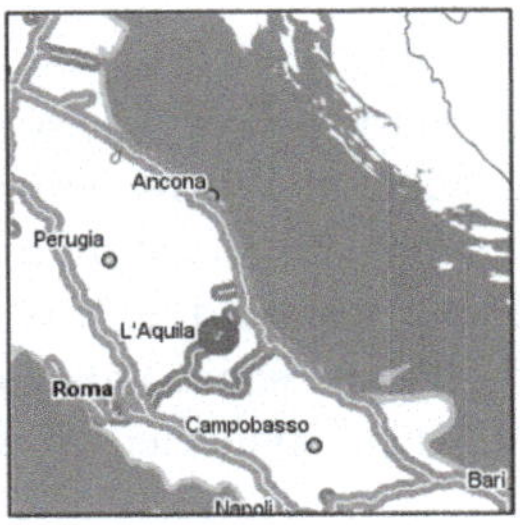

Fig. 2. L'Aquila city location

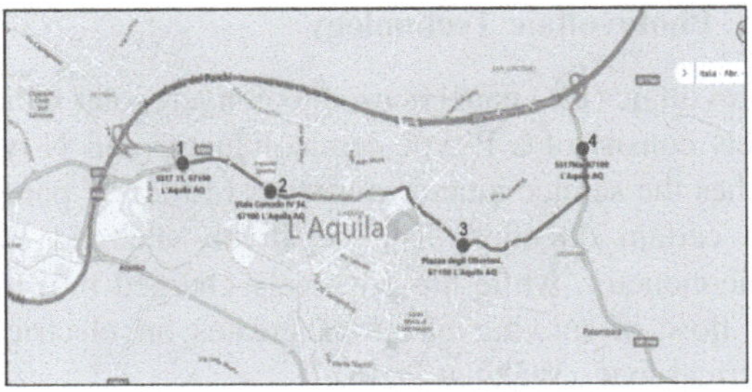

Fig. 3. SS 17 urban road in L'Aquila City [source: bing map]

The energy output calculation was performed by assuming an active lane width of 3.0 m.

3.1 Piezoelectric System Output

System energy output calculation was performed by taking into account the performance data of the Innowatech system: 200 kW/h per single lane for 1 km long stretch with a traffic flow of 600 vehicles/h (Shukla and Ansari 2018).

The calculations have been made assuming an optimal distribution of the piezoelectric modules on the road in line with the passage of the vehicle wheels.

Table 1 lists for each selected stretch, the tested traffic volume and the theoretical electric energy production per km of lane. Calculation results show that the weighted average specific energy output is 726 kWh/m^2 per year.

Yearly specific energy computation is evaluated assuming 24 h per day and 365 days per year.

Table 1. Road integrated piezoelectric system data and performance

Stretch	Lane length	Average traffic volume	Hourly energy per km of lane	Hourly total energy	Yearly total energy	Hourly specific energy	Yearly specific energy
	km	Vehic/h per lane	kWh/h/km	kWh/h	MWh/y	kWh/m^2	kWh/m^2/y
1-2	1.4	952	317	444	3892	0.106	927
2-1		927	309	433	3790	0.103	902
2-3	2.8	748	249	698	6116	0.083	728
3-2		752	251	702	6148	0.084	732
3-4	2,2	623	208	457	4002	0.069	606
4-3		612	204	449	3931	0.068	596
Tot	6.4			3183	27879		

3.2 Photovoltaic System Output

In this sub-section, is taken into consideration the installation of PV panel for road pavements in the urban segment (1-4) along the SS 17.

Electric energy (*E*) produced by a standard PV array integrated into a road pavement is calculated as:

$$E = [(\eta_m \cdot A \cdot G) \cdot (1 - \varepsilon_t)] \quad (1)$$

where:

- η_m is the PV system efficiency, considering the combined system losses due to temperature and low irradiance (using local ambient temperature), loss due to angular reflectance effects and other losses (cables, inverter etc.)
- A is the PV area
- G is the average sum of global irradiation per square meter received by the PV modules of the given system
- ε_t is the vehicular traffic effect parameter, that takes into consideration the shadowing of running cars on the PV panel surface. Its value is between 0 at no traffic condition and 1 at theoretical full traffic condition with zero distance among lined up vehicles.

A numerical method for ε_t calculation, which does not require experimental measurements of vehicle speed, is proposed below. For a given road with a vehicular flow (φ), the parameter $\varepsilon_t(\varphi)$ is defined as:

$$\varepsilon_t(\varphi) = D(\varphi) \cdot \bar{l} \cdot \frac{D(\varphi)}{D(\varphi_{max})} \quad (2)$$

where φ_{max} is the theoretical maximum vehicular flow of the road, $\bar{l}$ is the average length of vehicles, $D(\varphi)$ and $D(\varphi_{max})$ are the vehicular average density at (φ) and (φ_{max}) flow values, respectively. In the assumption of considering the breaking of the vehicles to take place with a uniformly decelerated motion, (φ_{max}) can be calculated by the Eq. (3):

$$\varphi_{max} = \sqrt{\frac{a'_m \cdot n}{2 \cdot k \cdot S}} \cdot 3600 \qquad \text{(vehic/h)} \quad (3)$$

where a'_m is constant average deceleration, n is the number of vehicles per convoy, S is the average spatial spacing between cars and k is a safety coefficient.

The φ_{max} value is reached at v_{max} speed, that can be calculated from Eq. (4):

$$v_{max} = \sqrt{\frac{2 \cdot a'_m \cdot S \cdot n}{k}} \cdot 3,6 \qquad \text{(km/h)} \quad (4)$$

The average density along a stretch of road of length L is given by the product of the reciprocal of the spatial dimension and the ratio between the sum of the times (t_i), in which each vehicle remains in road stretch and the relative temporal dimension of observation *T*.

$$D(\varphi) = \frac{1}{L} \cdot \frac{\sum_{i=1}^{n} t_i}{T} \qquad \text{(vehic/km)} \tag{5}$$

A φ_{max} value of 2078 (vehic/h) is obtained in the hypothesis of taking into account in Eqs. (3), (4) and (5) the following parameters: $a'_m = 3$ m/s^2, $S = 5$ m, $n = 1$ and $k = 1.2$. The PV performance was calculated at L'Aquila location (42°21'37" North, 13°22'43" East, Elevation: 651 m a.s.l.) by using PVGIS5 tool (PVGIS-CMSAF 2018) with the following environmental and technological assumptions: (i) yearly in-plane average solar irradiation of 1500 kWh/m^2 and (ii) crystalline silicon PV system potential efficiency of 11%. Figure 4 illustrates the theoretical specific monthly energy output. Calculation results show a specific yearly energy output of 1134 kWh/kW$_{pk}$ which equals to about 178 kWh/m^2 per year at no traffic condition ($\varepsilon_t = 0$).

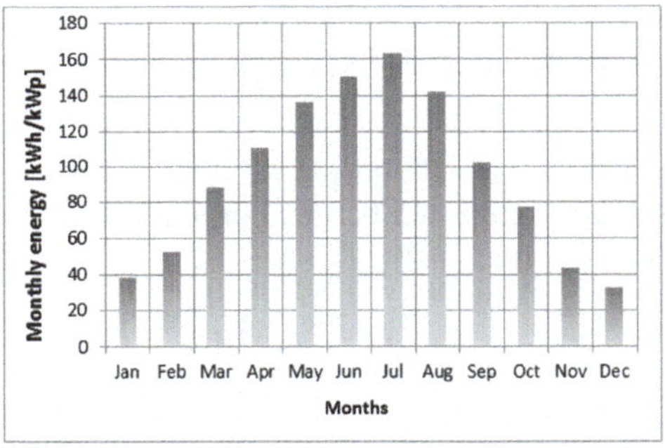

Fig. 4. Monthly energy output @ $\varepsilon_t = 0$

Table 2. Road integrated photovoltaic system data and performance

Stretch	Length	Average traffic volume	Yearly specific energy output @ no traffic	ε_t	Yearly specific energy output @ tested traffic	Yearly energy
	km	Vehic/h per lane	kWh/m^2/y	(%)	kWh/m^2/y	MWh/y
1-2	1.4	952	178	18.32	145.4	611
2-1		927		17.84	146.2	614
2-3	2.8	748		14.40	152.4	1280
3-2		752		14.47	152.2	1279
3-4	2.2	623		11.99	156.7	1034
4-3		612		11.78	157.0	1036
Tot	6.4					5854

Table 2 lists the road integrated photovoltaic system data. In the fifth column, the calculated ε_t values are reported. Calculation results show that the weighted average yearly specific energy output is 152 kWh/m^2 per year.

4 Conclusions

Piezoelectric and photovoltaic road pavements integrated technologies for energy harvesting were reviewed and examined. A preliminary performance simulation of the two systems was performed over an urban road located in L'Aquila city (central Italy) with real traffic and own environmental (irradiation) conditions.

Simulation results show the following weighted average energy theoretical outputs:

- piezoelectric system: 726 kWh/m^2 per year;
- photovoltaic system: 152 kWh/m^2 per year.

The substantial difference between the two examined systems consists in the fact that the piezoelectric devices generate electric energy at the expense of the vehicular energy while the photovoltaic devices uses the sun as primary source and does not interfere with energy of vehicles.

Results of the analysis show that the urban road, when properly designed and re-qualified through the integration of specific systems, can become a valuable energetic source. Future research efforts will focus on the system durability analysis of the integrated road electric energy harvesting solutions, investment costs in order to calculate the energy production cost by varying environmental and working conditions.

References

Agarwal A, Tong N, Hill D (2014) Assessment of Piezoelectric Materials for Roadway Energy Harvesting. California Energy Commissions, California

Chari A, Kour, R (2015) Piezoelectric Roads: Energy Harvesting Method Using Piezoelectric Technology

Alark A, Kulkarni (2013) Solar roadways – rebuilding our infrastructure and economy. Int J Eng Appl 3(3):1429–1436

Beck M, Gerasimchuk I, Harris M (2015) IISD Report on The End of Coal: Ontario's coal phase-out. International Institute for Sustainable Development (Canada)

Ciancetta F, Ometto A, et al (2016) Analysis of flywheel-fuel cell system for mini electrical bus during an urban route. In: SPEEDAM 2016, Capri, Italy. 978-1-5090-2067-6/16

Ciancetta F, Ometto A, et al (2016) Energy storage system comparison for mini electrical bus. In: SPEEDAM 2016, Capri, Italy. 978-1-5090-2067-6/16

Colagrande S, D'Ovidio G (2018) Electric energy harvesting solutions review from roads pavements. In: Proceedings of Transport Means 2018 International Conference, Trakai, Lithuania October, vol 1, pp 5–10

Dikshit T, Shrivastava D, Gorey A, Gupta A, Parandkar P (2010) Energy harvesting via piezoelectricity. Int J Inf Technol 2:265–270

D'Ovidio G (2007) Design of a hybrid power plant for city buses. World Electr Veh Assoc J 1:68–72

D'Ovidio G, Crisi F, Lanzara G (2008) On the magnetic resistance of YBaCuO bulk superconductor dynamically interacting with troubled flux of iron-homopolar magnetic track. J Optoelectron Adv Mater 10(5):1011–1016
D'Ovidio G, Crisi F, Lanzara G (2011) Design and optimization of UAQ4 experimental maglev module. Mater Sci Forum 1058(670):42–47
D'Ovidio G, Masciovecchio C, Totani G (2014) Sizing of hydrogen hybrid power unit for city bus. In: Proceedings of the 17th Transport Means 2014 International Conference, Lithuania, October, pp 50–53
D'Ovidio G, Masciovecchio C, Rotondale A (2016) Hydrogen fuel cell and kinetic energy recover system technologies for powering urban bus with zero emission energy cycle. J IET Intell Transp Syst 10(9):573–578
D'Ovidio G, Carpenito A, Masciovecchio C, Ometto A (2017) Preliminary analysis on advanced technologies for hydrogen light-rail train application in sub-urban non electrified routes. Ingegneria Ferroviaria 72(11):865–878
Edery-Azulay L (2010) Innowattech: harvesting energy and data; a standalone technology. In: First International Symposium. The Highway to Innovation, Israel national roads company Ltd, Tel Aviv
European Commission Clean Power for Transport (2013) A European alternative fuels strategy, COM (2013) 17 final
Hill D, Agarwal A, Tong N (2014) Assessment of Piezoelectric Materials for Roadway Energy Harvesting. DNV Kema
Ibrahim S, Ali W (2012) Power enhancement for piezoelectric energy harvester. In: Proceedings of the World Congress on Engineering, London, vol 2, pp 1018–1023
Kour R, Charif A (2016) Piezoelectric roads: energy harvesting method using piezoelectric technology. Innov Energy Res 5(1):1000132
Kumar P (2013) Piezo-smart roads. J Enhanced Res Sci Technol Eng 2:65–70
Kurzweilai R (2011) Innowattech attach harvests mechanical energy from roadways. http://www.kurzweilai.net/innowattech-harvests-mechanical-energy-from-roadways. Accessed 10 Apr 2018
Lanzara G, D'Ovidio G, Crisi F (2014) UAQ4 levitating train: Italian maglev transportation system. IEEE Veh Technol Mag 9(4):71–77
Mighall T, Phillips M (2013) Society and Exploitation Through Nature. Routledge, Abingdon
Nelson W (2010) Piezoelectric Materials: Structure, Properties and Applications. Nova Science Publishers, New York
PhysOrg.com (2009) Solar roadways awarded DOT contract to pave roads with solar cells. https://phys.org/news/2009-09-solar-roadways-awarded-dot-pave.html. Accessed 07 Sept 2009
PVGIS-CMSAF tool (2018) http://re.jrc.ec.europa.eu/pvg_tools/en/tools.html. Accessed 08 June 2018
Shukla A, Ansari SA (2018) Energy harvesting from road pavement: a cleaner and greener alternative. Int Res J Eng Technol (IRJET) 05(02):1612–1616
Songsukthawan P, Jettanasen C (2015) Performance analysis of maximum power transfer in piezoelectric energy harvesting. In: Proceedings of the International Multi Conference of Engineers and Computer Scientists, vol 2, pp 670–673
Stephy J, Keerthi SJ (2017) A review on solar roadways: the future of roads. Int J Recent Innov Eng Res 2(3):104–108
Wakulat RJ (2016) Photovoltaic (PV). Salem Press, Pasadena
Wattway in France (2016) https://www.theverge.com/solar-panel-road-electricity-france-normandy. Accessed 22 Dec 2016
Xiong et al (2012) New technologies for development of renewable energy in the public right-of-way. DTFH61-10-C-00016. FHWA 9th Quarterly Report, Virginia Tech. October

Environmental Sustainability and Energy Assessment of Bituminous Pavements Made with Unconventional Materials

Marinella Giunta[1(✉)], Marina Mistretta[2], Filippo Giammaria Praticò[3], and Maria Teresa Gulotta[4]

[1] Department of Civil, Energy, Environmental and Material Engineering, University Mediterranea of Reggio Calabria, via Graziella Feo di Vito, 89100 Reggio Calabria, Italy
marinella.giunta@unirc.it
[2] Department of Heritage-Architecture-Urbanism, University Mediterranea of Reggio Calabria, Salita Melissari, 89100 Reggio Calabria, Italy
marina.mistretta@unirc.it
[3] Department of Information Engineering, Infrastructure and Sustainable Energy, University Mediterranea of Reggio Calabria, via Graziella Feo di Vito, 89100 Reggio Calabria, Italy
filippo.pratico@unirc.it
[4] Department of Engineering, University of Palermo, viale delle Scienze, 90128 Palermo, Italy
teresa.gulotta@deim.unipa.it

Abstract. The increasing sensibility towards the questions related to climate change has made the scientists and technicians aware of the need to incorporate the principles of sustainable development into the road construction sector. In this view, recycled materials in road pavements and technologies to decrease the overall carbon footprint have become more and more important. Based on the ISO14040 series, the overall purpose of this study is calculating, by means of a systemic approach, the life-cycle energy and the carbon footprint of road pavement solutions. To this aim, several alternatives in terms of bituminous mixtures (warm mix asphalt, reclaimed asphalt pavements, waste plastics) were considered. All the phases of pavement life from material production to end-of-life stages were taken into account. Results demonstrate that even if there is not a best alternative from all the angles of the analysis, however, the WMA technology, combined with the use of RAP, improves the pavement energy and environmental performance, involving a reduction in primary energy consumption and raw materials, and avoiding impacts for disposal.

Keywords: Road pavement · Sustainability · Recycled materials

1 Introduction

Road pavement technicians and scientists are becoming more and more aware of the need to improve materials and practices of construction and maintenance with the purpose to minimize the burdens on society and environment, with particular regard to carbon footprint and energy consumption.

M. Pasetto et al. (Eds.): ISAP APE 2019, LNCE 48, pp. 123–132, 2020.
https://doi.org/10.1007/978-3-030-29779-4_12

Traditionally, pavements are constructed using hot mix asphalt (HMA) produced at a temperature of around 170 °C. Such high temperature consumes large amounts of energy and produces a lot of waste gases and solid dusts (Capitão et al. 2012). Furthermore, a great amount of non-renewable resources is required.

To reduce and mitigate the quoted environmental impacts associated with the use of HMA, several materials (e.g., crumb rubber, CR, reclaimed asphalt pavement, RAP, waste plastic, WP, industrial wastes and by products, W), and technologies (e.g., warm mix asphalt, WMA, half warm mix asphalt, HWMA, in-place recycling, HMA containing RAP), recognized as more environmental sustainable, have been recently developed and applied (Rodríguez-Alloza et al. 2015; Santos et al. 2015; Lee et al. 2010; Mladenovič et al. 2015; Praticò et al. 2013).

The abovementioned solutions are surely promising but the evaluation of how much they contribute to the pavement sustainability is a relevant matter to be investigated, because their effectiveness depends on the context in which they are applied and on the way the sustainability is measured and evaluated (Santos et al. 2019).

To this purpose, Life Cycle Assessment (LCA), carried out based on ISO14040 series, represents a suitable methodology recognized and applied by road engineers and industry to assess energy and environmental burdens associated to the road lifespan (Matthews et al. 2014). It includes all the phases of the life of roads: construction (manufacturing and transportation of the construction materials and components to the construction site), operations, maintenance, renovation, disassembly, and waste management. The outcomes of a LCA study can provide enough information to facilitate eco-design and knowledge-based comparative assertions (Bressi et al. 2018).

The goal of this study is to apply the life cycle approach, according to the international standards of series ISO 14040-14044 (International Organization for Standardization, 2006), to assess the energy consumption and the carbon footprint, that arise from the production, transportation and installation steps of road pavements, in order to identify the main hotspots along the pavement life-cycle.

2 Sustainable Asphalt Materials and Technologies: Strengths and Weaknesses

In this section a brief description of the strengths and weaknesses of WMA, WMA with RAP, and WP is reported.

Warm-mix asphalts (WMAs) are a set of technologies in which the asphalt mixture is produced, transported and compacted at a temperature lower than the one of traditional HMAs (20–30 °C less, cf. Capitão et al. 2012). WMA allows to have a bitumen mixture with suitable viscosity, without undermining the performance and the expected life of pavement layer. WMA technologies exhibit environmental benefits related with the reduction of the energy consumption and carbon footprint. The technical benefits are: (i) better compaction of the road; (ii) ability to haul paving mix for longer distances; (iii) extension of the paving season, (iv) higher durability of the pavement due to the lower aging of the binder during production, (v) less production of odours and fumes that makes safer the construction phase for the workers. Weaknesses of WMA pertain to rutting, moisture susceptibility (which may result in the premature rutting of pavement surface), and higher initial costs due to the additives.

The advantages of WMA technologies can be improved if they are blended with RAP. Studies about the feasibility of the combined application of WMA-high RAP percentage demonstrated that WMA technologies could be used with a content of RAP up to 100% (Mallick et al. 2008). WMA can present good mechanical properties when RAP is applied properly. Note that high percentages of RAP enhance the rutting resistance but result in higher moisture susceptibility (Praticò et al. 2011). Relevant issues to be addressed in a comprehensive evaluation of the performance of WMA-high RAP mixtures are: (i) the effect of different WMA technologies; (ii) the evaluation of rutting resistance and moisture susceptibility, (iii) the assessment of cracking and fatigue resistance for the presence of high RAP content (Zhao et al. 2013). Form an environmental standpoint, the use of the RAP is underscored by the need to reduce the exploitation of quarries and the disposal of the wastes resulting from rehabilitation projects of road pavements.

However, the evaluation of the benefits has to take into account also methods, technologies, emissions, and temperature/energy requirements for milling/removing, pre-treatment of RAP, RAP line in an asphalt plant, and use of rejuvenators.

The relevant production of waste plastic and the issues related to the landfill have elicited a great interest in the use of these wastes in sustainable pavement constructions, thanks to the indubitable environmental advantages. Indeed, landfill and incineration, the usual techniques to dispose plastic wastes, produce a leachate with negative effects on land, water and air (Bansal et al. 2017). In addition to the environmental benefits, many studies and applications demonstrate the technical appropriateness of the use WP, such as polypropylene and low-density polyethylene, in plain bituminous concrete mixtures, because of improved durability and fatigue life. WP, added to the bitumen in the proportion of 2–8% by bitumen weight, modifies binder properties, lead to an increase of the softening point. WP can be added also using the dry process, in which plastics are first blended with aggregates and after mixed with bitumen. WPs improve viscoelastic properties, Marshall quotient, and binding, and stripping properties of the mixture.

3 Energy and Carbon Footprint of Different Road Pavement Solutions

3.1 Goal and Scope Definition

The LCA methodology is applied to assess the energy and environmental impacts, arisen from the life cycle of the asphalt pavement in a typical urban road, according to the international standards of the series ISO 14040. In detail, five different types of bituminous mixtures are defined for the pavements under analysis, and a scenario analysis is carried out in order to identify the less impacting construction techniques from the energy and environmental point of view.

The contributions of each life-cycle step to the total impacts and the energy and environmental hotspots are identified in order to define suitable options of improvement. The models used for life cycle analyses are those available as a part of SimaPro 7.3.

The Cumulative Energy Demand method (Wernet et al. 2016) is used to quantify the life-cycle energy consumption, expressed in terms of Global Energy Requirement (GER). The environmental impact is assessed in terms of Carbon Footprint (CF), as contribution to global climate change, including emissions from fossil and biogenic carbon sources, emissions caused by land use change and carbon uptake by plants over a 100-year time horizon. To calculate the carbon dioxide equivalent (CO_{2eq}) of all non-CO_{2gases} (CH4, N2O, SF6, HFCs and CFCs) the environmental characterization factor is based on the ILCD 2011 impact assessment method (Van Oers 2016).

3.2 Functional Unit and System Boundary

The selected functional unit (FU) is 1 m^2 of road pavement. The study is developed "from cradle to grave", including all the processes and the activities that encompass raw materials sourcing, composite materials production, laying operations, maintenance and end-of-life works during pavement service life. The manufacturing of production equipment, buildings and other capital goods were not taken into account, because not included in the technical system (The International EPD® System 2013).

3.3 Scenario Definition for the Asphalt Pavement

The case study proposed in this paper is a two-lane, single carriageway road, 1 km length and 9.5 m width, with a pavement thickness of 320 mm. The pavement structure, which lays on the subgrade, is composed of the friction course (FC, 50 mm), the binder course (BIC, 70 mm), and the unbound base course (UBC, 200 mm).

The Benchmark Scenario (B) is characterized by:

- a FC (porous asphalt concrete), which contains: (i) HMA composed of modified bitumen (5% by mix weight), containing the five percent of Styrene-Butadiene-Styrene Polymer (SBS); (ii) quicklime (QL); (iii) cellulose fibres (FB), to avoid the binder drainage; (iv) mineral filler (FI); (v) mineral aggregates; (vi) in-place residual air voids (18%).
- a BIC (dense-graded asphalt concrete) that includes: (i) neat bitumen (5% by mix weight); (ii) mineral filler; (iii) mineral aggregates; (iv) in-place residual air voids (6%).
- an UBC composed of a given gradation of mineral aggregates and moisture content.

Starting from the Benchmark Scenario, four different scenarios of asphalt pavement, based on different construction techniques, are defined. The scenario analysis aims at assessing the variations induced, in terms of energy consumption and carbon footprint, by the use of recycled materials (Re-claimed Asphalt Pavement, waste plastics, and crumb rubber), and by asphalt plant characteristics and technology (Hot Mix Asphalt and Warm Mix Asphalt).

Scenario 1 is characterized by the use of neat bitumen and the use of waste plastics (WP), and crumb rubber (CR), from end-of-life tires. It is assumed that the selected WP derive from municipal solid wastes and CR from waste tires, thus addressing is-sues that concern land use reduction for disposal, non-renewable resource saving, and climate change mitigation.

In Scenario 2, a porous and warm mix asphalt (PAWMA) is considered as a friction course, including: (i) mineral aggregates and filler; (ii) modified bitumen (as for Benchmark Scenario); (iii) additive in a standard dosage (0.5% based on bitumen weight); (iv) residual air voids (18%). The BIC is manufactured using the WMA technology. It includes the same components reported above for the PAWMA, where the gradation of mineral filler and aggregates is different and a different asphalt binder percentage is given. The technologies above appear to allow the production of WMA by reducing the viscosity of the asphalt binder at a given temperature.

Scenario 3 includes PAWMA with the addition of 45% of RAP in FC and BIC. Furthermore, the UBC layer is mixed with 45% of RAP.

Scenario 4 is defined as Scenario 3, but the FC, BIC and UBC are assumed to contain 30% of RAP

The use of RAP in bituminous mixtures leads to save virgin materials and to avoid impacts for landfills.

3.4 Life Cycle Inventory and Data Quality

Table 1 reports the quantities per FU that refer to the different materials that compose the layers and to the average haul distances from production/supply sites to the construction sites. For each scenario, starting from the data in Table 1, Life Cycle Inventory (LCI) is performed to quantify the inputs and outputs of the examined system, by means of mass and energy balances of the selected FU.

The eco-profiles of energy sources, raw materials, transports, and waste treatments are included in the analysis based on international environmental databases (International Organization for Standardization 2006). Reference is made to International Organization for Standardization, 2006 to derive LCIs of bituminous materials. With regard to the aggregates, the authors assume to derive data from Calabrian and Sicilian quarries. The eco-profile of electricity is referred to the Italian electricity mix, while the eco-profiles of input materials are mainly referred to the European context.

The energy for the production of the bituminous mixtures are reported in Table 2 per kg of materials. The data used were collected by European case studies (Aurangzeb et al. 2014; Farina et al. 2017; Walch 2015).

Table 1. Quantities per FU and transport distances of employed component materials

	Scenario									
	Benchmark		1		2		3		4	
	kg/FU	km	kg/FU	km	kg/FU	km	kg/FU	km	kg/FU	km
AG	548.43	196	460.70	196	549.05	196	257.55	196	356.22	196
BIT	13.54	348	10.41	348	12.63	348	9.11	348	10.08	348
CR	–	–	20.89	100	–	–	–	–	–	–
FIB	0.29	205	0.25	205	0.32	205	0.29	205	0.29	205
FIL	47.37	196	45.93	196	47.37	196	48.40	196	48.06	196
QL	7.27	460	5.85	460	7.26	460	7.25	460	7.29	460
RAP	–	–	–	–	–	100	301.52	100	200.26	100
REJ	–	–	–	–	–	348	0.34	348	0.23	348
SBS	0.29	348	–	–	0.21	348	0.24	348	0.25	348
WP	–	–	20.89	100	–	–	–	–	–	–
WAT	44.00	–	44.00	–	44.00	–	45.16	–	44.74	–
Z	–	–	–	–	0.02	348	0.02	348	0.02	348

AG = Mineral Aggregates; BIT = Bitumen; CR = Crumb Rubber; FIB = Cellulose Fibres; FIL = Mineral Filler; QL = Quick Lime; RAP = Reclaimed Asphalt Pavement; REJ = Rejuvenating Agent; SBS = Styrene-Butadiene-Styrene Polymer; WP = Waste Plastic; WAT = Water; Z = Synthetic zeolites

Table 2. Quantities of energy sources required per types of asphalt

	Natural gas	Electricity
	m^3/kg	kWh/kg
HWA	0.0098	0.00425
HWA+cr+wp	0.0098	0.004675
WMA	0.01078	0.00575
WMA+rap 45%	0.009702	0.00645
WMA+rap 30%	0.010241	0.00625

3.5 Life Cycle Energy and Environmental Impact Assessment Results

The LCA model is used to evaluate the GER and CF associated to the life-cycle steps of the road pavement, including the production of materials, construction of pavement, transport, maintenance and end of life, in the scenarios described above.

GER is derived as the total primary energy consumed during the whole life cycle of the pavement. Figure 1 shows the contribution to GER of each life-cycle step in each scenario.

Outcomes show that the Benchmark Scenario presents the most significant value of GER (nearly 1,893 MJ/m^2). Scenarios 3 and 4, which include PAWMA with RAP, involve the lowest GER values (about 1,535 MJ/m^2 and 1,618 MJ/m^2 respectively), essentially due to the lower energy consumption in WMA production and to the use of RAP in the pavement layers, which implies a reduction of virgin raw materials requirement.

Production is the step with the highest contribution to GER, accounting for about 67% in all the assessed scenarios. With regard to the construction step, even if the contribution is negligible with respect the other steps, it was found that the highest share comes from RAP scenarios (about 7.03 MJ/m^2), due to the higher energy required for the milling process.

The maintenance step accounts for around 20% of the total GER in all the scenarios, due to the one replacement of the friction course layer during the pavement lifespan. With regard to transport and end-of-life, GER is about 6% in all scenarios.

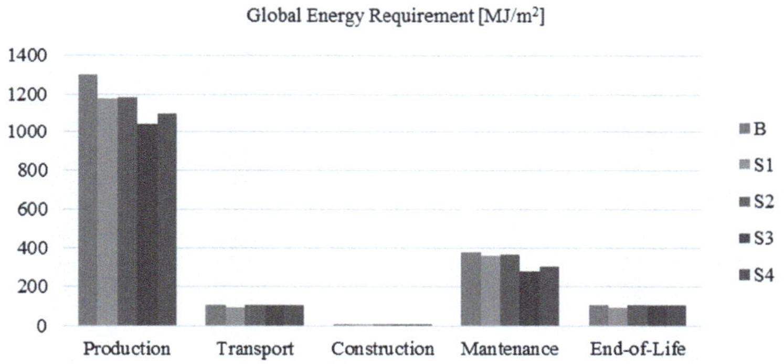

Fig. 1. GER of road pavement per FU: contribution of life-cycle steps for each assessed scenario (MJ/m^2).

A GER breakdown of the production step is reported in Fig. 2. It shows that the main contribution to GER comes from virgin bitumen production (between 46% to 56%), followed by the quicklime production (3–4%).

The global energy requirement for the mineral aggregates accounts for about 1–2% in the production step in all the scenarios. About 22–33% of the GER used in the production step is associated to the energy (EUsed) required (as electricity and natural gas) to produce the asphalt, while the transportation of the raw materials to the plant (TrP) accounts for about 15–17%.

Figure 3 shows CF results for all the assessed scenarios, expressed in kg of CO2eq per FU (m2). RAP scenarios involve the lowest total CF in comparison with the others. Scenario 3 involves the lowest contribution, with a reduction of 5% respect to Scenario 1 and of 8% respect to the Base Benchmark Scenario, which shows the highest contribution to CF.

The largest CF impacts refer to the production step (nearly 64% in Scenarios 1, 63% in Base Benchmark Scenario, and 62% in Scenarios 3 and 4), due to the transportation of materials and the energy used for the asphalt production (as reported in

Fig. 4). In fact, the production step includes the extraction and transportation of raw materials, the production of materials used in the pavement construction, and the energy sources required for the treatment.

With regard to transport from the plant to the site of construction, it accounts for nearly 8–9% in all the scenarios.

The CF associated to the construction phase is also in this case negligible with respect to the other stages. In any case, results show that the RAP scenarios involve the highest contribution in comparison to the other scenarios (0.45 kgCO2eq/m2), reaching the 1% on the total CF in all scenarios. With regard to the maintenance step, referred to the replacement of the friction course after 10 year of lifespan, the contribution to CF varies from 15% (Scenario 3) to 16% (Scenario 1). The end-of-life involves a share of 13% on the total CF in Base Benchmark scenario, Scenarios 2 and 3. Scenario 1 involves the lowest contribution, due to the material recycling.

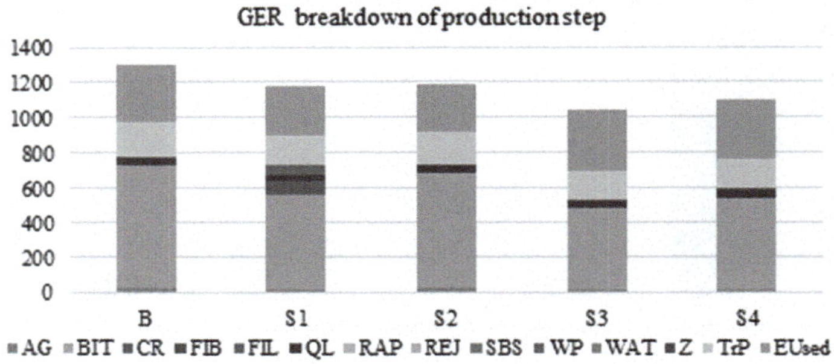

Fig. 2. GER breakdown of road pavement per FU: contribution of production steps for each assessed scenario (MJ/m^2).

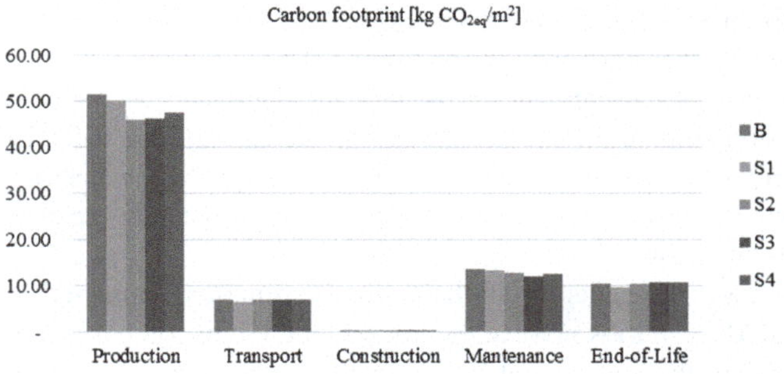

Fig. 3. Carbon Footprint (CF) of road pavement per FU: contribution of life-cycle steps for each assessed scenario (kgCO_{2eq}/m^2)

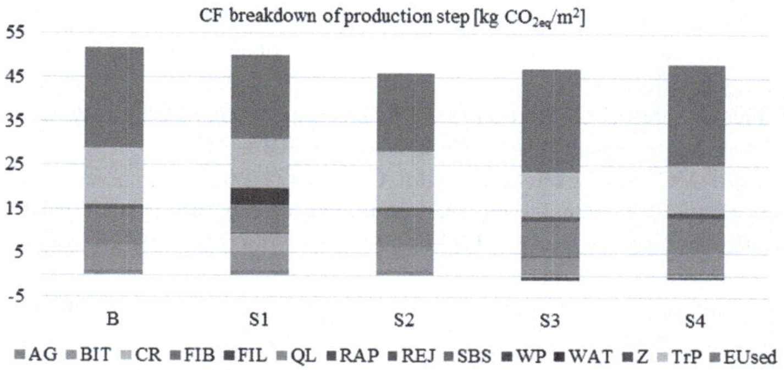

Fig. 4. CF breakdown of road pavement per FU: contribution of production steps for each assessed scenario ($kgCO_{2cq}/m^2$)

4 Conclusions

This study focuses on the life-cycle energy and carbon footprint of different road pavements, including materials production, transportation, construction, maintenance, and end-of-life, according to the international standards of ISO 14040 series. The results highlight that Scenario 3 (45% of RAP) represents the most sustainable pavement, followed by the Scenario 4 (30% of RAP). In fact, the most important result is that the WMA technology, combined with the use of RAP, improves the pavement energy and environmental performance, involving a reduction in primary energy consumption and raw materials, and avoiding impacts for disposal.

Furthermore, it is worth noting that material production step, including the raw material extraction and resource supply, involves the largest primary energy consumption and environmental impacts, mainly due to the production of bitumen.

Thus, planning a careful selection of 'greener' materials to be used in the production stage is of paramount importance to improve the eco-profile of the road pavement.

Despite the methodological limitations due to the use of secondary data for modelling the life cycle of a number of production materials and the lacking availability of process-specific data for such materials, outcomes show that LCA is a suitable methodology to provide a systemic approach for energy and environmental assessment for the sake of all stakeholders. This supports the development of the eco-design in the road sector, in the pursuit of low-carbon and low-energy products.

References

Aurangzeb Q, Al-Qadi IL, Ozer H, Yang R (2014) Hybrid life cycle assessment for asphalt mixtures with high RAP content. Resour Conserv Recycl https://doi.org/10.1016/j.resconrec.2013.12.004

Bansal S, Kumar Misra A, Bajpai P (2017) Evaluation of modified bituminous concrete mix developed using rubber and plastic waste materials. Int J Sustain Built Environ 6(2):442–448

Bressi S, Santos J, Giunta M, Pistonesi L, Lo Presti D (2018) A comparative life-cycle assessment of asphalt mixtures for railway sub-ballast containing alternative materials. Resour Conserv Recycl 137:76–88

Capitão SD, Picado-Santos LG, Martinho F (2012) Pavement engineering materials: Review on the use of warm-mix asphalt. Constr Build Mater 36:1016–1024

Farina A, Zanetti MC, Santagata E, Blengini GA (2017) Life cycle assessment applied to bituminous mixtures containing recycled materials: crumb rubber and reclaimed asphalt pavement. Resour Conserv Recycl 117:204–212. https://doi.org/10.1016/j.resconrec.2016.10.015

International Organization for Standardization. (2006) ISO 14040-Environmental management - Life Cycle Assessment - Principles and Framework, vol 3

Lee JC, Edil TB, Tinjum JM, Benson CH (2010) Quantitative assessment of environmental and economic benefits of recycled materials in highway construction. Transp. Res. Rec.: J. Transp. Res. Board 2158(1):138–142

Mallick RB, Kandhal PS, Bradbury RL (2008) Using warm-mix asphalt technology to incorporate high percentage of reclaimed asphalt pavement material in asphalt mixtures. Transp Res Rec: J Transp Res Board 2051(1):71–79

Matthews H, Hendrickson C, Matthews D (2014) Life Cycle Assessment: Quantitative Approaches for Decisions that Matter. https://www.lcatextbook.com/. Accessed 11 Feb 2019

Mladenovič A, Turk J, Kovač J, Mauko A, Cotič Z (2015) Environmental evaluation of two scenarios for the selection of materials for asphalt wearing courses. J Clean Prod 87:683–691

Praticò FG, Vaiana R, Giunta M (2011) Recycling pems back to innovative, silent, permeable road surfaces. In: Proceedings of 8th international conference on environmental engineering, ICEE 2011, March 2014, pp 1186–1192

Praticò FG, Vaiana R, Giunta M (2013) Pavement sustainability: permeable wearing courses by recycling porous European mixes. J Archit Eng 19(3):186–192

Rodríguez-Alloza AM, Malik A, Lenzen M, Gallego J (2015) Hybrid input–output life cycle assessment of warm mix asphalt mixtures. J Clean Prod 90:171–182

Santos J, Bryce J, Flintsch G, Ferreira A, Diefenderfer B (2015) A life cycle assessment of in-place recycling and conventional pavement construction and maintenance practices. Struct Infrastruct. Eng. 11(9):1199–1217

Santos J, Bressi S, Cerezo V, Lo Presti D (2019) SUP&R DSS: a sustainability-based decision support system for road pavements. J Clean Prod 206:524–540

The International EPD® System (2013) Product Category Rules (PCR) of the Environmental Product Declaration (EPD): "Highways, streets and roads (except elevated highways)"

Van Oers L (2016) CML-IA database, characterisation and normalisation factors for midpoint impact category indicators, Version 2015;4:5. Accessed 14 May 2018

Walch J (2015) A study on Lower Temperature Asphalts Commercialisation in the UK

Wernet G, Bauer C, Steubing B, Reinhard J, Moreno-Ruiz E, Weidema B (2016) The ecoinvent database version 3 (part I): overview and methodology. Int J Life Cycle Assess 21(9):1218–1230

Zhao S, Huang B, Shu X, Woods M (2013) Comparative evaluation of warm mix asphalt containing high percentages of reclaimed asphalt pavement. Constr Build Mater 44:92–100

Reflectivity and Durability Assessment of Solar Heat-Blocking Pavement

Masahiko Iwama[1(✉)], Tamotsu Yoshinaka[1], Shunsuke Nishioka[2], and Hiroshi Murakami[2]

[1] Research Institute, NIPPO Corporation, 6-70 Mihashi, Nishi-ku, Saitama, Japan
iwama_masahiko@nippo-c.jp
[2] Technical Planning Division, NIPPO Corporation, 1-19-11 Kyobashi, Chuo-ku, Tokyo, Japan

Abstract. In recent years, there is growing awareness of environmental issues, such as global warming and climate change. As asphalt pavement covers approximately twenty per cent of urban areas, it is considered to be a factor in the "urban heat island phenomenon". In order to tackle this problem from a paving perspective, solar heat-blocking pavement has been developed and applied as cool pavement technology to achieve the following benefits: a reduction in surface temperature and mitigation of urban heat. This paper describes a reflectivity and durability assessment of solar heat-blocking pavement in the laboratory and field, with the following conclusions being drawn from this study. With regard to the albedo characteristics, laboratory and field results show that the reduction in surface temperature is significantly related to the albedo characteristics. In terms of pavement durability in the laboratory and field, the surface of the solar heat-blocking pavement is strong enough for traffic loading, compared to dense-grade asphalt surfaces. Finally, the relationship between reflectivity and durability of solar heat-blocking pavement is highlighted through field monitoring results.

Keywords: Albedo · Surface temperature · Reflectivity · Durability

1 Introduction

There is growing awareness that temperatures have been rising worldwide due to the emerging issue of global warming and climate change. This kind of problem has become more pronounced, especially in Japan. Also, as asphalt pavements cover approximately twenty per cent of urban areas, it is considered to be a factor in the "urban heat island phenomenon", which significantly affects the thermal comfort of pedestrians.

According to Yoder and Witzak (1975), rising temperatures from the paving surface affect the properties of the surface layer. Therefore, bearing in mind these problems, reducing the surface temperature has become increasingly important in terms of sustainability as well as the environment. Pomerantz *et al.* (2000) indicated that making urban surfaces whiter to reflect both visible and infrared rays (i.e. sunlight) is the most

M. Pasetto et al. (Eds.): ISAP APE 2019, LNCE 48, pp. 133–143, 2020.
https://doi.org/10.1007/978-3-030-29779-4_13

practical way of mitigating surface heat. However, out of consideration for driver visibility, darker surfaces (e.g. grey) would be preferable based on familiarity. In addition, taking into account workability of existing pavement, surface treatment with a painted coating is one way to deal with the issue for existing surfaces. For these reasons, a new surface treatment technology called "solar heat-blocking pavement", which aims to reduce surface temperatures, has been developed in Japan (Yoshinaka *et al.* 2003).

The aim of this paper is to demonstrate a reflectivity and durability assessment of solar heat-blocking pavement and to address the potential for mitigation of Urban Heat-Island.

2 Solar Heat-Blocking Pavement

2.1 Basic Concept

Solar heat-blocking pavement was originally developed by utilizing the application of solar reflective technology to building roofs (Kinoshita 1998). Pomerantz *et al.* (2000) suggested that sealing a reflective material onto the surface layer contributes to both a reduction in surface temperature and the mitigation of surface damage. Therefore, a reduction in surface temperature during summer was expected by applying this technology to asphalt paving.

The function of this technology is based on the higher reflectivity of near-infrared rays and lower reflectivity of visible rays. In practice, the reflectivity of solar rays and infrared rays are represented by albedo. Albedo is defined as the ratio between incoming and reflected solar rays. A higher albedo means that the surface layer has a higher reflectivity of infrared rays, whereas a lower albedo indicates that infrared rays are absorbed into the surface layer, thus increasing the surface temperature. In order to prevent the surface layer from absorbing infrared rays, paint-based materials with a higher albedo were coated onto the existing paving (see Fig. 1).

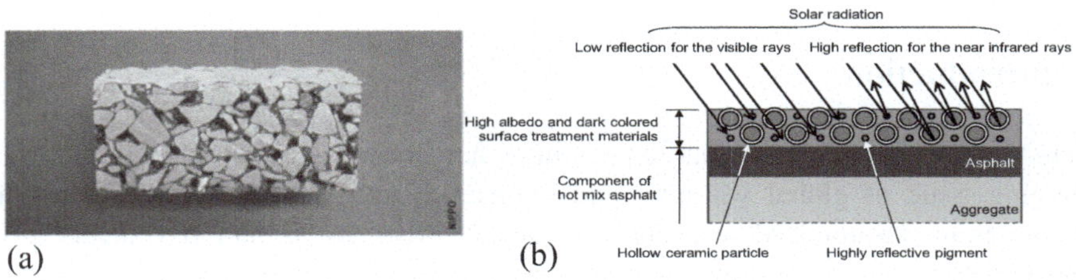

Fig. 1. Solar heat-blocking pavement (a) Photograph, and (b) Schematic image

2.2 Application Procedure

Solar heat-blocking pavements can be constructed by applying the developed solar reflective pigment to existing surfaces. In practice, two-pack resins, such as Methyl

Methacrylate (MMA), are used. The coating layer is about 1.0 mm thick. The coating layer consists of three components: prime layer, second layer and non-skid sand (see Fig. 2a). Firstly, the prime layer is applied to cover the existing surface; then the non-skid sand (i.e. artificial sand) is sprayed on immediately after the primary coating to ensure skid resistance; and finally, the second layer is applied as the coloured surface and to sandwich the non-skid sand. After curing for an hour, the site can be reopened to traffic. The coating work is normally performed with a specialist spray gun as shown in Fig. 2b. The material specification generally used for this pavement is as follows:

- Size of non-skid sand particles: 0.5–1.7 mm;
- Density of non-skid sand: 0.5 kg/m^2;
- Density of coating: 0.4–0.6 kg/m^2 in each layer.

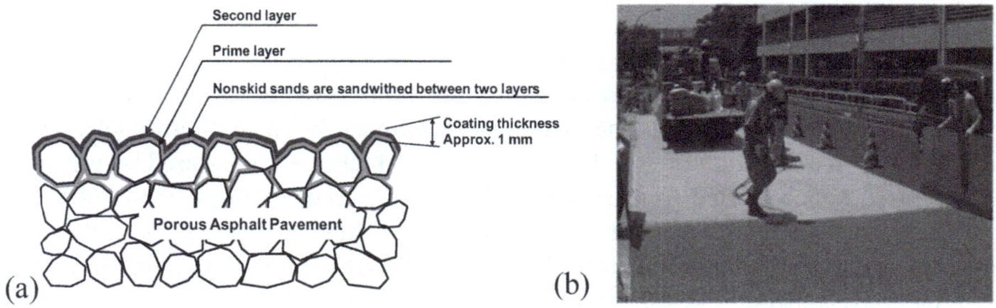

Fig. 2. Overview of the pavement (a) Structure, (b) Construction method

3 Performance Testing

3.1 Reflectivity

In order to examine the albedo characteristics, a comparison was made between three surfaces: solar heat-blocking pavement, conventional pavement (i.e. dense-graded asphalt pavement) and normal paint material. The test was conducted in accordance with the Japanese standard, JIS A 5759. In this case, the colour for both the solar heat-blocking pavement and the normal paint is grey, whilst that of the conventional pavement is black.

Figure 3 shows the comparison results. As shown in the figure, there are clear differences between the solar heat-blocking pavement and conventional painting materials. Conventional painting material has almost the same or a less reflective ratio across the entire wavelength. However, in the case of the solar heat-blocking pavement, the reflective rate for near-infrared rays in the wavelength is much higher than the others. This indicates that the solar heat-blocking pavement has higher albedo, despite the fact that the normal paint material is the same colour as the heat-blocking pavement.

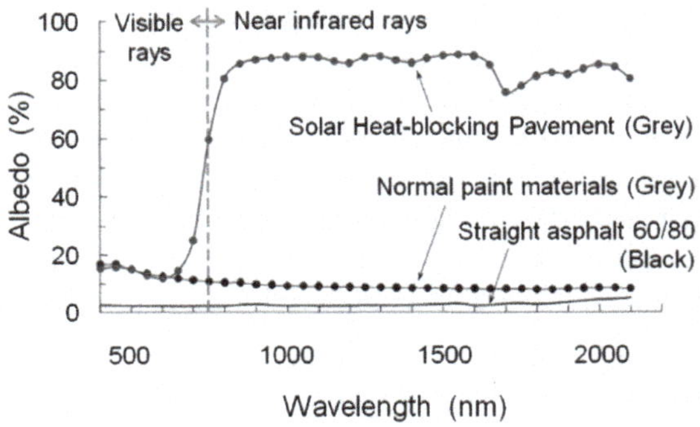

Fig. 3. Albedo characteristics of solar reflective pigment

3.2 Retro-Reflection

Solar heat-blocking pavement also has retro-reflection characteristics. Figure 4 shows the measurement results for solar heat-blocking pavement. In this test, artificial sunlight is radiated from a direction of 90°. The retro-reflection ratio was measured by receiving near-infrared rays reflected from solar heat-blocking pavement from 0° to 180°. The results clearly demonstrated that the near-infrared rays were reflected back in the same direction as those of sunlight. Therefore, it is evident that solar heat-blocking pavement has a recursive reflection property.

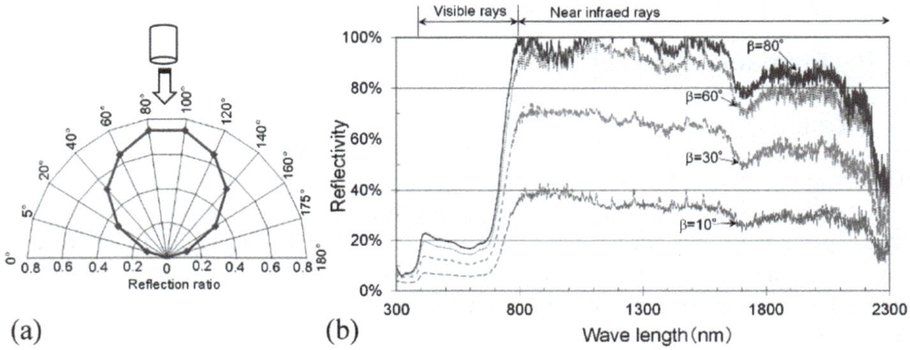

Fig. 4. Retro-reflection characteristics of solar heat-blocking pavement (Sunlight direction $\alpha = 90°$) (a) Retro-reflection characteristics, (b) Albedo characteristics in each reflected direction

3.3 Laboratory Lamp Test

In order to investigate the reduction in temperature when using solar heat-blocking pavement, laboratory lamp irradiation experiments were carried out, as per the Japanese pavement performance reference (Japan Road Association, 2008). The schematics and experimental results are shown in Fig. 5b.

In this experiment, porous asphalt mixture is used for both the solar heat-blocking pavement specimen and the conventional specimen. A special lamp with a similar wavelength to sunlight was used to evaluate the performance. The detailed experimental conditions are as follows:

- Experiment temperature: 30 °C (temperature control cabinet is preferable);
- Lamp height: A height at which the surface temperature of the conventional specimen reaches 60 °C in about three hours, is applied for the test;
- Measurement method: Thermocouples are used to measure surface temperature.

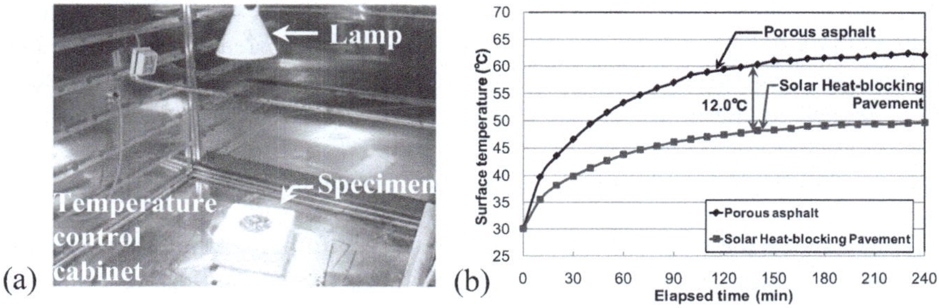

Fig. 5. Laboratory lamp test (a) Equipment, (b) Test result

As can be seen from Fig. 5, the results clearly show the difference between the two specimens. Although the result is not necessarily the same as the site temperature, laboratory lamp tests are often used to confirm the performance of solar heat-blocking pigment before site application to predict the on-site performance.

3.4 Torque Resistance

Considering the service state, the coating layer of solar heat-blocking pavement should be strong enough and have good bonding to the original asphalt surface. In order to investigate these factors, laboratory stripping-resistance tests were conducted. A schematic representation of the test equipment is shown in Fig. 6.

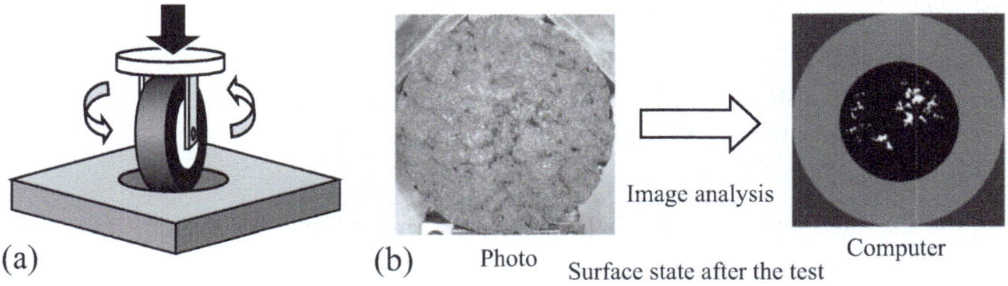

Fig. 6. Torque resistance test (a) Test equipment, and (b) Image analysis

A test load was conducted, following the laboratory stripping-resistance test method for solar heat-blocking pavement material. A test load was applied to the specimen by turning the tire left and right (Minegishi *et al.* 2010). A summary of the test parameters is presented below:

- Loading condition: Turning the tire left and right;
- Test temperature: 20 °C;
- Test load: 686 N;
- Number of cycles: 650.

After the test, a digital image of the specimen's surface was taken with a digital camera. Then, the stripping resistance of the solar heat-blocking layer is evaluated using computer image analysis which can record the stripped area of the surface. In this case, the degree of stripping is evaluated as a "stripped area ratio". The stripped area ratio is defined by the following equation:

$$\text{Stripped Area Ratio } (\%) = \text{Stripped Area/Wheel Contact Area} \tag{1}$$

Table 1 shows the torque test result. As can be seen from the table, compared to the performance criteria set by one organization (Minegishi and Ueno 2010), the coated layer demonstrates strong adhesion to the existing surface. As a result, a longer life cycle can be expected for solar heat-blocking pavement.

Table 1. Torque resistance test result

Stripping area rate (%)	
Performance requirement	≤40%
Solar heat-blocking pavement	9.9%

Note: Performance requirement (Minegishi and Ueno 2010)

3.5 Ravelling Test

The ravelling test was carried out in accordance with the testing procedure for solar heat-blocking pavement materials (Ueno *et al.* 2014). The test is the same as a conventional ravelling test (see Fig. 7). However, it was conducting under the following conditions:

- Temperature: 20 °C;
- Test duration: 180 s;
- Test speed for wheel: 200 cycles/min;
- Test speed: 66 cycles/min.

After the test, the surface of the specimen is captured by digital camera. The stripped area is analysed using the same procedure as the torque test.

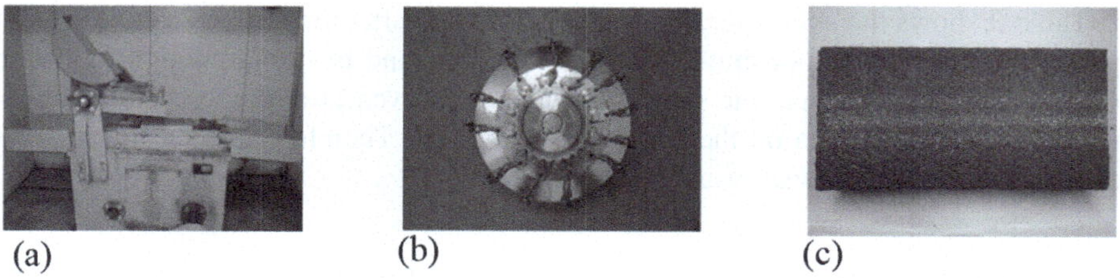

Fig. 7. Ravelling test (a) Overview, (b) Wheel with chain, (c) Specimen after the test

Table 2 shows the stripped area ratio of solar heat-blocking pavement after the ravelling tests. As can be seen in the table, the results clearly meet the performance criteria for both torque and ravelling tests. Therefore, it can be said that solar heat-blocking pavement has a strong coating layer.

Table 2. Ravelling test results for solar heat-blocking pavement

Test	Stripped area rate	Performance requirement
Labelling test	7%	$\leq$20%

Note: Stripped area rate = (stripped area/tire contact area) × 100
: Performance requirement (Ueno *et al.* 2014)

3.6 Accelerated Wear Test

Skid resistance is an important parameter for road users, because it is directly related to the drivers' safety. For solar heat-blocking pavement, there are some concerns that non-skid sands laminated by the coated layers would wear off due to repeated traffic load. In order to understand the skid resistance during service life, a laboratory accelerated wear test was conducted, then changes in skid resistance value were measured using the British Pendulum Number (BPN) (Minegishi *et al.* 2010). The details of the test conditions are as follows (Fig. 8):

- Test tire pressure: 320 ± 10 kPa;
- Test load: 1.96 kN;
- Contact pressure: 0.29 MPa;
- Test Speed: 20 km/h;
- Test temperature: 20 °C;
- Number of passes: 200,000.

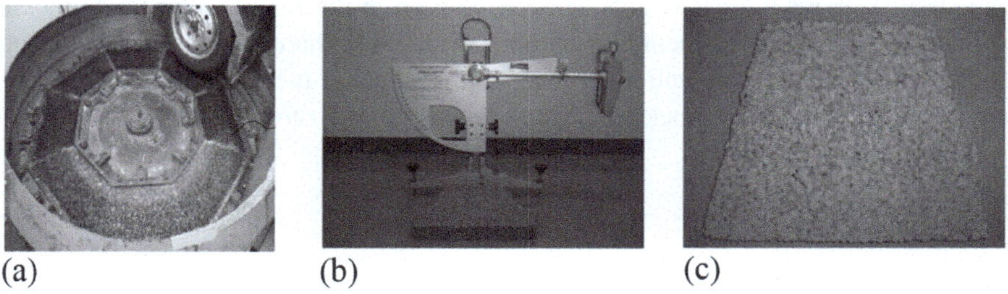

Fig. 8. Accelerated wear test: (a) Overview, (b) British pendulum, (c) Specimen after the test

Table 3 shows the skid resistance value before and after the test. As can be seen in the table, the results demonstrated that the non-skid sand is strong enough to ensure skid resistance, since it met the standard requirement even after the tests. In addition, there are no stripped area on the surface after the test. Therefore it is clear that solar heat-blocking pavement can resist wear.

Table 3. British pendulum number after the accelerated wear tests (AWT)

Test results	Specimen No.	Before AWT	After AWT
Measured value	1	82	65
	2	82	66
	Average	82	66
Standard requirement		≥ 60	≥ 55

Note: Performance requirement (Minegishi and Ueno 2010)

3.7 Aggregate Pop-Out

The bonding between aggregate is an important factor affecting the durability of pavement. Considering the strong coating layer, solar heat-blocking pavement might be a countermeasure for this problem. In order to confirm this effect, laboratory rotational tracking tests were conducted, in accordance with the Japanese Standard (see Fig. 9).

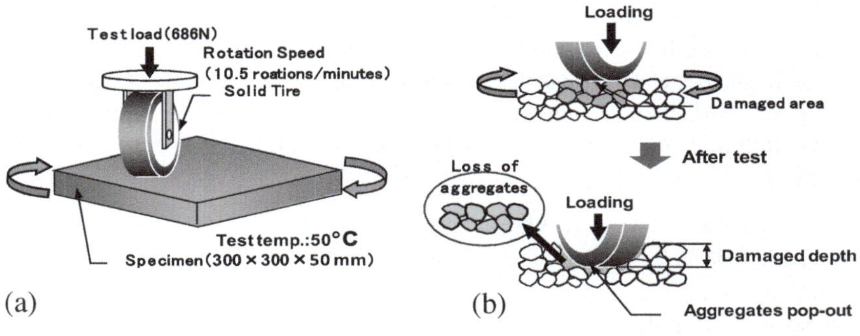

Fig. 9. Schematic of rotational shear tracking (a) Overview, (b) Loss of aggregate

Figure 10 shows the test results. As can be seen in the figure, solar heat-blocking pavement is much more resilient to rotational shear stress than conventional pavement. Particle loss rate for conventional pavement gradually increased during the test, whereas that of solar heat-blocking pavement was almost zero after the test.

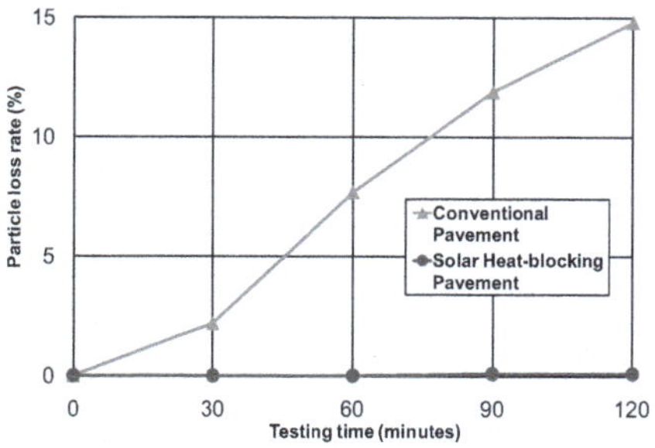

Fig. 10. Rotational shear tracking results

4 Performance in the Field

4.1 Reduction in Surface Temperature

Figure 11 compares the surface temperatures of both solar heat-blocking and dense graded pavements through a thermographic image at a site. As can be seen from the image, the surface temperature of the solar heat-blocking pavement was 35 °C, whilst that of dense graded pavement was 48.3 °C. Therefore, it is immediately evident that solar reflective coating can effectively reduce the surface temperature.

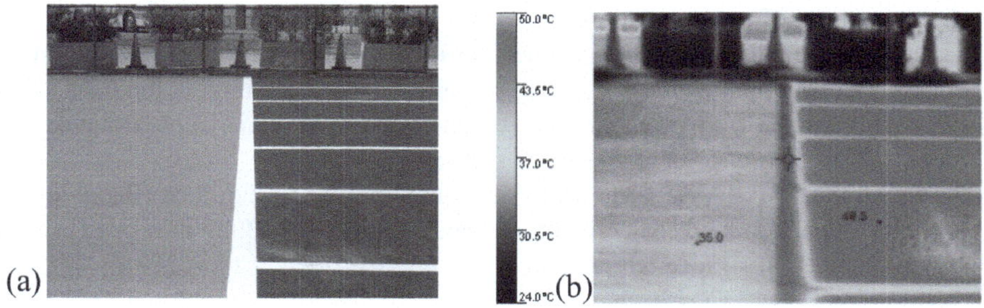

Fig. 11. Surface temperature (a) Photographic image, (b) Thermographic image Left: solar heat-blocking Pavement, Right: dense graded pavement

4.2 Improvement of Pavement Durability

Rut depth for the two surfaces (i.e. dense graded pavement and solar heat-blocking pavement) was also measured in regular monitoring in parallel with surface temperature measurements in the field. Figure 12 compares the maximum rut depth of the two pavements during the four-year monitoring (Hayakawa *et al.* 2009).

As can be seen from the figure, the results indicate that for both solar-heat blocking and dense graded pavement, the rut depth increases up to year two, thereafter it remains rather stable. However, in terms of rut depth, the solar heat-blocking pavement can reduce the maximum rut depth by a half, as compared to that of dense graded pavement. As a result, a clear difference is evident between solar heat-blocking pavement and dense graded pavement, in terms of rut depth.

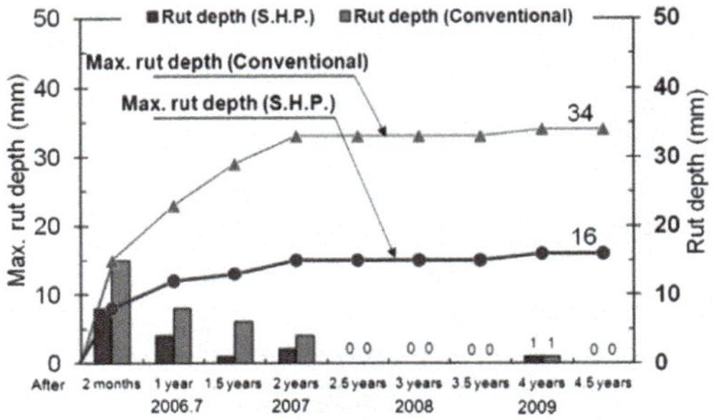

Fig. 12. Changes in the maximum rut depth over four years

5 Conclusions

This paper presents the reflectivity and durability assessment of solar heat-blocking pavement in the laboratory and field. Based on the results of laboratory experiments and field applications, the following conclusions can be drawn:

- With regard to reflectivity, the results obtained from laboratory experiments proved that solar heat-blocking pavement has significantly high albedo with unique retro-reflection characteristics; in the field, the surface temperature is reduced by approximately 13 °C compared to conventional dense graded asphalt paving due to the prevention of solar radiation.
- In terms of durability, the laboratory test results demonstrated that the surface of the solar heat-blocking pavement is strong enough for traffic loading; in the field, it can effectively reduce rutting, as the rate (based on rut depth) was approximately half compared to the dense graded asphalt surface.
- This technology is likely to be useful in mitigating the “urban heat island” effect, since a comparison between conventional and solar heat-blocking pavements shows the advantages in surface temperatures.

References

Yoder EJ, Witzak MW (1975) Principles of Pavement Design. Wiley, Hoboken

Pomerantz M, Pon B, Akbari H, Chang S-C (2000) The Effect of Pavements’ Temperature on Air Temperature in Large Cities, Lawrence Berkeley National Library

Yoshinaka T, Kinouchi T, Fukae N (2003) Study on high performance of solar radiation reflective pavement. In: Proceedings of 25th Japan road conference, No. 09P04. (in Japanese)

Kinoshita K (1998) A solar radiation reflective paint to improve the efficiency of cooling and heating. Ind Mater 46(5):35–38 (in Japanese)

Japan Industrial Standard (1994) JIS A 5759 Adhesive films for glazing

Japan Road Association (2008) Surface Temperature Measurement Method for Laboratory Specimen by lamp Irradiation, Reference book for Evaluation Method of Pavement Performance, pp 81–84. (in Japanese)

Minegishi J, Ueno S (2010) Performance requirement for durability on surface of solar heat-blocking pavement. Annual Report of Civil Engineering Support & Training Center, Tokyo Metropolitan Government, pp 41–50. (in Japanese)

Ueno S, Hashimoto Y, Minegishi J (2014) Evaluation method for Flaking resistance performance for solar heat-blocking pavement. Annual Report of Civil Engineering Support & Training Center, Tokyo Metropolitan Government, pp 31–42. (in Japanese)

Hayakawa I, Uematsu S, Sekita M, Yoshinaka T (2009) The application of solar heat-blocking pavement to airport taxiway targeting the prevention of rutting. J Pavement 44–8:9 (in Japanese)

Supply Curves Using LCA and LCCA for Conceptual Evaluation of Proposed Policies to Improve the Environment

John T. Harvey[1(✉)], Alissa Kendall[2], Ali Butt[1], Arash Saboori[1], Mark Lozano[3], and Maryam Ostovar[1]

[1] Department of Civil and Environmental Engineering, University of California Pavement Research Center, University of California, Davis, One Shields Avenue, Davis, CA 95616, USA
jtharvey@ucdavis.edu

[2] Department of Civil and Environmental Engineering and Energy and Efficiency Institute, University of California, Davis, One Shields Avenue, Davis, CA 95616, USA

[3] Energy and Efficiency Institute, University of California, Davis, One Shields Avenue, Davis, CA 95616, USA

Abstract. Many changes are being proposed to reduce greenhouse gas emissions by a multitude of sources, with the proposals based to varying degrees on science, economics, the potential to grow markets or shrink the markets of competitors, regulatory strategies, and attractiveness based on the ability to easily communicate the idea to the general public. Identifying, quantifying, and then selecting among the many possible strategies to achieve GHG reductions is difficult, especially without a standardized approach for comparison. A promising approach, supply curves, that has been used at a national level for developing abatement strategies for GHG reduction is proposed for use in this paper. Some of the critiques of past use of supply curves are being addressed through the use of the principles of consequential life cycle assessment and life cycle cost analysis. Pilot studies currently underway for a large state road agency and local governments will provide initial feedback on the ability to use this approach at a conceptual level for initial prioritization of alternatives. Initial results indicate that sufficient data can be gathered in a reasonable amount of time to compare alternatives and that the results can be compared on a much more consistent basis than has occurred previously.

Keywords: Greenhouse gas emissions · Life cycle assessment · Life cycle cost analysis · Supply curve · Benefit-cost · Conceptual analysis

1 Introduction

California's 2006 Climate Change Solutions Act (Assembly Bill 32) tasked many government entities, including local governments and government agencies, with reducing greenhouse gas (GHG) emissions to 1990 levels by 2020 (a 30% reduction), and 80% below 1990 levels by 2050. There is no single change that will achieve these

M. Pasetto et al. (Eds.): ISAP APE 2019, LNCE 48, pp. 144–152, 2020.
https://doi.org/10.1007/978-3-030-29779-4_14

ambitious goals, instead multiple changes must be made in the state's economy by many actors. Many changes are being proposed by a multitude of sources, with the proposals based to varying degrees on science, economics, the potential to grow markets or shrink the markets of competitors, regulatory strategies, and attractiveness based on the ability to easily communicate the idea to the general public. Identifying, quantifying, and then selecting among the many possible strategies to achieve GHG reductions is difficult, especially without a standardized approach for comparison.

UC Davis researchers at the University of California Pavement Research Center (UCPRC) and the National Center for Sustainable Transportation (NCST) listened to state and local policy leaders and transportation system operators over the last five years lament the difficulty of prioritizing the tens of strategies and tactics that are being proposed for changes in how they should design and operate systems to reduce greenhouse gas emissions to meet the requirements. Having worked extensively supporting state and local government in California with "life cycle thinking" data and tools for implementation of life cycle assessment for environmental impacts (including social impacts) and life cycle cost analysis for financial impacts, they believed that a process that considered the full system and the life cycle was important to provide the most beneficial and sustainable solutions while minimizing the likelihood of unintended negative consequences.

As an example, the California Department of Transportation (Caltrans) has many possible strategies to achieve greenhouse gas (GHG) emission reductions in Caltrans' operations of the state highway network to help meet the state's climate change mitigation goals. However, although many of the ideas for change appear to be attractive, simple and positive, the following is true for many of them:

- The net GHG reduction if fully successful has often not been quantified
- It has not been determined whether or not the proposed changes produce net GHG reductions, or might be found instead to cause potential increases when the full system in which they occur and the full life cycle are considered
- The time it will take to make the change happen has not been estimated,
- The process and difficulty of making the change have not been estimated, and
- Most importantly, the costs of making the change, both initial and life cycle, have often not been estimated.

A life cycle perspective is required for GHG accounting because benefits achieved during one stage of strategy's life cycle may be reduced or reversed by carbon-intensive upstream or downstream stages. Similarly, if an incomplete system view is taken benefits in one part of the system may be reduced or reversed (i.e. more carbon is emitted than business as usual) in another part of the system that was not considered. In some cases, two or more potential changes in operations are incompatible with each other in ways that will negate the benefits, and a full system view can help identify these conflicts.

The last point in the bullet list above is considered equally important with the calculation of emissions, because state government and the state's overall economy have finite capacity and political will to pay for change. The approach used in the studies described in this paper is that the greatest and fastest GHG reduction will occur if there is a prioritization in terms of GHG reduction benefit to cost. Prioritization based on benefit to cost will result in the most efficient use of existing funds to achieve the maximum reduction possible, in other words the most "bang for the buck". Unless there is this type of a prioritization, then the capacity of the public and the state's economy to implement the needed GHG reductions may be exceeded before the goals are reached. It is also considered important to be able to demonstrate to the public that efforts are being made to achieve GHG reduction goals in the most cost-effective ways possible in order to help maintain public support for those goals.

The ability to quantify the full-system, life cycle effects of decisions and changes in systems is advancing and improving using the life cycle assessment (LCA) approach and related analysis processes. The limitations and problems with LCA are also being identified so that more robust and trustworthy results can be produced. The methodology for life cycle cost analysis (LCCA) is already mature and used within Caltrans for support of decision-making regarding infrastructure choices.

The timeframe for change is also important because emission reductions that occur sooner will have greater beneficial impact than emission reductions that occur later or are spread out over a longer period of time. This is not accounted for in current global warming potential (GWP) calculations. Time-adjusted warming potential (Kendall 2012) should be used to account for the timing of emission reductions. Use of time-adjusted warming potential will help identify strategies providing the "fastest bang for the buck".

This study discusses and shows early examples of use of a GHG mitigation "supply curve" framework to support decision-making by Caltrans. The supply curve, as used in these studies, provides a method for selecting the most cost-effective strategies for mitigation by undertaking the following process for each strategy: it (1) quantifies the net effects on GHG quantity over the strategy's lifecycle, (2) considers the time required to make the change happen, (3) explores the process and difficulty of making the change happen, and (4) calculates the initial and lifecycle costs of the strategy.

This approach is being used for two studies:

1. To evaluate possible changes that Caltrans can make in its operations to reduce greenhouse gas emissions
2. To evaluate proposed actions for transportation in climate action plans that have been developed by cities and counties in California to reduce greenhouse gas emissions

2 The Approach

The approach used is to support strategic prioritization of approaches for reducing GHG emissions using what are called "marginal abatement curves", "supply curves", or "McKinsey curves" after the company that has made extensive use of them (Creyt et al. 2007). Supply curves illustrate the economics associated with changes and policies made for climate change mitigation. In particular, the work done by Lutsey and Sperling (2009) demonstrated how alternatives within the transportation sector can be quantified and compared using available information, and also compared with alternatives in other sectors of the economy. Transportation is particularly important in

California because it is responsible for approximately 41% of annual GHG emissions in the state. This percentage has increased and actual transportation emissions have increased as other sectors of the economy, particularly generation and use of electrical energy, have decreased (CARB 2018).

A generic example of a supply curve, adapted from Lutsey, is shown in Fig. 1. To implement the development of supply curves, a set of questions are answered and calculations are completed using the best available information about the proposed changes box (the complete set of questions and calculations are described later in the paper). The supply curve uses the best estimate of the benefit on the x-axis, with each box representing a proposed change and the width of the box indicating the size of the benefit. Reduction of greenhouse gas emissions is shown in the example, however this could be a performance metric for other environmental goals, such as air pollution which is a major concern in California, as well.

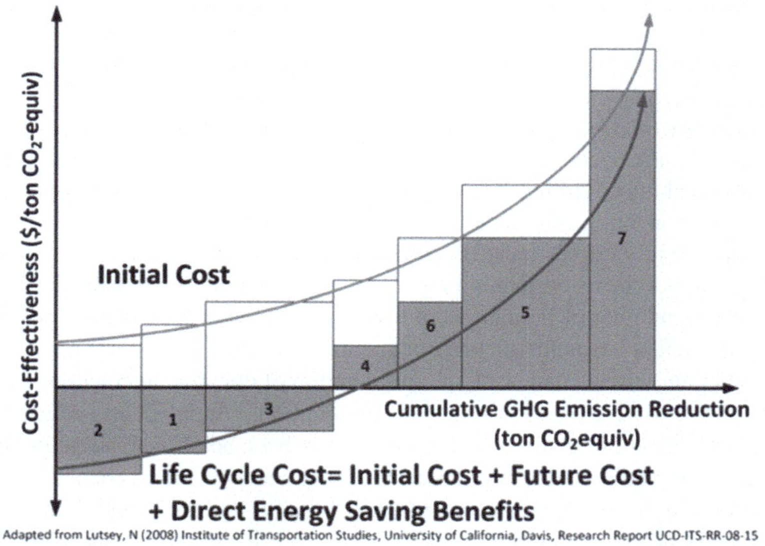

Fig. 1. Generic supply curve considering initial cost and life cycle cost.

In the approach being used, LCA is used to estimate the benefit by comparing GHG emissions from the proposed change over the life cycle analysis period versus current practice. The LCA is performed using the best available information, which can range from very poor to very good based on ISO 14044 (2006) data quality parameters as discussed related to pavements in the Federal Highway Administration Pavement LCA Framework (Harvey et al. 2016): time-related coverage, geographical coverage, technology coverage, precision, completeness, representativeness, consistency, and reproducibility. The documentation of the LCA for the supply curve needs to include a data quality assessment, which must be taken into consideration when comparing alternative proposed changes on the supply curve.

The y-axis of the supply curve shows the cost of the change per unit of benefit. Two values are calculated for each proposed change using the best available information: the initial cost of implementation and the long-term or life cycle cost. As with the LCA information, the economic analysis of the proposed changes for the supply curve is developed with the best available information and documentation is required of the assumptions, calculations and quality of the information used.

The proposed changes are put in rank order of cost effectiveness, with color coding to identify the level of uncertainty of the information used for the analysis (not shown in the example in Fig. 1). All changes have an implementation cost, but some changes will potentially result in a life cycle cost savings. Those changes that are to the left on the curve should be considered for implementation first, because they provide the most improvement for the least cost. Those that have negative life cycle costs are what Lutsey refers to as "no regrets" choices because they reduce costs over the life cycle. Moving to the right along the x-axis of the curve identifies the cumulative effect of changes towards the overall GHG reduction goal, and the increasing cost of achieving that goal. As with all economic analyses regarding public policy, the economic analysis should consider not only the overall costs, but who pays the costs or receives the savings, and whether those costs or savings are equitable.

The purpose of developing supply curves to review alternatives is to bring full system analysis, life cycle thinking, and above all, quantification, to their development in a decision-making environment where they are often absent, and to support decision-making for prioritization that includes consideration of economics.

However, supply curves must be used with caution, and are only one of the tools available to support decision-making regarding GHG and other pollutant reduction, not the only one. A number of limitations of supply curves have been identified, including omission of ancillary benefits of greenhouse gas emission abatement, poor consideration of uncertainty in the data, lack of consideration of dynamic interactions over time, and lack of transparency concerning their assumptions. Supply curves based on the individual assessment of abatement measures suffer from additional shortcomings such as not considering interactions, non-economic costs, and behavioral changes, as well as incorrect counting of benefits, and inconsistent baselines (Kesicki and Akins 2012). It has been suggested that supply curves be used more for comparisons of alternatives than for quantifying cumulative progress to abatement (Huang et al. 2016). The ability of supply curves to predict future abatement has been critiqued because of the lack of considerations of longer-term changes in markets driven by consumer changes, the timing of policy actions, actions taken by other actors in the market, and changes in future technologies (Morris et al. 2012). Most of these critiques have focused on national-level supply curves, rather than more granular and often less complex curves for agency- and local-level curves, but they must be kept in mind when using supply curves to support decision-making.

These critiques are intended to be addressed somewhat by the use of LCA and LCCA approaches by the additional information that is intended to be gathered as part of the development of the supply curves, and in particular the use of consequential LCA which assumes that decisions will result in changes in the market rather than attributional LCA which assumes that market will not change.

The full set of questions that for which information is being gathered for the supply curve studies are as follows:

1. Define the action intended to create change in GHG emissions.
2. Define the system in which the change occurs.
3. Estimate whether the market will change or the action only changes market share.
4. State who the change will impact
5. State who is responsible for implementing the change
6. State who pays for costs of the change or benefits from savings
   a. Government, level of government
   b. Producers without pass through to consumers
   c. Consumers
7. State what the method used to create the change will be:
   a. Market
   b. Market incentives
   c. Regulation
   d. Legislation
   e. Public programs incentivizing change
   f. Education
8. Show estimates or calculations of what effects the change intended to reduce GHG emissions will have on these these other environmental and resource use indicators:
   a. Air pollution
   b. Water pollution
   c. Energy use
      i. Renewable
      ii. Non-renewable
      iii. Renewable energy source used as material
      iv. Non-renewable energy source used as material
   d. Water use
   e. Use of other natural resources
9. State how the effectiveness of the change in reducing GHG reductions (the performance indicators) will be measured, modeled or estimated once implemented.
10. State who will be responsible for measuring, modeling or estimating the performance metrics.
11. Supply curve calculation development questions:
    a. Expected change in GHG emissions per unit of change in the system.
    b. Expected maximum units of change in the system.
    c. Time to reach maximum units of change.
    d. Expected shape of change rate:
       i. Linear
       ii. Increasing to maximum
       iii. Decreasing to maximum
       iv. S-shaped
    e. Estimated initial cost per unit of change
    f. Estimated life cycle cost per unit of change

The information used to develop the answers to all questions needs to be fully documented, including:

- Citations
- Development of optimistic, best and pessimistic estimates to the extent possible to permit sensitivity analysis
- Identification of the level of disagreement between different sources of information
- A ranking of the data and estimation quality such as Excellent, Good, Fair, Poor, Completely Unknown

The recommendation is to submit supply curves and their documentation to outside critical review by interested stakeholders before using them for decision-making and documentation of the critiques and responses by the supply curve developers, following ISO LCA principles.

3 Applications in Studies Currently Underway

This approach is currently be piloted for proposed changes in the operations of the California Department of Transportation (funded by Caltrans), and for alternative strategies being included in climate action plans under development by California local and regional planning agencies (funded by NCST).

For the Caltrans study, the above methodology is currently being applied as a pilot for six mitigation strategies that could be implemented by Caltrans. These strategies were selected to provide a wide range of topics with which to test the evaluation process:

1. Efficient maintenance of pavement roughness
2. Energy harvesting through piezoelectric technology
3. Automating bridge tolling systems
4. Increased use of reclaimed asphalt pavement
5. Electrification for light vehicles and use of bio-based diesl as alternative fuels for the Caltrans fleet, and
6. Installing solar and wind energy technologies within the state highway network right-of-way

A description of one of these potential changes and initial findings from the study that is currently underway are described below. The analysis period is from 2019 to 2050, which is the state's target year for achieving GHG reduction goals.

Efficient Maintenance of Pavement Roughness

Pavement condition affects the fuel use of vehicles and therefore both greenhouse gas (GHG) emissions and the cost of transportation, while maintaining pavement condition is a direct cost to road agencies. Pavement condition affects the fuel use of vehicles through rolling resistance, i.e. energy losses due to interaction between vehicles and the pavement. The relative impact of the three elements of rolling resistance (roughness, texture and structural deflection) on fuel economy and GHG emissions from on-road vehicles depends primarily on the level of pavement roughness in California.

Alternative maintenance strategies and condition trigger levels for treatment are being considered using the Caltrans pavement management system for the full 80,000 lane-km state network, and using materials and construction emissions factors developed by Wang et al. (2014) and IRI progression models developed by Jeremy Lea and Ester Tseng of the UCPRC, which are implemented in the PMS. Current decision trees consider cracking first, and then an IRI of 2.7 m/km (170 inches/mile) to trigger treatment. Alternatives considered are use in the decision trees of the IRI trigger of 3.6 m/km (224 inches/mile) used prior to 2012, and a potential future alternative that focuses on keeping sections with higher traffic volumes smoother to maximize the reduction in GHG balancing greater material and construction emissions versus fuel savings per vehicle multiplied by the number of vehicles based on Wang et al. (2014). Triggers are 1.6 m/km (100 inches/mile) for the highest trafficked sections transitioning to the current 2.7 m/km for the rest of the network. Alternative budget scenarios are also being analyzed.

Because the calculation were already set up in the PMS, most of the work for consisted of hand testing the implementation in the code, and developing the scenarios. The results indicate that on the order of 1 to 2 million metric tons (MMT) of GHG emissions can be reduced on average with increased spending on maintenance and rehabilitation to maintain smoother pavement out of total emissions for the state of about 450 MMT. The benefit to direct agency cost ranges from about \$150 to more than \$600/MMT reduction. These numbers can be compared with the price of carbon on the California carbon market of between \$10 and \$20/MMT.

Inclusion of road user fuel savings dramatically reduces the life cycle costs, in some cases resulting in net savings considering agency and user, however, consideration must be given to increased fuel use from lower prices and smoother roads. The optimized IRI triggers are the remaining scenario to be tried.

4 Summary and Conclusions

Governments and road agencies have goals for reducing GHG emissions and other environmental impacts and also face cost constraints. In democracies, there is a need to maintain public support for policies and practices to achieve these critical environmental goals by choosing the most cost-effective alternatives, and honestly, transparently, and effectively communicating the approach used in decision making, the expected benefits and costs, and the metrics for measuring the performance of the decision makers in delivering the results. Many potential changes in the policies and practices of road agencies are being proposed, both internally and externally. However, there is often a lack of quantitative information regarding the benefits and costs of these proposals, and a lack of definition regarding how the changes will interact in a larger system in which they will occur and their long-term effects, and who they will affect which has equity implications.

A promising approach, called supply curves, that has been used at a national level for developing abatement strategies for GHG reduction is proposed for use in this paper. Some of the critiques of past use of supply curves are being addressed through the use of the principles of consequential life cycle assessment and life cycle cost analysis. Pilot studies currently underway for a large state road agency and local governments will provide initial feedback on the ability to use this approach at a conceptual level for initial prioritization of alternatives. Initial results indicate that sufficient data can be gathered in a reasonable amount of time to compare alternatives and that the results can be compared on a much more consistent basis than has occurred previously. It is apparent from work to date that a number of important assumptions need to be made, that need to be fully documented, and assessed for quality, for consideration in decision making.

Acknowledgements. This work was undertaken with funding from the California Department of Transportation (Caltrans) and from the National Center for Sustainable Transportation, which is greatly appreciated. The opinions and conclusions expressed in this paper are those of the authors and do not necessarily represent those of the State of California, NCST or the Federal Highway Administration.

References

California Air Resources Board (2018) California Greenhouse Gas Emissions Emissions for 2000 to 2016 Trends of Emissions and Other Indicators. https://www.arb.ca.gov/cc/inventory/pubs/reports/2000_2016/ghg_inventory_trends_00-16.pdf. Accessed 5 Mar 2019

Creyts J, Durkach A, Nyquist S, Ostrowski K, Stephenson J (2007) Reducing U.S. Greenhouse Gas Emissions: How Much at What Cost? McKinsey & Company for the Conference Board. U.S. Greenhouse Gas Abatement Mapping Initiative. https://www.mckinsey.com/business-functions/sustainability/our-insights/reducing-us-greenhouse-gas-emissions

Harvey J, Meijer J, Ozer H, Al-Qadi I, Saboori A, Kendall A (2016) Pavement Life-Cycle Assessment Framework. FHWA-HIF-16-014. http://www.fhwa.dot.gov/pavement/sustainability/hif16014.pdf

Huang S, Lopin K, Chou K-L (2016) The applicability of marginal abatement cost approach: a comprehensive review. J Clean Prod 127(20):59–71. https://doi.org/10.1016/j.jclepro.2016.04.013

International Standards Organization (2006) Environmental Management - Life cycle assessment - Requirements and Guidelines. ISO 14044. International Organization for Standardization, Geneva, Switzerland

Kendall A (2012) Time-adjusted global warming potentials for LCA and carbon footprints. Int J Life Cycle Assess 17(3):1042–1049. https://doi.org/10.1007/s11367-012-0436-5

Kesicki F, Akins P (2012) Marginal abatement cost curves: a call for caution. Clim Policy 12 (2):219–236. https://doi.org/10.1080/14693062.2011.582347

Lutsey N, Sperling D (2009) Greenhouse gas mitigation supply curve for the United States for transport versus other sectors. Transp Res Part D: Transp Environ 14(3):222–229. https://doi.org/10.1016/j.trd.2008.12.002

Morris J, Patlsev S, Reilly J (2012) Marginal abatement costs and marginal welfare costs for greenhouse gas emissions reductions: results from the EPPA model. J Environ Model Assess 17:325. https://doi.org/10.1007/s10666-011-9298-7

Wang T, Harvey J, Kendall A (2014) Reducing greenhouse gas emissions through strategic management of highway pavement roughness. Environ Res Lett 9:034007. https://doi.org/10.1088/1748-9326/9/3/034007

Marginal Materials for Asphalt Pavements

Cold Recycling with Bitumen Emulsion of Marginal Aggregates for Road Pavements

Marco Pasetto[1(✉)] and Nicola Baldo[2]

[1] Department of Civil, Environmental and Architectural Engineering (DICEA), University of Padua, Via Marzolo 9, 35131 Padua, Italy
marco.pasetto@unipd.it

[2] Polytechnic Department of Engineering and Architecture (DPIA), University of Udine, Via del Cotonificio 108, 33100 Udine, Italy

Abstract. The paper deals with a laboratory study of bitumen emulsion bound mixtures for road pavements with an aggregate structure totally composed of waste materials, i.e. reclaimed asphalt pavement (RAP), steel slag, coal ash and glass wastes, combined in different ratios. The investigation was divided into a preliminary environmental and physical analysis of each waste material and a subsequent mechanical characterization of the bitumen emulsion bound mixtures, by means of indirect tensile strength, stiffness modulus and repeated load axial tests. Indirect tensile strength tests were also performed in wet conditions to evaluate the moisture resistance of the mixes. The main outcomes of the trial (indirect tensile strength at 25 °C on dry samples up to 0.37 MPa; stiffness modulus at 25 °C and 2 Hz up to 4,266 MPa, depending on the mixture) were compared with the requisites for acceptance of the main Italian Contract Specifications, which demonstrated that the analyzed marginal materials are suitable for use as integral substitutes of natural aggregates in the production of bitumen emulsion bound mixtures for road pavements.

Keywords: Cold recycling · Bitumen emulsion · Steel slag · Reclaimed asphalt pavement · Coal ash · Glass wastes

1 Introduction

The reuse of civil and industrial by-products in road pavements can enhance their environmental sustainability, representing a potential solution to two important issues, namely the disposal of these wastes at the end of their life and the necessity to substitute conventional quarried aggregates with alternative materials (Pasetto and Baldo 2013; Pasetto and Baldo 2018).

The main goal of the research was to investigate "cold recycled" mixtures prepared with civil and industrial by-products, cement and bitumen emulsion. Leaching and physical-mechanical testing was performed on the raw materials to investigate their environmental compatibility and technical suitability in bitumen emulsion bound mixtures to be used in road foundation layers. Within the pavement engineering field, the bitumen (and cement) treatment of granular materials is considered as one of the most effective stabilization methods (Bocci et al. 2011).

M. Pasetto et al. (Eds.): ISAP APE 2019, LNCE 48, pp. 155–163, 2020.
https://doi.org/10.1007/978-3-030-29779-4_15

Four civil and industrial by-products were considered in the research, namely electric arc furnace (EAF) steel slags, reclaimed asphalt pavement (RAP), glass wastes (GW) and coal fly (CF) ash; these materials are also known in the literature as "marginal or succedaneum aggregates" (Pasetto and Baldo 2006; Pasetto and Baldo 2010). The investigation analyzed the by-products individually, as well as mixtures of the recycled aggregates, combined in different ratios.

2 Materials

The steel slags investigated are a by-product of the steel production process based on the electric arc furnace technology. The RAP came from milled highway pavements in Northern Italy, made of asphalt concretes with a similar composition. The fly ash was produced by the combustion of coal in boilers for the production of steam to be used in power stations and industrial plants. The glass wastes mainly derived from food and beverage containers. All the marginal materials considered are solid, odor-free and mainly greyish in color, with the exception of the RAP, that appears basically black. The pH is 11.2 for the coal ash, 9.6 for the EAF slags, 8.2 for the glass wastes. The four by-products were provided by different private companies located in Northern Italy.

For all the mixes, a commercial over-stabilized bitumen emulsion was used (C60B10); its main characteristics had been investigated by the producer. It was decided to use this emulsion because it ensures a reasonably delayed cement setting and allows the mixing of a high content of fine materials.

Portland cement CEM II/B LL 32.5R was chosen as "active filler", for all the mixtures considered in the investigation.

2.1 Environmental Analysis of the Marginal Materials

Table 1 reports the toxicological investigation results of the marginal materials considered in this research. Remarkable differences can be observed between the by-products, in terms of initial concentration of heavy metals. EAF slags presented higher contents of thallium and chromium (total), than the other materials. The coal ash had the highest initial content of cadmium, nickel, selenium, arsenic and beryllium. The highest values of lead, copper and zinc were recorded for the glass wastes.

Table 2 presents the data obtained by means of the leaching tests performed on the marginal materials. The toxic characteristic leachability procedure (TCLP) adopted needs end-over-end agitation, a 20:1 liquid-to-solid ratio and an equilibrium time set at 18 h. A sample of no less than 100 g has to be extracted with a proper leaching solution. After agitating it for 18 h, the extracts have to be separated from the solids using a glass fiber filter (Siddique et al. 2010). The extracts are subsequently analyzed by inductively coupled plasma–atomic emission spectrometer.

All data were within the legal limits in Italy (Legislative Decree 152/2006), thus demonstrating that the marginal aggregates present no toxicological issues. Therefore, according to Italian Law, the by-products analyzed can be classified as "non-hazardous, special non-toxic and non-noxious refuse". Consequently, it was considered unnecessary to perform specific tests on the specimens of the bitumen emulsion bound mixtures.

Table 1. Initial concentration of heavy metals (mg/kg)

Element	EAF slags	Glass wastes	Coal ash
Copper (Cu)	219.0	955.0	298.8
Cadmium (Cd)	<1.0	<0.5	2.1
Lead (Pb)	<1.0	217.3	53.4
Zinc (Zn)	165	884.6	212.9
Chromium total (Cr)	4275.0	244	288.6
Chromium Hexavalent (Cr)	<5.0	<5	1.7
Nickel (Ni)	9.1	12.9	139.1
Mercury (Hg)	<1.0	<0.5	<1.0
Selenium (Se)	8.8	<2.0	61.2
Arsenic (As)	<5.0	<2.0	22.4
Beryllium (Be)	0.6	0.8	4.4
Antimony (Sb)	33.5	1.9	5.4
Thallium (TI)	31.1	<0.5	<1.0

Table 2. Leaching concentration of heavy metals

Element	EAF slags	Glass wastes	Coal ash	Legal thresholds
Copper (Cu)	<0.001 mg/l	0.043 mg/l	<0.05 mg/l	<0.05 mg/l
Cadmium (Cd)	<1.0 μg/l	<1.0 μg/l	<5.0 μg/l	<5 μg/l
Lead (Pb)	<5.0 μg/l	<5.0 μg/l	<50.0 μg/l	<50 μg/l
Zinc (Zn)	<0.001 mg/l	<0.001 mg/l	<3.0 mg/l	<3.0 mg/l
Chromium (Cr)	8.0 μg/l	<1.0 μg/l	<50.0 μg/l	<50 μg/l
Nickel (Ni)	<3.0 μg/l	<3.0 μg/l	<10.0 μg/l	<10 μg/l
Mercury (Hg)	<1.0 μg/l	<1.0 μg/l	<1.0 μg/l	<1 μg/l
Selenium (Se)	<10.0 μg/l	<5.0 μg/l	<10.0 μg/l	<10 μg/l
Arsenic (As)	<5.0 μg/l	<5.0 μg/l	<50.0 μg/l	<50 μg/l
Barium (Ba)	0.94 mg/l	0.01 mg/l	<1.0 mg/l	<1 mg/l

The chemical characteristics of the steel slags were further investigated in terms of oxides composition by means of X-ray fluorescence; high contents of FeO (30.3%) and CaO (27.9%), as well as SiO_2 (17.4%), MgO (6.4%) and Al_2O_3 (4.9%) were observed.

2.2 Physical-Mechanical Characterization of the Marginal Materials

The grading curves of the marginal materials, obtained following the EN 933-1 Standard, are presented in Fig. 1; as expected, the steel slag was the coarsest aggregate, whereas the finest one was the coal ash. The glass wastes showed a particle size distribution quite close to that of RAP, but with the latter resulting coarser.

Table 3 presents the main physical-mechanical characteristics of the marginal materials, along with the specific standard used.

Although the HRB-AASHTO classification method was originally studied for conventional soils, it can be still considered interesting in order to obtain a qualitative equivalence between by-products and conventional stone materials, with regard to some important physical characteristics, i.e. water sensitivity and grading size.

It was not possible to determine the Plastic Limit (according to EN ISO 17892-12) for all the marginal materials. Considering also the percentages passing through sieves of 2, 0.4 and 0.075 mm (Table 1), it was possible to classify the marginal materials, on the basis of the HRB-AASHTO method, as A1 soils, which can be utilized in road construction (GW can correspond to an A1-b soil, while steel slags and RAP to an A1-a soil), except for the coal ash, which can be classified as A2-4 soil.

For the CF ash, bulk and dry densities values were well below than those of steel slags, RAP and glass wastes.

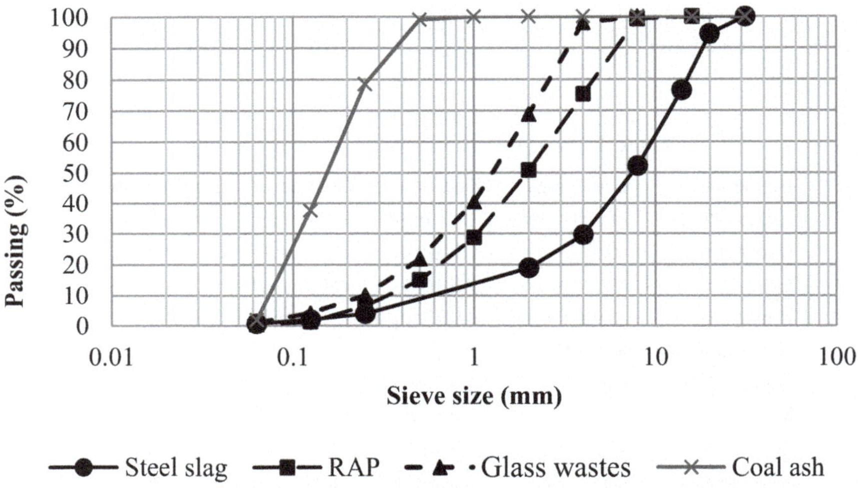

Fig. 1. Grading curves of the aggregates

Table 3. Physical and mechanical characteristics of the marginal aggregates

Physical ÷ mechanical properties	EAF slag	RAP	Glass wastes	Coal ash
Los Angeles coefficient (%) EN 1097-2	24	34	–	–
Equivalent in sand (%) EN 933-8	77	92	89	–
Grain bulk density (g/cm^3) CNR 64/78	3.53	2.57	2.63	2.20
Grain dry density (g/cm^3) CNR 63/78	3.41	2.49	2.56	2.07
Plasticity Index (-) EN ISO/TS 17892-12	0	0	0	0
ASTM 10 sieve passing (%)	20.0	40.5	70.4	100.0
ASTM 40 sieve passing (%)	6.5	11.0	20.6	99.5
ASTM 200 sieve passing (%)	0.9	0.9	3.2	26.2

The Los Angeles test allowed a satisfactory resistance of the steel slag to abrasion and friction to be verified; the LA coefficient was lower than the acceptance threshold of ANAS - Italian National Road Authority (2016), set at 25%. Conversely, the RAP materials presented an LA coefficient of 43%, higher than that of the steel slags.

The cleanliness of the aggregates, represented by the Equivalent in Sand value, was very good for all the marginal materials and higher than the minimum ANAS threshold, set at 50%.

The volumetric stability test performed on the steel slags, based on the Standard EN 1744/1 part 15.3, demonstrated a null expansion after the 168 h prescribed by the test protocol.

3 Mixtures

3.1 Composition and Grading Curves of the Mixes

The investigation focused on five different mixes, all characterized by a constant rate of glass wastes (15%) and coal ash (5%). The other two marginal aggregates, namely EAF slags and RAP, were used in 5 different and complementary proportions within the range 0–80%, as reported in Table 4.

Table 4. Composition of the mixtures (%)

Aggregate type	Mix 1	Mix 2	Mix 3	Mix 4	Mix 5
EAF slags	80	60	40	20	0
RAP	0	20	40	60	80
Glass wastes	15	15	15	15	15
Coal ash	5	5	5	5	5

The steel slag quantity in the five mixtures (Mix 1, Mix 2, Mix 3, Mix 4, Mix 5) was progressively substituted by 20% with an equal amount of RAP, therefore Mix 3 was designed with an equal content of steel slags and RAP, both at 40%. This was done to evaluate the possibility of using the two main marginal materials with different ratios, in order to take into account their availability in the production plant. In fact a crucial aspect for the potential reutilization of these wastes in the pavement engineering field, is the possibility of ensuring, for all the time required by the road construction works, the necessary quantity of the mixture, even if a temporary lower availability of one type of marginal material could occur.

In this investigation, the aggregate structure was entirely made with civil and industrial by-products, in order to satisfy specific construction needs of the highway network in North-Eastern Italy. More precisely, all the mixtures have been designed for pavement foundation layers. Another relevant research goal was to rework the aggregates provided by the production plant to the minimum extent; the aggregates were therefore used during the mixing phase with their original particle size assortment.

For each mixture the design grading curve was obtained from the integration of the raw by-products in different ratios.

The design grading curves, obtained from the combinations of marginal materials, were compared with the Asphalt Academy reference envelope (2009) in order to evaluate their suitability (Fig. 2). Only the grading curves of Mix 1 and Mix 2 resulted entirely within the conventional reference envelope, whereas for the other three mixes it was necessary to consider also the envelope's "acceptable limit" (Fig. 2).

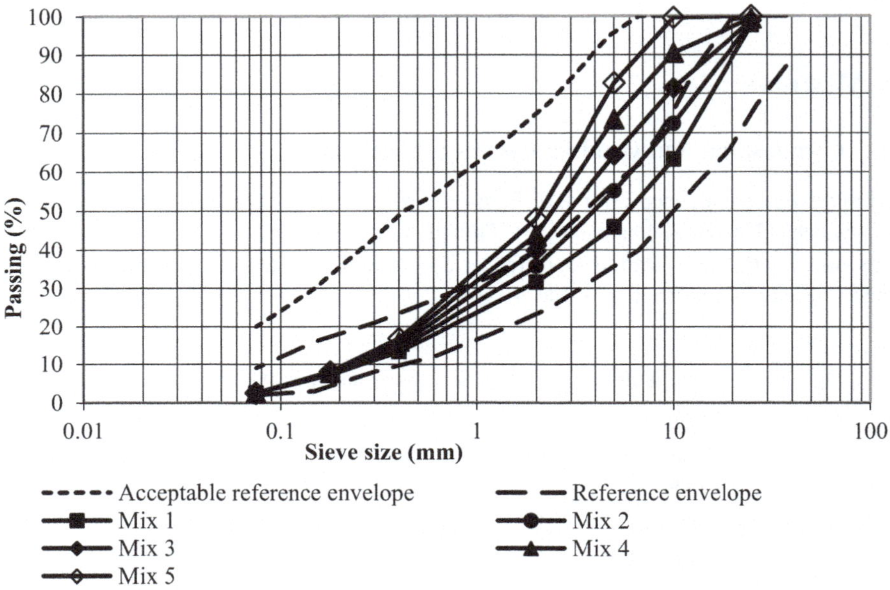

Fig. 2. Grading curves of the mixtures

3.2 Optimization of the Mixtures

The optimal binder content was identified by means of an experimental procedure based on the indirect tensile strength test, performed on a series of cylindrical specimens for each of the 5 mixtures analyzed, in which, maintaining the type and quantity of marginal materials fixed, the bitumen emulsion content was changed at steps of 0.5% by weight of the aggregate, in the range 2.0–4.0%.

According to previous investigations (Bocci et al. 2011), the active filler, namely the cement, was used at a constant content, set at 2%. This percentage should guarantee an adequate moisture resistance of the mixtures, without any fragility issue (Wirtgen 2012). On the basis of preliminary tests, for all the mixtures a water content was used equal to 6.0% by weight of the aggregate.

The specimens were compacted by means of a gyratory compactor, using a 150 mm diameter mold, applying a constant pressure of 600 kPa, speed of 30 rotations per minute and angle of gyration equal to 1.25° (EN 12697-31). Each specimen was compacted applying 180 gyrations to achieve the maximum density.

According to the methodology proposed by the Asphalt Academy (2009) the curing conditions were set at 40 °C for 72 h. The specimens were conditioned at the test temperature in an environmental conditioning chamber. Indirect Tensile Strength (ITS) was evaluated at 25 °C, based on the specifications in Standard EN 12697-23. The indirect tensile strength tests were performed on both dry and soaked specimens, to determine the moisture susceptibility of the mixtures. The soaked specimens were submerged in water at 25 °C for 24 h prior to testing (Thanaya 2003).

The Optimal Binder Content (OBC) for each mix was determined as the minimum percentage that ensures the minimum dry ITS prescribed in the ANAS specifications.

The dry and soaked ITS data, along with the Tensile Strength Ratio (TSR) values, are presented in Table 5. The TSR was determined as the ratio between the ITS values obtained for soaked specimens and those of dry specimens.

Table 5. Mix design results

Property	Mix 1	Mix 2	Mix 3	Mix 4	Mix 5
ITS dry (MPa)	0.33	0.37	0.37	0.34	0.33
ITS soaked (MPa)	0.22	0.25	0.30	0.28	0.28
TSR (%)	67	68	81	82	85

With respect to the acceptance requisites of the main Italian specifications (ANAS 2016), related to the dry ITS, each of the mixes passed the minimum threshold of 0.32 MPa at 3% of bitumen emulsion, on the weight of the aggregate. Hence, an identical OBC was found for all the mixes, equal to the intermediate bituminous binder percentage considered. With regard to moisture resistance, a satisfactory behavior was verified for all the mixes; the minimum TSR value suggested in the literature, equal to 50%, (Thanaya 2003), was achieved for all combinations of marginal materials, with a bitumen emulsion content of 3%.

4 Performance Characterization

4.1 Stiffness Characterization

Indirect Tensile Stiffness Modulus (ITSM) tests, according to Annex C of the EN 12697-26 standard, were conducted on the mixes, in order to evaluate their mechanical behavior, with respect to repeated load cycles. Cylindrical specimens, prepared by gyratory compactor like those used for the indirect tensile strength optimization, underwent ITSM tests at 10, 25 and 40 °C.

The results of the Stiffness Modulus investigation are shown in Fig. 3. Given the visco-elastic nature of the bitumen emulsion, it is possible to observe the expected reduction in stiffness with the raising of the testing temperature, for all the mixes analyzed. The highest Stiffness Modulus was recorded for Mix 3, at each of the testing

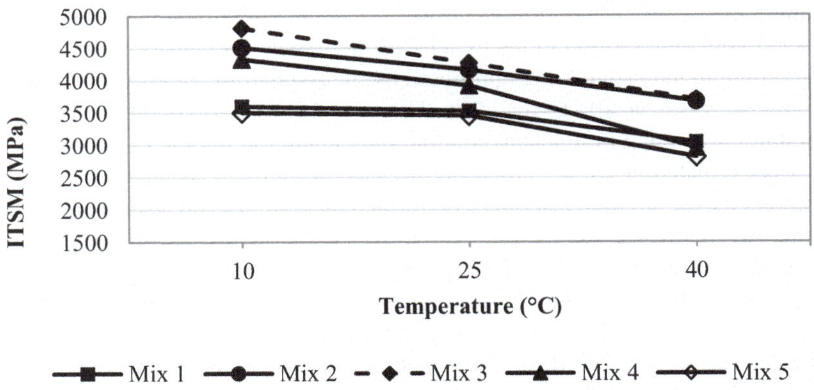

Fig. 3. ITSM tests results

temperatures, with values in the range 3,695–4,812 MPa, depending on the temperature. The lowest stiffness was observed for the mixes made only with steel slags or RAP, namely Mix 1 and Mix 5, respectively.

4.2 Permanent Deformation Analysis

The permanent deformation resistance of the mixes was evaluated through a Repeated Load Axial Test (RLAT) without confinement, applying 1,800 pulses of 100 kPa stress, with loading and unloading times set at 1 s, to cylindrical specimens prepared by means of the gyratory compactor, as was done for those undergoing the ITSM tests. The testing temperature was set at 30 °C, considering the weather conditions in Northern Italy and the depth of the foundation layers (for which a higher temperature would not be reasonable). The permanent deformation value accumulated at the end of the test on each mixture is represented in Fig. 4; a much higher resistance was verified of Mix 3 with respect to the other mixes.

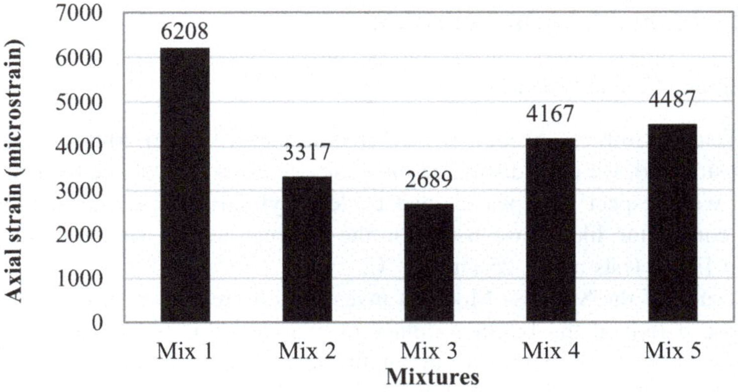

Fig. 4. RLAT results

5 Conclusions

The toxicological, physical and mechanical characterization of the marginal materials investigated, i.e. EAF steel slags, RAP, glass wastes and coal ash, has demonstrated the feasibility of reusing such materials in pavement foundation layers and thereby enhancing the environmental sustainability of road infrastructures.

The acceptance requisites specified by the main Italian road technical standards for cold mixtures, in terms of dry indirect tensile strength, were satisfied for a content of 2% of cement and 3% of bitumen emulsion, for all combinations of marginal materials considered.

With respect to durability, all the optimized mixtures demonstrated a satisfactory moisture resistance, with tensile strength ratios even higher than 80%, depending on the mix composition.

The best laboratory results were obtained for the mixture with a balanced percentage of steel slags and RAP (40% of both), which developed a dry indirect tensile strength of 0.37 MPa.

The comparative ranking of the mixtures resulting from the indirect tensile strength analysis, was also observed with the performance characterization, conducted by means of stiffness modulus tests and repeated load axial tests.

References

ANAS S.p.A. (2016) Capitolato speciale d'appalto - Norme Tecniche. Azienda Nazionale Autonoma delle Strade, Roma, IT (in Italian)

Asphalt Academy (2009) Technical Guideline: Bitumen Stabilised Materials, A Guideline for the Design and Construction of Bitumen Emulsion and Foamed Bitumen Stabilised Materials, Second Edn. Asphalt Academy

Bocci M, Grilli A, Cardone F, Graziani A (2011) A study on the mechanical behaviour of cement–bitumen treated materials. Constr Build Mater 25(2):773–778

Pasetto M, Baldo N (2006) Electric arc furnace steel slags in "high performance" asphalt mixes: a laboratory characterization. In: 2006 TMS fall extraction and processing division: sohn international symposium, 27–31 August, San Diego, CA

Pasetto M, Baldo N (2010) Recycling of steel slags in road foundations. Environ Eng Manag J 9 (6):773–777

Pasetto M, Baldo N (2013) Cement bound mixtures with metallurgical slags for road constructions: mix design and mechanical characterization. Inzynieria Mineralna 14(2):15–20

Pasetto M, Baldo N (2018) Re-use of industrial wastes in cement bound mixtures for road construction. Environ Eng Manag J 17(2):417–426

Siddique R, Kaur G, Rajor A (2010) Waste foundry sand and its leachate characteristics. Resour Conserv Recycl 54:1027–1036

Thanaya INA (2003) Improving the performance of cold bituminous emulsion mixtures incorporating waste materials. Ph.D. thesis, School of Civil Engineering, The University of Leeds, UK

Wirtgen GmbH (2012) Wirtgen Cold Recycling Technology. Windhagen, Germany

Experimental Investigation of Performance Properties of Asphalt Mixture Designed with the Re-recycled RAP and EAFSS

Di Wang[1], Augusto Cannone Falchetto[1,2,3(✉)], Ki Hoon Moon[4], Chiara Riccardi[1], and Michael P. Wistuba[1]

[1] Department of Civil Engineering – ISBS, Technische Universität Braunschweig, 38106 Brunswick, Germany
{di.wang, a.cannone-falchetto, chiara.riccardi, m.wistuba}@tu-bs.de

[2] Department of Civil and Environmental Engineering, University of Alaska Fairbanks, Fairbanks, AK 99775, USA
afalchetto@alaska.edu

[3] Department of Civil and Environmental Engineering, Nagaoka University of Technology, 1603-1 Kamitomiokamachi, Nagaoka, Niigata Prefecture 940-2188, Japan

[4] Korea Expressway Corporation Pavement Research Division (KECPRD), Dongtan-Myeon, Hwaseong-Si, Gyeonggi-Do 445-812, South Korea
zetamkh@ex.co.kr

Abstract. In this paper, the combined use of re-recycled reclaimed asphalt pavement (RAP) and Electric Arc-Furnace Steel Slag (EAFSS) is experimentally investigated and evaluated. First, a virgin asphalt mixture for surface layers is prepared and next aged in the laboratory to reproduce artificial RAP. Then, this material is used to prepare a set of first generation recycled mixtures which further undergoes the same procedure to finally produce RAP and mixtures presenting a second generation of recycling (re-recycling). Meanwhile, the combination of EAFSS and RAP of different generations is also incorporated in the mix design. Fatigue, low temperature creep, and fracture tests are conducted to evaluate the mixtures' mechanical properties. Both recycled and re-recycled mixtures prepared with or without slags shows better fatigue behavior than virgin material. Higher stiffness and poorer relaxation response are exhibited by re-recycled RAP mixtures compared with the recycled materials, although within levels commonly observed in the literature. Fracture tests indicate a complex behavior for re-recycled mixtures showing a similar response to the virgin materials for a higher amount of RAP. Only limited differences can be found when slag is used. The present experimental study seems to provide evidence on the possibility of combining re-recycled RAP and EAFSS to designing asphalt mixtures for pavement application.

Keywords: Asphalt mixture · Reclaimed asphalt pavement · Re-recycling · Electric arc-furnace steel slag (EAFSS) · Mechanical properties

M. Pasetto et al. (Eds.): ISAP APE 2019, LNCE 48, pp. 164–173, 2020.
https://doi.org/10.1007/978-3-030-29779-4_16

1 Introduction

In view of economic and environmental aspects, recycled and industrial by-product materials have seen increasing use in pavement construction in the last decades by both road authorities and research institutions (Cannone Falchetto et al. 2017b; Moon et al. 2017). This includes Reclaimed Asphalt Pavement (RAP) (Poulikakos et al. 2017), recycled asphalt shingle (RAS) (You et al. 2011), Constructions and Demolition Waste (CDW) (Tahmoorian et al. 2018), crumb rubber (Arabani et al. 2018) and steel slags (Grönniger et al. 2017), among others. Within this set of materials, the use of RAP and by-product such as slags has experienced a significant increment over the past years (Grönniger et al. 2017; Moon et al. 2017; EAPA 2017).

For RAP, approximately 72.5 and 47.3 million tons are reclaimed annually in the U.S., and Europe (including Turkey), respectively (EAPA 2017). Among those more than 90% are reused for pavement application, making RAP the most recycled pavement material both in the U.S. and Europe (EAPA 2017). Due to the rapid increase of the heavy traffic load, maintenance and rehabilitation are needed before the end of the service life, eventually leading to the need for pavement recycling (Brantley and Townsend 1999). This massive operation on the road networks is prompting the need to manage a large amount of recycled material (Heneash 2013). However, most of the pavements were originally designed with a certain amount of RAP, raising the critical question and challenge of the second generation of recycling or re-recycled of RAP.

An attempt to evaluate the possibility of using re-recycling RAP was first conducted in Japan in 1995 (Yoshikane 1995), followed with a number of studies in U.K., Switzerland and Germany (Heneash 2013; Hugener and Kawakami 2017; Cannone Falchetto et al. 2018a; Wang et al. 2019). In these previous efforts, when a maximum of 40% re-recycled RAP was used, comparable rutting, abrasion, fatigue, water sensitivity, thermal cracking, and low temperature fracture properties could be observed between the mixture prepared with re-recycled materials and those designed with virgin binder and aggregate. This suggested that re-recycling of RAP could be a feasible option for the next generation of asphalt pavements.

With respect to industrial by-products, Electric Arc-Furnace Steel Slag (EAFSS) is obtained from the iron and steel industry during the rapid cooling process. It is mainly composed of calcium, iron, aluminum, magnesium, and silicon oxides, which together account for an average of approximately 90% of the weight of the material (Sofilić 2010). Despite the volume expansion caused by hydration, better performance properties were found compared with natural aggregates also in the different phases of bituminous materials, such as asphalt mortar and asphalt mixture (Ahmedzade and Sengoz 2009; Grönniger et al. 2017; Moon et al. 2017).

Although significant effort has been devoted to studying both re-recycled RAP and steel slags, not many studies can be found on the combined use of these materials (Moon et al. 2017). In this paper, the combined effect of re-recycling RAP and EAFSS on the mechanical properties of asphalt mixtures is experimentally investigated. Fatigue behavior is evaluated by the Indirect Tensile (IDT) cylindrical configuration test (EN 12697-24 2018). Meanwhile, low temperature creep and fracture properties are investigated by the Bending Beam Rheometer (BBR) (Marasteanu et al. 2009) and the

Semi-Circular Bending (SCB) tests (Cannone Falchetto et al. 2017a; Cannone Falchetto et al. 2018a). Finally, the corresponding parameters are used to evaluate the possibility of combined use of re-recycled RAP and EAFSS.

2 Materials and Experimentation

2.1 Material Preparations

A set of thirteen asphalt mixture is prepared for this study. First, an unmodified asphalt binder with Performance Grade of PG 58-28 (AASHTO M320-16 2016) and limestone aggregates with a nominal maximum aggregate size NMAS = 12.5 mm are selected to prepare a mixture for surface layer according to the conventional mix design used in Germany (Asphalt-StB 2007). This material is identified as reference. Then, the fresh materials are artificially aged (AASHTO R30 2012; Heneash 2013) in a laboratory environment and then crushed to obtain the first generation of RAP (RAP source A) to further produce the set of first generations of recycled mixtures with the same gradation curve of the virgin material. Similarly to the procedure previously used, these materials are crushed and aged to recreate the re-recycled RAP which is included in the second-generation recycled mixtures. At this stage, the materials prepared with 20% RAP is defined as RAP source B, while the one produced by 40% RAP is RAP source C. Replace of fractions was performed by corresponding proportions for each size. However, due to the binder in the RAP, small particles are clumped together, hence, fine aggregates and fillers are difficult to evaluate and sieve. Therefore, the RAP is replaced directly by weight. A single target air voids content of 7% and a PG 58-28 binder with a single proportion of 7% are used for all asphalt mixtures, while different percentages of EAFSS are incorporated. It should be noted that no rejuvenator is used in this study. Two additional mixtures were designed with virgin material and EAFSS for comparison purposes. Asphalt mixture slabs are firstly prepared in a laboratory environment, and then, cylindrical (FGSV 2009; EN 12697-26 2012) and SCB (Cannone Falchetto et al. 2018a) specimens are cored while the related BBR (ASTM D6816-11 2016) samples are cut from the slabs. Table 1 provides a summary of the asphalt mixtures used in this study.

2.2 Fatigue Testing

The fatigue properties of asphalt mixtures are evaluated with the cylindrical indirect tensile (CIDT) (ASTM D3497-79 2003) test with a diameter of 150 mm (FGSV 2009; EN 12697-26 2012). Cyclic loading is applied to the mixture sample at T = 20 °C with a frequency of 10 Hz (EN 12697-24 2018). A decreasing stiffness modulus, $|E|_N$, due to fatigue phenomena can be observed during the test together with the evolution of the energy ratio, *ER* as a function of the load cycles, N: $ER(N) \sim N \cdot |E|_N$.

The allowable number of load cycles, N_{macro}, and the corresponding fatigue life is determined by the peak of the *ER* curve as it is associated with the beginning of macro cracking. Conventionally, the Wöhler line (Walther and Wistuba 2014) can be used to express the material's fatigue function (Eq. 1). Hence, a minimum of three different stress amplitudes is required for each testing condition.

Table 1. Asphalt mixture

ID	Recycling level	EAFSS (%)	Recycled material (%)	RAP source
A	Virgin	-	0	-
B	Recycled RAP	0	20	A
C	Recycled RAP	0	40	A
D	Virgin	20	0	-
E	Virgin	40	0	-
F	Re-recycled RAP	0	20	B
G	Re-recycled RAP	0	40	B
H	Re-recycled RAP	10	10	B
I	Re-recycled RAP	20	20	B
J	Re-recycled RAP	0	20	C
K	Re-recycled RAP	0	40	C
L	Re-recycled RAP	10	10	C
M	Re-recycled RAP	20	20	C

$$N_{Macro} = C_1 \cdot \varepsilon_{el}^{C_2} \tag{1}$$

where ε_{el} is horizontal elastic initial strains, C_1 and C_2 are fitting constants. More detailed information can be found in previous research conducted at the Braunschweig Pavement Engineering Centre (ISBS) (Walther 2016).

2.3 Low Temperature Creep and Fracture Testing

Low temperature creep tests are performed on small asphalt mixture beams (l = 102 mm, b = 12.5 mm, and h = 6.25 mm) (Marasteanu et al. (2009), with a Bending Beam Rheometer (BBR) (ASTM D6816-11 2016) where air is used as cooling medium (Fig. 1a). A higher constant loading force equal to 4 N and extended testing time of 1000 s are imposed. Creep stiffness, $S(t)$, and m-value can be calculated as follows:

$$S(t) = \frac{1}{D(t)} = \frac{\sigma(t)}{\varepsilon(t)} = \frac{P \cdot l^3}{4 \cdot b \cdot h^3 \cdot \delta(t)} \quad \text{and} \quad m(t) = \left| \frac{d \log S(t)}{d \log t} \right| \tag{2}$$

where D (t) is the creep compliance; σ is the bending stress; $\varepsilon(t)$ is the bending strain and $\delta(t)$ is the beam deflection. Based on Eq. 2, thermal stress, $\sigma(T)$, can be computed with the Laplace transformation method (Cannone Falchetto et al. 2017c).

The low temperature fracture tests are performed with the Semi-Circular Bend (SCB) configuration (Cannone Falchetto et al. 2018a; Al-Qudsi et al. 2019). This test is conducted on a specimen with a diameter of 150 mm, a thickness of 25 mm and a straight notch 15 mm long and 1.5 mm wide (Fig. 1b). A Load Line Displacement (LLD) and a Crack Mouth Opening Displacement (CMOD) were used together for displacement measurements. CMOD signal is used to obtain a stable crack growth with

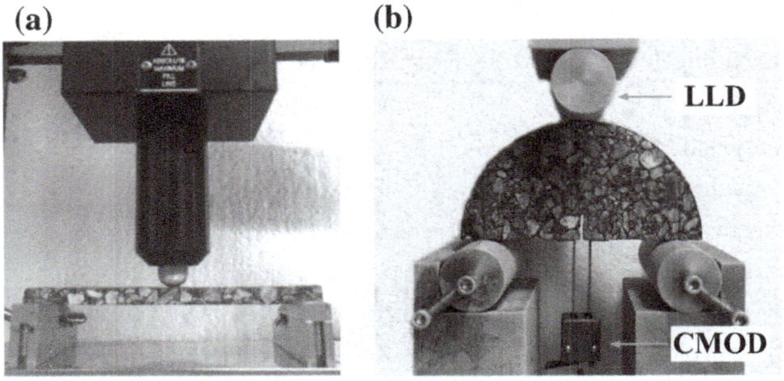

Fig. 1. (a) BBR testing setup (b) SCB testing device setup

a loading rate of 0.0005 mm/s, then the peak load, P_N, is measured at the same time. Load versus LLD curve can be generated from the measured SCB data and two main fracture parameters, fracture energy, G_F, and Mode I fracture toughness, K_{Ic}, are then calculated according to the following expressions:

$$K_{Ic} = [P_N/(2 \cdot r \cdot t)] \cdot \sqrt{\pi \cdot a} \cdot \left[Y_{I(S_0/r)} + (\Delta s_0/r) \cdot B\right] \tag{3}$$

$$G_F = W_f/A_{lig} \tag{4}$$

where P_N is the peak load in SCB test; t and r are thickness and radius of SCB specimen; a is a notch length with 15 mm; W_f is the work of fracture; A_{lig} is the area of the ligament; $Y_{I(SO/r)}$ is the normalized stress intensity factor.

3 Results and Analysis

3.1 Fatigue Behavior

In this study, CIDT is used to evaluate the fatigue response of each material, all the samples are tested under a stress-control loading at a constant temperature of 20 °C. In the purpose of determining the fatigue lines, the number of loading cycles to fatigue failure N_{macro} (fatigue life) are recorded and then plotted versus the initial strain level ε_{el} in a log-log plane against at the loading frequency of 10 Hz (Fig. 2).

In Fig. 2, the Wöhler curves are shifted upwards in an almost parallel way. Very good linear correlations can be observed for the entire set of mixtures with a coefficient of determination (R^2) higher than 0.95. A slight improvement in fatigue life can be found when mixtures are prepared with fresh materials and slags with respect to the virgin reference material. The mixtures prepared with the first generation of artificial RAP performed the best fatigue behavior in the entire set of mixtures, followed with the mixture prepared with re-recycled RAP and with or without slags. In Fig. 2a, only a limited difference can be observed among mixtures designed with different amounts of recycled RAP and slags. Even a closer observation of the mixtures prepared with RAP sources B and C (Fig. 2b) is still unable to visually show significant differences

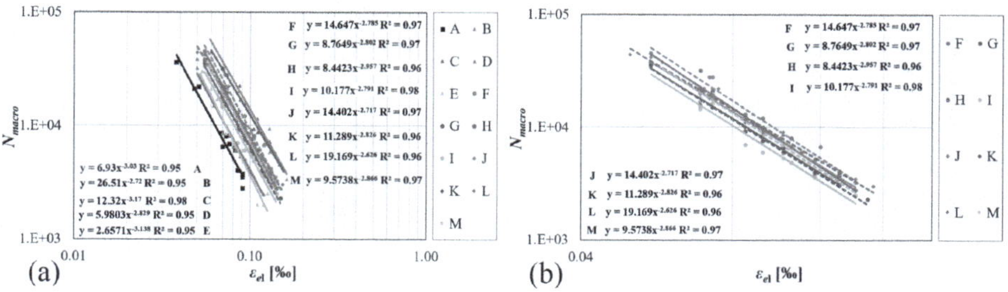

Fig. 2. Loading cycles to fatigue failure N_{macro} vs. initial strain level ε_{el} for three different loading amplitudes at a test temperature of 20 °C and test frequency 10 Hz. (a) comparison between the entire set of mixtures; (b) comparison between materials prepared with re-recycled RAP

between material prepared only with re-recycled RAP and those containing re-recycled RAP mixed with slags. This suggests that the mixtures prepared with re-recycled RAP with or without slags present comparable fatigue response at the selected testing temperature of 20 °C. This trend is not entirely surprising, as a similar result can be found in a previous research effort (Walther and Wistuba 2014). An opposite material behavior may be expected for fatigue tests conducted at lower temperatures (Walther 2016).

3.2 Low Temperature Creep and Fracture Tests Results

Figure 3 summarizes the results of creep stiffness and thermal stress for the entire set of asphalt mixtures. The experimental results obtained from BBR creep tests indicate an increased stiffness associated with a higher percentage of RAP and re-recycled RAP (Fig. 3a). In addition, materials prepared with a combination of slags and re-recycled RAP show even higher stiffness. Based on Fig. 3b, similar thermal stress is obtained for recycled mixtures in comparison to the virgin one, while a substantially higher $\sigma(T)$ can be observed for material containing slags and re-recycled mixtures. This suggests poorer relaxation capabilities. Nevertheless, the observed thermal stress is consistent with a previous study and can be considered as acceptable (Cannone Falchetto et al. 2018b). This appears to support the idea of combining re-recycled RAP and slags for pavement construction (Yoshikane 1995; Heneash 2013; Hugener and Kawakami 2017).

The combined effect of re-recycled RAP and slags on the low temperature fracture properties of the asphalt mixture is investigated with a simple SCB fracture test. In Table 2, the SCB results of peak load P_N, fracture energy, G_F, and fracture toughness, K_{Ic} are presented, while Fig. 4 illustrates the Load vs LLD curves.

An increase in the peak load, P_N, can be observed when recycled RAP and slags are used, while a very moderate decreasing trend is experienced for the mixture containing re-recycled RAP with or without slags. With respect to recycled RAP, a very similar pattern can be observed for fracture energy and fracture toughness (Table 2). Moreover, the increase in material brittleness, for recycled RAP is highlighted by the shape

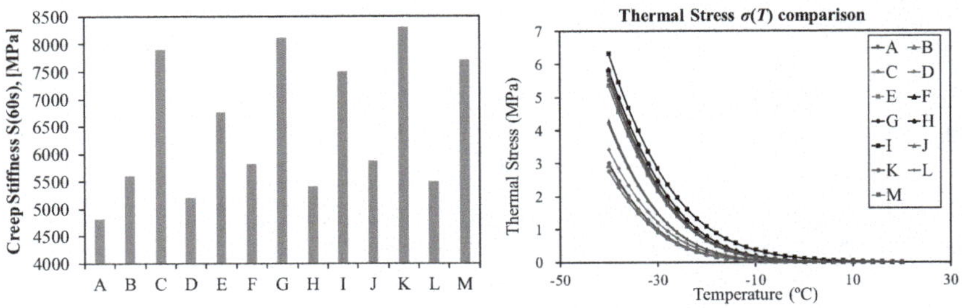

(a) (b)

Fig. 3. (a) BBR creep stiffness at t = 60 s at −6 °C; (b) $\sigma(T)$ comparison

Table 2. SCB test results

Mix ID	P_N (kN)	G_F (kN/m)	K_{Ic} (MPa * $m^{0.5}$)	Mix ID	P_N (kN)	G_F (kN/m)	K_{Ic} (MPa * $m^{0.5}$)
A	3.77	0.411	1.119	H	3.78	0.375	1.122
B	3.96	0.432	1.175	I	3.75	0.369	1.113
C	4.15	0.452	1.232	J	3.86	0.413	1.146
D	4.14	0.404	1.229	K	3.77	0.421	1.119
E	3.88	0.381	1.152	L	3.91	0.415	1.161
F	3.82	0.376	1.134	M	3.85	0.429	1.143
G	3.81	0.372	1.131				

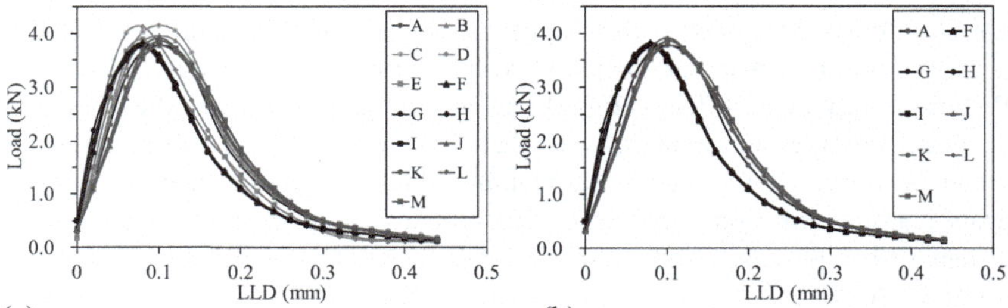

(a) (b)

Fig. 4. Load vs. LLD curves for all the different mixtures (a) comparison between the entire set of mixtures; (b) comparison between materials prepared with re-recycled RAP

of the softening part of the load-displacement curve that is squeezed toward the y-axis (Fig. 4a). This overall trend suggests that the presence of the recycled RAP is improving the mixture strength, although this is associated with reduced fracture energy which is dissipated when the material fails. A similar trend is experienced also for mixture prepared with slags without RAP.

The presence of re-recycled RAP appears to introduce a twofold behavior (Fig. 4b). While for source B (20% recycled RAP) there is more evident embrittlement, this is not true for source C (40% recycled RAP). This may be due to the higher amount of recycled RAP in the latter case which is only minimally blending with the virgin binder used in the mixture, ultimately resulting in a more plastic behavior overall similar to the virgin mixture due to the softer consistency of the virgin binder. A very moderate improvement in fracture properties appears to be experienced when EAFSS are introduced in the mixture. This fracture behavior seems to be partially in contrast with the higher thermal stress observed in the previous section, prompting the need for further investigation to support these observations.

4 Summary and Conclusions

In this paper, the effect of the combined use of re-recycled RAP and EAFSS slags on fatigue, low temperature creep and fracture properties of asphalt mixture was experimentally investigated based on the cylindrical testing geometry, on BBR mixture creep stiffness and SCB fracture test. Asphalt mixtures designed with different amounts of recycled materials and with or without EAFSS are prepared for this purpose. Based on experimental work and data analysis the following conclusions can be drawn:

- The mixtures prepared with recycled materials and slags shows better fatigue properties compared to the virgin mixture. Only slight differences can be observed between mixtures prepared with re-recycled RAP with or without slags.
- The Results of low temperature creep test indicate that mixtures prepared with re-recycled RAP, with and without EAFSS, have inferior relaxation properties with respect to virgin mixtures. However, this is within values previously observed in the literature. Moderately better performance is exhibited when EAFSS are used.
- Relatively similar low temperature fracture performances were found for mixture prepared with re-recycled RAP in comparison to the conventional fresh mixture when higher amount of RAP was used. Similarly to creep tests, better performance is found when slags are used. This is in partial disagreement with the creep tests results.

The present experimental study provides evidence on the possibility of the combined use of re-recycled of RAP and slags. Nevertheless, it should be remarked that only limited asphalt mixtures were prepared, while tests were conducted only in a laboratory environment. Additional materials and experimental work need to be performed to further evaluate the effect of recycled materials and industrial by-products on pavement mixtures.

Acknowledgments. The support of Institut für Straßenwesen at TU Braunschweig, Germany, and Korea Expressway Corporation is acknowledged. Augusto Cannone Falchetto would like to acknowledge the support of the Japan Society for Promotion of Science – JSPS international research in Japan program.

References

AASHTO M320-16 (2016) Standard specification for performance-graded asphalt binder. American Association of State Highway and Transportation Officials

AASHTO R 30 (2012) Standard Practice for Mixture Conditioning of Hot Mix Asphalt (HMA). American Association of State Highway and Transportation Officials

Ahmedzade P, Sengoz B (2009) Evaluation of steel slag coarse aggregate in hot mix asphalt concrete. J Hazard Mater 165(1–3):300–305

Al-Qudsi A, Cannone Falchetto A, Wang D, Büchler S, Kim YS, Wistuba MP (2019) Finite element simulation of semi-circular bending test using self-affine fractal cracks. In: 8th European asphalt technology association conference

Arabani M, Tahami SA, Hamedi GH (2018) Performance evaluation of dry process crumb rubber-modified asphalt mixtures with nanomaterial. Road Mater Pavement Des 19(5):1241–1258

Asphalt-StB TL (2007) Technische Lieferbedingungen für Asphaltmischgut für den Bau von Verkehrsflächenbefestigungen. Köln: Forschungsgesellschaft für Straßen -und Verkehrswesen (FGSV). (in German)

ASTM D3497-79 (2003) Standard test method for dynamic modulus of asphalt mixtures. ASTM International, West Conshohocken, PA

ASTM D6816-11 (2016) Standard practice for determining low-temperature performance grade (PG) of asphalt binders. ASTM International, West Conshohocken, PA

Brantley AS, Townsend TG (1999) Leaching of pollutants from reclaimed asphalt pavement. Environ Eng Sci 16(2):105–116

Cannone Falchetto A, Moon KH, Lee CB, Wistuba MP (2017a) Correlation of low temperature fracture and strength properties between SCB and IDT tests using a simple 2D FEM approach. Road Mater Pavement Des 18(S2):329–338

Cannone Falchetto A, Moon KH, Wang D, Riccardi C, Kang MS, Wistuba MP (2017b) Investigation on the combined recycling of reclaimed asphalt pavement and steel slag in asphalt mixture at low temperature. In: International conference on mechanisms of cracking and debonding in pavements, vol 13, pp 75–80

Cannone Falchetto A, Moon KH, Wang D, Riccardi C, Wistuba MP (2018a) Comparison of low-temperature fracture and strength properties of asphalt mixture obtained from IDT and SCB under different testing configurations. Road Mater Pavement Des 19(3):591–604

Cannone Falchetto A, Moon KH, Wang D, Riccardi C, You ZP, Wistuba MP (2018b) Experimental investigation on double recycling of asphalt mixture for pavement applications. In: 6th international conference on sustainable solid waste management, pp 1–10

Cannone Falchetto A, Moon KH, Wistuba MP (2017c) An alternative method for computing thermal stress in asphalt mixture: the laplace transformation. Road Mater Pavement Des 18 (Suppl 2):226–240

EAPA, European Asphalt Pavement Association (2017) Asphalt in Figures. https://eapa.org/wp-content/uploads/2018/12/AIF_2017.pdf. Accessed 11 Oct 2018

EN 12697-24 (2018) Bituminous mixtures. Test methods for resistance to fatigue. European Committee for Standardization, Brussels, Belgium

EN 12697-26 (2012) Bituminous Mixtures. Test methods for Hot Mix Asphalt stiffness. European Committee for Standardization, Brussels, Belgium

FGSV (2009) Arbeitsanleitung zur Bestimmung des Steifigkeits- und Ermüdungsverhaltens von Asphalten mit dem Spaltzug-Schwellversuch als Eingangsgröße in die Dimensionierung von Asphaltbefestigungen. AL-SP – ASPHALT 09, Forschungsgesellschaft für Straßen- und Verkehrswesen, Köln. (in German)

Grönniger J, Cannone Falchetto A, Isailović I, Wang D, Wistuba MP (2017) Experimental investigation of asphalt mixture containing Linz-Donawitz steel slag. J Traffic Transp Eng (Engl Ed) 4(4):372–379

Heneash U (2013) Effect of the repeated recycling on hot mix asphalt properties. Doctoral dissertation, University of Nottingham, UK

Hugener M, Kawakami A (2017) Simulating repeated recycling of hot mix asphalt. Road Mater Pavement Des 18:76–90. Issue sup2: EATA 2017

Marasteanu MO, Velasquez RA, Cannone Falchetto A, Zofka A (2009) Development of a simple test to determine the low temperature creep compliance of asphalt mixtures. NCHRP-133 Final Report

Moon KH, Cannone Falchetto A, Wang D, Riccardi C, Wistuba MP (2017) Mechanical performance of asphalt mortar containing hydrated lime and EAFSS at low and high temperatures. Materials 10(7):743

Poulikakos LD, Papadaskalopoulou C, Hofko B, Gschösser F, Cannone Falchetto A, Bueno M, Arraigada M, Sousa J, Ruiz R, Pettit C, Loizidou M, Partl M (2017) Harvesting the unexplored potential of European waste materials for road construction. Resour Conserv Recycl 116:32–44

Sofilić AMT (2010) Characterization of the EAF steel slag as aggregate for use in road construction. In: CISAP4, 4th international conference on safety & environment in process industry, vol 19, pp 117–123

Tahmoorian F, Samali B, Yeaman J, Crabb R (2018) The use of glass to optimize bitumen absorption of hot mix asphalt containing recycled construction aggregates. Materials 11 (7):1053

Walther A (2016) Performance evaluation of HMA with high RAP contents using Rejuvenators. International Society for Asphalt Pavements, USA

Walther A, Wistuba MP (2014) Auswirkung der Wiederverwendung von Ausbauasphalt in Asphalttragschichten auf mechanische Eigenschaften. Project Report, TU Braunschweig ISBS. (in German)

Wang D, Cannone Falchetto A, Moon KH, Riccardi C, Pei JZ, Wen Y (2019) Experimental investigation on the use of re-recycled reclaimed asphalt pavement (RAP) in asphalt mixtures. Environ Sci Pollut Res. (under review)

Yoshikane T (1995) Investigation of deterioration of recycled hot-mixed asphalt concrete pavement and a trial re-recycling of asphalt concrete. In: Disposal and recycling of organic and polymeric construction materials: proceedings of the international RILEM workshop, vol 27. CRC Press, pp 210–221

You Z, Mills-Beale J, Fini E, Goh SW, Colbert B (2011) Evaluation of low-temperature binder properties of warm-mix asphalt, extracted and recovered RAP and RAS, and bioasphalt. J Mater Civil Eng 23(11):1569–1574

Influence of Crumb Rubber Added by Dry Process on Linear Viscoelastic Properties and Tensile Strength of Bituminous Mixtures

Yasmina Mahmoudi[1(✉)], Salvatore Mangiafico[1], Cédric Sauzéat[1], Hervé Di Benedetto[1], Simon Pouget[2], and Jean-Philippe Faure[3]

[1] Université de Lyon/ENTPE/LTDS (UMR CNRS 5513), 3 Rue Maurice Audin, 69120 Vaulx-en-Velin, France
yasmina.mahmoudi@entpe.fr
[2] Eiffage Infrastructures, 8 Rue du Dauphiné, 69960 Corbas, France
[3] Aliapur, 71 Cours Albert Thomas, 69003 Lyon, France

Abstract. In this study, the thermomechanical behaviour of bituminous mixtures produced with CR by dry process was evaluated. Two bituminous mixtures were produced and tested: a reference one without CR and a second one with 3% of CR on the mixture weight. The used CR had a continuous 3/6 mm grain size distribution. Complex modulus was determined using tension/compression tests with sinusoidal loading, at nine temperatures from −25 °C to 50 °C and eight frequencies from 0.003 to 10 Hz with an imposed strain amplitude of 50 µm/m. For both mixtures, Time-Temperature Superposition Principle was validated. At high frequency/low temperatures, lower values of norm of complex modulus were found for the mixture containing CR. Tensile strength was determined using direct tensile tests on two samples at 10 °C and 3.125×10^{-2} %/min imposed strain rate.

Keywords: Crumb rubber · Dry process · Bituminous mixture · Linear viscoelastic behaviour · Complex modulus · Tensile strength

1 Introduction

One of the techniques to recycle end-of-life tires consists in recovering rubber from them and incorporating it into bituminous mixtures. Two main processes may be used.

In the wet process, Crumb Rubber (CR), usually in the form of fine particles, is added to the bitumen to act as a modifier (Caltrans 2006; Lo Presti 2013). In the dry process, CR, usually in the form of coarse particles (in order to reduce its specific surface area), is added as an aggregate and interactions between the binder and the particles of CR are neglected. However, some studies highlighted the possibility that interactions occur even in the dry process (López-Moro et al. 2013; Shen, Li and Xie 2017). In the literature, previous researches show contrasted results about mechanical performances of bituminous mixtures with CR added by dry process. Xie and Shen (2016) observed higher fatigue and rutting resistances of mixtures with CR added by dry process in comparison with mixtures without CR. However, the reported performances remain poorer than those of bituminous mixtures modified with polymers or

M. Pasetto et al. (Eds.): ISAP APE 2019, LNCE 48, pp. 174–182, 2020.
https://doi.org/10.1007/978-3-030-29779-4_17

with CR added by wet process. On the contrary, other authors observed an improvement of several performances such as resistances to rutting, fatigue, long and short-term ageing and low temperature cracking with respect to bituminous mixtures modified with polymers or bituminous mixtures with CR added by wet process (Cao 2007; Feiteira Dias et al. 2014; Moreno et al. 2013; Neto et al. 2005; Rahman et al. 2005).

The objective of the presented work is to analyse the effect of the incorporation of CR by dry process on the linear viscoelastic properties of a bituminous mixture. This study is part of a PhD thesis in the context of a collaboration between University of Lyon/École Nationale des Travaux Publics de l'État (ENTPE) and two companies: Aliapur and Eiffage Infrastructures.

2 Tested Materials

Two different bituminous mixtures were produced. The first one, used as the reference material, is a BB5®-type mixture (NF EN 13108-1 2007). The second mixture is the same BB5® in which 3% of coarse aggregates (on the mixture weight) are replaced by 3% of CR (on the mixture weight) with a 3/6 mm grain size distribution and a specific weight of 1200 kg/m^3. The same 35/50 pen grade bitumen modified with SBS was used for both mixtures. Figure 1 shows the volumetric grading curves of all aggregates

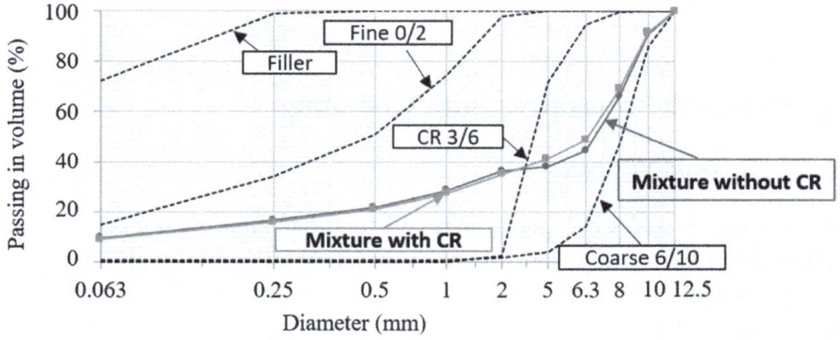

Fig. 1. Volumetric aggregate gradation curve for the two studied bituminous mixtures.

Table 1. Composition of the two studied bituminous mixtures with and without CR.

		Coarse aggregates 6/10 mm	Fine aggregates 0/2 mm	Filler	Bitumen	CR 3/6 mm
Mixture without CR	Mass content (% of total mass)	62.0	27.0	5.90	5.10	/
	Volume content (% of total volume)	56.5	25.4	5.51	12.6	/
Mixture with CR	Mass content (% of total mass)	59.0	27.0	5.90	5.10	3.0
	Volume content (% of total volume)	52.0	24.5	5.30	12.1	6.1

fractions as well as the volumetric grading curves obtained for the two bituminous mixtures. The proportions of every component used in the production of each mixture are presented in Table 1.

To produce the mixture without CR, aggregates and bitumen were heated at 180 °C and mixed during 120 s. A similar procedure was followed to produce the mixture with CR, except that CR was added to the aggregates prior to the addition of bitumen. A LPC-type wheel compactor was used to produce a 600 × 400 × 150 mm^3 slab of each mixture, according to NF EN 12697-33+A1 (2007). Twelve cylindrical samples were cored and sawn from each slab. Their dimensions were 150 mm in length and 75 mm in diameter. Air void content of each specimen was geometrically determined. Four specimens with similar air void contents were chosen for complex modulus tests: two specimens for the mixture without CR, called S1 and S2, and two specimens for the mixture with CR, called CR1 and CR2. Two samples with similar air void contents were chosen for tensile tests: sample S3 for the mixture without CR and sample CR3 for the mixture with CR. Table 2 shows air void contents of tested specimens.

Table 2. Air void content of tested specimens.

Tested specimen	CR1	CR2	S1	S2	CR3	S3
Air void content (%)	6.06	5.67	6.25	5.94	4.90	5.05
	Complex modulus tests				Tensile tests	

3 Experimental Procedures and Modelling

3.1 Complex Modulus Tests

Complex modulus tests were performed by applying sinusoidal cyclic loading in Tension/Compression with a 50 μm/m imposed axial strain amplitude, using a hydraulic press associated with a thermal chamber for temperature control. Samples were loaded at nine temperatures (−25 °C, −15 °C, −5 °C, 5 °C, 15 °C, 25 °C, 35 °C, 45 °C and 50 °C) and eight frequencies (0.003 Hz, 0.01 Hz, 0.03 Hz, 0.1 Hz, 0.3 Hz,

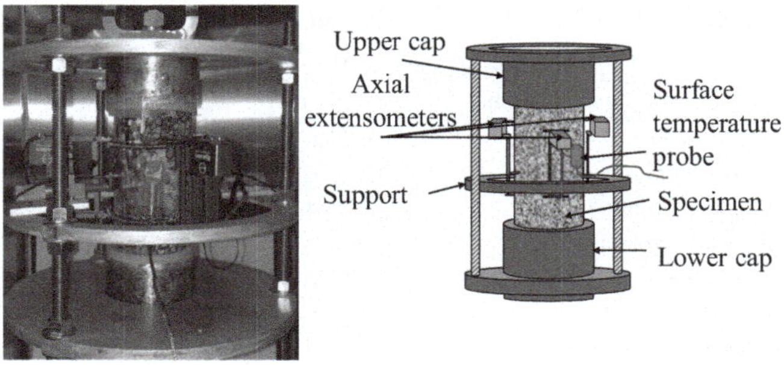

Fig. 2. Picture and scheme of a specimen instrumented.

1 Hz, 3 Hz and 10 Hz). Three extensometers were placed at 120° around the samples and used to monitor axial strain ε_1. Axial stress σ was calculated based on the force measured by a load cell embedded in the press. A surface temperature probe measured the temperature of surface of specimens. Figure 2 shows a picture and a scheme of an instrumented sample.

Expressions of axial strain ε_1, axial stress σ are presented as follows in Eqs. (1)–(2):

$$\varepsilon_1(t) = \varepsilon_{01}\sin(\omega t) = Im\left[\varepsilon_{01}e^{i(\omega t)}\right] \tag{1}$$

$$\sigma(t) = \sigma_0\sin(\omega t + \varphi_E) = Im\left[\sigma_0 e^{i(\omega t + \varphi_E)}\right] \tag{2}$$

where ε_{01} is the axial strain amplitude, ω is the angular frequency linked to the frequency f by $\omega = 2\pi f$, i is the imaginary unit defined by $i^2 = -1$, σ_0 is the stress amplitude, φ_E the phase shift between axial strain and axial stress.

Complex modulus E^* is defined as in Eq. (3):

$$E^* = \frac{\sigma_0 e^{i(\omega t + \varphi_E)}}{\varepsilon_{01}e^{i(\omega t)}} = |E^*|e^{i\varphi_E} = E_1 + iE_2 \tag{3}$$

3.2 2S2P1D Model

Experimental data were fitted with 2S2P1D model (Olard and Di Benedetto 2003), an analogical linear viscoelastic model composed of two springs, two parabolic elements and one linear dashpot (Fig. 3).

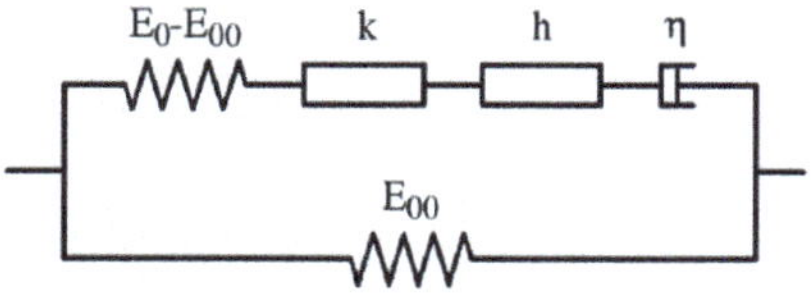

Fig. 3. 2S2P1D analogical model.

Equation (4) gives the expression of complex modulus according to 2S2P1D model for a given combination of temperature T and ω:

$$E^*(\omega, T) = E_{00} + \frac{E_0 - E_{00}}{1 + \delta(i\omega\tau(T))^{-k} + (i\omega\tau(T))^{-h} + (i\beta\omega\tau(T))^{-1}} \tag{4}$$

where E_0 is the glassy asymptotic modulus for $\omega \to \infty$, E_{00} is the static asymptotic modulus, for $\omega \to 0$, k, h and δ are dimensionless constants, $\tau(T)$ is a characteristic time depending on temperature and β is a constant linked with the Newtonian viscosity η, defined by Eq. (5):

$$\eta(T) = (E_0 - E_{00})\beta\tau(T) \tag{5}$$

If the Time-Temperature superposition principle is validated, master curves of E^* can be obtained at a given reference temperature T_{REF} by shifting isotherms according to temperature shift factors a_T, calculated as shows Eq. (6):

$$a_T(T) = \frac{\tau(T)}{\tau(T_{REF})} \tag{6}$$

Temperature shift factors can be modelled as functions of temperature using the WLF equation Eq. (7), depending on constants C_1 and C_2 and on reference temperature T_{REF} (Williams, Landel and Ferry 1955):

$$\log(a_T) = -\frac{C_1(T - T_{REF})}{C_2 + T - T_{REF}} \tag{7}$$

where a_T is a shift factor at temperature T for the reference temperature T_{REF}.

3.3 Direct Tensile Tests

Direct tensile tests were performed on samples CR3 and S3 in order to determine tensile strength at 10 °C and 3.125×10^{-2} %/min imposed strain rate. The same equipment and test configuration as for complex modulus tests was used. The procedure of the tests consisted in applying a constant strain rate until failure of the sample, while monitoring the corresponding tensile stress.

4 Results and Analysis

4.1 Complex Modulus

Experimental data of complex modulus obtained for all the tested samples are presented in Figs. 4, 5 and 6. A good repeatability of experimental results of complex modulus is observed for both mixtures.

Table 3. 2S2P1D and WLF parameters for tested samples.

At any T_{REF}							At T_{REF} = 15 °C		
Sample	E_{00} (MPa)	E_0 (MPa)	δ (-)	k (-)	h (-)	β (-)	τ (s)	C_1 (-)	C_2 (°C)
S1	28	3.23×10^4	1.95	0.174	0.56	300	1.19×10^{-1}	28.1	180.3
S2	28	3.23×10^4	1.95	0.174	0.56	300	1.31×10^{-1}		
CR1	21	2.95×10^4	1.82	0.158	0.52	1000	8.28×10^{-2}	31.9	202.9
CR2	21	2.95×10^4	1.82	0.158	0.52	1000	9.16×10^{-2}		

Time-Temperature Superposition Principle (TTSP) is verified for both mixtures, as unique curves are obtained for complex modulus in Cole-Cole plane and Black space (Fig. 4). Very close values of shift factors were used for both samples of each mixture at all temperatures. Therefore, only one set of WLF constants was used to fit a_T values of both samples of the same material (Table 3). Furthermore, similar shift factors were obtained for the two mixtures, with and without CR. As an example, shift factors obtained for samples S1 and CR1 are shown in Fig. 5. Master curves of $|E^*|$ and φ_E at 15 °C (obtained using shift factors calculated according to the WLF equation) are plotted in Fig. 6a and b for all tested samples.

At high frequencies/low temperatures, the mixture with CR shows lower values of $|E^*|$ than the mixture without CR. Similar values of φ_E are observed for both mixtures. 2S2P1D model fits well experimental data of E^*. For each mixture, identical values of 2S2P1D constants were used for the two tested specimens, except for τ, which shows very slight differences. Values of glassy modulus E_0 reflect the previous observation, as lower values are obtained for the mixture with CR compared to the mixture without CR.

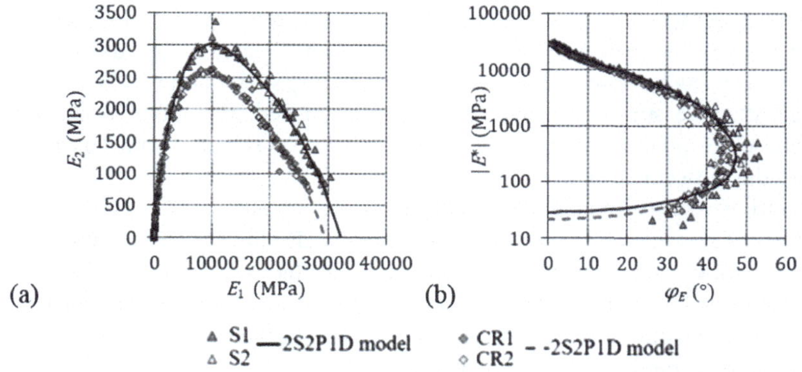

Fig. 4. Experimental results and 2S2P1D fitting of **E*** for all tested samples (S1 and S2 without CR, CR1 and CR2 with CR): (a) Cole-Cole plane, (b) Black space.

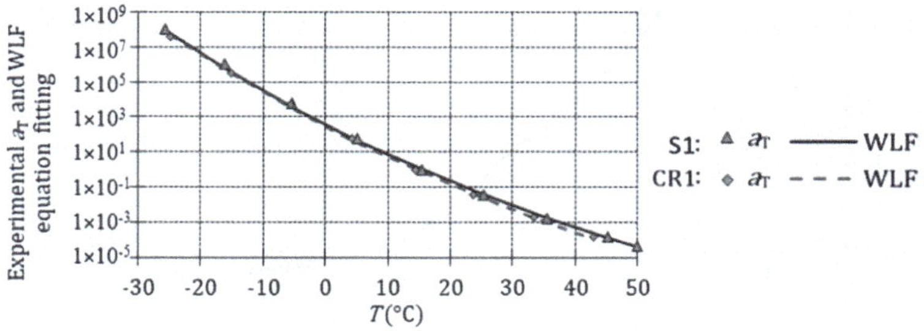

Fig. 5. Examples of experimental temperature shift factors and WLF modelling for samples S1 (at $\mathbf{T_{REF}}$= 15.49 °C) and CR1 (at $\mathbf{T_{REF}}$ = 14.48 °C).

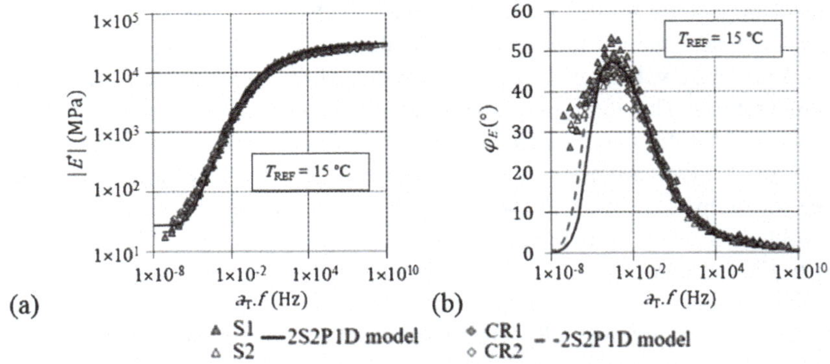

Fig. 6. Experimental results and 2S2P1D fitting of $\mathbf{E}^*$ for all tested samples (S1 and S2 without CR, CR1 and CR2 with CR): Master curves of (a) $|\mathbf{E}^*|$ and (b) φ_E at $\mathbf{T_{REF}}$ = 15 °C.

4.2 Tensile Strength

Results of direct tensile tests on samples CR3 and S3 are shown in Fig. 7 and Table 4. Sample CR3 suffered from a premature detachment from the cap, during the tests. However, as it can be observed from the corresponding stress-strain curve, this problem occurred after the stress peak had already been reached. For this reason, only peak stress and corresponding strain were used as tensile resistance parameters in order to compare the performances of the two samples.

According to experimental data, peak stress of specimen CR3 is 14.2% lower than peak stress of specimen S3. Specimen CR3 exhibits also lower values of corresponding strain with respect to specimen S3. Hence, these results suggest that tensile strength of the specimen with CR is lower in comparison with the specimen without CR. Considering the premature detachment of sample CR3, further tests will be necessary to validate the repeatability of experimental results.

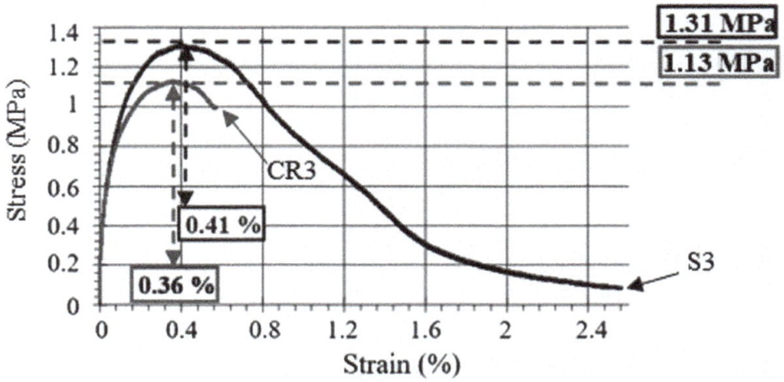

Fig. 7. Experimental results of direct tensile tests performed on specimens CR3 and S3.

Table 4. Values of peak stress and corresponding strain obtained by direct tensile tests on specimens CR3 and S3.

Tested specimen	Peak stress (MPa)	Corr. strain (%)
CR3	1.13	0.36
S3	1.31	0.41

5 Conclusions

The aim of this study is to investigate the effect of CR addition by dry process on LVE behaviour and tensile strength of a bituminous mixture. Two mixtures were produced, a reference one without CR and a similar one, with 3% CR (by weight of the mixture) replacing the same mass of aggregates. Complex modulus tests were carried out on two samples of each mixture, at several temperatures and loading frequencies. From the experimental results obtained, the following conclusions can be drawn. The time-temperature superposition principle is validated for both mixtures, with and without CR. At low temperature/high frequency, lower values of norm of complex modulus were found for the mixture with CR compared to the mixture without CR. 2S2P1D model successfully fitted experimental data of complex modulus of both mixtures. Lower values of asymptotic glassy modulus E_0 were obtained for the mixture with CR. Direct tensile tests were performed on two samples, one of each mixture, at 10 °C and with 3.125×10^{-2} %/min imposed strain rate. Experimental results show lower tensile strength for the specimen with CR in comparison with tensile strength of the specimen without CR.

Acknowledgements. This work is supported by the French Environment and Energy Management Agency (ADEME).

References

Caltrans (2006) Asphalt Rubber Usage Guide: State of California Department of Transportation, Materials Engineering and Testing Services

Cao W (2007) Study on properties of recycled tire rubber modified asphalt mixtures using dry process. Constr Build00 Mater 21(5):1011–1015. https://doi.org/10.1016/j.conbuildmat.2006.02.004

Feiteira Dias JL, Picado-Santos LG, Capitão SD (2014) Mechanical performance of dry process fine crumb rubber asphalt mixtures placed on the Portuguese road network. Constr Build Mater 73:247–254. https://doi.org/10.1016/j.conbuildmat.2014.09.110

Lo Presti D (2013) Recycled tyre rubber modified bitumens for road asphalt mixtures: a literature review. Constr Build Mater 49:863–881. https://doi.org/10.1016/j.conbuildmat.2013.09.007

López-Moro FJ, Moro MC, Hernández-Olivares F, Witoszek-Schultz B, Alonso-Fernández M (2013) Microscopic analysis of the interaction between crumb rubber and bitumen in asphalt mixtures using the dry process. Constr Build Mater 48:691–699. https://doi.org/10.1016/j.conbuildmat.2013.07.041

Moreno F, Sol M, Martín J, Pérez M, Rubio MC (2013) The effect of crumb rubber modifier on the resistance of asphalt mixes to plastic deformation. Mater Des 47:274–280. https://doi.org/10.1016/j.matdes.2012.12.022

NF EN 12697-33+A1 (2007) Bituminous mixtures - Test methods for hot mix asphalt - Part 33: specimen prepared by roller compactor

NF EN 13108-1 (2007) Bituminous mixtures - Material specifications - Part 1: asphalt concrete

Olard F, Di Benedetto H (2003) General 2S2P1D model and relation between the linear viscoelastic behaviours of bituminous binders and mixes. Road Mater Pavement Des 4 (2):185–224. https://doi.org/10.3166/rmpd.4.185-224

Rahman M, Airey GD, Collop AC (2005) The mechanical properties of the dry process CRM asphalt mixtures following short-term and long-term ageing. Presented at the 7th international conference on the bearing capacity of roads and airfields, Trondheim, Norway

Neto SAD, Farias M, Mello LGR, Pereira PAA, Pais JC (2005) The use of crumb rubber in asphalt mixtures using the dry process. Presented at the international symposium on pavement recycling, Sao Paulo, Brasil

Shen J, Li B, Xie Z (2017) Interaction between crumb rubber modifier (CRM) and asphalt binder in dry process. Constr Build Mater 149:202–206. https://doi.org/10.1016/j.conbuildmat.2017.04.191

Williams ML, Landel RF, Ferry JD (1955) The temperature dependence of relaxation mechanisms in amorphous polymers and other glass-forming liquids. J Am Chem Soc 77 (14):3701–3707. https://doi.org/10.1021/ja01619a008

Xie Z, Shen J (2016) Performance properties of rubberized stone matrix asphalt mixtures produced through different processes. Constr Build Mater 104:230–234. https://doi.org/10.1016/j.conbuildmat.2015.12.063

A Preliminary Investigation into the Use of Alkali-Activated Blast Furnace Slag Mortars for High-Performance Pervious Concrete Pavements

Marco Bassani[1(✉)], Luca Tefa[1], and Paola Palmero[2]

[1] Department of Environment, Land and Infrastructure Engineering, Politecnico di Torino, 24, Corso Duca degli Abruzzi, 10129 Turin, Italy
marco.bassani@polito.it

[2] Department of Applied Science and Technology, Politecnico di Torino, 24, Corso Duca degli Abruzzi, 10129 Turin, Italy

Abstract. Pervious concrete (PC) pavements are water control systems for infiltrating storm water, reducing runoffs, contrasting heat-island effects and noise diffusion in urban environments. Their poor strength and stiffness, which stem from their porosity, limit their application for use in sidewalks and residential streets. To overcome this limitation, the research aimed at developing an alkali-activated blast furnace slag binder (AA-BFS) which, when added to the concrete mixture, would enable it to attain the characteristics required for application in trafficked road pavements. A Portland cement (CEM-I) and a blast furnace cement (CEM-III/C) were included for comparison purposes. Standard and high-performance pervious concrete (HPPC) mortars were tested with the same mix ratio of binder, sand, and admixtures. The effects of a latex polymer admixture were also analysed. After 2, 7, and 28 days of curing, AA-BFS mortars exhibited the greatest flexural and compressive strengths. As expected, CEM-III/C needed more time to achieve strengths comparable to those recorded for CEM-I. The addition of latex failed to result in any discernible benefits in terms of the stress-strain behaviour of mortars. The results confirm that AA-BFS is a sustainable option for HPPC production.

Keywords: Pervious concrete · Ordinary portland cement · Alkali-activated blast furnace slag · Geo-polymerization · Latex polymer admixture

1 Introduction

Porous concrete (PC) is enjoying increased popularity in urban pavement applications since it provides roadways with storm-water runoff retention capacity to counteract flooding, facilitates a replenishment of local groundwater sources, contrasts heat-island effects, and limits the diffusion of noise related traffic (Ahiablame et al. 2012; Bassani et al. 2017). A PC is made up of coarse aggregate particles wrapped in a binder paste or mortar which includes functional admixtures to enhance internal cohesion. Its properties are strongly affected by the volume of voids, which represents between 15 and

M. Pasetto et al. (Eds.): ISAP APE 2019, LNCE 48, pp. 183–192, 2020.
https://doi.org/10.1007/978-3-030-29779-4_18

35% of the content thanks to the low or negligible amount of sand in the material. Consequently, there are maintenance issues with the use of PC in urban road pavements due to ravelling and clogging (Gupta 2014).

The use of PC is consistent with policies promoting sustainable infrastructures (Dolan et al. 2006). However, its diffusion is limited to pedestrian paths, bikeways, residential streets, and parking lots (Ferguson 2005). Due to its pervious nature, PC has lower flexural and compressive strength values, and is less resistant to fatigue cracking when compared to conventional concrete. To overcome these limitations, some have tried to improve its properties so that it may be used in heavy traffic and cold weather conditions. This aim has been pursued by adding functional admixtures and fibres (Bhutta et al. 2012). Among admixtures, latex serves to improve the bonds between the cement paste and aggregates, to increase strength and to control the volume of permeable voids (Shu et al. 2011; Huang et al. 2010). The addition of solid latex to cement in a ratio of 10% in weight terms permits a reduction in the water/cement (w/c) ratio, which in turn increases the tensile strength (Wang et al. 2006). Latex slows down the hardening process enhancing the resistance to freeze-thaw action (Kevern et al. 2011).

A number of studies have demonstrated the high strength values of alkali-activated (AA) fly ash (FA), blast furnace slag (BFS), cement kiln dust, stone cutting powders, and other by-products (Khale and Chaudhary 2007; Palmero et al. 2017). The use of AA binders includes a thermal treatment during curing to accelerate the geo-polymerization process (Alonso and Palomo 2001) and to develop adequate strength (Görhan and Kürklü 2014). However, for sustainable construction procedures, thermal treatments should be avoided. With this in mind, Krizan and Zivanovic (2002) investigated mortars including BFS and alkaline solution (AS) in proportions reported in EN 196-1. After 28 days of curing at room temperature, the compressive strength ranged from 65.0 to 80.0 MPa depending on the concentration of the AS.

In the specific context of PC, the recourse to AA binders is gaining interest due to the benefits deriving from a large-scale reduction in CO_2 emissions, and the reuse of by-products otherwise destined for landfill sites (Provis 2014). However, most of the investigations into PC were conducted using ordinary Portland cements (Ferguson 2005), while only a few works considered AA binders (Sata et al. 2013; Zaetang et al. 2015). At present, none has compared the performance of ordinary and AA binders in PC. This investigation focused on the strength of mortars made with AA-BFS as a binder. CEM-I and CEM-III/C were considered for comparison purposes. The first is a Portland cement of 32.5 MPa class, the second is a blast-furnace cement. The pros and cons resulting from the addition of a latex admixture to the mortar were considered specifically in terms of strength. The mortars were designed with a binder, sand, water, and admixture mix ratio suitable for the preparation of a HPPC.

2 Materials and Methods

CEM-I 32.5 R includes more than 95% of clinker, while the CEM-III/C is a blast furnace cement obtained by mixing CEM-I (15 wt%) and ground granulated BFS (85 wt%) according to EN 197-1. CEM-I is widely used in the preparation of PC

(Ibrahim et al. 2014), while CEM-III/C was considered here to compare the properties of hydrated and AA-BFS.

Grain size distribution (determined by Fristch Analysette 22 laser granulometer) and chemical composition values for CEM-I, CEM-III/C and BFS are shown in Fig. 1a and Table 1 respectively. CEM-I has the finest distribution. The most relevant differences in oxide content are related to silica (SiO_2) and calcium oxide (CaO), the former higher in BFS than CEM-I, the second higher in CEM-I. BFS, unlike CEM-I, also contains magnesium carbonate (as indicated by the presence of MgO and CO_2). Figure 1b shows the grain size distribution of the natural river sand (mix-design) and the siliceous one for standard mortars (EN 196-1 sand).

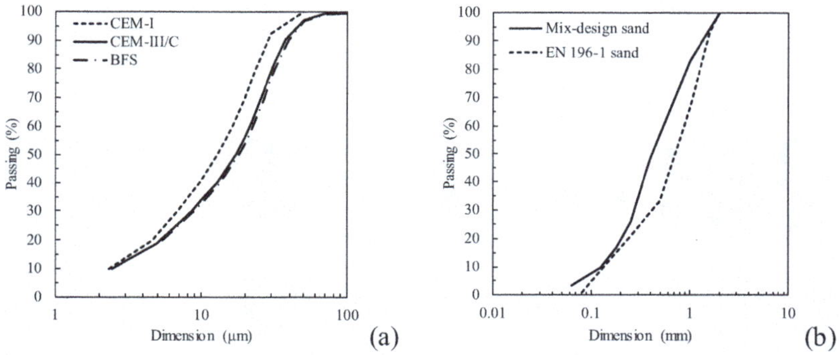

Fig. 1. (a) Particle size distribution of CEM-I, CEM-III/C, BFS; size distribution of mix-design and standard (EN 196-1) sands

The latex, a stable dispersion of a synthetic styrene-butadiene elastomer in water, was added at 7.0 wt% with respect to the binder. It has a density of 1.02 g/cm^3, a pH of 8, and a dry solid content of 36%. The alkaline solution (AS) was obtained by combining 80 wt% of sodium silicate (Na_2SiO_3) and 20 wt% of a sodium hydroxide (NaOH) solution (50 wt%, 25 M). The sodium silicate was already in liquid form and had a SiO_2/Na_2O mass ratio of 3.4 (i.e., SiO_2 = 28.1%, Na_2O = 8.4%, H_2O = 63.5%) and a pH of 11.6.

When deciding on the binder, aggregate and liquid mix ratio for the investigated PC, consideration was given to typical PC mixture proportions proposed in literature (Huang et al. 2010; Arhin and Madhi 2014). According to ACI 522R-10, typical ranges are: 270–415 kg/m^3 of cement, 1190–1480 kg/m^3 of coarse aggregate, a limited or zero amount of fine aggregate, a 0.27–0.34 w/c mass ratio, and a 4.0–4.5 aggregate/binder mass ratio (a/b).

The optimal mix ratio is when the mortar fully covers the aggregate but does not flow away. Tennis et al. (2004) indicated that this condition is obtained when a small amount of fresh PC mixture tightened by hand does not crumble or lose voids between grains, and the paste does not stick to the hand itself. Figure 2 illustrates the four PC mixture consistency levels adopted in this investigation.

Table 1. Chemical composition (mass percentage, only values larger than 1) of CEM-I, CEM-III/C and BFS (CEM III/C composition is the weighted average of its constituents, CEM-I and BFS)

Oxides	CEM-I	CEM-III/C	BFS
Al_2O_3	3.91	8.44	9.24
SiO_2	18.40	29.11	31.00
SO_3	3.71	1.99	1.69
K_2O	1.26	0.50	0.37
CaO	67.60	43.21	38.90
Fe_2O_3	5.07	1.67	1.07
SrO	0.04	0.07	0.08
CO_2	-	8.47	9.97
MgO	-	5.53	6.50

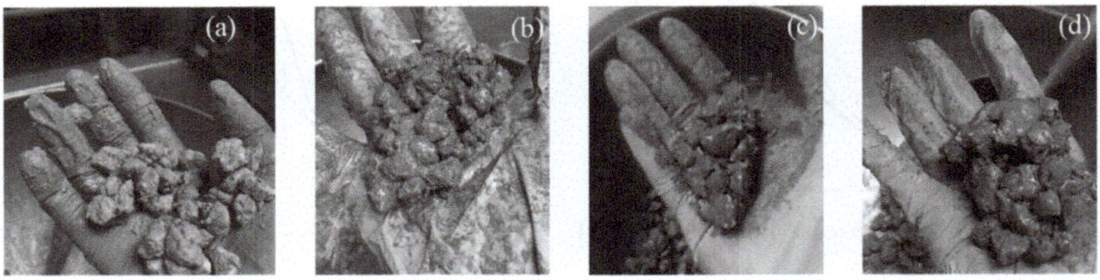

Fig. 2. (a) "poor", (b) "nearly good", (c) "good", and (d) "excellent" PC mixture consistencies

HPPC mortars were optimized for consistency by differentiating between mortars with (mix A) and without (mix B) latex. A viscosity modifying admixture (VMA) was added to provide cohesion and prevent segregation and bleeding (Aïtcin and Flatt 2016). A polycarboxylate-ether (PCE) based superplasticizer to improve workability, and a calcareous aggregate of size fraction 4–16 mm (particle density of 2.773 Mg/m^3, water absorption of 0.1% according to EN 1097-6) were used. Table 2 reports the CEM-I mix ratio results for several attempts depending on the water to binder (w/b), coarse aggregate to binder (a/b), sand to coarse aggregate (s/a), latex to binder (l/b), VMA to binder (VMA/b), and superplasticizer to binder (PCE/b) mass ratios. The optimal proportions were assessed for HPPC containing CEM-I binder only and adopted for the other two mortars. From the results in Table 2, it is evident that mixtures A-10 and B-3 have the optimal mix ratio. Table 3 summarizes both standard and HPPC mortar mix proportions. The 12 different mortars included two compositions (standard, HPPC), three binders (CEM-I, CEM-III/C, AA-BFS), and two mix types (A, B). As a result, a total of 180 samples were prepared and tested (12 mortars × 5 replicates × 3 curing times).

It is worth noting that mortars with CEM-III/C and AA-BFS would have needed different constituent mix ratios to comply with the consistency requirements (Fig. 2). However, for comparison purposes, the same proportions for CEM-I were adopted.

Table 2. Consistency of HPPC mixtures made with CEM-I (w = water, b = binder, a = coarse aggregate, s = sand, l = latex, VMA = viscosity modifying admixture, PCE = polycarboxylate-ether-based superplasticizer)

Mixture	w/b	a/b	s/a	l/b	VMA/b	PCE/b	Consistency
A-1	0.35	4.4	0.05	0.07	0.005	0.010	Poor
A-2	0.35	4.4	0.06	0.07	0.005	0.010	Poor
A-3	0.35	4.4	0.07	0.07	0.005	0.010	Poor
A-4	0.35	4.4	0.08	0.07	0.005	0.010	Nearly good
A-5	0.33	4.3	0.05	0.07	-	-	Poor
A-6	0.33	4.3	0.06	0.07	-	-	Poor
A-7	0.33	4.3	0.07	0.07	-	-	Nearly good
A-8	0.33	4.3	0.07	0.07	0.005	0.010	Poor
A-9	0.28	4.4	0.05	0.07	0.005	-	Good
A-10	**0.32**	**4.4**	**0.05**	**0.07**	**0.005**	**0.010**	**Excellent**
A-11	0.26	4.3	0.07	0.07	-	-	Poor
B-1	0.35	4.4	0.05	-	0.005	0.010	Good
B-2	0.35	4.3	0.07	-	0.009	0.009	Good
B-3	**0.35**	**4.4**	**0.06**	-	**0.005**	**0.010**	**Excellent**
B-4	0.35	4.4	0.07	-	0.005	0.010	Poor
B-5	0.33	4.3	0.05	-	-	-	Poor
B-6	0.33	4.3	0.05	-	-	0.010	Poor
B-7	0.33	4.3	0.05	-	0.005	0.010	Poor
B-8	0.33	4.3	0.06	-	0.005	0.010	Good
B-9	0.33	4.3	0.07	-	0.005	0.010	Nearly good

Table 3. Mix ratios for mortars designed for HPPC (for significance of symbols see Table 2)

Mortar type	Mix code	Latex	w/b	s/b	l/b
Standard (EN 196-1)	A	Yes	0.45	3.0	0.07
	B	No	0.50	3.0	0.00
Designed for HPPC	A	Yes	0.32	0.22	0.07
	B	No	0.35	0.26	0.00

VMA and PCE-based plasticizer were used only in the mix-design phase. Fresh mortars were cast in 80 × 20 × 20 mm prismatic moulds. All samples were demoulded after 24 h of curing (95% r.h, room temperature).

CEM-I and CEM-III/C mortars were immersed in water, while AA-BFS samples continued their curing in a humidity chamber at room temperature. In this work, AA-BFS samples were not immersed in water in order to avoid both the partial dissolution of unreacted species during the initial curing stages, and the lowering of the OH- concentration in the pore solution with a reduction in alkali activation potential (Naghizadeh and Ekolu 2017). Although different curing conditions could have had an influence on mechanical strength, each material underwent an optimal curing process.

Maximum flexural ($\sigma_{f,max}$) and compressive ($\sigma_{c,max}$) stress at failure values were determined at 2, 7, and 28 days. Flexural strength was measured in a three-point bending configuration, while compressive strength was tested on the two residual pieces according to EN 196-1 (constant strain rate of 0.25 mm/min) with a 50 kN testing machine.

3 Results and Discussion

Figure 3 includes the results for all flexural (Fig. 3a) and compressive (Fig. 3b) strength tests. Regarding standard mortars, CEM-I always provided the highest $\sigma_{f,max}$ value irrespective of curing time. For instance, 2-day cured specimens exhibited average values of 5.3 and 5.8 MPa, with or without latex respectively. At this curing time, mortars with CEM-III/C developed little flexural strength (0.5 MPa with latex and 0.9 MPa without latex). After 7 days, CEM-III/C achieved the $\sigma_{f,max}$ of AA-BFS mortars. In particular, those without latex exhibited slightly higher $\sigma_{f,max}$ values (4.4 MPa for CEM-III/C, 3.7 MPa for AA-BFS). The same behaviour was evident after 28 days: $\sigma_{f,max}$ values for standard mortar AA-BFS with latex were higher than CEM-I, independently of curing time. This trend is reversed when no latex is included in the mixture with mortars containing CEM-III/C always exhibiting the lowest $\sigma_{f,max}$.

Figure 3a provides evidence that HPPC mortars exhibited a different behaviour to standard mortars. Values of $\sigma_{f,max}$ for CEM-III/C and AA-BFS were, on average, higher than those for standard mortars. In contrast, the $\sigma_{f,max}$ for CEM-I HPPC were, on average, 18% lower than standard mortars.

The addition of latex to HPPC mortars did not lead to improvements in strength. HPPC mortars with CEM-I and latex showed lower $\sigma_{f,max}$ and $\sigma_{c,max}$ than those without latex, independently of curing time. In the case of CEM-III/C, differences in $\sigma_{c,max}$ (Fig. 3b) and $\sigma_{f,max}$ (Fig. 3a) between mix A and B are limited. 2-and 7-day cured specimens of HPPC mortars with latex had slightly lower strengths. Vice versa, 28-day cured mortars containing CEM-III/C and latex exhibited average $\sigma_{c,max}$ and $\sigma_{f,max}$ of 38.1 and 7.9 MPa, while those without latex reached 37.9 and 7.8 MPa respectively. A similar mechanical behaviour was recorded for AA-BFS mortars, with higher strengths in the case of mix A and mix B at 2 and 7 days of curing. On the 28th day the average value of $\sigma_{c,max}$ in AA-BFS mortars with latex exceeded the corresponding value for mortars without latex (70.8 MPa in mix A versus 68.4 MPa in mix B).

To determine whether the differences between the average strength ($\sigma_{c,max}$ and $\sigma_{f,max}$) values of different mortars were statistically significant, several t-tests were carried out with a significance level of 0.05. Results show that in almost all cases the null hypothesis H_0: $\mu_{\sigma X} = \mu_{\sigma Y}$ (with X and Y representing two generic set of samples) was rejected, thus demonstrating that two average values are statistically different.

The results demonstrate that AA-BFS with HPPC proportions attained the highest compressive and flexural strength values independently of curing time and the presence of latex. This is consistent with the findings of Krizan and Zivanovic (2002), who recorded higher compressive strength values in AA-BFS mortars with sodium silicate only and a w/b ratio of 0.43 compared to the values for reference mortars made with

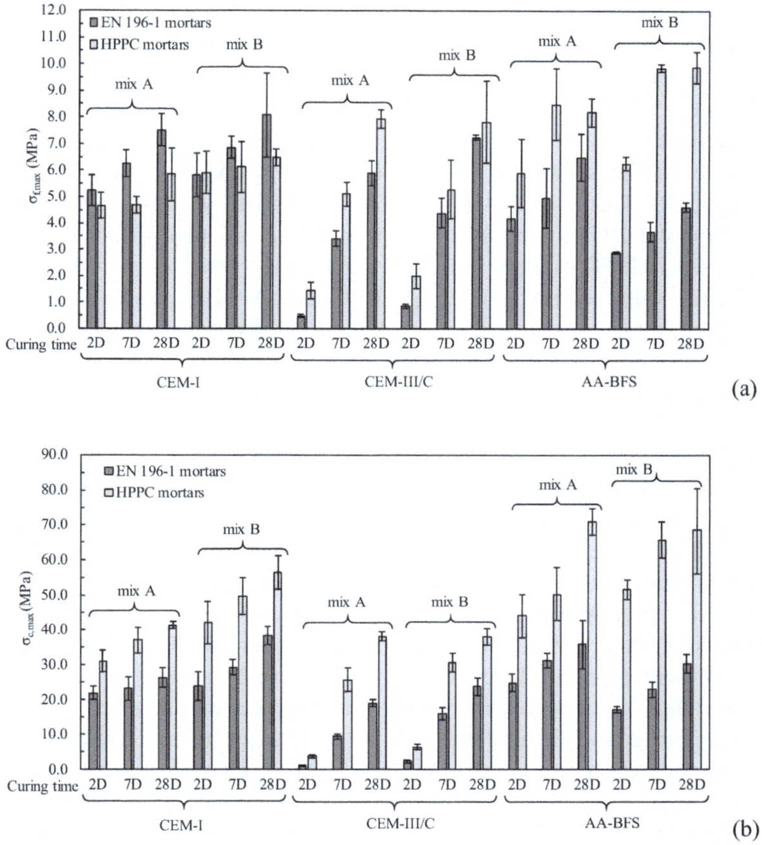

Fig. 3. Flexural (a) and compressive (b) strength values after 2, 7 and 28 days (D) of curing for the 12 mortars. Error bars indicate one standard deviation around the average

Portland cement. The results obtained here support the hypothesis that AA-BFS can be used to widen the field of PC applications to scenarios where ordinary Portland cements may fail due to its limited strength.

After 7 days of curing, the $\sigma_{f,max}$ values for AA-BFS mortars with latex were 45% and 40% higher than the values for CEM-I and CEM-III/C mortars. In the case of 7-day cured HPPC mortars without latex, those with AA-BFS exhibited an average value of $\sigma_{f,max}$ equal to 9.9 MPa in comparison to the 6.1 and 5.3 MPa values for mortars with CEM-I and CEM-III/C respectively. After 28 days of curing the gap between AA-BFS and CEM-III/C mortars decreased ($\sigma_{f,max}$ of AA-BFS 4% and 21% greater than CEM-III/C with and without latex respectively), while mortars with CEM-I showed the lowest values of $\sigma_{f,max}$. It is worth noting that the flexural strength of concrete is a fundamental parameter controlling pavement performance and durability. The higher the $\sigma_{f,max}$, the more durable and resistant the pavement with respect to traffic loads and temperature differentials.

All HPPC mortars without latex (mix B) exhibited a higher average $\sigma_{c,max}$ than those including latex (mix A), with the sole exception of AA-BFS cured for 28 days. The negative effect proved more marked in CEM-I and AA-BFS mortars, with only slight differences between mixes A and B noted in mortars containing CEM-III/C. The lower average density of the latex containing specimens is believed to be the main cause of lower strengths. Standard mortars of CEM-I, CEM-III/C, AA-BFS with latex showed an average density of 2181, 2087, and 2066 kg/m^3 in comparison to the values of 2230, 2196, and 1958 kg/m^3 for the corresponding mixtures without latex.

In the case of HPPC mortars in mix A, the average densities were equal to 2057, 1993, and 2079 kg/m^3 with CEM-I, CEM-III/C, and AA-BFS respectively, while the corresponding mortars without latex (mix B) exhibited density values equal to 2132, 2035, and 2096 kg/m^3.

Wang and Wang (2010) observed that both the apparent and dry bulk density of CEM-II 52.5 R mortars decreased with an increase in the content of the styrene-acrylic ester copolymer latex admixture. The authors pointed out that the addition of latex generally reduces both compressive and flexural strengths of cured specimens (3, 7 and 28 days), but improved the final toughness. Sumathy et al. (1997) also observed strength reductions in cement mortar specimens containing 10% (in weight) of latex. The authors explained the reduction in 19% of $\sigma_{c,max}$ in comparison with a reference mortar by citing the low probability of the polymer entering the matrix in the case of large amounts of binder. Zhong and Chen (2002) investigated CEM-II 52.5 R mortars including different types of latex: carboxylic styrene-butadiene rubber latex, vinyl chloride-vinylidene chloride copolymer latex, and styrene-acrylic ester latex. A reduction in compressive and flexural strength in comparison to a reference mortar was always observed. Ukrainczyk and Rogina (2013) estimated that the compressive strength of mortars containing calcium aluminate cement (a sand to cement ratio equal to 3) decreases by varying the amount of polymer latex admixture from 0% to 9% of cement mass. Lee et al. (2016) analysed both $\sigma_{c,max}$ and $\sigma_{f,max}$ of geo-polymeric mortars containing FA and BFS activated by an alkaline solution of NaOH and sodium silicate. They concluded that the addition of a styrene-butadiene latex failed to provide any benefits in terms of strength improvement. Latex causes a decrease of pH$^+$ in the fresh mixture, which resulted in a lower alkali-activation potential of raw powders and a reduction in $\sigma_{c,max}$.

4 Conclusions

In conclusion, the key findings of this investigation follows:

1. HPPC mortars always exhibited greater flexural and compressive strengths than standard mortars thanks to the lower w/b and s/b ratios;
2. mortars including AA-BFS exhibited higher compressive and flexural strengths than mortars made with CEM-I and CEM-III/C cements;
3. no benefits from the addition of latex in terms of compressive and flexural strength were observed, with the exception of AA-BFS mortars;

4. benefits associated with the addition of latex are not definitive, since its inclusion improves the cohesion between mortars and coarse aggregates. This fact needs to be properly investigated at the (upper) scale of HPPC mixtures.

One of the most important implications of this research is that there is still room for additional steps in the research on HPPC including AA-BFS, since the mix proportion which best complies with consistency requirements can be further modified and optimized. This shall be the aim of a future research investigation into the use of alternative binders suitable for the implementation and extension of PC technology.

This study is a preliminary laboratory work on the effects of binder type on the performance of pervious concrete (PC), with an emphasis on the strength properties of mortars following the inclusion of potential binders. Stiffness and strength of PC at the full concrete mix scale, as well as other material properties, are the subject of this work in progress.

Acknowledgements. The research performed by the two Departments of the Politecnico di Torino was partly funded by Regione Piemonte in the frame of WIN_STREET Project (*Water IN STReet design with Environmental Engineering Technologies for urbanized areas*, code F.E.S.R. 2007/2013).

References

Ahiablame LM, Engel BA, Chaubey I (2012) Effectiveness of low impact development practices: literature review and suggestions for future research. Water Air Soil Pollut 223(7):4253–4273

Aïtcin P-C, Flatt RJ (2016) Science and technology of concrete admixtures. Woodhead Publishing, Cambridge

Alonso S, Palomo A (2001) Alkaline activation of metakaolin and calcium hydroxide mixtures: influence of temperature, activator concentration and solids ratio. Mater Lett 47(1–2):55–62

Arhin SA, Madhi R (2014) Optimal mix designs for pervious concrete urban area. Int J Eng Res Technol 3(12):42–50

Bassani M, Tefa L, Comino E, Rosso M, Giurca F, Garcia Perez A, et al (2017) Environmental and engineering performance assessment of biofilters and retention systems for pavement stormwater. In: Transport infrastructure and systems: proceedings of the AIIT international congress on transport infrastructure and systems, Rome, Italy, 10–12 April 2017. CRC Press

Bhutta MAR, Tsuruta K, Mirza J (2012) Evaluation of high-performance porous concrete properties. Constr Build Mater 31:67–73

Dolan LM, Bohemen H, Whelan P, Akbar KF, O'Malley V, O'Leary G, Keizer PJ (2006) Towards the sustainable development of modern road ecosystems. In: The ecology of transportation: managing mobility for the environment. Springer, Dordrecht, pp 275–331

Ferguson BK (2005) Porous pavements. Taylor & Francis, Boca Raton

Görhan G, Kürklü G (2014) The influence of the NaOH solution on the properties of the fly ash-based geopolymer mortar cured at different temperatures. Compos B Eng 58:371–377

Gupta R (2014) Monitoring in situ performance of pervious concrete in British Columbia—a pilot study. Case Stud Constr Mater 1:1–9

Huang B, Wu H, Shu X, Burdette EG (2010) Laboratory evaluation of permeability and strength of polymer-modified pervious concrete. Constr Build Mater 24(5):818–823

Ibrahim A, Mahmoud E, Yamin M, Patibandla VC (2014) Experimental study on Portland cement pervious concrete mechanical and hydrological properties. Constr Build Mater 50:524–529

Kevern JT, Schaefer VR, Wang K (2011) Mixture proportion development and performance evaluation of pervious concrete for overlay applications. ACI Mater J 108(4):439–448

Khale D, Chaudhary R (2007) Mechanism of geopolymerization and factors influencing its development: a review. J Mater Sci 42(3):729–746

Krizan D, Zivanovic B (2002) Effects of dosage and modulus of water glass on early hydration of alkali–slag cements. Cem Concr Res 32(8):1181–1188

Lee NK, Kim EM, Lee HK (2016) Mechanical properties and setting characteristics of geopolymer mortar using styrene-butadiene (SB) latex. Constr Build Mater 113:264–272

Naghizadeh A, Ekolu SO (2017) Pozzolanic materials and waste products for formulation of geopolymer cements in developing countries: a review. Concr Beton 151:22–31

Palmero P, Formia A, Tulliani J-M, Antonaci P (2017) Valorisation of alumino-silicate stone muds: from wastes to source materials for innovative alkali-activated materials. Cement Concr Compos 83:251–262

Provis JL (2014) Geopolymers and other alkali activated materials: why, how, and what? Mater Struct 47(1–2):11–25

Sata V, Wongsa A, Chindaprasirt P (2013) Properties of pervious geopolymer concrete using recycled aggregates. Constr Build Mater 42:33–39

Shu X, Huang B, Wu H, Dong Q, Burdette EG (2011) Performance comparison of laboratory and field produced pervious concrete mixtures. Constr Build Mater 25(8):3187–3192

Sumathy CT, Dharakumar M, Saroja Devi M, Saccubai S (1997) Modification of cement mortars by polymer latex. J Appl Polym Sci 63(10):1251–1257

Tennis PD, Leming ML, Akers DJ (2004) Pervious concrete pavements. Portland Cement Association, Skokie

Ukrainczyk N, Rogina A (2013) Styrene–butadiene latex modified calcium aluminate cement mortar. Cement Concr Compos 41:16–23

Wang K, Schaefer VR, Kevern JT, Suleiman MT (2006) Development of mix proportion for functional and durable pervious concrete. In: Proceedings of the 2006 NRMCA concrete technology forum—focus on pervious concrete

Wang R, Wang P (2010) Function of styrene-acrylic ester copolymer latex in cement mortar. Mater Struct 43(4):443–451

Zaetang Y, Wongsa A, Sata V, Chindaprasirt P (2015) Use of coal ash as geopolymer binder and coarse aggregate in pervious concrete. Constr Build Mater 96:289–295

Zhong S, Chen Z (2002) Properties of latex blends and its modified cement mortars. Cem Concr Res 32(10):1515–1524

Experimental Study on Use of Recycled Polymer as Modifier in Mastic and Asphalt Mixture

Francesco Mazzotta, Andrea Simone(✉), Valeria Vignali, Claudio Lantieri, and Giulio Dondi

Department of Civil, Chemical, Environmental, and Materials Engineering (DICAM), University of Bologna, Viale Risorgimento 2, 40136 Bologna, Italy
andrea.simone@unibo.it

Abstract. Environment wastes caused by technological and industrial development are increasing. Whereas natural resources and disposal area for those wastes are decreasing day to day. So, recycling and reuse of waste materials has become crucial in terms of protection of environment and economy. In this research, a new polymer coming from the recycling of mixed plastic wastes has been used as fibres to increase asphalt mixture mechanical properties. In order to evaluate the rheo-mechanical effects of this new polymer the multiscale approach has been applied, evaluating mastic and mixture properties. The study has been started from the Asphalt Concrete scale through Indirect Tensile Stiffness Modulus (ITSM) test and Repeated Load Axial (RLA) test, in order to obtain stiffness and permanent deformation resistance of the mixtures. The scale has been reduced from mixture to mastic, obtaining complex modulus and phase angle of mastics using Dynamic Shear Rheometer (DSR). The results confirm as the new polymer increases mixture bearing capacity and rutting resistance.

Keywords: Bitumen · Waste plastic · Stiffness modulus · Phase angle

1 Introduction

In the last years a lot of research studies have shown that the neat bitumen is not able to provide the necessary performances of a road pavement in terms of heavy traffic resistance. Usually the bitumen acts as a binder for the different size aggregates and their mixing constitute the layers of a road pavement. Although aggregates angularity and shape have considered as primary factors that affect the development of the mechanical performance of asphalt pavements (Dondi et al. 2012), mainly due to large petrochemical industrial development, new studies were started adding modifiers into the asphalt mixture in order to improve its performances. In order to minimize the environmental impacts of roads construction and maintenance, moreover, it is necessary to quantify the energy and resources savings (Sangiorgi et al. 2016). This leads to the using of eco-friendly construction techniques, to reduce the use of virgin raw materials and to promote the recycling and reuse of the so-called secondary raw

M. Pasetto et al. (Eds.): ISAP APE 2019, LNCE 48, pp. 193–203, 2020.
https://doi.org/10.1007/978-3-030-29779-4_19

materials (materials which, after complete initial use, may be used repeatedly in production as starting material).

An increasing number of innovative and environmentally friendly materials has been launched on the market, and others are still under study. Some of them are: rubber (Sangiorgi et al. 2017, 2018; Vignali et al. 2016), bleaching clays from the food industry (Sangiorgi et al. 2014; Mazzotta et al. 2015), glass powder (Simone et al. 2017).

Disposal of plastic waste has become a major concern worldwide due to the considerable increase in volume and growth (Angelone et al. 2016), since it is not a biodegradable material and considered a major environmental pollutant (Moghaddam et al. 2013). Therefore, it would be beneficial if plastic waste could be reused in pavement construction.

Recent studies were conducted to evaluate utilization of plastic waste in asphalt mixes and its effects on the performance on flexible pavements (Sangita and Verinder 2011; Gawande et al. 2012; Swami and Jirge 2012; Chavan 2013; Abdo 2017). Results of these studies showed asphalt mixes containing plastic waste exhibited improvement in engineering properties (i.e., Marshall stability, resistant to water, and resistant to crack propagation). Several studies investigated using plastic waste as an asphalt binder modifier to enhance its performance (Hinislioglua and Agar 2014; Al-Humeidawi 2014), other studies evaluated the use of plastic waste as a partial replacement of aggregates (Hassani et al. 2007; Rajasekaran et al. 2013).

The object of this paper is the performances evaluation of an asphalt mixture modified with a new kind of polymer coming from urban and industrial recycled plastic waste. This technique permits to modify the mixture directly to the plant without changing the industrial production process. This method allows both to save money and reduce environmental impact, increasing the mixture performances. The mechanical properties of the studied mixture were evaluated through dynamic mechanical test. In particular complex stiffness modulus and creep modulus of the mixture have been extracted using three different modifier percentages. As studied by Hospodka et al., is possible to find correlations between asphalt mixture fatigue tests and mastic fatigue test with new DSR methods (Hospodka et al. 2018). Following the same field of investigation, in order to analyze mixture results in a smaller investigation scale, rheological test has been conducted on normalized mastics.

2 Material and Methodology

2.1 Aggregates

The aggregates used in this investigation were limestone crushed stones with the 20% of Reclaimed Asphalt Pavement (RAP). The selected gradation of aggregates is shown in Fig. 1. The choice of grain size was based on the one of the typical Italian specifications (ANAS) gradation limits of HMA mixtures for surface layers. The volumetric mass of the aggregates was calculated according to the EN 1097-6 standard.

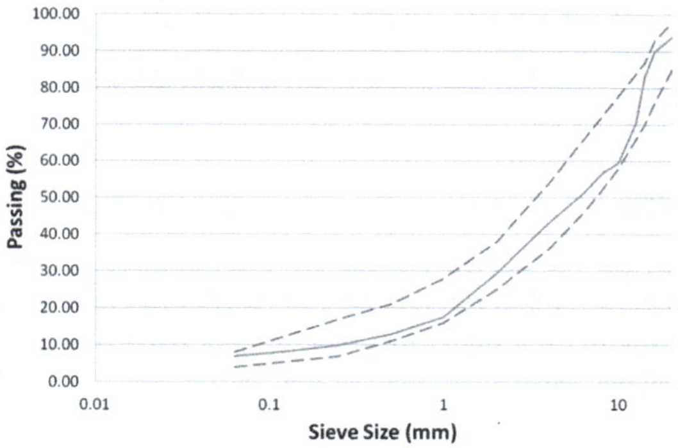

Fig. 1. Applied aggregate gradation. (ANAS, Italian specifications).

2.2 Bitumen

For this research, in order to analyze the polymer effects in the neat binder, a common Italian 50/70 pen bitumen has been used to made asphalt mixtures and mastics. The characteristics of the bitumen were evaluated and shown in Table 1. Considering 0.9% of bitumen recovered from RAP, the optimal percentage of bitumen 50/70 pen was equal to 4%.

Table 1. Physical properties of the binder.

Property	Unit	Pen grade 50/70	Standard
Penetration @ 25 °C	dmm	50	EN 1426
Softening point	°C	50	EN 1427
Dynamic viscosity @160 °C	Pa s	0.5	EN 12596

2.3 Fillers

With the aim to reproduce mastics and mixtures corresponding to real cases, virgin limestone filler and recovery limestone were used. The limestone filler is the product of limestone crushing and it is configured as a fine aggregate white in color and characterized by an amorphous structure. Limestone filler is one of the most suitable materials for the mastics production because of the mineralogical and chemical nature (proportional to its specific surface area and, therefore, the presence of very fine elements). Specifically, the used limestone filler complied with the following requirements:

- Calcium carbonate content $\geq 75\%$ by mass (typically $CaCO_3 \approx 95\%$);
- Clay content (adsorption of methylene blue) $\leq$ 1.20 g/100 g;
- Organic content (TOC) $\leq$ 0.20% by mass (for the type LL $\leq$ 0.50% by mass, $FeO_2 \approx 0.2\%$ to $SiO_2 \approx 0.4\%$).

The recovery filler has been extrapolated from an asphalt plant in Bologna, and it comes from the aspiration of powders generated by the coarse aggregates. The extracted filler was sieved to sieve 0.063 mm. The amount of filler in the asphalt mixtures was 5% in weight of aggregates, 1.2% is composed by recovery filler and the remaining 3.8% by virgin limestone filler.

2.4 PB25 Polymer

The PB25 is a thermoplastic polymer comprising a blend of polyolefins coming from urban and industrial waste. In this research PB25 waste plastic has been used as asphalt mixture modifier. It could be product as granule, pellet and powder. In this research it has been used in powder form as shown in Fig. 2.

Fig. 2. PB25 in granular form, pellet form and powder form (from left to right).

2.5 Mixture and Mastic Proportions

The aim of this study is to define PB25 effects analysing mastics rheological characteristics and mixture mechanical properties at different additive percentages. In particular the amount of PB25 considered was:

(a) 0 kg per 1 ton of asphalt mixture;
(b) 1.5 kg per 1 ton of asphalt mixture;
(c) 2.5 kg per 1 ton of asphalt mixture;
(d) 3.5 kg per 1 ton of asphalt mixture;

Starting from the gradation curve shown in Fig. 1, all the asphalt mixtures components have been kept constant for all specimens except for the plastic waste PB25. In this way, totally four sample have made and called PC0 without polymer and PC15, PC25 and PC35 when was applied respectively the cases (b), (c) and (d). The multi scale approach has been identified on (a) and (d) cases, obtaining two mastics called PC0 and PC35 with the single components propositions shown in Table 2.

Table 2. Mix Proposition-Mastic.

Material	PC0 (g)	PC35 (g)
Bitumen 50/70 pen	234.00	234.00
PB25	0	18.64
Limestone filler	238.70	128.77
Recovery filler	191.00	191.00

3 Experimental Work

3.1 Asphalt Mixture Tests

In the current research, the volumetric characteristics have been determined according to standard EN 12697-8 in addition to compaction curves analysis to investigate the rate of workability. The 150 mm diameter specimens were compacted by gyratory compactor under constant 600 kPa pressure and external angle of 1.25 °C for 100 gyrations. The obtained samples was tested according to UNI EN 12697-26 and UNI EN 12697-25 standards, to obtain both the indirect tensile stiffness modulus and the accumulated deformation. The ITSM test has been conducted to evaluate the mixture bearing capacity and its and thermo-sensitivity behaviour at 10–20, and RLAT test was important to understand the rutting mixture phenomena at 40 °C.

3.2 Mastic Tests

Amplitude Sweep (AS) tests were conducted according to EN 14470 to investigate mastics visco-elastic region at 10 °C, applying a constant frequency of 10 rad/s (1.59 Hz). In order to represent the complex modulus (G*) and phase angle (δ), Frequency Sweep (FS) test according to EN 14470 was performed. Master curves were calculated from the tests, finding the relationship between load frequencies, complex modulus G* and phase angle δ. The reference temperature for the master curves is 20 °C. The Williams-Landel-Ferry model was used to obtain the temperatures shift factors. The frequencies range was between 0.01 and 10 Hz. The Multiple Stress Creep and Recovery test (MSCR) was run according to the AASHTO TP 70–07 (2013). According to this standard, mastic sample is loaded at a constant creep stress for 1 s, followed by a zero stress recovery of 9 s. Ten cycles of creep and recovery are run at 0.1 kPa creep stress, followed by ten at 3.2 kPa creep stress. The non-recoverable compliance (J_{nr}) and the percent recovery after ten cycles at 0.1 and 3.2 kPa were studied. The J_{nr} value was calculated as the ratio between the average non recoverable strain for 10 creep and recovery cycles, and the applied stress for those cycles. The testing temperature of 58 °C was adopted (Fig. 3).

Fig. 3. Asphalt mixture and mastic tests configurations: ITSM (a), RLAT (b), AS and FS and MSCR (c).

4 Results and Discussion

4.1 Volumetric Analysis

Figure 4 shows the compaction curves for the studied mixtures. From the developed compaction curves, it is evident that the slope of the mixtures containing the polymer is the same of the slope of mixture PC0 without modifier (Table 3).

Table 3. Volumetric properties of asphalt mixtures (* Gyration number: 100).

Specimen	VMA (%)	VA (%)	VFA (%)
PC0	12.00	1.70	87.10
PC15	12.30	2.10	83.30
PC25	12.60	2.40	81.10
PC35	16.6	2.50	80.10

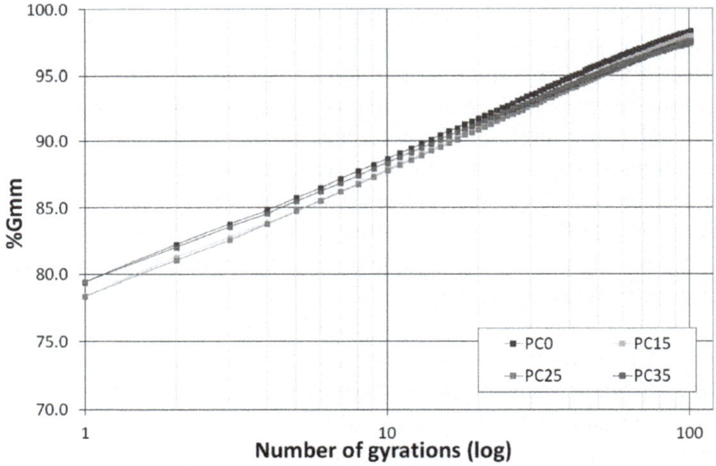

Fig. 4. PC0, PC15, PC25 and PC35 compaction curves comparison.

It can be deduced that the PB25 has no negative effect on the mixture compaction phase, that is, the presence of polymer does not affect the binder viscosity and therefore does not alter mixing and compaction temperatures.

4.2 Stiffness Properties

The dynamic mechanical characterisation was centred on the study of the Stiffness Modulus as set out in the EN 12697-26 (Annex C) standard. According to the standard, the Modulus was determined through a pulse loading with a 124 ms rise-time, to generate a predefined horizontal deformation of 2 μm in the core of the sample. This last device setting has been necessary because of the high materials stiffness. The effect of the waste filler was assessed, looking at the stiffness and at the AC thermal sensitivity. From the analysis of Fig. 5 it can be deduced that the use of waste plastic powder PB25 increases the mixture stiffness throughout the temperature range if compared to the mixture without polymer. At 10 °C for PC35 there is an increase of 33.8% of stiffness compared to PC0, at 20 °C the increase is 33.3% and at 30 °C there is a difference of 52.5%. The stiffness curves are parallel in all temperatures range, the polymer does not affect the mixture thermo-sensitivity but has significant effects on its bearing capacity.

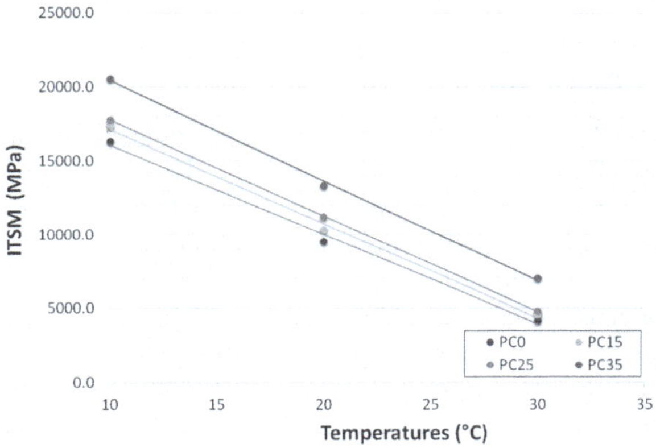

Fig. 5. Indirect Tensile Stiffness Modulus (ITSM) values

4.3 Permanent Deformation Evaluation

The repeated load axial test is used to determine permanent deformation resistance (EN 12697-25). In RLAT specimen is loaded with frequency of 0.5 Hz and a pressure of 100 ± 2 kPa. The specimen accumulated axial deformation was measured after 3600 loading cycles at test temperature of 40 °C. Testing was carried out with a uniaxial load of 100 kPa repeated for 3600 cycles at 40 °C. The Repeated Load Axial Test (RLAT) was implemented to evaluate resistance to permanent deformations in direct uniaxial compression test configuration. From the analysis of results, it is evident the different resistance to permanent deformation of the four mixture at 40 °C. The traditional sample PC0 has less creep stiffness of 10 MPa compare to all other modified sample mixture. PC35 modified with the highest percentage of PB25 has shown the maximum creep stiffness value, equal to 19.6 MPa (Fig. 6). This confirms that the resistance to permanent deformation vary according to the amount of polymer in the bituminous mixtures. Moreover from RLAT test has been obtained the accumulated strain, and also in this case is evident that for all specimen PC0 the degree of deformation is higher than the mixtures containing PB25. The accumulated strain became lower increasing the polymer percentage inside the mixture (Fig. 6).

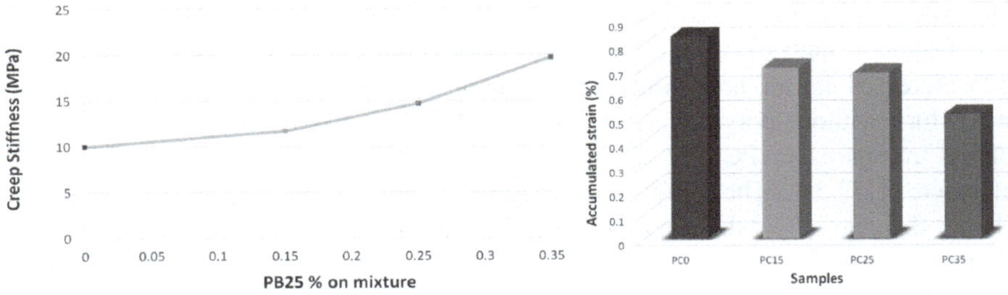

Fig. 6. RLA test results at 40 °C in terms of creep stiffness and of accumulated strain.

4.4 Rheological Properties

Rheological tests in linear and not linear range were fundamental to investigate the polymer effects on the bituminous matrix. Mastics visco-elastic properties in linear range are shown in Fig. 7. At high frequency G* tends to the same value for both mastic (5600 MPa at 10 Hz). Differences can be observed at low frequencies where PC35 mastic shows higher moduli. In particular G* value of PC35 at lowest frequency of 0.01 is 473 Pa and G* value of PC0 at the same frequency is 163 Pa. The PB25 increases the shear stress resistance at high temperatures; the phase angle trends at high temperatures shows an increase in the elastic response due to the polymer presence. However, the mastic response is predominantly viscous and the modification of the asphalt concrete cannot be compared to the rheological effects that occurs in the bitumen modified through the use of thermoplastic elastomers.

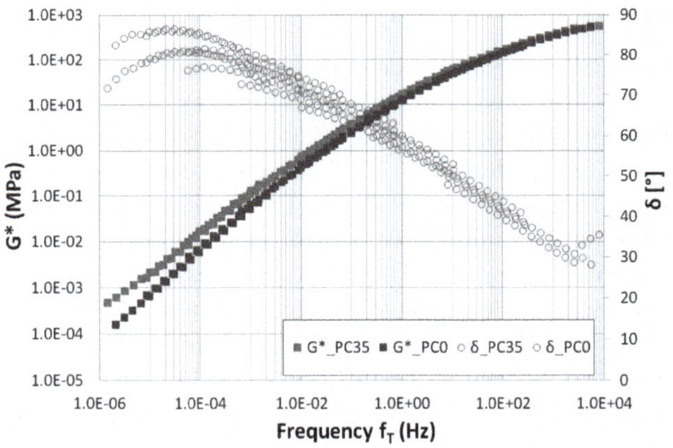

Fig. 7. Master curves comparison.

In order to study the rutting resistance and find correlations between RLA tests and rheological properties, mastics permanent deformations were characterized by means of Multiple Stress Creep Recovery (MSCR) at 58 °C. The standardized temperature was chosen in order to compare these results with RLAT results on asphalt mixture.

Figure 8 shows the MSCR curves at the end of the 3.2 kPa shear stress level cycles. Mastics containing PB25 exhibits the stiffer behavior, accumulating less deformation at the end of the test. The bitumen characteristics of the studied mastics exhibit significantly different response because of the polymer presence. PB25 has increased the resistance to deformation during the creep phase but this mastic exhibited a slight recovery during the recovery phase as confirmed on the FS sweep test analysis. The results of MSCR curves are confirmed studying the non-recoverable compliance values (Fig. 9). J_{nr} values were calculated for the four mastics under 0.1 and 3.2 kPa shear stresses at 58 °C. The PC35 mastic, obtained with unmodified bitumen, have lower values of non-recoverable compliance at both shear stress levels, showing less sensitivity to permanent deformations. This results confirms the RLA test results, in terms of creep stiffness increasing.

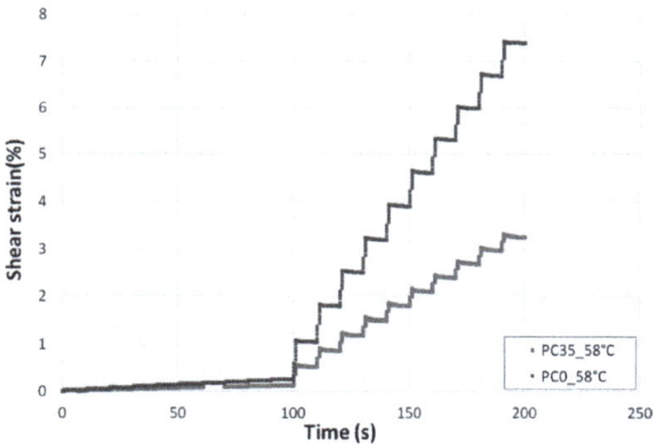

Fig. 8. MSCR test result for PC0 and PC35 at 58 °C under 0.1 kPa and 3.2 kPa.

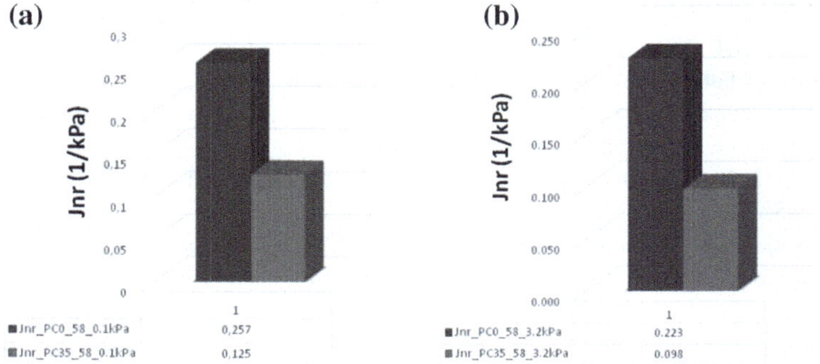

Fig. 9. Jnr results at 0.1 kPa (a) and 3.2 kPa (b).

5 Conclusions

The present research confirmed the applicability of the reuse of plastic urban waste as polyolefins powder to produce high performance asphalt mixtures, improving neat binders characteristic. The main purpose of this research was to characterize the PB25 as suitable material for the road pavements construction, with the opportunity to modify asphalt mixtures directly in the plant.

The results obtained from mechanical characterization have confirmed the suitability of the material to be used as mixture modifier. In terms of interaction between PB25 and unmodified bitumen the rheological properties of mastics was studied. In particular, from the results obtained performing FS and MSCR tests, it can be stated that the used polymer increases the mastic stiffness and exalts elastic shear stress response. From the asphalt mixture analysis the following conclusions can be stated:

- The Indirect Tensile Stiffness Modulus (ITSM) shows a significant stiffness mixture increasing, proportional to the presence of polymer;
- Permanent deformation potentiality analysis, by means of RLA test, revealed the higher performance of the mixtures made with PB25. In addition, the creep stiffness values were also in accordance to MSCR values at high temperature which validates the approaches.

The use of PB25 in the asphalt mixtures not only allows the recovery and reuse of a waste material, but also it allows to improve the mixtures performances in terms of resistance to permanent deformation at high temperatures and bearing capacity.

References

Abdo AM (2017) Investigation the effects of adding waste plastic on asphalt mixes performance. J Eng Appl Sci 12:15

Al-Humeidawi BH (2014) Utilization of plastic waste and recycle concrete aggregate in production of hot mix asphalt. J Eng Sci 7(4):322–330

Angelone S, Cauhape Casaux M, Martinez FO (2016) Green pavements: reuse of plastic waste in asphalt mixtures. Mater Struct 49(5):1655–1665

Chavan MAJ (2013) Use of plastic waste in flexible pavements. Int J Appl Innov Eng Manag 2(4):540–552

Dondi G, Simone A, Vignali V, Manganelli G (2012) Discrete particle element analysis of aggregate interaction in granular mixes for asphalt: combined DEM and experimental study. In: Proceedings of 7th RILEM international conference on cracking in pavements, 20–22 June, Delft, The Netherlands

Gawande A, Zamare G, Renge VC, Tayde S, Bharsakale G (2012) An overview on plastic waste utilization in asphalting of roads. J Eng Res Stud 3(2):1–5

Hassani A, Ganjidoust H, Amir AM (2007) Use of plastic waste (Poly-ethylene terephthalate) in asphalt concrete mix as aggregate replacement. J Waste Manag Res 23(4):322–327

Hinislioglua S, Agar E (2014) Use of waste high density polyethylene as bitumen modifier in asphalt concrete mix. Mater Lett 58:267–271

Mazzotta F, Sangiorgi C, Vignali V, Lantieri C, Dondi G (2015) Rheological characterization of bituminous mastics added with waste bleaching clays. In: Proceedings of 8th RILEM symposium, 7–9 October, Ancona, Italy

Moghaddam TB, Karim MR, Soltani M (2013) Utilization of waste plastic bottles in asphalt mixture. J Eng Sci Technol 8(3):264–271

Rajasekaran S, Vasudevan R, Paulraj S (2013) Reuse of plastic wastes coated aggregates-bitumen mix composite for road application - green method. Am J Eng Res 2(11):01–13

Sangiorgi C, Eskandarsefat S, Tataranni P, Simone A, Vignali V, Lantieri C, Dondi G (2017) A complete laboratory assessment of crumb rubber porous asphalt. Constr Build Mater 132:500–507

Sangiorgi C, Tataranni P, Simone A, Vignali V, Lantieri C, Dondi G (2014) Waste bleaching clays as fillers in hot bituminous mixtures. Constr Build Mater 73:320–325

Sangiorgi C, Tataranni P, Simone A, Vignali V, Lantieri C, Dondi G (2016) Assessment of waste bleaching clay as alternative filler for the production of porous asphalts. Constr Build Mater 109:1–7

Sangiorgi C, Tataranni P, Simone A, Vignali V, Lantieri C, Dondi G (2018) Stone mastic asphalt (SMA) with crumb rubber according to a new dry-hybrid technology: a laboratory and trial field evaluation. Constr Build Mater 182:200–209

Sangita GR, Verinder K (2011) A novel approach to improve road quality by utilizing plastic waste in road construction. J Environ Res Dev 5(4):1036–1042

Simone A, Mazzotta F, Eskandarsefat S, Sangiorgi C, Vignali V, Lantieri C, Dondi G (2017) Experimental application of waste glass powder filler in recycled dense-graded asphalt mixtures. Road Mater Pavement Des 20:592–607

Swami V, Jirge A (2012) Use of plastic waste in construction of bituminous road. Int J Eng Sci Technol 4(5):2351–2355

Vignali V, Mazzotta F, Sangiorgi C, Simone A, Lantieri C, Dondi G (2016) Incorporation of rubber powder as filler in a new dry-hybrid technology: rheological and 3D DEM mastic performances evaluation. Materials 9:842

Hospodka M, Hofko B, Blab R (2018) Introducing a new specimen shape to assess the fatigue performance of asphalt mastic by dynamic shear rheometer testing. Mater Struct 51:46

Preliminary Study on the Mechanical Properties of an Asphalt Mixture Containing RAR Modifiers

Christina Plati(✉), Brad Cliatt, and Andreas Loizos

Laboratory of Pavement Engineering, National Technical University of Athens, Athens, Greece
cplati@central.ntua.gr

Abstract. This study is an attempt to investigate a reacted and activated rubber (RAR) compound modifier, which is an elastomeric asphalt extender. This type of material is comprised mainly of fine crumb rubber granules with the addition of an activated mineral binder stabilizer; the compounds are typically introduced into the asphalt concrete mixture as a replacement of the bitumen binder. There are two rationales for the potential use of these modifiers within flexible asphalt concrete mixtures. The first rationale is the potential advantages related to long-term environmental sustainability, as inclusion of these rubber compounds can transform an industrial waste product (rubber tires) into a viable replacement for bitumen within asphalt concrete mixtures. The second rationale is the potential of enhancing various mechanical properties of the asphalt mixtures, such as to provide increased resistance to permanent deformation, enhanced pavement friction properties and/or increased durability and strength of the produced asphalt mixtures. With this mind the objective of the current investigation is to preliminary evaluate the performance of the addition of RAR modifiers into an open graded asphalt wearing course mixture. For the research varying percentages of the bituminous binder are replaced by a RAR modifier with the objective to define the most suitable percentage of RAR to be included within the investigated open graded asphalt mixture in terms of compaction degree and stiffness modulus.

Keywords: Asphalt mixtures · Modifier · Modulus · Sustainability

1 Background and Objectives

Pavements structures and roadway are an integral portion of the transportation infrastructure. When designing and/or maintaining current pavement infrastructure assets pavement engineers and other responsible authorities need to take into consideration multiple and often competing factors in order to design and build these projects. One of these issues in the concept of sustainability and the reutilization and valorization of up until now waste products that may have ended up in landfills. Rubbers vehicles tires are one of these waste products that has a great potential to be fully reutilized.

One of the techniques for this is through their potential incorporation as a constituent in the construction of asphalt flexible pavements. For a brief background on the

M. Pasetto et al. (Eds.): ISAP APE 2019, LNCE 48, pp. 204–213, 2020.
https://doi.org/10.1007/978-3-030-29779-4_20

subject, Lo Presti (2013) provided an extensive overview of rubber tires being recycled through their incorporation as a components of asphalt concrete mixtures. In the paper the potential value of rubber tire and their engineering properties are discussed. In addition, the main processes to turn waste tires into a rubber material that will be suitable for utilization in asphalt pavement mixtures. These processes include: ambient grinding of the tires at normal temperatures, cryogenic grinding of the tire at elevated temperatures and other procedure that are less utilized. In addition to describing the transformation processed for the waste tire, the paper also describes the basic procedures for inclusion of the rubber into asphalt mixtures.

Beyond descriptions of the general processes, extensive research has been undertaken over the years that determine the mechanical and rheological properties of these rubber modified mixtures. Multiple researchers have investigated the rheological properties of the mixture including Mturi et al. (2014). Moreno-Navarro and Rubio Gámez (2012) investigated strength characteristics, while other researchers such as Fontes et al. (2010) have investigated issues related to permanent deformation. Cong et al. (2013) following an extensive investigation of a large combination of bitumen types and rubber percentage concluded that with the addition of rubber to the asphalt mixture the softening point increases and better results are seen for the elastic recovery, viscosity and complex modulus, thus rutting tendency is decreased. Arabani et al. (2018) outlined the issues related to crumb rubber in regards to implementing the material in a dry process. In this type of process the addition of crumb rubber has issues related to bonding of the rubber materials to the bitumen and overall homogeneity of the mixture. The investigation introduced a nanomaterial in an attempt to improve upon these issues. Venudharan et al. (2016) provided an overview of the most recent asphalt mix design practices and the research included a review of current research and a look at the future of waste rubber materials in gap grade asphalt mixtures. They concluded that more research is needed to understand the effect of rubber materials in asphalt concrete mixtures.

Beyond available research on the mechanical and rheological characteristic of these modified mixtures more recent research has focused on the overall energy consumption and environmental impacts the incorporation these techniques may have and their potential for sustainable solutions. Wang et al. (2018) investigated the energy consumption of these mixture concluding that consumption is lower than conventional mixtures. It was stressed out that the technique is a green technology that lowers greenhouse gas emissions, reduces energy consumption, lower the levels required for virgin materials and emphasized that the idea should be utilized with the goal to achieve environmental sustainability.

With the positive results in regards to the potential advantages of rubber inclusion into asphalt mixtures, a newer technique has emerged in order to incorporate the waste rubber material into asphalt concrete mixtures. The technique involves a new material type called a reacted and activated rubber (RAR) compound. This material, which is an elastomeric asphalt extender, aims at overcoming known issues related to how to more effectively and homogenously incorporate these waste rubber tire materials into asphalt mixtures. In order to do this, the RAR modified material has been suggested to ensure that the bond between the rubber and bitumen is already formed before the mix procedures are commenced. To do this the RAR material is comprised of three different components: crumb rubber materials, bitumen and a filler with the percentage usually between 60–65%, 20–25% and 15–20% respectively and this combination is designed to mitigate bitumen and rubber bonding issues.

This type of RAR materials is newer and as such it has been less investigated and less in depth related research is currently available. Sousa et al. (2013) has provided a brief overview of both the suggested RAR material and technique in addition to explaining the potential benefits it may have over the traditional crumb rubber materials considering compositions with varying percentage of RAR. The investigation looked into changes in viscosity and the effect on fatigue and permanent deformation of mixtures containing RAR materials. Other researchers have looked into the rheological properties of asphalt mixtures containing RAR materials. Kedarisetty et al. (2016) and Kedarisetty et al. (2018) for example investigated the rheological properties of asphalt mixtures containing RAR materials. The more recent investigation (Kedarisetty et al. 2018) concluded that the mixture showed significant improvements in mixture performance and further recommended that more aggregate mixture gradations and types be investigated both within the laboratory and in-situ.

Following the previous researchers, the current study attempts to build more knowledge on the RAR technique in regards to an open graded wearing course layer with RAR materials incorporated. With this in mind the main objective is to carry out a preliminary investigation into the achieved compaction degree and stiffness modulus results, over a range of test temperature of mixtures containing varying percentages of a RAR modifier in comparison to a normal asphalt mixture for an open graded asphalt wearing course layer.

2 Materials Tested

2.1 Design Mixture

The asphalt concrete material investigated was designed for an open graded wearing course layer. More specifically the mixture was designed to fit within the specifications for a wearing course layer materials with a maximum aggregate size of 12.5 mm. A sieve analysis on aggregate materials and gradation is shown in Table 1.

Table 1. Mixture gradation

Sieve (mm)	% Passing	Design limits (% Passing)
37	100	100
31.5	100	100
20	100	100
12.5	92.6	84–100
9.0	61.9	52–82
4.00	25.3	18–46
2.00	19.5	5–24
1.12	13.9	3–19
0.063	1.0	0–9

Information concerning the bitumen and the initial mixture properties with no RAR modifiers is shown in Table 2. The maximum mixture density was 2520 kg/m3.

Table 2. Bitumen and AC mixture properties

Bitumen information			Mix design (no additives - 0%)		
Type	25/55–70	EN 13108	Stability (KN)	11.3	EN 12697-34
Penetration (25C)	44	EN 1426	Flow (mm)	4.4	EN 12697-34
Softening point(C)	75.8	EN 1427	Air voids (%)	10.9	EN 12697-8
Elastic recovery	94.8	EN 13398-10	Density (kg/m3)	2245	EN 12697-6
Density	1.03	EN 15326	Water Sensitivity	0.82	EN 12697-12

2.2 RAR Modifier

The RAR modifier is comprised of three materials, a fine crumb rubber, inert fillers and bitumen. The material in its physical state is a black/gray fine materials with a bulk density of 0.6gr/cm3 and a flash point of greater than 300 °C. The individual rubber modifier particles are less than 600 microns in size with the majority of the individual rubber modified particles are between 250–600 microns in size. A schematic representation of the material is shown in Fig. 1.

Fig. 1. Investigated reacted and activated rubber

3 Research Methodology

3.1 Mixture Preparation

For the investigation mixtures containing: 0, 10%, 20% and 40% (by weight of bitumen) RAR modifier were produced. The original asphalt mixture contained 5.4% bitumen by mass of aggregate. The RAR modifier material was added to the mixture to produce the investigated mixtures. The basic procedure is shown in Fig. 2.

Fig. 2. Mixture preparation

For the investigated asphalt mixture the RAR modifier was added first to the heated aggregates in a dry process, then the bitumen was added. The aggregates were heated to between 175–185 °C before the addition of the rubber modifier, in order to account for the addition of the RAR modifier at room temperature. The RAR modifier was added to the dry aggregates and then mixed for 10–20 s to ensure even distribution of the modifier in the mix. The following step was the addition of the 25/55–70 bitumen. After the addition of the bitumen the mixture was further mixed for an additional 30–45 s to ensure complete aggregate coverage and a homogenous asphalt mixture. The asphalt mixture was then, immediately transferred back to the oven and maintained at a temperature of 170–180 °C for one hour. The asphalt mixture was stirred multiple times during this period to ensure that RAR modifier was fully and homogeneously integrated into the asphalt mixture before proceeding to the next preparation stages.

3.2 Mixture Compaction

The heated RAR modified asphalt mixture was removed from the oven and transferred in order to be compacted in a steel rolling compactor. The rolling compactor is a device designed to compact asphalt material slabs in the laboratory under conditions that simulate in situ compaction. The device is comprised of a large curved steel roller that pivots and applies a moving force similar to in-situ rolling machines. For the investigation a static compaction load was applied onto the material until desired void levels were achieved. For the compaction a 305 mm x 305 mm with a 50 mm height mould was utilized. The material and mould were heated to 170–180 °C and the compaction of the materials was carried out with a temperature between 160–170 °C. Batches of the prepared asphalt mixture were removed from the oven and placed within the steel moulds required for the compaction procedures. The asphalt mixture introduced into the slab was based on the asphalt mixture maximum density and the target voids. The compaction process was a static process and the compaction procedures were in accordance to EN 12697 - Bituminous mixtures - Test methods for hot mix asphalt - Part 33: Specimen prepared by roller compactor. A schematic representation of the compaction process is shown in Fig. 3.

Fig. 3. Compaction procedure

3.3 Production of Test Specimens

The slabs produced in the rolling compaction process were allowed to cool to room temperature before being extracted from their metallic mould. The extracted slab was then placed under a coring machine where four 100 mm cores per slab were extracted. The distance of between the cores and from the edge of the slab were maximized. A schematic representation of the coring process is shown in Fig. 4. Final dimensions of the cores were 39–42 mm in height and a diameter of 100 mm.

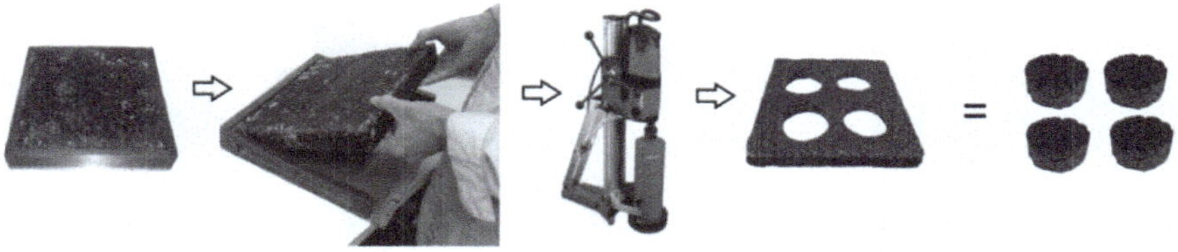

Fig. 4. Coring of the specimens

3.4 Specimen Testing

For the investigation the strength of the cores was tested via the EN standard 12697 Part 26: Stiffness. More specifically Annex C Indirect tension to cylindrical specimens (IT-CY) was utilized to determine the measured stiffness modulus. Each of the prepared test specimens was tested along two axes separated by 90°. The average of the two was determined as the average modulus. For the testing a target rise time of (124 ± 4) ms was set. The target peak transient horizontal deformation was set to 0,005% of the specimen diameter or in the range of 4.8–5.2 microns. Specimens that did not meet these specifications were either retested later or excluded from further analysis. A schematic representation of the testing process is shown in Fig. 5.

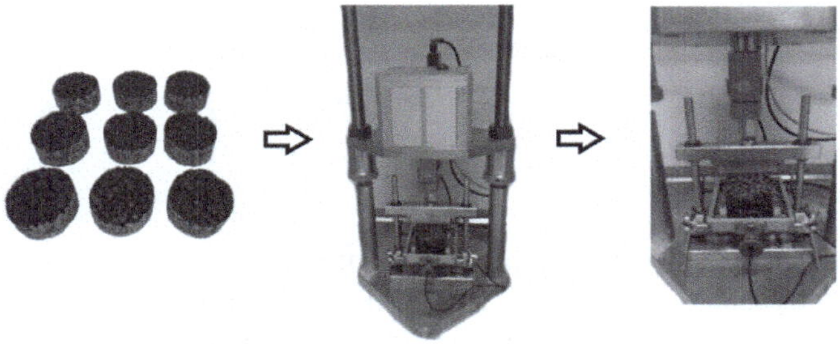

Fig. 5. Testing of the specimens

Each of the specimens was tested at four separate temperatures of 15, 20, 25 and 30 °C. For each of the test temperatures the specimens were acclimated in a climate controlled chamber in order to reach test temperature for a period of 4–6 h. The tolerance was set at ±0.5 °C of the test temperature. A dummy specimen of the same materials was utilized to ensure the correct testing temperature within the chamber. After test specimens reached the target temperature they were transferred immediately to climate control testing chamber and immediately tested.

4 Testing Results and Analysis

4.1 Compaction Results

Figure 6 below shows the achieved bulk density during the compaction procedures.

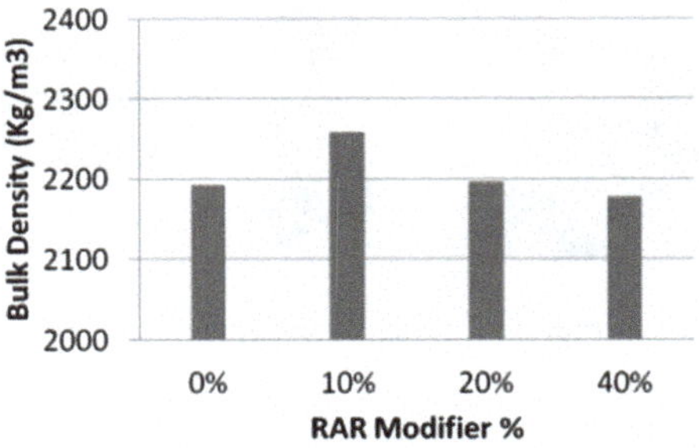

Fig. 6. Bulk density (kg/m3)

Figure 7 below shows the achieved air voids during the compaction procedures.

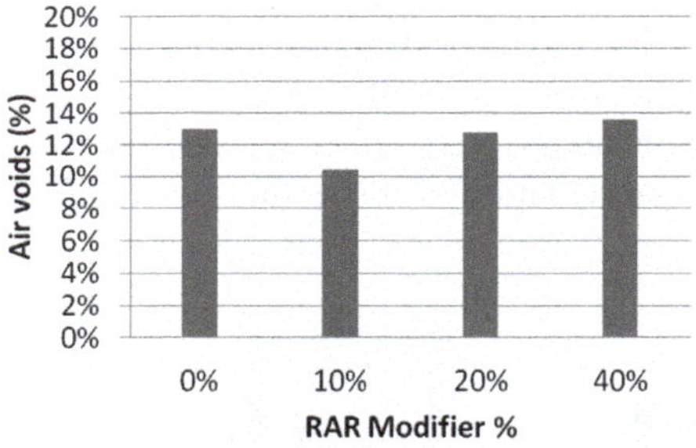

Fig. 7. Air voids (%)

From both, the bulk density compactions results and the achieved air voids % it can be seen that the asphalt mixture with 10% RAR exhibited the best overall results and exceeded those of the design mixture. The 20% RAR mixture produced results similar to the non RAR mixture.

4.2 Modulus Results

Figure 8 below shows the stiffness results according to the EN standard 12697-26 testing procedure.

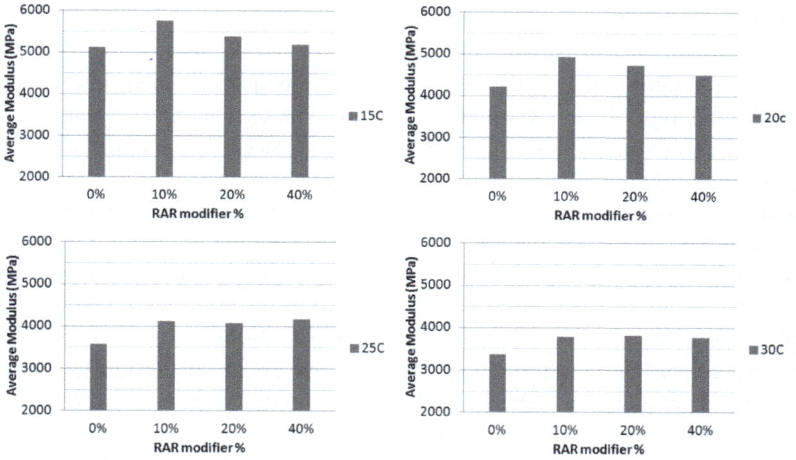

Fig. 8. Modulus results (15–30 °C)

The results presented in Fig. 8 concern the four test temperatures of 15, 20, 25, and 30 degrees Celsius. As can be seen from the results for 15 °C, as the RAR percentage is increased from 0–10% there is a 12% increase in the modulus values. For 20 °C the corresponding increase is 17%, with a 16% increase at 25 °C and a 12% at 30 °C. So for all temperatures the increase is in the range from 12–17% when the RAR % is increased to 10%. However when the RAR% is increased from 10–20% there is a −6% to −1% decrease in modulus values for 15–25 °C. While for 30 °C there is a slight increase of 1%. Similar results are found when increasing the RAR% from 20–40% with the modulus value decreasing up to 5%.

Figure 9 below provides a graphic representation of the influence of both the temperature and the percentage of RAR modifier added to the asphalt mixture. It can be seen that as the temperature increases to 30% the RAR mixture modulus values are nearly identical. While at the same time all RAR mixture are 12–13% greater than the non RAR asphalt mixture. This increase is valid for all temperatures and shows that the RAR modifier has a positive effect on the modulus values.

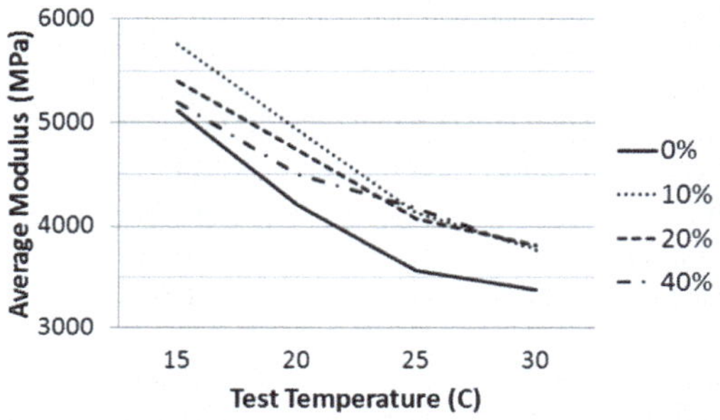

Fig. 9. Effect of test temperature

5 Conclusions

Based on the results of the preliminary investigation it was shown that the asphalt mixtures containing the RAR modifier had an average modulus greater than the original asphalt mixture at all temperatures. At the lowest temperature of 15 °C the difference range from +17% for the 10% mixture to +1% for the 40% mixture. While as the temperatures increased to 25–30 °C the differences in modulus of the RAR mixtures diminished. More specifically as the test temperature reached 30 °C all RAR mixtures had a stiffness modulus of 12–13% greater than the original asphalt mixture. Based on the results of the achieved bulk density and the air voids it can be seen that the asphalt mixture exhibited the highest bulk density and the lowest air voids in the mixture. Based on the above mentioned results and based on the results of the preliminary investigation it can be concluded that the optimal percentage of RAR modifier to be added to the investigated asphalt mixture is 10% by weight of bitumen. The current investigation was a preliminary investigation to determine the optimal percentage of the RAR modifier to be added based on the results of the achieved compaction degree and stiffness modulus. More investigation is required to determine other properties of the RAR modifier mixture regarding cracking susceptibility, rutting potential sustainability issues etc. The investigation is ongoing.

References

Arabani M, Tahami SA, Hamedi GH (2018) Performance evaluation of dry process crumb rubber-modified asphalt mixtures with nanomaterial. Road Mater Pavement Des 19(5):1241–1258. https://doi.org/10.1080/14680629.2017.1302356

Cong P, Xun P, Xing M, Chen S (2013) Investigation of asphalt binder containing various crumb rubbers and asphalts. Constr Build Mater. https://doi.org/10.1016/j.conbuildmat.2012.11.063

Fontes PL, Trichês G, Pais J, Pereira P (2010) Evaluating permanent deformation in asphalt rubber mixtures. Constr Build Mater. https://doi.org/10.1016/j.conbuildmat.2009.12.021

Kebria DY, Moafimadani SR, Goli Y (2015) Laboratory investigation of the effect of crumb rubber on the characteristics and rheological behaviour of asphalt binder. Road Mater Pavement Des. https://doi.org/10.1080/14680629.2015.1042015

Kedarisetty S, Biligiri KP, Sousa JB (2016) Advanced rheological characterization of Reacted and Activated Rubber (RAR) modified asphalt binders. Constr Build Mater https://doi.org/10.1016/j.conbuildmat.2016.06.043

Kedarisetty S, Gourab S, Krishna Prapoorna BK, Sousa J (2018) Reacted and Activated Rubber (RAR)-modified dense-graded asphalt mixtures: design and performance evaluation. J Test Eval 46:20170211. https://doi.org/10.1520/JTE20170211

Lo Presti D (2013) Recycled tyre rubber modified bitumens for road asphalt mixtures: a literature review. Constr Build Mater. https://doi.org/10.1016/j.conbuildmat.2013.09.007

Moreno-Navarro F, Rubio Gámez MC (2012) Influence of crumb rubber on the indirect tensile strength and stiffness modulus of hot bituminous mixes. J Mater Civ Eng. https://doi.org/10.1061/(ASCE)MT.1943-5533.0000436

Mturi GAJ, O'Connell J, Zoorob SE, De Beer M (2014) A study of crumb rubber modified bitumen used in South Africa. Road Materials and Pavement Des. https://doi.org/10.1080/14680629.2014.910130

Sousa JB, Vorobiev A, Rowe GM, Ishai I (2013) Reacted and Activated Rubber: Elastomeric Asphalt Extender. Transp Res Rec: J Transp Res Board. https://doi.org/10.3141/2371-04

Venudharan V, Biligiri KP, Sousa JB, Way GB (2016) Asphalt-rubber gap-graded mixture design practices: a state-of-the-art research review and future perspective. Road Mater Pavement Des. https://doi.org/10.1080/14680629.2016.1182060

Wang T, Xiao F, Zhu X, Huang B, Wang J, Amirkhanian S (2018) Energy consumption and environmental impact of rubberized asphalt pavement. J Clean Prod. https://doi.org/10.1016/j.jclepro.2018.01.086

Influence of the Production Temperature on the Optimization Process of Asphalt Mixes Prepared with Steel Slag Aggregates Only

Emiliano Pasquini[1](✉), Giovanni Giacomello[1], Marta Skaf[2], Vanesa Ortega-Lopez[3], Juan Manuel Manso[3], and Marco Pasetto[1]

[1] Department of Civil, Environmental and Architectural Engineering (ICEA), University of Padua, Via Marzolo 9, 35131 Padua, Italy
emiliano.pasquini@unipd.it
[2] Department of Construction, Escuela Politécnica Superior, University of Burgos, Calle Villadiego s/n, 09001 Burgos, Spain
[3] Department of Civil Engineering, Escuela Politécnica Superior, University of Burgos, Calle Villadiego s/n, 09001 Burgos, Spain

Abstract. This paper presents a comprehensive research aimed at evaluating the feasibility of preparing bituminous mixes using slag aggregates only (EAF slag and LF slag from electric steelmaking), i.e. without any natural aggregate. Here, the second step of the experimental study, currently in progress, is presented, focusing on investigating possible countermeasures to the issues related to a reduced compactability of steel slag materials, such has been evidenced by the first phase of the research. To this aim, the effect of using a higher production temperature to prepare the steel slag mixtures has been analyzed in terms of volumetric properties as well as basic mechanical performance and durability. A reference mixture, prepared with natural virgin aggregate and cement as filler, was also investigated for comparison purposes. Basically, the main experimental findings showed that the mixes prepared with steel slags still suffered poor workability even when higher production temperature was used. This led to a higher strength but more brittle mechanical behavior whereas promising results were obtained in terms of water and aging resistance. Anyway, it is worth noting that the measured performance for the slag mixtures can be mostly considered acceptable according to the requirements usually adopted for material acceptance.

Keywords: EAF steel slag · LF steel slag · Mechanical properties · Durability

1 Introduction

In this century, the Society has finally realized the urge of saving natural resources by reusing or recycling the wastes, reducing also the increasing problem of landfilling. In the construction industry, where enormous amounts of natural resources are employed worldwide, this conviction has brought together Scientists, Industry and Governments to seek and validate the alternatives to the environment exploitation in a safe and reliable framework.

M. Pasetto et al. (Eds.): ISAP APE 2019, LNCE 48, pp. 214–223, 2020.
https://doi.org/10.1007/978-3-030-29779-4_21

Road construction mainly uses as raw materials big amounts of natural aggregates that are traditionally extracted from quarries or pits. Many different residues have been for a long time tested as substitutes of natural aggregates in bituminous mixtures, such as reclaimed asphalt pavement (Frigio et al. 2015), coal bottom ash (Pasetto and Baldo 2008), recycled concrete (Arabani et al. 2013), construction and demolition wastes (Pérez et al. 2012) and of course, steel slags (Pasetto et al. 2017).

On the other side, steel industry, in a continuously growing trend, last year produced 1.6 billion tons of steel (World_Steel_Association 2018). Steel is a highly recyclable product, but during its production, some by-products are generated that the industry is fighting to turn into valuable resources (Craig Heidrich et al. 2017). In the electric cycle of carbon steel production, mainly two by-products are produced: the Electric Arc Furnace (EAF) slag and the Ladle Furnace slag (LFS), whose generation is estimated to exceed 10 million tons a year, only in Europe (EUROSLAG 2013).

The EAF slag has been used for some time as a substitute for natural gravel in construction materials, especially in the manufacture of bituminous road mixtures. After conditioning (stabilization, crushing, sieving and washing) it shows properties that adapt perfectly to be used as a coarse aggregate in granular layers and bituminous mixtures, being able in some cases even to be used in high-demanding mixes, i.e. for heavy traffic in wearing courses (Skaf et al. 2017).

The reuse of the other by-product, the LFS is more complicated, due to its dusty appearance and potential expansion (Moliné et al. 2018; Ortega-López et al. 2017). Some initial uses are being explored, mainly in the preparation of Portland cement clinker (Adolfsson et al. 2011) or to replace cement in construction applications (García-Cuadrado et al. 2018; Manso et al. 2011; Ortega-López et al. 2014). Very recently, its particle size and slight hydraulicity are making it be valued as a potential candidate to be used as filler in bituminous mixtures (Bocci 2018; Skaf et al. 2016)

2 Background and Goal

Based on the abovementioned state of the art, a comprehensive research program aimed at evaluating the feasibility of preparing bituminous mixes combining Electric Arc Furnace and Ladle Furnace steel slag aggregates only (i.e. without natural aggregates) is currently in progress. The first steps of the research showed promising performance of the steel slag mixes even if clear issues related to excessive air void content of the mixtures made with LFS slag as fine aggregate were observed (Skaf et al. 2018). Thus, further research activities were planned to assess both the causes and the possible solutions of such a lower workability.

In particular, the present paper reports the results obtained analyzing the opportunity of adopting higher production (i.e. mixing and compaction) temperatures in order to overcome the difficulty in effective compaction of bituminous mixes prepared with LFS as fine aggregate. To accomplish this objective, volumetric properties as well as main mechanical (indirect tensile strength and particle loss resistance) and durability (water and aging susceptibility) characteristics of different mixtures prepared combining LFS (as fine aggregates and filler) and EAF (as coarse aggregates) slags were evaluated. Such properties were then compared with those of the corresponding reference mixtures (i.e. produced using virgin aggregates). Other variables potentially affecting the mixture volumetrics (e.g. softer bitumen, different gradations, etc.) were also investigated. The results will be reported in other papers.

3 Materials and Mixes

A plain 50/70 penetration grade bitumen was used to prepare the tested mixtures.

As in the previous phase of the research (Skaf et al. 2018), the reference mixture (hereafter named SSC) was obtained by using ordinary Portland cement CEM I/42.5 R as typical filler (particle size < 0.063 mm) as well as natural siliceous aggregates as fine (0.063 mm < particle size < 2 mm) and coarse (particle size > 2 mm) fractions. Then, two different steel slag based mixtures were investigated, both prepared using LFS as filler and fine aggregate. In this case, the coarse fraction of one mixture (hereafter named SLL) consisted of natural siliceous aggregates whereas the other material (hereafter named ELL) was prepared using only steel slag aggregates (i.e. EAF steel slag was used as the coarse aggregate). Such mixture compositions were chosen because an imperfect compaction was observed in the previous steps of the research when LFS slag replaced the fine aggregate fraction of the mixture (Skaf et al. 2018). Table 1 summarizes the aggregate composition of the tested mixtures. Further details can be found elsewhere (Skaf et al. 2018).

Table 1. Tested mixtures

Tested mix	Coarse aggregate	Fine aggregate	Filler	Production temperature
SSC	Silice	Silice	Cement	155 °C
SLL	Silice	LFS	LFS	155 °C
SLL_h	Silice	LFS	LFS	170 °C
ELL	EAF	LFS	LFS	155 °C
ELL_h	EAF	LFS	LFS	170 °C

All the mixtures were prepared adopting the same aggregate gradation for surface layers, i.e. a dense graded granulometric distribution with a nominal maximum aggregate size of 16 mm. Starting from a previous Marshall mix design (Skaf et al. 2018) and in order to avoid too many variables to take into account, all the mixes were manufactured using 12.5% bitumen content by the volume of the aggregates, properly considering the different densities of the aggregates.

Test specimens were compacted through the Marshall hammer (EN 12697–30) by applying 50 or 75 blows per face depending on the subsequent testing procedure. 155 °C was fixed as reference temperature for mixing and compaction operations. Moreover, as anticipated, steel slag based mixtures were also prepared at an arbitrary higher temperature (170 °C) in order to verify the possibility of improving the compactability of such materials while maintaining their mechanical properties and durability. Such mixes were coded using the suffix "_h".

4 Test Methods

4.1 Volumetric Properties Assessment

The volumetric properties of all the manufactured specimens (12 for each mix) in terms of air void content (AVC), voids in the mineral aggregates (VMA) and voids filled with bitumen (VFB) were determined according to EN 12697-8. In particular, the bulk density of test samples was measured through the saturated surface dry procedure reported in EN 12697-6 whereas the maximum density of the mixes was calculated using the mathematical procedure defined in EN 12697-5. To this aim, the particle density of the different aggregates was obtained following the pycnometer method described in EN 1097-6.

4.2 Mechanical Characterization

The basic mechanical characterization of the tested materials was based on indirect tensile strength (ITS) and Cantabro tests.

In order to assess the fracture resistance of the tested materials, ITS tests on cylindrical samples compacted through the Marshall procedure were carried out at 25 °C according to EN 12697-23 by applying a deformation rate of 0.85 mm/s. In this study, three replicates for each tested mixture were carried out.

On the other hand, the so-called Cantabro test was performed according to EN 12697-17 in order to estimate the particle loss (PL) resistance of the selected materials. Also in this case, three replicate specimens were tested for each material.

4.3 Durability Evaluation

Durability of the tested mixes was evaluated in terms of water and aging effects.

As suggested by EN 12697-12, the well-known Indirect Tensile Strength Ratio (ITSR) was used to evaluate the water resistance of the selected materials. To this aim, wet conditioning of selected specimens was carried out by submerging the samples in a water bath at 40 °C for 72 h (EN 12697-12). Then, three ITS test repetitions were carried out for each selected material both for dry and wet samples.

On the other hand, the increase in particle loss PL (calculated through the Cantabro test described above) after a laboratory aging process was selected as the key parameter to evaluate the aging resistance of the tested asphalt mixes. To achieve this goal, a sample of the compacted specimens was aged in forced draft oven at 85 °C for 120 h according to what suggested by AASHTO R30. Finally, three Cantabro tests were carried out both for unaged and aged samples for each material.

5 Experimental Findings

5.1 Volumetric Characteristics

Figure 1 summarizes the volumetric properties of all the tested mixtures in terms of air void content (AVC), voids in the mineral aggregates (VMA) and voids filled with bitumen (VFB). In particular, the graph shows the average values of AVC, VMA and VFB along with their corresponding maximum and minimum measured values that are represented by the error bars.

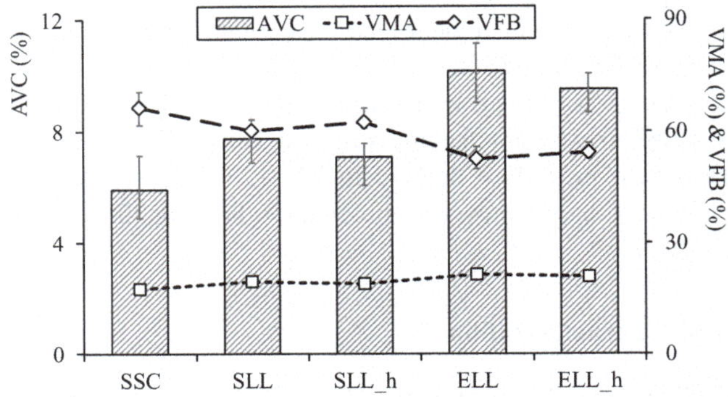

Fig. 1. Volumetric properties (voids, VMA, and VFB) of the tested mixtures

As it can be noted, the mixtures prepared with the slags (SLL and ELL) demonstrated some compactability issues, especially in the case of ELL mixture prepared with steel slag aggregate only (EAF for the coarse fraction and LFS for the fine fraction and the filler). In particular, the use of LFS as fine aggregate and filler (SLL material) led to a less pronounced (but still significant) decrease in workability of the mixture likely due to a stiffening effect of the asphalt mortar, which made the compaction more difficult (Skaf et al. 2018). A further substantial worsening of the volumetric properties occurred when EAF aggregates were introduced to replace the reference siliceous coarse fraction (ELL mix); in this case, it is hypothesized that the higher angularity of the EAF aggregates negatively contributed to the unacceptable measured void content. The statistical significance of AVC differences among SSC, SLL and ELL was assessed through a one-way Analysis of Variance (ANOVA) at a 95% confidence level whose results are reported in Table 2.

Table 2. Main ANOVA test results for air void content (AVC)

Comparison	Significant?	p-value
SSC vs. SLL	YES	1.62E-07
SLL vs. ELL	YES	7.87E-11
SLL vs. SLL_h	YES	1.22E-03
ELL vs. ELL_h	YES	7.20E-03
SSC vs. SLL_h	YES	1.12E-04

To overcome the abovementioned issues, an attempt in enhancing the production temperatures of the steel slag mixtures was proposed. To this aim, both SLL and ELL materials were produced at 15 °C higher temperature (SLL_h and ELL_h, respectively) with respect to the reference SLL and ELL mixes. The experimental findings reported in Fig. 1 demonstrate that a clear decrease in the air void content can be achieved thanks to the enhancement of the production temperature, even if this does not allow to attain the same volumetric properties of the reference SSC mixture. Again, the statistical significance of the reduction of the air void content for the "_h" mixtures is demonstrated through an ANOVA at a 95% confidence level (Table 2).

5.2 Mechanical Properties

Indirect tensile strength (ITS) and particle loss (PL) resistance were selected as key parameters to represent the basic mechanical properties of the investigated materials. The average experimental results are synthetized in Fig. 2 where the error bars show the corresponding maximum and minimum measured values.

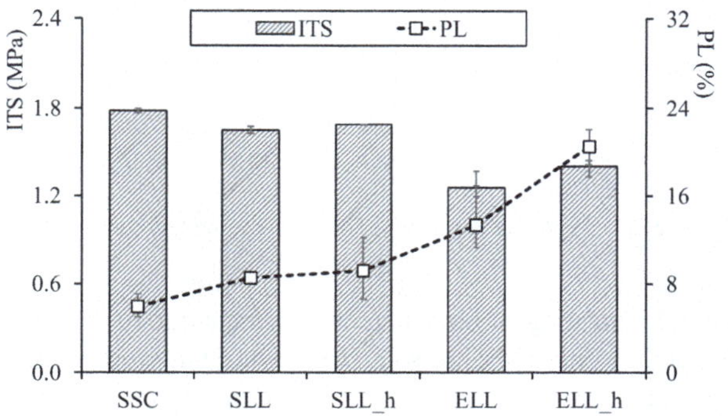

Fig. 2. Basic mechanical properties (ITS and PL) of the tested mixtures

As it can be observed, the presence of LFS aggregates led to a certain reduction of strength and internal cohesion that further decreased in the case of mixtures prepared with steel slags only (ELL). These experimental findings could be ascribed to the poor volumetric properties discussed above, rather than to an actual negative contribution of the slags to the mechanical behavior of the materials. Moreover, it is worth noting that ITS and PL of slag mixes are still satisfactory based on usual acceptance requirements (typically fixed at 0.8 MPa and 20%, respectively).

On the other hand, the effect of the higher preparation temperature on the mechanical properties of the tested materials seems to be related to the oxidative hardening suffered by the bitumen due to such higher mixing and compaction temperature. In fact, it is worth noting that both the indirect tensile strength and the particle loss of SLL_h and ELL_h exhibited a slight increase with respect to the corresponding SLL and ELL mixes, respectively, thus denoting both a stronger and more brittle behavior of the former "_h" materials.

5.3 Durability

The durability of the selected steel slag materials was studied in terms of moisture susceptibility and aging resistance. To this aim, the indirect tensile strength ratio

(ITSR) and the particle loss (PL) increase were determined after specific water and aging conditioning procedures, respectively.

In this sense, Fig. 3 shows the average indirect tensile strengths measured both in dry and wet condition (i.e. after a dry or wet conditioning period) along with their minimum and maximum measured values (error bars) and the corresponding ITSR values.

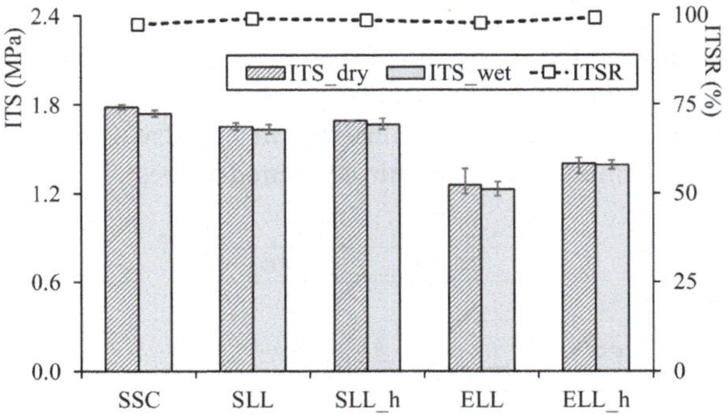

Fig. 3. Water resistance (ITSR) of the tested mixtures

The experimental data suggest no detrimental effects due to the addition of the slags both for the fine fraction only (SLL mix) and the whole aggregate skeleton (ELL mix) with respect to the reference SSC mixture notwithstanding the higher voids of the slag based materials. Thus, a good affinity seems to exist between the selected binder and steel slag aggregates.

Moreover, the preparation temperature does not seem to significantly affect the moisture susceptibility of the mixes prepared with steel slag aggregates since the ITSR evidenced by SLL_h and ELL_h mixtures were substantially coincident with those of the corresponding reference SLL and ELL mixtures, respectively. However, the overall results highlight the bland water action imposed by the water conditioning method suggested by the European standard (EN 12697-12) that often does not allow a clear differentiation among different material behaviors.

Finally, the aging effect on the particle loss resistance of the investigated materials can be observed in Fig. 4 where average PL values of unaged and aged samples (i.e. before and after an aging conditioning period) are reported also highlighting the minimum and maximum measure values and the related percentage increase of PL due to the laboratory aging procedure.

First of all, the results depicted in the figure show a progressive reduction of the aging resistance (enhancement of the PL percentage increase) passing from the reference SSC mix to the steel slag based SLL and ELL materials. This increasing effect of aging could be related to the increasing air void content of SLL and ELL mixes that probably enhanced the detrimental action of the high temperature conditioning on the oxidative aging degree of the asphalt binder. Such probable higher embrittlement of the bitumen film led to the observed reduced raveling resistance of the steel slag mixes.

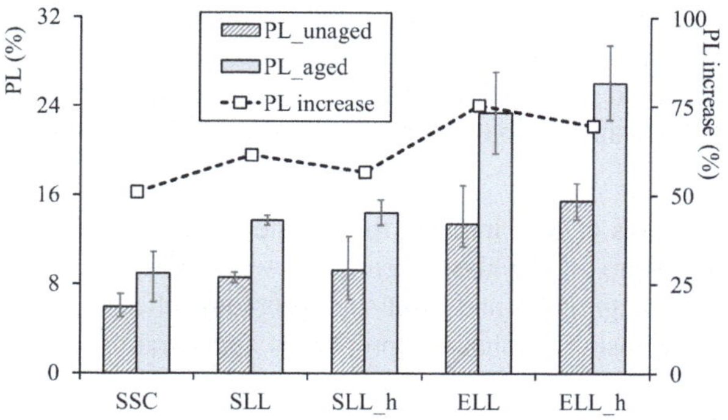

Fig. 4. Aging resistance (PL increase) of the tested mixtures

On the other hand, SLL_h and ELL_h materials, prepared adopting higher mixing and compaction temperatures, seemed to be characterized by an enhanced aging resistance (i.e. lower increase of particle loss due to aging) with respect to the corresponding SLL and ELL mixes, respectively. Two different factors maybe contributed to such observed behavior: i) "_h" mixes were characterized by significantly lower air voids with respect to the corresponding reference mixtures thus reducing their oxidative aging level; ii) the materials prepared at higher temperatures already suffered a higher aging degree during the preparation, thus reducing the effect of the aging during the conditioning period.

6 Conclusions and Further Studies

The present paper illustrates the effect of different production temperatures on the volumetric characteristics, the mechanical behavior and the durability properties of different steel slag based asphalt mixtures. Electric arc furnace slags were used for the coarse aggregate fraction whereas ladle furnace slags were selected for the fine aggregate fraction and the filler.

Based on the experimental findings, the following main conclusions can be drawn:

- The mixtures prepared with steel slags demonstrated compactability issues (especially in the case of the mixture prepared with steel slag aggregates only) likely due to a stiffer asphalt mortar and higher angularity of steel slags. Such poor volumetric characteristics negatively affected the mechanical properties (strength and cohesion) and the aging resistance of the steel slag materials without affecting their moisture susceptibility. This latter finding suggests a good affinity between the selected binder and steel slag aggregates;
- A significant improvement of the volumetric properties of the steel slag mixtures was achieved thanks to the enhancement of their production temperature; however, such materials did not attain the same volumetric properties of the reference mix. The higher production temperature led to a higher oxidative hardening of the bitumen during the preparation of the samples that involved an increase in tensile

strength and aging resistance coupled with a reduction of particle loss resistance without significant changes in moisture susceptibility;

- Overall, the observed performance of the steel slag based materials can generally be considered still satisfactory based on common technical requirements for material acceptance.

These main findings clearly indicate that further studies are needed to both prove the hypothesized reasons of the abovementioned workability issues due to the extensive use of steel slag aggregates and limit such problems effectively while maintaining or enhancing the promising mechanical properties and durability.

Acknowledgements. This work was supported by the Junta de Castilla y León and FEDER Funds through grant BU119P17 awarded to the group UIC-231; by the University of Burgos through grant Y135.GI awarded to the SUCONS group and by the University of Padua through grant BIRD182754 awarded by the Department of Civil, Environmental and Architectural Engineering (ICEA).

References

Adolfsson D, Robinson R, Engström F, Björkman B (2011) Influence of mineralogy on the hydraulic properties of ladle slag. Cem Concr Res 41:865–871

Arabani M, Moghadas Nejad F, Azarhoosh AR (2013) Laboratory evaluation of recycled waste concrete into asphalt mixtures. Int J Pavement Eng 14:531–539

Bocci E (2018) Use of ladle furnace slag as filler in hot asphalt mixtures. Constr Build Mater 161:156–164

Craig Heidrich KK, Reiche T, Merkel T (2017). Iron and steel slags: global perspective on the circular economy. In: 9th EUROSLAG conference, Metz

EUROSLAG (2013) Position paper on the status of ferrous slag

Frigio F, Pasquini E, Partl MN, Canestrari F (2015) Use of reclaimed asphalt in porous asphalt mixtures: laboratory and field evaluations. J Mater Civ Eng 27(7):1–9

García-Cuadrado J, Santamaría-Vicario I, Rodríguez A, Calderón V, Gutiérrez-González S (2018) Lime-cement mortars designed with steelmaking slags as aggregates and validation study of their properties using mathematical models. Constr Build Mater 188:210–220

Manso JM, Hernández D, Losáñez MM, González JJ (2011) Design and elaboration of concrete mixtures using steelmaking slags. ACI Mater J 108:673–681

Moliné MN, Calvo WA, Martinez AGT, Galliano PG (2018) Ambient weathering of steelmaking ladle slags. Ceram Int 44:18920–18927

Ortega-López V, Manso JM, Cuesta II, González JJ (2014) The long-term accelerated expansion of various ladle-furnace basic slags and their soil-stabilization applications. Constr Build Mater 68:455–464

Ortega-López V, Skaf M, Santamaría A (2017) The reuse of ladle furnace basic slags in clayey soil-stabilization applications. In: Soil Stabilization: types, methods and applications, pp 231–271

Pasetto M, Baldo N (2008) Laboratory evaluation of bituminous mixtures with bottom ash from municipal solid waste incineration (MSWI). In: 2008 global symposium on recycling, waste treatment and clean technology REWAS 2008, Cancun

Pasetto M, Baliello A, Giacomello G, Pasquini E (2017) Sustainable solutions for road pavements: a multi-scale characterization of warm mix asphalts containing steel slags. J Clean Prod 166:835–843

Pérez I, Pasandín AR, Medina L (2012) Hot mix asphalt using C&D waste as coarse aggregates. Mater Des 36:840–846

Skaf M, Manso JM, Aragón Á, Fuente-Alonso JA, Ortega-López V (2017) EAF slag in asphalt mixes: a brief review of its possible re-use. Resour Conserv Recycl 120:176–185

Skaf M, Ortega-López V, Fuente-Alonso JA, Santamaría A, Manso JM (2016) Ladle furnace slag in asphalt mixes. Constr Build Mater 122:488–495

Skaf M, Ortega-López V, Manso J, Pasquini E, Pasetto M (2018) Mix design and preliminary validation of sustainable asphalt concrete manufactured with electric arc and ladle furnace steel slags. In: Road and rail infrastructure V –5th international conference on road and rail infrastructure CETRA2018, Zara

World_Steel_Association (2018). https://www.worldsteel.org/

Long-Term Aging Behaviour of Asphalt Mixtures Modified with Crumb Rubber Using the Dry Process

Israel Rodríguez-Fernández[1,2(✉)], Maria Chiara Cavalli[1], Lily D. Poulikakos[1], and Moises Bueno[1]

[1] Empa, Swiss Federal Laboratories for Material Science and Technology, 8600 Duebendorf, Switzerland
[2] GITECO Research Group, University of Cantabria, 39005 Santander, Spain
israel.rodriguez@unican.es

Abstract. The use of crumb rubber from waste tires as additive for asphalt mixtures has several environmental benefits in comparison with traditional options such as landfill disposal and energy recovery. Moreover, several studies have shown better performance in asphalt mixtures modified with crumb rubber compared to conventional asphalt. In this sense, crumb rubber can be incorporated by wet or dry processes. In the wet process, crumb rubber is added and blended into an asphalt binder prior to adding the heated aggregates. The major advantage of the dry process is related to the manufacturing procedure. Here, the crumb rubber is added into the mixer together with the aggregates and no additional equipment is required. Afterwards, once the asphalt mixture is prepared, a conditioning time must be allowed to ensure a proper interaction between crumb rubber and the binder. Many research and practical works have shown that this conditioning time along with the type, size and quantity of the crumb rubber added, bitumen content or air voids content has a significant influence on short-term performance of the modified mixtures. Our study aims to further investigate the ageing effect and the long-term performance of asphalt mixtures modified with crumb rubber using the dry process. To do this, semi-dense asphalt (SDA) mixtures with 12% air void content were prepared incorporating two different types of crumb rubber. Then, the performance of these mixtures was evaluated before and after a long-term aging treatment and compare to a conventional polymer modified mixture. The obtained results confirm that the incorporation of crumb rubber using the dry process can lead to mixtures with adequate long-term performance as well as less aging than a conventional polymer modified mixture.

Keywords: Crumb rubber · Dry process · Aging

1 Introduction

Waste tires are a global problem and an increasing risk to the environment, because they are often incorrectly stored and disposed of. It is estimated that every year almost 1,000 million tires end their service life and more than 50% are discarded without any

M. Pasetto et al. (Eds.): ISAP APE 2019, LNCE 48, pp. 224–232, 2020.
https://doi.org/10.1007/978-3-030-29779-4_22

treatment (Thomas et al. 2016). The practice of disposal by burning has proved to create serious fire hazards and environmental pollution, and the practice of disposal by land filling has become difficult because of the depletion of the available sites. In addition, these stockpiles present the threat of uncontrolled fires and other environmental hazards (Lo Presti 2013; Svoboda et al. 2018; Thomas et al. 2016; Xue and Shinozuka 2013).

One of the alternatives to recycle tires is to use the rubber components in asphalt concrete production. This application has several environmental benefits not only related to the reduction of used tires, but also replacement of raw materials and an extension of the pavement service life are achieved. In terms of energy consumption, despite production of asphalt with tire rubber is a high-energy consuming process (higher mixing temperature and longer mixing duration), the consumed energy was found generally lower than that of conventional asphalt mixture during the construction and maintenance phases, implying that asphalt with tire rubber had evident energy saving advantages in the life cycle. In addition, this recovery process had great benefits in energy saving compared with the landfill disposal and energy recovery (Farina et al. 2017; Wang et al. 2018).

A recovery process known as “granulate recovery” generates the material suitable to use in asphalt production. In this process, the waste tires are cut up into small pieces and the steel and fabric components are removed. Then, by further reducing the size, it is possible to produce Crumb Rubber (CR), which are suitable to be used in the asphalt industry. There are several technologies to produce CR, being the most habitual the ambient grinding, cryogenic grinding, wet-grinding and hydro jet size reduction. Nowadays, all of these techniques result in a highly controlled material, being the resulting rubber particles consistently sized and very clean (Lo Presti 2013). The shape and size of the final CR particles will vary depending on the process employed.

Two different processes can be used to incorporate CR into the asphalt mix. These processes are known as “wet process” and “dry process”. In the wet process, CR (0.075 to 1.2 mm) is mixed with bitumen at elevated temperatures before mixing with the aggregate. In the dry process, CR (0.4 to 10 mm) replaces a small portion of the fine aggregate (typically 1% to 3% by mass of the total aggregate in the mixture) and is blended with the aggregate before addition of the bitumen (Rahman et al. 2010). Wet process is more popular and has been more frequently used. The less popularity of the dry process is caused by the poorer results that it produces, especially in the early years (e.g. poor reproducibility and the premature failure of road surfacing) (Buncher 1995; Shook 1990; Lawrence et al. 1991). These limitations generated a certain lack of confidence in the technique, and the dry process has not been developed as much as the wet process (Moreno et al. 2011). The main problem associate to CR mixes is their lack of cohesion, which is primarily due to a poor interaction between CR and bitumen. This causes lower resistance to moisture, raveling and a reduction in the bearing capacity of the pavement. This interaction is conditioned by the properties of both materials, CR and bitumen. Also, the characteristics of the mix has an influence in this interaction. The mixing time, the temperature or the conditioning time (time during the CR-bitumen interaction occurs), among others, have an influence in the mix performance (Moreno et al. 2011).

Several laboratory studies have been conducted in order to determine a design methodology that ensure a good performance of a dry process originated mixture (Abdul Hassan et al. 2014). In this type of mixtures, as said before, CR performs as part of the aggregate of the mix, but also it partially modifies the binder properties through a CR-bitumen interaction. These two functions have been found to improve the mixture performance, improving the resistance to fatigue cracking, permanent deformation and cracking at low temperature, reducing cracking propagation and increasing durability (Cao 2007; da Silva et al. 2018; Feiteira Dias et al. 2014; Hernández-Olivares et al. 2009; Moreno et al. 2012; Xie and Shen 2013). The research in last years has been focused on determining the influence on mixture performance of different variables such as the conditioning time, manufacturing time, size and quantity of the CR added, bitumen content or air voids content (Abdul Hassan et al. 2014; Cao 2007; da Silva et al. 2018; Farouk et al. 2017; Feiteira Dias et al. 2014; Hernández-Olivares et al. 2009; Moreno et al. 2012, 2011; Xie and Shen 2013).

These works demonstrate the suitability of the dry process which has now the potential to be a common practice in future years. However, detailed studies to optimize the technique are still required. In this sense, the evaluation of their long-term performance should be considered.

Xie and Shen (2013) evaluates the long-term performance of three years in service testing pavement of stone matrix asphalt (SMA) pavements with CR added in dry process. They did not find any distresses related to cracking, rutting, raveling, bleeding, pushing or potholes. In addition, the samples taken from the field showed slightly higher stability and lower flow than control mixtures. Rahman et al. (2005) evaluated the mechanical properties of modified asphalt mixtures following short-term and long-term ageing. They concluded that the influence of short-term ageing on mechanical properties was far greater compared to long-term ageing. The two performance indicators used in this study, fatigue and resistance to permanent deformation, marginally improved following short-term ageing, but generally deteriorated after long-term ageing.

This study aims to further investigate the properties of asphalt mixtures modified with CR using the dry process after the long-term ageing. To do this, semi-dense asphalt (SDA) mixture was prepared incorporating two different types of CR using the dry process. Then, the performance of these mixtures was evaluated before and after a long-term aging treatment. The results were compared with a reference mixture.

2 Experimental Design

2.1 Asphalt Mix Manufacture and Characterization

Three SDA 4-12 mixtures were manufactured (Table 1). Two of them were experimental mixtures modified with CR using the dry method and the other was a reference mixture. Moreover, a conventional 50/70 penetration grade binder was used for the experimental mixtures whereas a conventional polymer modified binder (PMB 45/80-65) was used for the reference mix (binder content = 6.2% by weight of mixture).

Table 1. Particle size distribution of aggregates used for SDA 4-12 mixtures design

Sieve size (mm)	Fraction (% by mass)
5.6	100.0
4.0	90.0
2.0	23.0
1.0	14.0
0.5	11.0
0.25	9.0
0.125	8.0
0.063	7.0

In addition, two different types of CR were used. Both were obtained by using ambient grinding method and the resulting particles had a maximum particle size of 800 μm. One of them was modified using polymers (henceforth CR_1) whereas for the other, not additional treatment was applied (henceforth CR_2) (Loderer et al. 2018).

The dry process was used to prepare the mixtures modified with CR. First, the preheated aggregates (185 °C) were mixed with the CR (1% by mass of mixture) during 1.5 min. Then, the preheated binder (160 °C) was added and mixed with the aggregates and CR during another 2 min. Once mixed, the mixture was placed in the oven at 165 °C during 120 min (conditioning time). After this time, the samples were compacted at 155 °C. It is important to note that CR was incorporated to the mixture without modify the granular composition.

In order to assess the final characteristics of the asphalt mixtures, volumetric properties (EN 12697-5; EN 12697-6; EN 12697-8) were measured. As SDA is a type of mixture with a high porosity, the geometric method is used to calculate the bulk density. Finally, Marshall (EN 12697-34) and water sensitivity (EN 12697-12) tests were conducted.

2.2 Aging Methodology

After conditioning, an artificial accelerated aging process was applied to the mixtures. In this case, the conditioning process was assumed as short-term aging method. Then, the long-term aging was simulated using the SHRP long-term oven aging (LTOA) method. This method establishes that the mixtures should be placed in the oven at 85 ° C for 120 h. This accelerated long-term aging protocol is meant to represent 15 years of field ageing in a Wet-No-Freeze climate and 7 years in a Dry-Freeze climate (Airey 2003). The performance of aged mixtures was evaluated through the same laboratory tests performed before (volumetric properties, Marshall test and water sensitivity test).

3 Results and Discussion

The volumetric properties measured for asphalt mixtures prepared with and without addition of CR are shown in Fig. 1. All the mixtures meets the requirement established by the Swiss standard for this kind of mixtures in terms of air void content (10-14%). Note that the CR-modified mixtures show lower air void content than reference mixture. This is expected as the mix design was not adjusted to account for this additional material and its swelling.

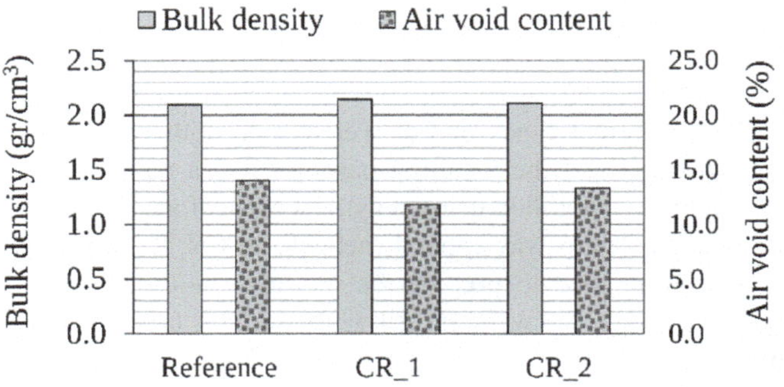

Fig. 1. Volumetric properties

Comparing CR-modified mixtures, which only differ in the CR type used, there a significant difference in the air voids content. This difference could be related with a different interaction between CR and binder. A different interaction could affect the CR particles size, the binder properties or the resultant binder content, and therefore the volumetric composition. The differences in the treatment commented above or differences in the specific surface of CR particles could be the main reasons that justify a different interaction with the binder.

Regarding mechanical performance, as said before, aging produces a change in the composition and rheological behavior of bitumen, resulting in an increment in the mixture stiffness. This behavior could be indirectly evaluated by the Marshall test. After aging, an increment in the Marshall stability join to a decrement in the Marshall flow could indicated that the mixtures is stiffer than before. The Marshall Quotient (MQ), calculated as the ratio between stability and flow, is recognized as an indicator of the mixture stiffness. Thus, the aging effect could be evaluated throng the MQ values before and after long-term aging.

The Marshall test results obtained before and after long-term aging are plotted in Figs. 2. It can be observed that the reference mix shows the behavior described before where the stability increases (+10%) and the flow reduces (−11%) their values, resulting in a higher MQ (+23%). However, CR-modified mixtures do not show this behavior. For these mixtures, both stability and flow increases after long-term aging in the same proportion (ca. +20%), resulting in no variations for the MQ. This fact could be related with the reactivation of the rubber during the aging process. These results could indicate that CR-modified mixtures increases the load-bearing capacity due to increased cohesion after aging (higher stability) without reduce their elastic behavior (same MQ). In conclusion, the CR-modified mixtures seems to be less susceptible to change their mechanical performance due to de long-term aging effect.

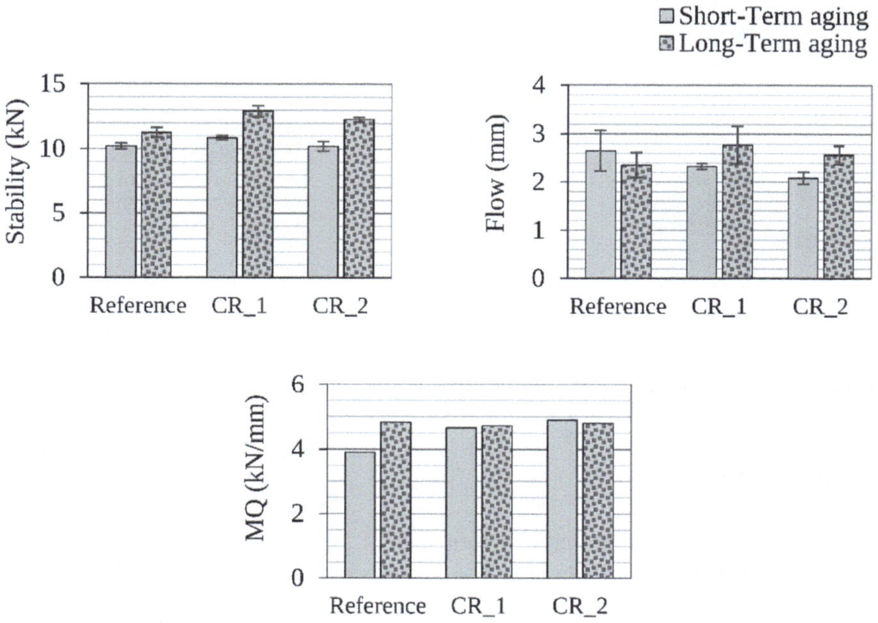

Fig. 2. Marshall test results before and after long-term aging

Regarding water sensitivity test, the normal tendency is to increase indirect tensile strength (ITS) after long-term aging (Fig. 3). This increment is higher in the conditioned samples (ITS wet) than in the unconditioned samples (ITS dry), therefore the ITSR also increases after long-term aging. In this case, the CR-modified mixtures show similar behavior than the reference mixtures. The only differences are in the magnitude of changes. A higher increment in the ITS values could be related with a lower aging resistance. According to this analysis, the mixture modified with CR_2 is the mixture with the best aging performance. The mixture modified with CR_1 and the reference mixture show similar performance.

The minimum value required by the Swiss standard for SDA mixtures for ITSR is 70%. In this case, reference mixture is the only mixture that fulfil this requirement before long-term aging. However, the ITSR values obtained in the CR-modified mixtures are close to the reference mixture. Authors would recommend to adjust the mixture recipe in order to improve its performance (e.g. increase the bitumen content) or the use of agents to promote a better adhesion between mineral and binder. It is important to note that the bitumen used in these experimental mixtures was a conventional bitumen whereas the technical manuals recommend the use of polymer

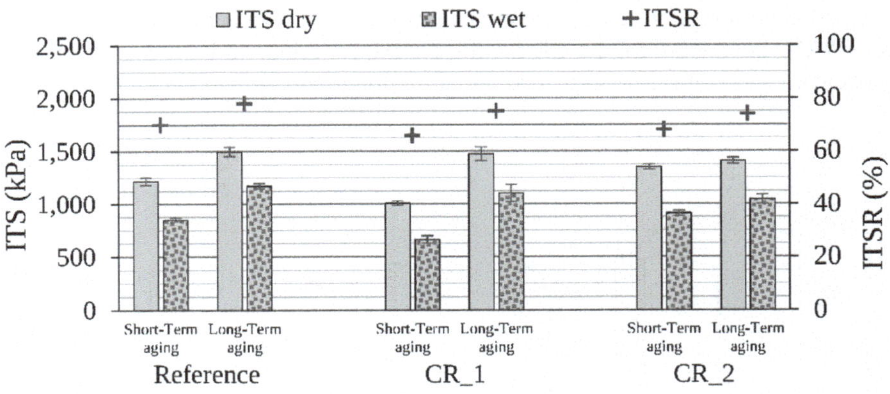

Fig. 3. Water sensitivity test results before and after long-term aging

modified binders. After long-term aging, all the mixtures show a higher ITSR value, being this increment similar for all of them (ca. +10%).

4 Conclusions

In the present study, the long-term performance of asphalt mixtures modified with CR using the dry process was evaluated. The findings are summarized as follows:

- Based on Marshall test, conventional bitumen mixtures modified with CR using the dry method improves the long-term aging susceptibility compared to polymeric modified mixture.
- CR-modified mixtures do not increase the Marshall Quotient value after long-term aging. This results could indicate that mixture stiffness keeps constant after aging.
- Water susceptibility of CR-modified mixtures should be improved. Despite these mixtures get similar results than the reference mixture, they do not reach the minimum value required by the standard. Authors would recommend to adjust the mixture recipe or the use of agents to promote a better adhesion between mineral and binder.
- The incorporation of CR by the dry method does not cause any problem during mixture manufacture neither during compaction after 120 min of conditioning time.

Acknowledgements. This research project is partially funded by Innosuisse project number 25880.1 PFIW-IW. Industrial support from the companies TRS, Ammann, Weibel and FBB is gratefully acknowledged. Israel Rodríguez-Fernández was supported by the FPU Programme of the Spanish Ministry of Education, Culture and Sport [grant number FPU-14/06997].

References

Abdul Hassan N, Airey GD, Putra Jaya R, Mashros N, Aziz MA (2014) A review of crumb rubber modification in dry mixed rubberised asphalt mixtures. Jurnal Teknologi 70(4):127–134. https://doi.org/10.11113/jt.v70.3501

Airey GD (2003) State of the art report on ageing test methods for bituminous pavement materials. Int J Pavement Eng 4(3):165–176

Buncher M (1995) Evaluating the effects of the wet and dry processes for including crumb rubber modifier in hot mix asphalt

Cao W (2007) Study on properties of recycled tire rubber modified asphalt mixtures using dry process. Constr Build Mater 21(5):1011–1015. https://doi.org/10.1016/J.CONBUILDMAT.2006.02.004

da Silva L, Benta A, Picado-Santos L (2018) Asphalt rubber concrete fabricated by the dry process: laboratory assessment of resistance against reflection cracking. Constr Build Mater 160:539–550. https://doi.org/10.1016/J.CONBUILDMAT.2017.11.081

Farina A, Zanetti MC, Santagata E, Blengini GA (2017) Life cycle assessment applied to bituminous mixtures containing recycled materials: crumb rubber and reclaimed asphalt pavement. Resour Conserv Recycl 117:204–212. https://doi.org/10.1016/J.RESCONREC.2016.10.015

Farouk AIB, Hassan NA, Mahmud MZH, Mirza J, Jaya RP, Hainin MR, Yaacob H, Yusoff NIM (2017) Effects of mixture design variables on rubber–bitumen interaction: properties of dry mixed rubberized asphalt mixture. Mater Struct 50(1):12. https://doi.org/10.1617/s11527-016-0932-3

Feiteira Dias JL, Picado-Santos LG, Capitão SD (2014) Mechanical performance of dry process fine crumb rubber asphalt mixtures placed on the Portuguese road network. Constr Build Mater 73:247–254. https://doi.org/10.1016/J.CONBUILDMAT.2014.09.110

Hernández-Olivares F, Witoszek-Schultz B, Alonso-Fernández M, Benito-Moro C (2009) Rubber-modified hot-mix asphalt pavement by dry process. Int J Pavement Eng 10(4):277–288. https://doi.org/10.1080/10298430802169416

Shook JF (1990) Experimental construction of rubber-modified asphalt pavements in New York State

Lawrence C, Killackey B, Lynch D (1991) Experimental hot mix pavement with scrap tire rubber at Thamesville, Ontario

Lo Presti D (2013) Recycled tyre rubber modified bitumens for road asphalt mixtures: a literature review. Constr Build Mater 49:863–881. https://doi.org/10.1016/j.conbuildmat.2013.09.007

Loderer C, Partl MN, Poulikakos LD (2018) Effect of crumb rubber production technology on performance of modified bitumen. Constr Build Mater 191(2018):1159–1171. https://doi.org/10.1016/j.conbuildmat.2018.10.046

Moreno F, Rubio MC, Martinez-Echevarria MJ (2011) Analysis of digestion time and the crumb rubber percentage in dry-process crumb rubber modified hot bituminous mixes. Constr Build Mater 25(5):2323–2334. https://doi.org/10.1016/J.CONBUILDMAT.2010.11.029

Moreno F, Rubio MC, Martinez-Echevarria MJ (2012) The mechanical performance of dry-process crumb rubber modified hot bituminous mixes: the influence of digestion time and crumb rubber percentage. Constr Build Mater 26(1):466–474. https://doi.org/10.1016/J.CONBUILDMAT.2011.06.046

Rahman M, Airey GD, Collop A (2005) The mechanical properties of the dry process CRM asphalt mixtures following short-term and long-term ageing. Retrieved from. https://www.dora.dmu.ac.uk/handle/2086/6881

Rahman MM, Airey GD, Collop AC (2010) Moisture susceptibility of high and low compaction dry process crumb rubber-modified asphalt mixtures. Transp Res Rec: J Transp Res Board 2180(1):121–129. https://doi.org/10.3141/2180-14

Svoboda J, Vaclavik V, Dvorsky T, Klus L, Zajac R (2018) The potential utilization of the rubber material after waste tire recycling. In: IOP conference series: materials science and engineering, vol 385, no 1, p 12057

Thomas BS, Gupta RC, Panicker VJ (2016) Recycling of waste tire rubber as aggregate in concrete: durability-related performance. J Clean Prod 112:504–513. https://doi.org/10.1016/J.JCLEPRO.2015.08.046

Wang T, Xiao F, Zhu X, Huang B, Wang J, Amirkhanian S (2018) Energy consumption and environmental impact of rubberized asphalt pavement. J Clean Prod 180:139–158. https://doi.org/10.1016/J.JCLEPRO.2018.01.086

Xie Z, Shen J (2013) Long-term performance of SMA mixtures added with crumb rubbers in dry process. In: Airfield and highway pavement 2013, pp 1145–1155. American Society of Civil Engineers. https://doi.org/10.1061/9780784413005.096

Xue J, Shinozuka M (2013) Rubberized concrete: a green structural material with enhanced energy-dissipation capability. Constr Build Mater 42:196–204. https://doi.org/10.1016/J.CONBUILDMAT.2013.01.005

Hot, Warm and Cold Recycling

100% Recycling of Low-Temp Asphalt for Minor Roads – Lab Compaction and Traffic Simulation

Christiane Raab[1(✉)], Manfred N. Partl[1], and Cédric Bensa[2]

[1] Road Engineering/Sealing Components, EMPA, Swiss Federal Laboratories for Material Science and Technology, Dübendorf, Switzerland
christiane.raab@empa.ch

[2] Civil Engineering, Polytech Clermont-Ferrand, Aubière, France

Abstract. Triggered by early experience in Sweden, the international project "Optimal recycling of reclaimed asphalt pavement" (ORRAP) for minor roads in the Upper Rhine region aims at developing a new strategy for 100% reclaimed asphalt pavement (RAP) at ambient temperature without adding bituminous binders. The still ongoing research involves laboratory experiments as well as in situ test sections. The link between small scale laboratory experiments and in situ testing is provided by medium scaled traffic simulation in the laboratory. The paper describes first results on compaction evaluation in the laboratory using different methods as well as traffic simulation with a medium scaled mobile traffic load simulator. The results show that compaction in laboratory at ambient temperature is very difficult to achieve and that the technology from Sweden as such is not practicable for the Upper Rhine Region, because of the required long curing times. However, it was found that compaction at low temperatures (60 °C) appears possible and provides promising results regarding stability and rutting.

Keywords: Reclaimed asphalt pavement RAP · 100% recycling · Low-temperature asphalt · Minor roads · Compaction · Rutting test · Traffic simulator

1 Introduction

The Upper Rhine region combining regions of Germany, France and Switzerland along there corner area between Basel and Strasbourg has a densely populated and busy regional road network. Renewal and maintenance operations generate large quantities of RAP (recycled asphalt pavement) piling up to voluminous stockpiles of ecological, economic and political relevance. Nowadays, RAP is either reused for hot or low temperature mixtures by adding new material, either in form of single material components or a certain percentage of new asphalt mixture (Hugener and Seeberger 2015); (Aurangzeb et al. 2012). In case of hot recycling, bituminous binders and/or rejuvenators are added whereas for low temperature recycling emulsions, foam bitumen and other components are added (Zaumanis et al. 2013), (Cannone Falchetto et al. 2019). By applying 100% recycling technology for minor roads using RAP at ambient

M. Pasetto et al. (Eds.): ISAP APE 2019, LNCE 48, pp. 235–245, 2020.
https://doi.org/10.1007/978-3-030-29779-4_23

temperature without adding new bituminous binder, the current RAP recycling rate could be greatly increased, minimizing environmental impacts (CO_2 emissions, energy consumption and consumption of natural resources) and reducing road rehabilitation costs. Thus, significant ecological and economic gains could be realized for the management bodies of low traffic communal roads, which account for about 50% of the road network in the Upper Rhine region.

The ongoing project “Optimal recycling of reclaimed asphalt pavement” (ORRAP) (ORRAP 2006–2020) is based on successful experiences in Sweden where the method was carried out in field trials on municipal roads and on low traffic highways several years ago (Jacobson 2002), (European Project FP7th DIRECT-MAT 2007–2011). Here, 100% RAP aggregates were used at ambient temperature for base courses, which were later covered with hot asphalt surface layers. According to Swedish experience the base courses built at ambient temperature require post-compaction by traffic as well as curing to settle and gain their ultimate strength. Therefore, they need to be left uncovered for at least 6 months. However, during this time serious restrictions on the traffic regime, such as speed limits, must be imposed to the drivers. Since this long maturing process was not considered a feasible strategy for the 3 countries involved in ORRAP, the question of slightly increasing compaction temperature was one proposal to solve this problem.

The ORRAP project is composed of 3 phases. In the first phase small scale laboratory testing is done. In a second phase a medium scale traffic simulation in the laboratory is carried out providing the link to the third phase which is the construction of an in situ low traffic test section with 100% RAP. Since the research is a collaboration between 3 different country regions, in all phases RAP from the 3 different locations in France, Germany and Switzerland are chosen and tested.

2 Objective

The objective of this paper is to present the first results and conclusions of the small and medium scale compaction evaluation in the laboratory with different compaction methods using the RAP material from Switzerland. Further, results from the traffic simulation with the medium scaled traffic load simulator and the large wheel rutting tester are shown and discussed.

3 Material

The material, discussed in the following, was provided by an associate Swiss partner. It was taken from an existing unprotected open-air pile of stored RAP. The material was processed to the size of 0/16 mm on site and transported to the Empa laboratory in two big bags.

Prior to the material investigation and testing, the asphalt aggregate was homogenized and portioned by pouring it through a riffle box as described in the European standard (EN 932-1 1996) in the laboratory. Figure 1 and Table 1 present the material characteristics of the Swiss RAP. The presented values in Fig. 1 are mean values of two material characterizations called RAP1 and RAP2 in the depiction of the grading curve.

Table 1. RAP material characteristics (mean values)

Bitumen content [mass-%]	Density [kg/m3]	Water content [%]
6.2	2.371	6.0

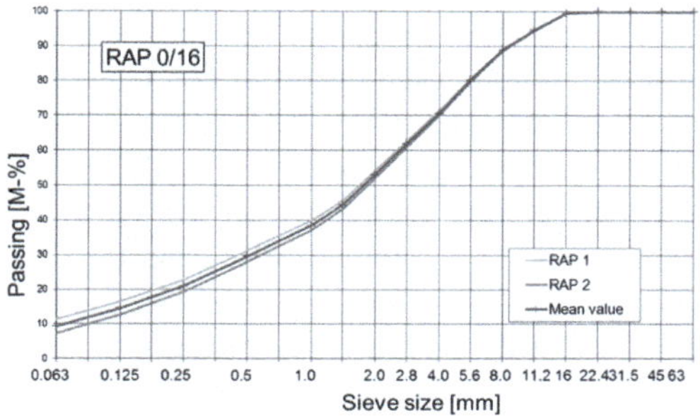

Fig. 1. Gradation of aggregates in 0/16 mm RAP (22.4 RA 0/16 according EN 13108-8)

4 Testing

Testing was conducted in two steps: In a first step the compatibility of small laboratory specimens was investigated, while the second step consisted of the compaction of medium size specimens. Further, the medium seize specimens were used for an investigation with the standardized, so-called, large wheel rutting tester as well as the laboratory medium scaled mobile traffic simulator for determining their stability and rutting performance.

4.1 Compaction of Laboratory Small Size Specimens

For the compaction of small laboratory size specimens, two different compaction methods were used: Marshall compaction and gyratory compaction (specimen mould size Ø150 mm).

In both cases, different compaction efforts (number of blows or gyrations) and temperatures were applied after trials with standard compaction effort and ambient temperature (ca. 20 °C). The RAP was not dried having the determined water content of about 6% as mentioned in Table 1 and the residual water content after heating and compaction was not determined.

Table 2 provides an overview of compaction methods, effort and RAP temperature.

Table 2. Compaction method, effort and temperature

Compaction method	Compaction effort [blows/gyrations]	RAP Temperature [°C]			
Marshall	50/100	20	60	80	-
Gyratory	204/410	20	60	80	100

From Marshall compaction at ambient temperature (20 °C), applying 50 blows on each side, it was not possible to achieve stable specimens, which can be unmolded without falling apart and tested. Using 100 blows on each side increased the specimen stability but, still, did not prove sufficient. The gyratory compaction results were similar leading to a collapse of the specimens during unmolding.

On the other hand, Fig. 2 demonstrates that compaction efforts at higher temperatures resulted in an immediate improvement of the specimens' stability in both compaction cases.

Using temperatures above ambient temperature even made it possible to determine Marshall stability and flow. Although the results at 60 °C are far below the values of regular asphalt mixtures, they show that 100% RAP specimens without adding any new material can be compacted achieving at least some stability. At a compaction temperature of 80 °C Marshall stability was found in the range of regular hot asphalt mixtures for low volume roads. The volumetrically determine air void content of the Marshall specimens was 10.8 vol-%. Table 3 presents the achieved Marshall characteristics for 3 specimens each (M1 to M3) at compaction temperatures of 60 °C and 80 °C and compaction efforts of 50 and 100 blows.

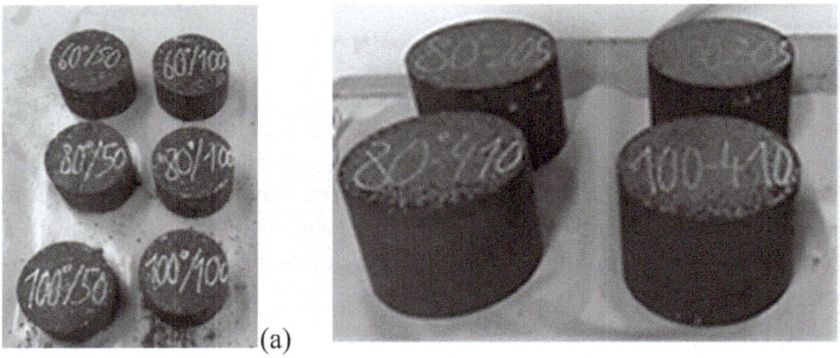
(a) (b)

Fig. 2. Marshall (a) and Gyratory (b) specimens compacted at different temperatures (first number) and efforts (second number)

Table 3. Marshall values with 50 and 100 blows at 60° and 80 °C

	50 blows			100 blows		
	M1	M2	M3	M1	M2	M3
60°						
Height [mm]	70.5	68.3	70.0	67.5	68.3	68.2
Stability [kN]	3.2	4.2	4.0	7.1	7.8	7.9
Flow [mm]	3.9	4.0	6.8	3.0	3.4	3.1
80°						
Height [mm]	65.9	64.8	65.3	65.6	64.7	65.2
Stability [kN]	11.5	12.0	11.8	13.0	14.0	13.5
Flow [mm]	3.6	3.4	3.1	3.9	2.7	2.8

Figure 3 gives an overview on the heights of the Marshall specimens for different compaction efforts and temperatures. According to Table 3 and Fig. 3 Marshall stability clearly increases with increasing compaction effort and temperature. This means that low temperature may be somehow compensated by increasing compaction effort.

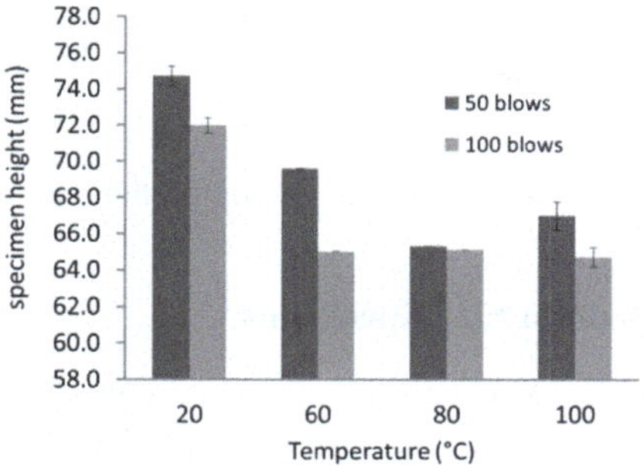

Fig. 3. Height of Marshall specimens in function of compaction effort and temperature

Similar observations can be made for the gyratory compaction. Figure 4 indicates a significant difference between compaction at ambient and elevated temperatures up to 100 °C. No significant fume emission was observed. From Fig. 4a it appears that a compaction already at 60 °C results in similar compaction over time as at 80 °C and 100 °C. Figure 4b presents the parameters K as a function of H for the linear logarithmic approach of the different compaction curves

$$H(N) = H(N = 1) - K \ln(N) \quad (1)$$

where, N denotes the number of gyrations, H the height of the specimen and K a constant describing the compaction rate.

Figure 4b clearly demonstrates that a large height H(N = 1) at the beginning of the gyrations results in a comparatively steep, i.e. fast compaction. Interestingly, whereas the mixtures compacted at 60 °C and 80 °C have identical compaction characteristics, the H(=1) and K values of the 100 °C mixture are slightly trending back towards those of the 20 °C mixture. In spite of the fact that the presented gyratory results are confirmed by only one replica, this observation may be an indication that heating RAP too much may further increase the viscosity and stiffening behavior of the existing RAP binder, leading to a compaction behavior that seems closer to the 20 °C mixture.

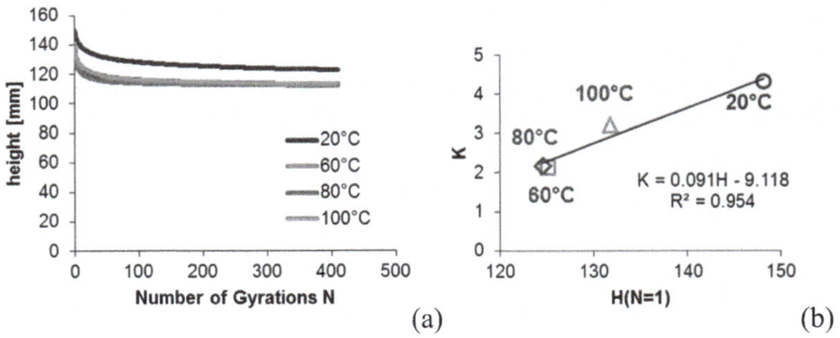

Fig. 4. Height of Gyratory specimens vs. compaction effort and temperature

4.2 Compaction of Medium Size Specimens

Parallel to the specimens for Marshall and gyratory compaction, medium size specimens (300 mm x 180 mm, height 100 mm) were produced. In a first version, specimens were produced with the rubber tire compaction as normally required for the large wheel rutting tests (European Standard EN 12697-22). In a second version specimens of the same size were made with the same compaction machine but now using an Empa-designed steel roller, as depicted in Fig. 5 (Raab et al. 2017). As for the compaction of small size specimens, the temperature was varied between ambient temperature (20 °C) and 80 °C for both versions of compaction.

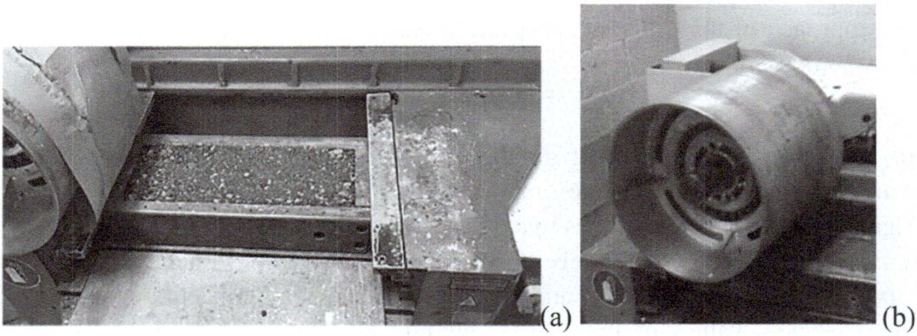

Fig. 5. Large rutting test compactor modified with steel roller (a) during compaction, (b) detail

Again, 20 °C was found insufficient for producing stable samples when demolding (Fig. 6a). In contrast, self-supporting samples could be fabricated already with compaction temperatures of 60 °C (Fig. 6b).

Fig. 6. Large rutting test specimens compacted with modified steel roller. (a) Compaction at 20 °C, (b) compaction at 60 °C

In a further step the compaction of large specimens for the laboratory scale traffic simulator was evaluated.

First compaction experiments at ambient temperature with a plate compactor, as used for the sidewalk compaction, were not very successful (Fig. 7). In these experiments specimens with the size of 1800 × 400 × 80 mm were tried to be compacted laterally confined by a wooden frame (Fig. 7).

Fig. 7. Compaction of specimen for use with traffic simulator using a plate compactor

Through additional heating of the surface with a torch, it appeared possible to achieve more or less compacted specimens. However, the result could not be considered for traffic simulation since the surface showed large areas of segregation.

In a next step the construction of an asphalt slab 1300 mm × 430 mm × 65 mm in a special wooden frame was carried out using a roller compactor (Fig. 8). For this purpose RAP was heated up to a temperature of 60 °C before being filled into the wooden frame in Fig. 8.

The compaction was done manually by moving the roller back and forth in static compaction mode without vibration. Figure 8 depicts the specimen during compaction. It was found that the total aggregate loss was only about 0.1% by weight and the specimen surface showed no in plane segregation (see Fig. 8).

Fig. 8. Specimen during compaction,

4.3 Rutting Testing with Laboratory Scaled Mobile Traffic Simulator

After construction, the specimen was tested with the laboratory scaled traffic simulator MMLS3 (Model Mobile Load Simulator), developed by Hugo (Hugo and Epps 2004) applying a scaled tire load of 2.1kN in one trafficking direction with four 1.05 m distant pneumatic 300 mm wheels that were inflated to 600 kPa (see Fig. 9). The machine (length × width × height = 2.4 × 0.6 × 1.2 m^3) enables about 7200 load applications per hour, i.e. at a speed of 2.6 m/s. This corresponds to a loading frequency of about 4 Hz for a measured tread length of 0.11 m. Testing was performed during hot summer month at ambient temperature (between 22 °C and 28 °C) up to 80’000 load passings without lateral wandering of the loading tires.

Rutting was measured with an automatic profilometer at 3 different points within the wheel path of the MMLS3, one in the middle and two others in the same distance of 300 mm from the middle.

Fig. 9. Traffic Load Simulator MMLS3 situated on top of the constructed specimen

The rutting test results are depicted in Fig. 10. Overall, rut depth is very low with a maximum of 1.6 mm after 80’000 load passings. The difference between profile 1 and profiles 2 and 3 is due to the fact that rutting is more profound at the location were the MMLS3 wheel touches down (profile 1).

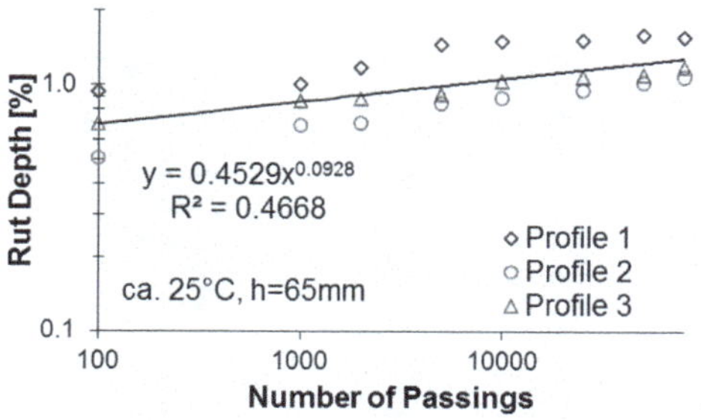

Fig. 10. Rut depth measurements at the 3 profiles

4.4 Rutting Testing with Large Wheel Rutting Tester

For rutting testing with the standardized, so-called, large wheel device, samples were compacted at 60 °C in steel molds as described in Sect. 4.2. In accordance to the European standard (European Standard EN 12697-22, 2003), rutting tests were conducted at 60 °C up to 30’000 cycles as required for heavy trafficked roads. In order to evaluate the influence of specimen height and the application for different layers as described in the standard, two different kind of 500 mm x 180 mm specimens were fabricated: Type A: surface and binder course with a height of 50 mm; type B: base course with a height of 100 mm. The rutting test results of all specimens under laterally confined conditions are shown in Fig. 11. Specimens of different heights after testing are depicted in Fig. 12.

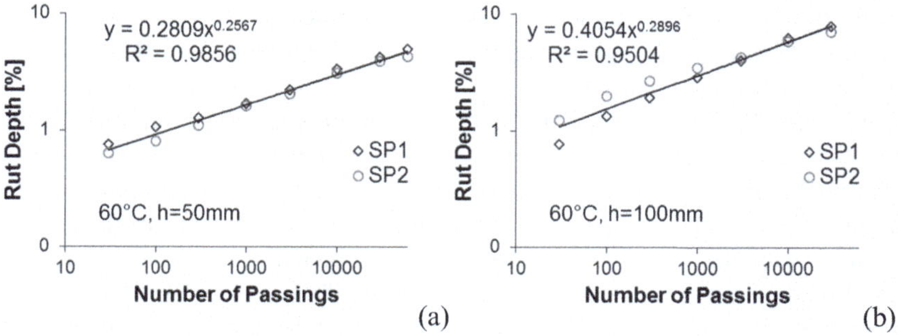

Fig. 11. Rut depth measurements of two specimen each, height 50 mm (a), height 100 mm (b)

From Figs. 11 and 12 follows that rut depth are low, especially considering the fact that the material should only be used for low volume roads, eventually covered with a surface course and being aware of the unlikeliness, that the whole layer under service would ever reach temperatures of 60 °C. Values for specimens of both heights even do not exceed the limit for road category H (very heavy traffic) with <10% of height at 30'000 cycles according to the Swiss standard Annex (SN 640431, 2014), (EN 18103-1. (2008). As expected, rutting is more prominent for the specimens with 100 mm height.

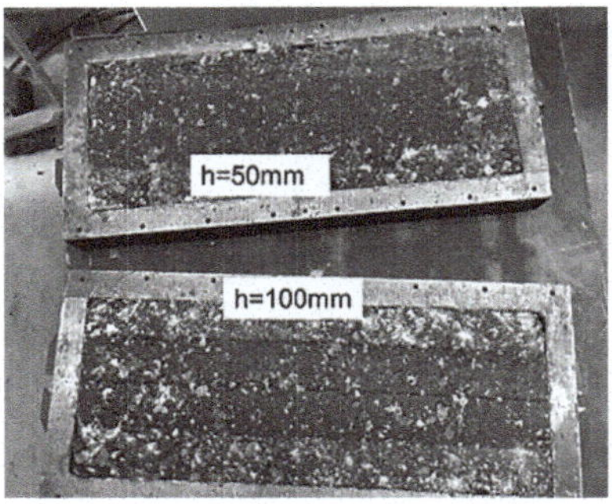

Fig. 12. Specimens after rutting test of 30'000 cycles.

5 Conclusions

In the present study, a laboratory compaction evaluation of 100% recycled asphalt aggregate (RAP) using different methods and specimens' sizes as well as a study on the rutting behavior was carried out to evaluate both the potential of compacting RAP at ambient temperatures and its effect on mechanical characteristics and selected performance properties. The study clearly shows the potential of using asphalt aggregates at low compaction temperatures and is therefore very promising for a successful in situ installation of these materials as planned in the ORRAP project.

Although compaction of small and medium size specimens at ambient temperature of 20 °C turned out not sufficient, it was found that increasing the temperature by only very little (60 °C) changed compaction behavior drastically and, with this, the possibility of constructing stable and rutting resistant asphalt aggregate samples without long curing times as proposed in a Swedish study. However, the study also suggests a certain risk that heating RAP too much (100 °C) may even influence the compaction behavior in a negative way. Investigation of rutting resistance with the large wheel rutting tester at 60 °C as well as the scaled traffic simulator MMLS3 at ambient temperature, both under laterally confinement of the specimens, generally produced low rut depth. Specimens investigated with the large wheel device even fulfilled the requirements for heavy trafficked roads. From the investigation and findings it seems feasible to construct low volume roads with 100% of RAP at low compaction temperature as it is the goal of the ORRAP project. If possible, the compaction effort should be increased. Raising the compaction temperature by only 40 °C might also be a way to accelerate in situ compaction, thus avoiding RAP base courses uncovered for a considerable amount of time before placing the hot mix surface layers.

References

Aurangzeb Q, Al-Qadi IL, Abuawad IM, Pine WJ, Trepanier JS (2012) Achieving desired volumetrics and performance for mixtures with high percentage of reclaimed asphalt pavement asphalt materials and mixtures 2012. Transp Res Rec: J Transp Res Board 2:34–42

Cannone Falchetto A, Porot L, Riccardi C, Hugener M, Tebaldi G, Dave E (2019) Effects of rejuvenator on reclaimed asphalt binder: an exploratory study of the RILEM TC 264-RAP Task Group 3. In: Rilem Bookseries, RILEM 252-CMB symposium. Chemo-mechanical characterization of bituminous materials, vol 20, pp 195–200. https://doi.org/10.1007/978-3-030-00476-7_31

European Standard EN 12697-22 (2003) Bituminous mixtures - Test methods for hot mix asphalt - Part 22: Wheel tracking, CEN, Brussels

European Project: FP7th project DIRECT-MAT Dismantling an RECyling Techniques for Road MATerials (2007-2011)

European Standard EN 932-1 (1996) Tests for general properties of aggregates - Part 1: methods for sampling, CEN, Brussels

Hugener M, Seeberger M (2015) Optimum RAP content in hot mix asphalt recycling. In: 4th ISAP APE symposium, 20–21 November, Tokyo, 11 p

Jacobson T (2002) Reclaimed crushed asphalt – experiences from field studies examples from Jönkönping, Västeras, Goteborg municipal roads. (Asfaltgranulat som obundet bär- och förstärkningslager. Återvinning av krossad asfalt som bär- och förstärkningslager. Del 2 – Erfarenheter från fältstudier) VTI notat 32

Hugo F, Epps, A (2004) Significant findings from accelerated pavement testing, NCHRP Synthesis 325, TRB, Washington

ORRAP (2016–2020) https://www.orrap.org/

Raab C, Arraigada M, Partl MN, Schiffmann F (2017) Cracking and interlayer bonding performance of reinforced asphalt pavements. Eur J Environ Civ Eng (EJECE) 21(sup1):14–26. https://doi.org/10.1080/19648189.2017.1306462

Swiss Standard Annex SN 640431 to European Standard EN 13108-1 (2014) Mischgutanforderungen – Teil 1: Asphaltbeton

Zaumanis M, Mallick R, Frank R (2013) Evaluation of rejuvenator's effectiveness with conventional mix testing for 100% reclaimed asphalt pavement mixtures. TRB Rec: J Transp Res Board 2370(1):17–25

Impacts of Recycling Agent on Superpave Mixture Containing RAP

Sujit Kumar Pradhan[1(✉)] and Umesh Chandra Sahoo[2]

[1] Civil Engineering Department, Indira Gandhi Institute of Technology, Sarang 759146, Odisha, India
skp14@iitbbs.ac.in

[2] School of Infrastructure, Indian Institute of Technology, Bhubaneswar 752050, Odisha, India
ucsahoo@iitbbs.ac.in

Abstract. This research presents the laboratory outcomes in which the influence of pongamia oil, a locally available non-edible oil as rejuvenator on the performance of hot mix asphalt containing reclaimed asphalt pavement was assessed. A research study has been taken up in three stages to assess the potential of rejuvenator. First stage focussed on determining optimum dosage of rejuvenator. The aged binder was then mixed with the rejuvenator at varying rates and the rheological properties were studied in terms of rutting, and fatigue using Dynamic Shear Rheometer. Second stage, thermal stability of the binder determined with the help of Thermo gravimetric analysis, which shows binder samples are stable up to 230 °C. From the results, it was concluded that 5% of pongamia oil can successfully rejuvenate the aged binder. Third stage, performance of HMA prepared with varying RAP content (30%, 40%, 50%, 60% and 70%) rejuvenated with Pongamia oil and control mixture was assessed according to superpave mix design guidelines. The performances of the rejuvenated HMAs were evaluated in terms of volumetric properties, indirect tensile strength, rutting and fatigue parameters. By comparing rejuvenated mixtures, 60% rejuvenated RAP mixture performs were found to perform superior than the control mix.

Keywords: Reclaimed · Asphalt pavement · Recycling · Rejuvenator · Pongamia oil · Rutting · Fatigue

1 Introduction

With the incessantly intensified material costs and environmental consciousness, the usage of RAP in pavement construction and rehabilitation has fascinated recognition attentions in the past few years as it reduces the utilization of virgin binder and aggregate. Higher RAP has the possibility to contribute fatigue, thermal cracking, and moisture damage leading to pavement failure. This may be due to the degradation of rheological properties of aged binder due to oxidation and volitization. Research shows rejuvenators have the ability to reinstate the properties of the aged binder. Researchers have used different types of oils, acids and other bio products as rejuvenators and studied their impact on the aged binder in terms of rheology, chemical analysis and

M. Pasetto et al. (Eds.): ISAP APE 2019, LNCE 48, pp. 246–255, 2020.
https://doi.org/10.1007/978-3-030-29779-4_24

thermal analysis such as waste cooking oil (Azahar et al. 2016), waste vegetable oil (Cao et al. 2018), waste engine oil (Qiu et al. 2018; Jia et al. 2014), soybean oil, maize oil (Portugal et al. 2017; Elkashef and Williams 2017), pine tree bio rejuvenator (Borghi et al. 2017), sunflower oil (Some et al. 2016). The dosage of rejuvenator plays a key role in effective restoring the properties of aged bitumen binder. To decide the quantity of RA required to attain a certain grade various criteria have been used, such as rheological properties (Yu et al. 2014), performance properties (Im et al. 2014; Shen et al. 2007). If the RAP content in the asphalt is higher, inclusion of rejuvenator improves the cracking resistance, rutting resistance and moisture susceptibility of the recycled mixtures (Im et al. 2014; Porot et al. 2017). Other studies have reported some degradation (Shen et al. 2007). The purpose of this research was to explore the effect of Pongamia oil as a rejuvenator for the higher RAP content in the hot mix asphalt mixes.

2 Materials and Methods

2.1 Materials

Virgin aggregate collected locally from Khurda district in the state of Odisha. VG30 grade of virgin binder was used in this research. RAP material was gathered from a local road rehabilitation project in the territory of Odisha, India. Pongamia oil utilized in the study from a local source in Odisha. Physical properties & chemical composition of pongamia oil are presented in Tables 1 and 2 respectively.

Table 1. Physical properties of Pongamia oil

Property	Value
Density	0.924 g/cm^3
Specific gravity	0.925
Viscosity	40.2 mm^2/sec
Fire point	230 °C
Flash point	225 °C

Table 2. Chemical composition of Pongamia oil

Fatty acid	Percentages
Palmitic	3.7%–7.9%
Stearic	2.4%–8.9%
Oleic	44.5%–71.3%
Linoleic	10.8%–18.3%
Linolenic	2.6%

2.2 Methodology

The first stage of the study consists of binder characterization based on determining the optimum dosage range of the rejuvenator. The extracted RAP binder mixed with the rejuvenator at varying percentages within dosage range. Second stage of the study includes the evaluation of rheological properties such as rutting (conducted at a frequency of 1.59 Hz within temperature range 40 °C to 70 °C) and fatigue parameter (AASHTO TP101, 2014) conducted on the binder sample at 10 °C to determine the fatigue lives by using DSR and thermal analysis by using TGA which characterize materials that exhibit mass loss due to decomposition, evaporation, desorption, reduction and weight gain due to oxidation and absorption. Third stage, to prepare a mix design to find the optimum binder content of control mix and RAP mixture by using rejuvenator. A superpave mixture design N_{design} = 109 was conducted according to with SHRP 1987, SP-2 for 19 mm nominal size of the aggregate. As virgin aggregate and RAP aggregate available in different stock piles gradation of control mix and for each percentages of RAP was adjusted to be close as possible to the target gradation according to superpave criteria. Total 6 nos of mixtures were designed to perform the performance testing. The first mixture was the control mixture. The second mixture (R30PON), third mixture (R40PON), fourth mixture (R50PON), fifth mixture (R60PON), sixth mixture (R70PON) contained 30%, 40%, 50%, 60%, 70% RAP and 5% rejuvenator by weight of aged binder respectively. Different mix performances were characterized which is specified in Fig. 1. Figure 2a presented the particle size distribution of various mix which satisfy the requirement according to the superpave specification.

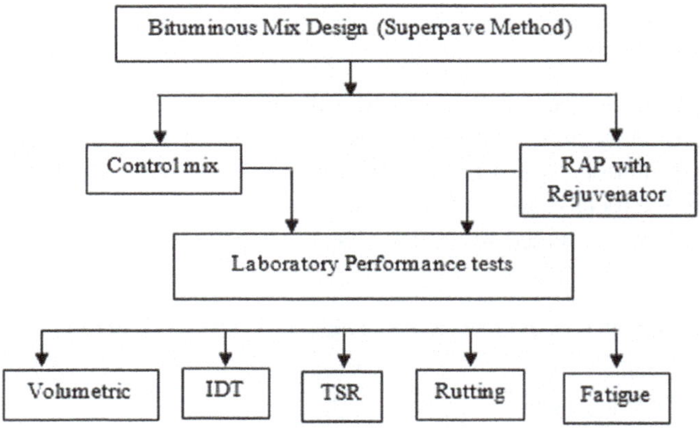

Fig. 1. Experimental plan

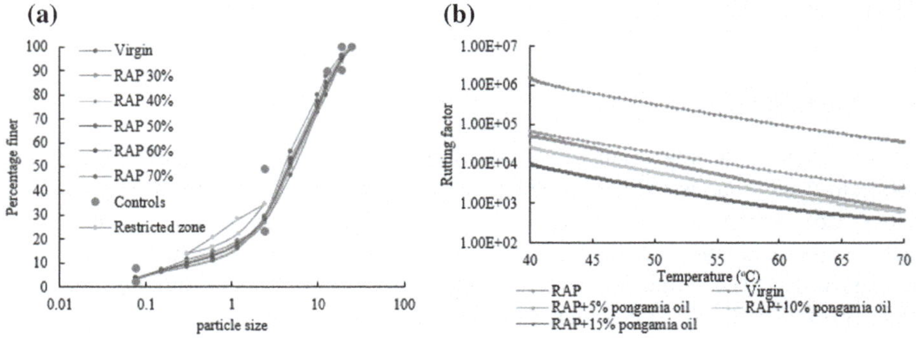

Fig. 2. (a) Particle size distribution curve of different mixtures (b) Rutting factor of binders

3 Results and Discussions

3.1 Rejuvenator Dosage

Rejuvenator dosage determination listed in Table 3.

Table 3. Rejuvenator dosage determination

State of the binder	Criteria	Rejuvenator dosage
Original state	At temp 70 °C	RC $\leq$ 17.10%
	G*/sinδ $\geq$ 1 kPa	RC $\leq$ 15.3%
RTFO residue	G*/sinδ $\geq$ 2.2 kPa	RC $\leq$ 18.02%
RTFO+PAV residue	G*sinδ $\leq$ 5000 kPa	RC $\geq$ 4.86%

From the above table it is seen that, rejuvenator dosage found between 4.86% to 15.3% so that pongamia oil mixed with the RAP binder at a rate of 5%, 10%, and 15% by weight of aged binder to study the performance properties.

3.2 Rutting

Rutting means the irretrievable deformation of asphalt during the loading. It is one of the major distresses in the asphalt pavements, especially at higher summer temperatures and under heavy axle loads. Binder with a higher G*/sinδbut a smaller flow deformation at a high temperature has a high rutting resistance. With increase in temperature, the bitumen in the mix becomes softer and hence rutting takes place. Asphalt with higher G*/sinδ value, has a higher rutting resistance. Figure 2b shows the variation of rutting parameter with temperature for rejuvenated RAP binders. This temperature sweep test was run with a temperature range of 40 °C to 70 °C, using oscillatory shear in strain-controlled mode at a frequency 1.59 Hz. The rutting resistance factor decreases as the temperature and rejuvenator dosage increases. So, rejuvenated binders that have an excessive addition of pongamia oil have a poor rutting resistance. It may be observed from the figure that the rutting resistance factor decreases with increase in temperature. RAP binders show highest rutting resistance as compared to the other binders. Rejuvenated binders with 5% pongamia oil showed better rut resistance in comparison with virgin, 10%, and 15% rejuvenated binders.

3.3 Fatigue

Figure 3a shows that variation of load cycles to fatigue failure for virgin, RAP, and rejuvenated RAP binder at 2.5% strain level. Rejuvenator significantly increases the fatigue life. For example, 5% pongamia oil rejuvenated RAP binder having fatigue life better than virgin binder. Figure 3b presents the plot between integrity parameter (C) versus damage intensity from the linear amplitude sweep test for virgin, RAP, and rejuvenated RAP binder. The estimation of C equivalent to 1 indicates the highest integrity with no damage, whereas the values of C equal to 0 represent the complete damage to the binder indicating a softer material. For better execution under fatigue, lower values of C_1 and higher value of C_2 are desirable, prompting higher C. The value of B (= 2α) reflects the inclination of fatigue life. It may be seen from Table 4 that with the inclusion of rejuvenator decrease C_1 and increase C_2 indicating that rejuvenator softens the binder and decrease the damage growth. This is due to addition of pongamia oil into aged binder decrease the α, specifying the fatigue life would be less empathetic to applied strains under traffic condition. So, inclusion of pongamia oil highly effective on fatigue resistance of binder.

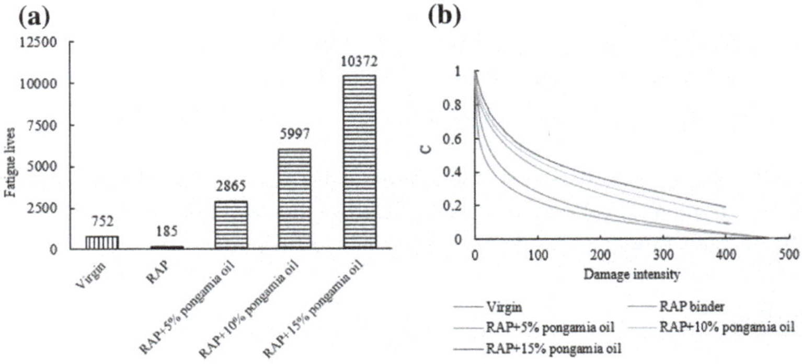

Fig. 3. (a) Fatigue lives of binders (b) Integrity parameter vs damage intensity of binders

3.4 Mass Loss

Figure 4a shows the mass loss curve of binders. Generally, aging of the binder is mainly due to the loss of volatiles and oxidation. It is a symptom of binder excretion during construction and long-time usage of pavement. The thermo gravimetric curves of virgin, RAP, and rejuvenated RAP show two distinct sections. The initial section showing minimum mass loss. After that decomposition occurs where the residue falls to a value of about 35%. The intense mass loss takes place at a temperature of 390 °C to 500 °C. We may conclude that rejuvenator aged binder thermally stable up to 230 °C.

From the rheological characterization, it was observed that 5% of Pongamia oil by weight successfully rejuvenated the aged binder. Therefore the same dosage was adopted to assess the performance of the mixtures as presented in the following paragraphs.

3.5 Volumetric

The summary of all mix designs parameters are presented in Table 4. The average optimum binder content (OBC) of the rejuvenated mixture decreases as the RAP content increases. The quantity of binder to be incorporated to a bituminous mixture cannot be too excessive or too little. The concept of determining the optimum binder content is to include adequate amount of binder so that the aggregate are fully coated with bitumen and the voids within the bituminous material are sealed up. Incorporated in the text and revised accordingly. The OBC of rejuvenated RAP was lower than control mix using soft binder. %VMA of the mixtures containing rejuvenated RAP were lower than virgin mixtures. It decreased with the increase of the rejuvenator content in mixes. As the compaction efforts for individual mixtures were the same, the variability of VMA due to combined effects of aged binder and pongamia oil. This tendency of volumetric properties of mixture may be due to softened binder that decreases the VMA and VFB. So, pongamia oil reduce the viscosity which enhances the workability of asphalt mixture. All the mixtures such as control and rejuvenated RAP mixture fulfil the volumetric according to superpave specifications.

Table 4. Volumetric of the mixtures

Specification	Virgin	RAP mixed with Pongamia oil				
		30%	40%	50%	60%	70%
%Max density at N_{ini}	85.5	85.45	85.3	85.3	85.48	85.33
%Max density at N_{des}	95.7	95.42	95.46	95.3	95.21	94.92
%Max density at N_{max}	97.58	97.48	97.19	97.2	96.82	96.88
OBC	5.5	5.29	5.27	5.24	5.22	5.2
Binder, RAP	Nil	1.1	1.21	1.36	1.48	1.92
Binder, Virgin	5.5	4.19	4.06	3.88	3.74	3.28
Gmm	2.562	2.571	2.552	2.54	2.546	2.549
%VMA	13.55	13.23	13.21	13.17	13.15	13.12
%VFB	70.42	69.64	69.59	69.48	69.39	69.27
$P_{0.075}$	2.95	3.21	3.39	3.54	3.73	3.9
P_{be}	3.39	3.45	3.52	3.57	3.61	3.71
Dust proportion (DP)	0.86	0.93	0.96	0.99	1.03	1.05

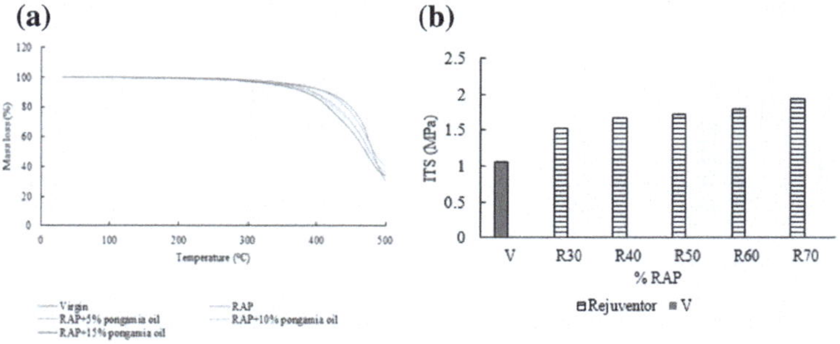

Fig. 4. (a) Mass loss versus temperature plot of binders (b) Indirect tensile strength of various mixture

3.6 Indirect Tensile Strength

Figure 4b shows the average measured indirect tensile strength (ASTM D6931, 2017) for the control and rejuvenated RAP mixtures. It was observed that inclusion of rejuvenator in the RAP resulted in an increase in ITS values. This suggests a decrease in the brittleness and a possible increase in fatigue life. When 30% RAP was used, the ITS value increased by 44.76% in comparison with virgin mix. Similarly, this trend was more prominent when 40%, 50%, 60%, 70% RAP was used, the ITS value increased by 59.04%, 64.76%, 71.42%, 85.71% respectively when compared with virgin mix. This showed that rejuvenated RAP mixtures having ITS results better than the control mixture consisting softer binder.

3.7 Moisture Sensitivity/Tensile Strength Ratio (TSR)

TSR (AASHTO T283. 2014) determined by taking the ratio between tensile strength of the conditioned samples and tensile strength of the control samples. TSR was used to assess the resistance of the asphalt mixtures to moisture-induced damage. A minimum TSR value of 80% is commonly specified. Figure 5a represents the TSR results of the rejuvenated RAP and control mixtures. The results show that all the mixture satisfies the minimum TSR value. As shown in Figure the control mix exhibited a satisfactory TSR value of 82.65%. With inclusion of pongamia oil in the mixture, it was observed that 60% rejuvenated RAP mixture have highest TSR value of about 93.13%. But TSR value increases from 30% to 50% RAP and 70% RAP mixture having value even less than control mixture. Mixture containing RAP 60% has preferred anti stripping properties better over different mixtures. So, we may say rejuvenated RAP have the almost better resistance of moisture susceptibility than virgin mixtures.

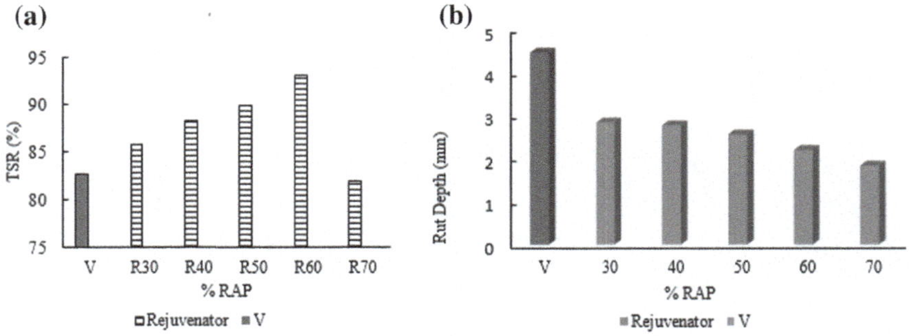

Fig. 5. (a) Tensile strength ratio of mixtures (b) Rut depth of bituminous mixtures

3.8 Rutting

Rutting (BS EN 12697-22, 2003) performed on the mixture for the nominal particle size of 19 mm. Specimens of size 305 mm × 305 mm and thickness of 50 mm were prepared at 7% air voids, using a roller compactor with a temperature of 60 °C, using 700 N wheel load. To determine the rutting susceptibility of the asphalt mixtures, a wheel tracking device was used following the European guidelines. Wheel-tracking machine, enable the test specimen in its cradle to be moved backwards and forwards under the loaded wheel in a fixed horizontal plane. The centre-line of the tyre track shall be not more than 5 mm from the theoretical centre of the specimen. The centre of the contact area of the tyre shall describe simple harmonic motion with respect to the centre of the top surface of the test specimen with a total distance of travel of 230 mm and a frequency of 26.5 load cycles per 60 s for the test device with a temperature of 60 °C within the chamber. The test was conducted for 20,000 passes (or 10,000 cycles) of wheel and the deformation was recorded using an internal data acquisition and control system. Figure 5b shows the rut depth of all mixture less than specified limiting value of 12.5 mm after 20000 passes or 10000 cycles. The control mixture has the maximum deflection of 4.46 mm due to the low stiffness values of the mixture. Inclusion of pongamia oil increased the rutting resistance and the concentrations of pongamia oil significantly decrease the rutting depth. All the mixture such as control and rejuvenated RAP mixtures are within the specified limits.

3.9 Fatigue

Figure 6 presenting the outcomes of indirect tensile fatigue test (according to (BS DD ABF, 2003). This test method is generally adopted because of the ease of sample preparation and testing. Constant stress mode of testing was opted and was conducted at 25 °C temperature. Fatigue testing frame is sufficiently rigid nature so that all the applied load is transmitted to the specimen and with top and bottom loading strip. It also consists of linear variable differential transformer having gauge length of at least 10 mm. A repeated diametrical load was applied at a frequency of 1 Hz, with 0.1 s loading and 0.9 s rest period. Occurrence of either 9 mm vertical deformation or total failure of the sample was considered as failure criteria. The sample size used here is cylindrical with thickness of 38 mm and diameter of 100 mm. Using pongamia oil enhanced the fatigue lives of the mixes in contrast to control mixture. Thus, their incorporation successfully reduces the stiffness of RAP. So, decreased stiffness of the rejuvenated RAP mixtures lead to increased fatigue life. Rejuvenator is effective in bringing down the embrittlement of the aged binder in RAP and consequently improving fatigue life performance. But rejuvenated 70% RAP mix showing fatigue life lower than virgin mixtures due to the high stiffness.

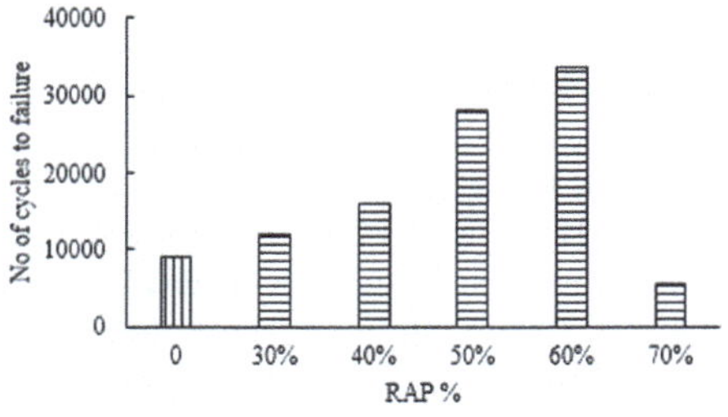

Fig. 6. Fatigue lives of mixtures

4 Conclusion

Based on the results the following conclusions were made. Overally, 5% pongamia oil was observed to successfully rejuvenate the aged binder in terms of long term performance such as rutting and fatigue. The TGA mass loss curves showed rejuvenated binders are thermal stable up to 230 °C. Volumetric results shows decrease in optimum binder content with increase in RAP containing pongamia oil, provide an economic benefit by decreasing the binder content. The average IDT strength increases with increase in RAP content even more than virgin mixture and rutting decreases as compared to control mixture satisfying minimum criteria of less than 12.5 mm for 20,000 passes. So, increase rutting resistance achieved from the addition of pongamia oil in RAP. In terms of durability, all the mixture show better resistance to moisture induced damage. In consideration of fatigue life rejuvenated RAP mixture show better fatigue lives up to 60% but 70% RAP falls below the control mixture. By considering the finding of this research it can be deduced that it is feasible to design the HMA up to 60% RAP with rejuvenator. The utilization of rejuvenated RAP may be an aggressive solution for sustainable ecological effect identified with virgin material and waste disposal.

Acknowledgements. The author's would like to thank School of Infrastructure, Indian Institute of Technology Bhubaneswar for laboratory support for carried out this research work.

References

Azahar WNA, Jaya RP, Hainin MR (2016) Chemical modification of waste cooking oil to improve the physical and rheological properties of asphalt binder. Constr Build Mater 126:218–226

Borghi A, Carrion AJB, Presti DL (2017) Effects of laboratory aging on properties of bio rejuvenated asphalt binder. J Mater Civ Eng 29(10):04017149:1-10

Cao X, Wang H, Cao X (2018) Investigation of rheological and chemical properties asphalt binder rejuvenated with waste vegetable oil. Constr Build Mater 180:455–463

Elkashef M, Williams RC (2017) Improving fatigue and low temperature performance of 100% RAP mixtures using a soybean-derived rejuvenator. Constr Build Mater 151:345–352

Im S, Zhou F, Lee R, Scullion T (2014) Impacts of rejuvenators on performance and engineering properties of asphalt mixtures containing recycled materials. Constr Build Mater 53:596–603

Jia X, Huang B, Bowers BF (2014) Infrared spectra and rheological properties of asphalt cement containing waste engine oil residues. Constr Build Mater 50:683–691

Porot L, Broere D, Wistuba M, Gronniger J (2017) Asphalt and binder evaluation of asphalt mix with 70% reclaimed asphalt. Road Mater Pavement Des 18:66–75

Portugal ACX, Lucena LCFL, Lucena AEFL (2017) Rheological properties of asphalt binders prepared with maize oil. Constr Build Mater 152:1015–1026

Qiu Y, Ding H, Rahman A (2018) Damage characteristics of waste engine oil bottom rejuvenated asphalt binder in the non-linear range and its microstructure. Constr Build Mater 174:202–209

Shen J, Amirkhanian S, Miller JA (2007) Effects of rejuvenating agents on superpave mixtures containing reclaimed asphalt pavement. J Mater Civ Eng 19:376–384

Some SC, Pavoine A, Chailleux E (2016) Evaluation of the potential use of waste sun flower rap seed oils-modified natural bitumen as binders for asphalt pavement design. Int J Pavement Res Technol 9:368–375

Yu X, Zaumanis M, Dos Santos S (2014) Rheological, microscopic, and chemical characterization of the rejuvenating effect on asphalt binders. Fuel 135:162–171

Evaluation of Reliability of RILEM Fragmentation Test

Francesco Preti, Stefano Noto(✉), Beatriz Chagas Silva Gouveia, and Gabriele Tebaldi

Department of Engineering and Architecture, Università di Parma, Parco Area delle Scienze, 181/A, 43124 Parma, Italy
stefano.noto@studenti.unipr.it

Abstract. Black Curve and White curve are a well-known method to fully characterize a RAP source, however involves a major effort in terms of laboratory equipment and time-consuming process. RILEM has validated two protocols for a fast RAP characterization: Fragmentation test and Cohesion test. In this study, the authors show the results of a reliability evaluation of the Fragmentation test. Considering this test evaluates the stability of aggregates gradation of the RAP under compaction, the authors have induced an artificial variation in the RAP gradation source to create different RAPs that can be considered just apparently the same. A certain amount of RAP aggregates has been gradually replaced with steel slags to assess the sensitivity of this test. One source of RAP has been used and three different batches were selected: 100% RAP, 50% RAP + 50% steel slags and finally 100% steel slags. Steel slag aggregates have been chosen because of its low Los Angeles abrasion coefficient and its increasingly use on hot mix asphalt with RAP. A series of laboratory tests have been performed using the modified Proctor, as required by the investigated RILEM protocol, and the percentage of material passing through a control sieve has been assessed. The obtained results are encouraging, as they fit with the initial assumption that RAP sources with different percentages of RAP clusters are differentiated by the Fragmentation test. Thus, a simple job site laboratory can provide a key information on the quality of the material to be used on field applications.

Keywords: RAP characterization · Fragmentation test · Protocol validation

1 Introduction

With regard to pavement maintenance and rehabilitation procedures, an important amount of Reclaimed Asphalt Pavement (RAP) is produced worldwide due to the milling operations of old pavements. RAP is a composite material which consists roughly in a blend of aggregates covered by the aged bitumen, and its particles may be either a single piece of stone aggregate coated with aged bitumen, or a cluster formed of fine coated aggregates bounded together by the bitumen. As a consequence of environmental, economical and/or technical reasons, RAP material has been largely used to produce recycled asphalt mixtures, either on hot or cold mixing processes.

M. Pasetto et al. (Eds.): ISAP APE 2019, LNCE 48, pp. 256–263, 2020.
https://doi.org/10.1007/978-3-030-29779-4_25

On hot recycling techniques, the amount of added RAP is usually never higher than 40%, and the material must be heated before mixed with the heated natural aggregate and the new bitumen. Although the use of RAP contributes to reduce the use of virgin material, possible problems related to cracking behavior and a shorter fatigue life should be taken into account on the mix design, as the aged bitumen content on the RAP, when heated, partially blends with the new bitumen, increasing the stiffness of the final mixture (West et al. 2013).

On the other hand, cold recycling techniques are mostly used as base course in flexible pavement rehabilitation, and the RAP material is mixed together with foamed bitumen or bituminous emulsion without previous heating. The cold recycled mixtures may have up to 100% of RAP aggregates (Asphalt Academy 2009; Wirtgen GmbH 2012).

A series of studies on RAP characterization were conducted on several different laboratories around the world, supervised by the Technical Committee 237-SIB of RILEM, entitled "Testing and characterization of sustainable innovative bituminous materials and systems" (Partl et al. 2018). One of the goals of this committee was to find or develop simple, quick and repeatable tests to characterize RAP directly on job sites, and without the need to proceed with the extraction of the bitumen. As a result, some simplified test procedures to characterize RAP were proposed by the committee. According to the referred RILEM committee, the fragmentation test (Perraton et al. 2016) was developed to measure the resistance of the RAP particles to breaking down after carrying a modified Proctor Test (EN 13286-2). Therefore, RAP can be quickly evaluated on field, by verifying the amount of material passing on a control sieve after subjected to the induced strokes. This information can be very useful to understand how the gradation curve of the RAP may change during mixing and compaction on field, due to the breaking of fine aggregates clusters.

2 Objective and Scope

This case study has as main objective the investigation of the Fragmentation test proposed by the RILEM Technical Committee 237-SIB, concerning its sensitivity to distinguish RAP with different characteristics. More specifically, this study aims to verify if the fragmentation test can differentiate RAPs with different compositions, such as different percentages of old bitumen and different amount of fine clusters, accessing, indirectly how much their grading curves may change due to mixing and compaction operations.

To reach the proposed objective, the scope of the study was delimited on analyzing one type of RAP, artificially mixed with steel slag on different proportions, to assess if the fragmentation test is able to classify differently those materials.

3 Materials and Methods

This study characterized one type of RAP, produced in north Italy, from milling operations taken place on local roads. Black and white gradation curves were determined, and are shown on Fig. 1. The percentage of old bitumen was also determined, and is in average 4.3% of the total dry aggregate mass.

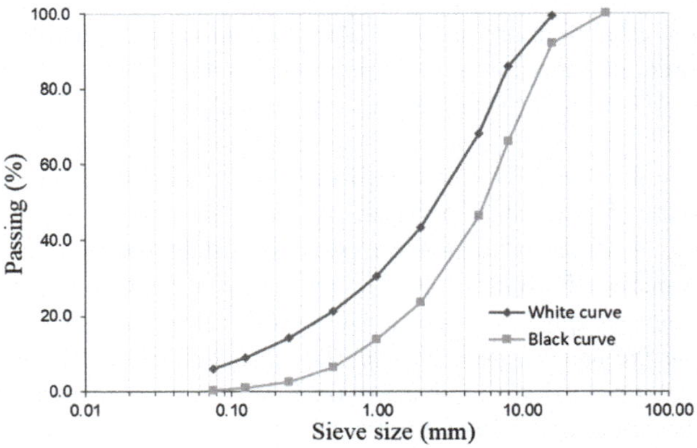

Fig. 1. White and black curve of the investigated RAP

The selected RAP was mixed in laboratory with steel slag, at a proportion of 50%, in order to artificially create a material with similar gradation characteristics but in a different composition. An initial evaluation showed that the original RAP contained no steel slag. The adoption of steel slag is justified to artificially simulate RAP mixtures with different characteristics in terms of compositions, such as different percentages of old bitumen and especially different percentage of RAP cluster in the mix. The used steel slags were provided already sieved and divided in separated bags, each one corresponding to one size class, as established by the fragmentation test procedure (5/10 mm; 10/15 mm; 15/20 mm; and 20/30 mm). As a consequence, a total of 3 different types of material, were tested: 100%RAP, 50%RAP/50% steel slag, and 100% steel slag. One replicate of each class and material was tested, assuming the RAP is a highly heterogeneous material.

3.1 The Fragmentation Test

According to the RILEM protocol, in general terms, the fragmentation test consists on the measurement of the material passing on a control sieve after a series of strokes of a normalized falling weight, as illustrated on Fig. 2.

The adopted procedures, as prescribed on the protocol, initiated with the sieving of the material on four different size classes: 5/10 mm; 10/15 mm; 15/20 mm; and

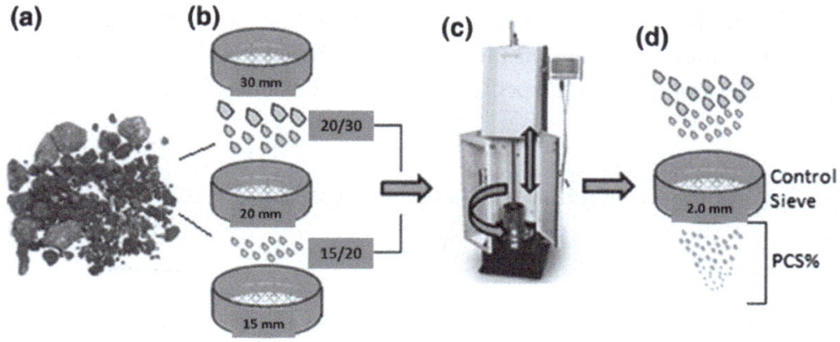

Fig. 2. Illustrative procedure and apparatus of the fragmentation test, adapted from Tebaldi et al. (2018)

20/30 mm. Each sample was tested under the modified Proctor procedure, which establishes 56 blows per each the five layers. After the compaction, the material was sieved on a control sieve of 2,0 mm. The coefficient of fragmentation was determined by the percentage of the weight of the material passing on the control sieve related to the total weight of the material before compaction.

The test was performed on three different temperatures, 5, 20 and 40 °C, for the 100% RAP and the 50%RAP/50% steel slag. The 100% steel slag was only tested at room temperature. The absence of binder on the steel slag indicates that the material should not be temperature dependent, supported by previous fragmentation tests performed on natural aggregates, which showed results do not vary with testing temperature (Tebaldi et al. 2018). Tested materials were conditioned on the required temperature for at least 4 h before running the modified Proctor.

As a final result, RAP materials can be classified on four different categories, considering the percentage of passing material at control sieve (%PCS), with focus on 5/10 mm fraction at 5 °C. These classification categories (B, C, D, and E) were originaly used by the RILEM Committee, based on the definitions of the normative NF EN 12620 standard. The Table 1 presents the categories and its limits, as proposed by the RILEM Committee (Tebaldi et al. 2018).

Table 1. RAP categories and limits proposed by RILEM (Tebaldi et al. 2018)

Category	B	C	D	E
Limits on %PCS class 5/10 mm @5 °C	≤ 7	≤ 9	≤ 11	≤ 14

4 Results and Discussions

The results of the investigation are presented on the Figs. 3, 4, 5, 6 and 7. The Fig. 3 shows the percentage of passing material at control sieve (%PCS) from the sample of the 5/10 mm fraction, at 5 °C. These results allow to access the classification of the material, as proposed by the RILEM Committee. The results indicate that the Fragmentation test was able to differentiate the materials (100% RAP, 50% RAP + 50% steel slag and 100% steel slag), as the samples were respectively classified as class D, C and B.

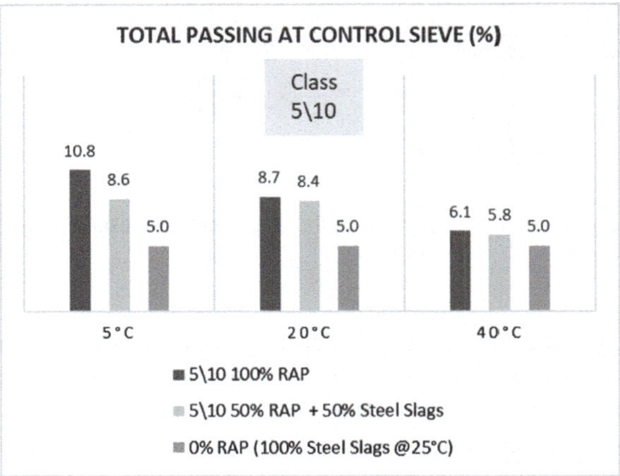

Fig. 3. %PCS by temperature, at class 5/10 mm

The results shown on Fig. 3 for all three tested temperatures also converge to the expected behavior of the artificially changed material (50% RAP + 50% steel slag), which present smaller variation of %PCS between tested temperatures, as compared to the original material (100% RAP), given the fact that the decrease of %PCS is due to the smaller amount of RAP on the material.

The Figs. 4, 5 and 6 report the final %PCS for each aggregates size ranges (classes) at three different required temperatures, 5 °C, 20 °C and 40 °C respectively. A higher %PCS for the two bigger size classes have been noticed, for each experimented mixture, at each tested temperature, indicating a higher amount of clusters for higher sized particles. With a focus on the total %PCS of the three mixtures, it is possible to observe a reduction of the fines passing on the control sieve, as the percentage of RAP aggregates decreases in the mixtures. This trend is detected at all three tested temperatures.

The diagram presented on Fig. 7 shows the variation of the total %PCS as a function of temperature, for both 100% RAP and 50% RAP + 50% Steel Slag. The values are standardized referring to the %PCS at 5 °C for each mixture. This graph is able to highlight a better understanding of the temperature susceptibility of the two analyzed mixtures, which represent different percentages of bitumen.

Looking at the slope of the 100% RAP material, the blue line, it is possible to notice a dramatic reduction in %PCS as the temperatures increases above 20 °C. This reduction may indicate that some amount of old bitumen on the RAP may soften as the temperate increases, holding together more fines, as the fragmentation test applies some compaction energy the material. On the other hand, the slope of the 50% RAP + 50% Steel Slag mixed material, the orange line, shows a softer decrease as a function of temperature. This behavior reflects the amount of bitumen inside the mixture, therefore, smaller amount of pre-existing clusters of fine materials, endorsing the ability of the test to distinguish two different mixtures.

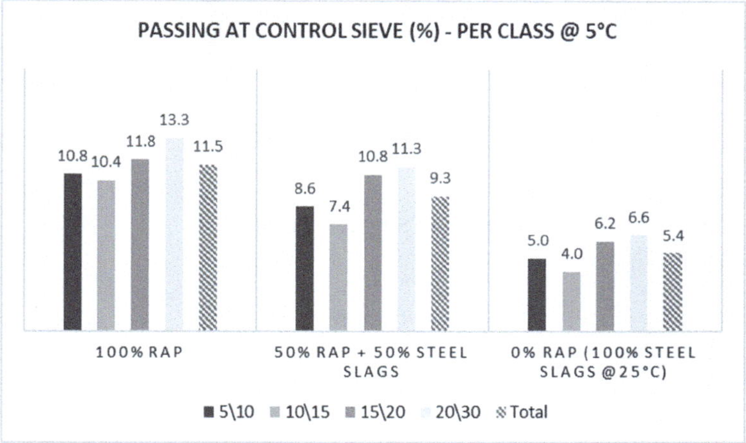

Fig. 4. %PCS by class, at 5 °C

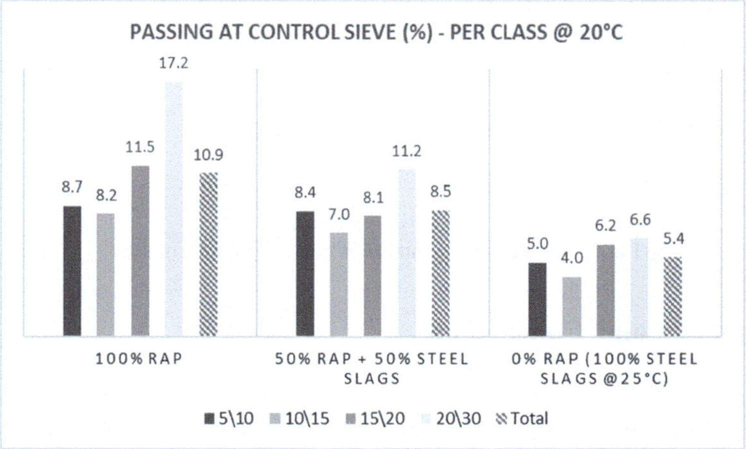

Fig. 5. %PCS by class, at 20 °C

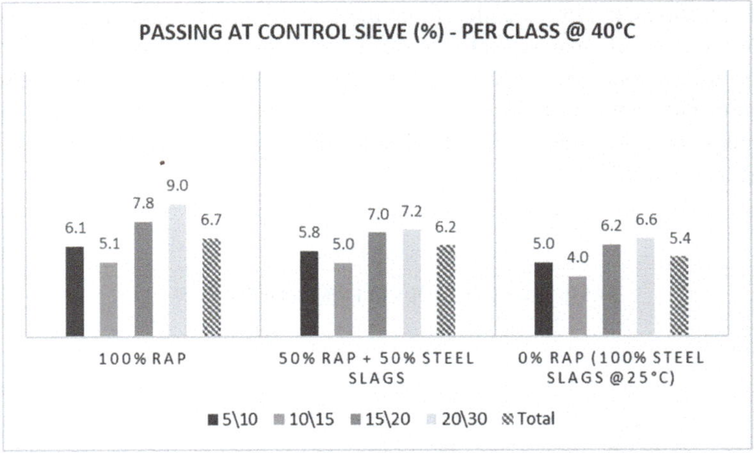

Fig. 6. %PCS by class, at 40 °C

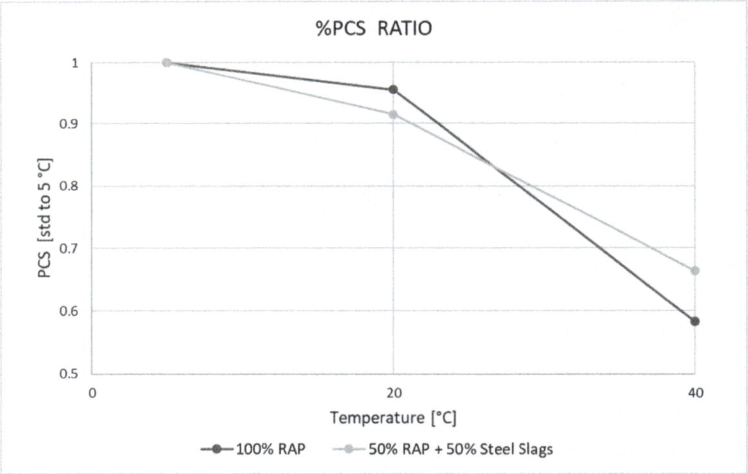

Fig. 7. Standardized total %PCS (Color figure online)

5 Summary and Conclusions

In the present study, a method has been presented to assess the reliability of Fragmentation Test, proposed by RILEM to quickly characterized RAP materials on job sites. A single source of RAP was evaluated. The composition of mixture was artificially and arbitrary modified, by adding an amount of steel slags equal to 50% of total aggregates weight, to each aggregate size range. The analysis of the data brings the authors to consider the test is sensitive to different type of RAP source, as it was able to differentiate the studied materials, giving them different classifications. In addition, the test was also able to show that materials with less amount of clusters (i.e. 50% RAP + 50% steel slag) have smaller variations on their grading curves.

On the other hand, results of the test made at the temperature of 40 °C may indicate that the mixing and compaction temperature has an important influence on the changings of the gradation curve of the materials containing RAP, and the variations are dependent on the RAP content in the mixture.

The results showed in this paper agreed with the results published by Perraton et al. (2016) and by Tebaldi et al. (2018), in the framework of the RILEM Technical Committee 237-SIB, they allow to say that the fragmentation test is able to differentiate RAP with different compositions, despite the analysis is run on equal size ranges.

References

Asphalt Academy (2009) Technical guideline: Bitumen stabilised materials. A guide for the design and construction of bitumen emulsion and foamed bitumen stabilised materials

EN 13286-2 (2002) Unbound and hydraulic bound mixtures – Part 2: test methods for laboratory reference density and water content – Proctor compaction

Partl MN, Porot L, Di Benedetto H, Canestrari F, Marsac P, Tebaldi G (2018) Testing and characterization of sustainable innovative bituminous materials and systems

Perraton D, Tebaldi G, Dave E, Bilodeau F, Giacomello G, Grilli A, Graziani A, Bocci M, Grenfell J, Muraya P, Pasetto M, Kuna K, Apeagyei A, Lo Presti D, Airey G, Jenkins K, Hajj E, Hugener M, Marsac P (2016) Tests campaign analysis to evaluate the capability of fragmentation test to characterize recycled asphalt pavement (RAP) material. In: 8th RILEM international symposium on testing and characterization of sustainable and innovative bituminous materials, pp 965–976. Springer, Dordrecht

Tebaldi G, Dave E, Hugener M, Falchetto AC, Perraton D, Grilli A, Presti DL, Pasetto M, Loizos A, Jenkins K, Apeagyei A (2018) Cold recycling of reclaimed asphalt pavements. In: Testing and Characterization of Sustainable Innovative Bituminous Materials and Systems, pp 239–296. Springer, Cham

West R, Willis JR, Marasteanu M (2013) NCHRP Report 752: improved mix design, evaluation, and materials management practices for hot mix asphalt with high reclaimed asphalt pavement content. Transportation Research Board of the National Academies, Washington, DC

Wirtgen GmbH (2012) Wirtgen cold recycling technology, 1st edn. Windhagen, Germany

Development of a Soybean-Based Rejuvenator for Asphalt Mixtures Containing High Reclaimed Asphalt Pavement Content

Mohammad Reza Pouranian, Reyhaneh Rahbar-Rastegar(✉), and John E. Haddock

Lyles School of Civil Engineering, Purdue University, West Lafayette, IN 47907, USA
rrahbar@purdue.edu

Abstract. In spite of its economic and environmental benefits, the use of reclaimed asphalt pavement (RAP) is limited due to the impact of aged binder on the performance of asphalt pavements. Currently, one of the most popular methods to mitigate the aged RAP binder is the addition of a rejuvenating agent. Such an agent acts to restore life to the aged binder thus making it less brittle. The main objective of this study is to evaluate a soybean-based byproduct as a RAP bio-rejuvenating agent. To achieve this objective, asphalt mixtures were designed containing various RAP types and content, and three different percentages of bio-rejuvenator. Test specimens of the various mixtures were produced and tested to determine complex modulus, flow number, and fatigue cracking characteristics. The results indicate that increasing the percentage of bio-rejuvenator improves the viscoelastic parameters and results in a significant improvement in fatigue cracking properties of asphalt mixtures containing RAP, but also slightly decreases rutting performance.

Keywords: Reclaimed asphalt pavement · Soybean acidulated soapstock · Rejuvenator · Cracking performance · Rutting performance

1 Introduction

Using reclaimed asphalt pavement (RAP) in the production of asphalt mixtures has increased in recent years due to economic and environmental benefits. The national asphalt pavement association (NAPA) reported that usage of RAP in asphalt mixtures reached 76.9 million tons in 2016 in the United States, a 37% increase compared to 2009 (Hansen and Copeland 2017). However, as RAP contents continue to rise, it becomes more challenging to counteract the RAP's asphalt binder brittleness, a result of oxidation during pavement life. The oxidation process results in decreased asphalt flexibility, adhesive capability (bonding between asphalt and aggregate), and cohesive capability (bonding inside asphalt). These effects have a significant impact on the cracking properties of mixtures. One method for counteracting this brittleness is the addition of a rejuvenating agent (Boyer 2000; Shen et al. 2007; Seidel and Haddock 2014; Elkashef et al. 2018; Zhang et al. 2019). In most studies associated with the application of rejuvenators in asphalt mixtures containing RAP, the main objective has

M. Pasetto et al. (Eds.): ISAP APE 2019, LNCE 48, pp. 264–273, 2020.
https://doi.org/10.1007/978-3-030-29779-4_26

been to attain an increase in the cracking performance without jeopardizing other mixture properties (Epps Martin et al. 2017). Also, with the aid of better rejuvenators, RAP percentages can be pushed higher, allowing the conservation of virgin materials and increasing economic benefits.

Findings from previous research show that soybean products, specifically soybean acidulated soapstock (SAS) can be incorporated as a rejuvenating agent (Seidel and Haddock 2014; Portugal et al. 2018). By adjusting the chemical fractions, it has been reported that SAS can effectively soften the aged RAP asphalt binder and do so in a less costly manner than the many current rejuvenation practices. The objective of this study is to evaluate asphalt mixtures containing RAP and SAS rejuvenator to determine if the addition of SAS can increase high-RAP asphalt mixture cracking performance at intermediate temperatures (fatigue) without overly impacting high temperature performance (rutting).

2 Mixture and Materials

To evaluate combinations of RAP and SAS rejuvenator in asphalt mixtures, the laboratory performance of asphalt mixtures containing SAS-rejuvenated RAP were compared to those asphalt mixtures containing RAP, but no rejuvenator. The experimental design incorporated three rejuvenator contents (0, 1, 3% of total weight of RAP binder), two RAP contents (25 and 45%), and two RAPs of varying age (RAP A, RAP B). Results of binder extraction for RAP A and B revealed that the binder performance grades of RAPs A and B were PG 94-4 and PG 94-10, respectively. Table 1 shows the experimental plan. The "x" in the table indicates the combinations tested. Table 2 shows the aggregate and RAP stockpile gradations along with the percentage of each component in the various mixtures. The volumetric properties of the mixtures are shown in Table 3.

Table 1. Experimental design

Rejuvenator content, %	0		1		3	
RAP content, %	RAP type					
	A	B	A	B	A	B
25	x	x	x	x	x	x
45	x	x	x	x	x	x

All mixtures were fabricated using a PG 70-22 virgin binder and 100 gyrations of Superpave Gyratory Compactor (SGC) for design traffic of 10 to 30 million equivalent single axle load (ESAL). To prepare specimens for laboratory performance testing, the rejuvenator was added directly to the RAP, prior to incorporating the RAP into the mixture.

Table 2. Aggregate gradations (percent passing) and specific gravities

	#11 Dolomite	Blast furnace slag	#24 Crushed sand	Natural Sand	RAP A	RAP B	Baghouse fines
Sieve size (mm)	Aggregate and RAP stockpile						
12.5	100	100	100	100	100	100	100
9.5	90.1	89.3	100	100	93.5	92.8	100
4.75	19.4	22.8	99.9	99.9	64.4	44.2	100
2.36	3.5	8.7	92.6	89.1	47.9	29.5	100
1.18	1.7	6.7	60.9	70.1	34.1	22.2	100
0.600	0	0	37.0	47.7	23.1	18.0	100
0.300	0	0	21.0	17.0	12.3	10.5	100
0.150	0	0	10.2	2.2	7.0	6.0	100
0.075	0	0	4.9	0.8	4.2	3.8	98.0
Bulk specific gravity	2.681	2.442	2.781	2.612	2.651	2.691	2.800
Mixture designation	Percent component in the mixture						
RAP A-25%	15	17	30	11	25	0	2
RAP A-45%	11	13	22	7	45	0	2
RAP B-25%	15	17	30	0	11	25	2
RAP B-45%	11	12	22	0	8	45	2

Table 3. Mixture volumetric properties

Mixture type	Pb (%)	Pbe (%)	Va (%)	VFA (%)	VMA (%)	Dust ratio
RAP A-25%	4.8	4.16	4.1	69.9	14.3	1.1
RAP A-45%	5.3	5.76	4.0	76.8	15.2	0.9
RAP B-25%	4.4	5.28	4.1	74.6	15.7	0.8
RAP B-45%	4.8	4.81	3.9	73.1	14.9	1.0

3 Performance Tests

3.1 Viscoelastic Properties

The performance of asphalt mixtures is highly influenced by viscoelastic properties. The lower the temperature or faster the vehicle speed, the stiffer the asphalt mixture. It is well accepted that a higher stiffness improves the rutting resistance, while it might deteriorate the cracking behavior. On the other hand, the higher relaxation capability of asphalt mixtures is desired for better cracking performance, while it might worsen the resistance to permanent deformation. Complex modulus testing is used to determine the stiffness (dynamic modulus) and relaxation capability (phase angle, δ) of asphalt mixtures over a range of temperatures and loading rates. A sinusoidal load is applied and the values of resultant strain (deformation) are measured to determine the stress-strain relationship. In this study, the test was performed at six frequencies (25, 10, 5, 1, 0.5, and 0.1 Hz) and three temperatures (4.4, 21.1 and 37.8 °C), in accordance with AASHTO, T 342-11. The dynamic modulus and phase angle master curves were obtained using the time-temperature superposition principal.

3.2 Fatigue Resistance

The stiffness and brittleness of asphalt mixtures affect their fatigue cracking behavior. The more brittle a mixture, the more likely fatigue cracking will occur. Laboratory testing can be performed to predict the fatigue cracking behavior of asphalt mixtures. The Semi-Circular Bend (SCB) test (AASHTO TP 124) is a relatively new method for evaluating the fatigue resistance of asphalt mixtures (Al-Qadi et al. 2015). In this test, a vertical load is applied to a pre-notched semi-circular specimen at a displacement rate of 50 mm/min and the crack propagates along the pre-cut notch. The fatigue resistance of an asphalt mixture is evaluated using two cracking indices, fracture energy (G_f) and flexibility index (FI). The definition of these indices is shown in Fig. 1. In general, the higher the values of G_f and FI, the better a mixture's cracking resistance.

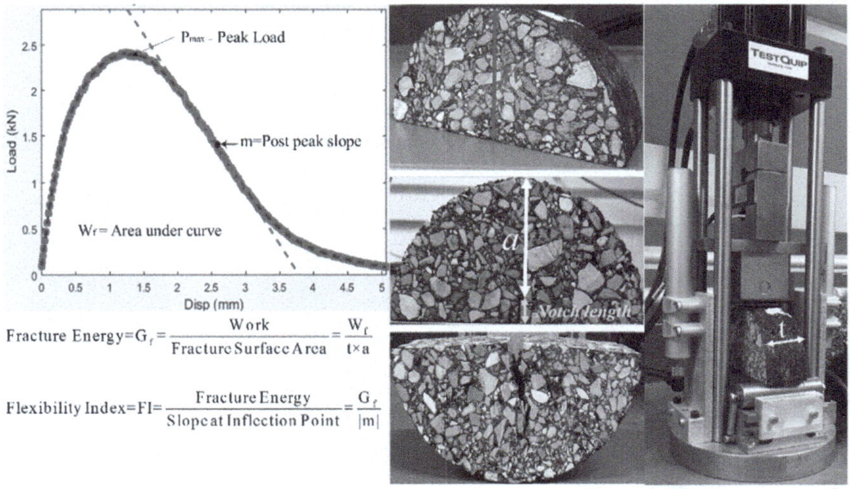

Fig. 1. Typical fracture test load-displacement curve and definition of fracture cracking indices

3.3 Rutting Resistance

Asphalt mixtures can deform at high temperatures when loaded. The Flow Number (FN) test is conducted to evaluate asphalt mixture resistance to permanent deformation (rutting). The test uses a vertical pulsed cyclic load (0.9 s) to simulate the load of a heavy vehicle that passes repeatedly over an asphalt pavement. In this study, the FN test was performed at a temperature of 50.5 °C based on the AASHTO, T 79-10 specification.

4 Results and Discussion

The dynamic modulus and phase angle results for all mixtures at reference temperature (21 °C) are shown in Figs. 2 and 3, respectively. Generally, a higher phase angle (relaxation capability) is desired for better cracking performance, while a higher stiffness helps improve a mixture's resistance to deformation. The mixtures containing RAP A appear to be stiffer than mixtures with RAP B. Furthermore, as expected, the mixtures with 45% RAP have higher dynamic modulus than 25% RAP mixtures.

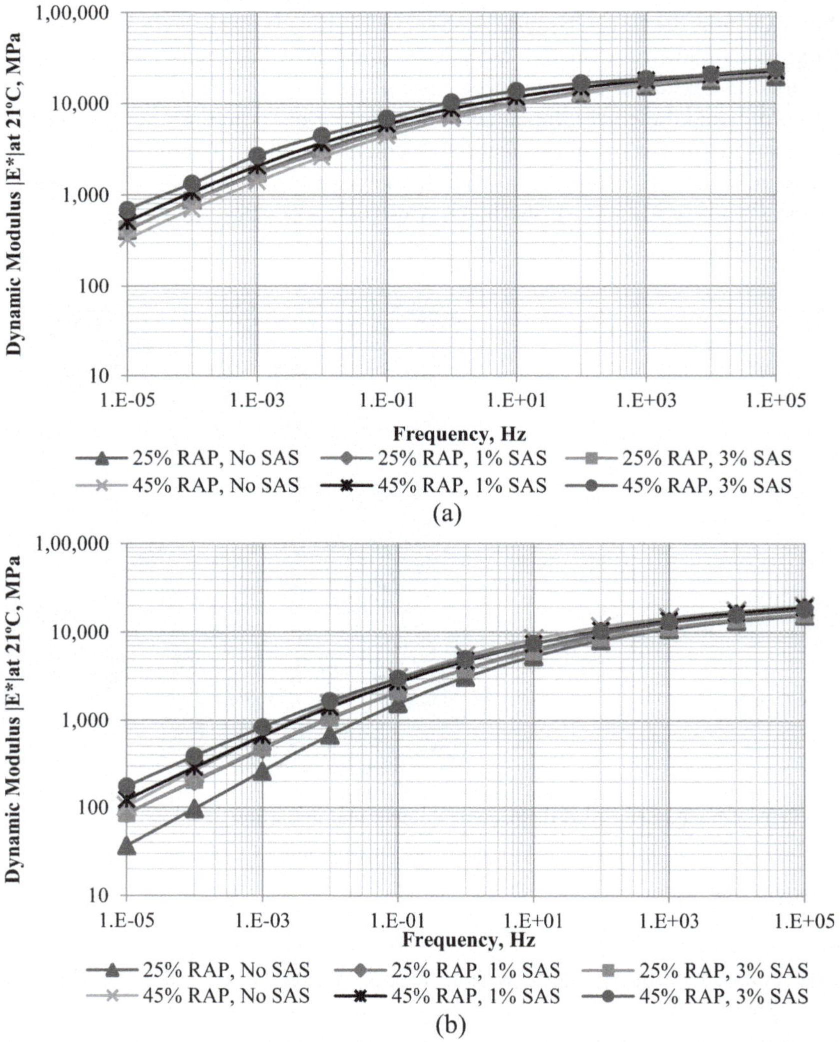

Fig. 2. Dynamic modulus master curve for mixtures containing RAP: (a) Type A; and (b) Type B.

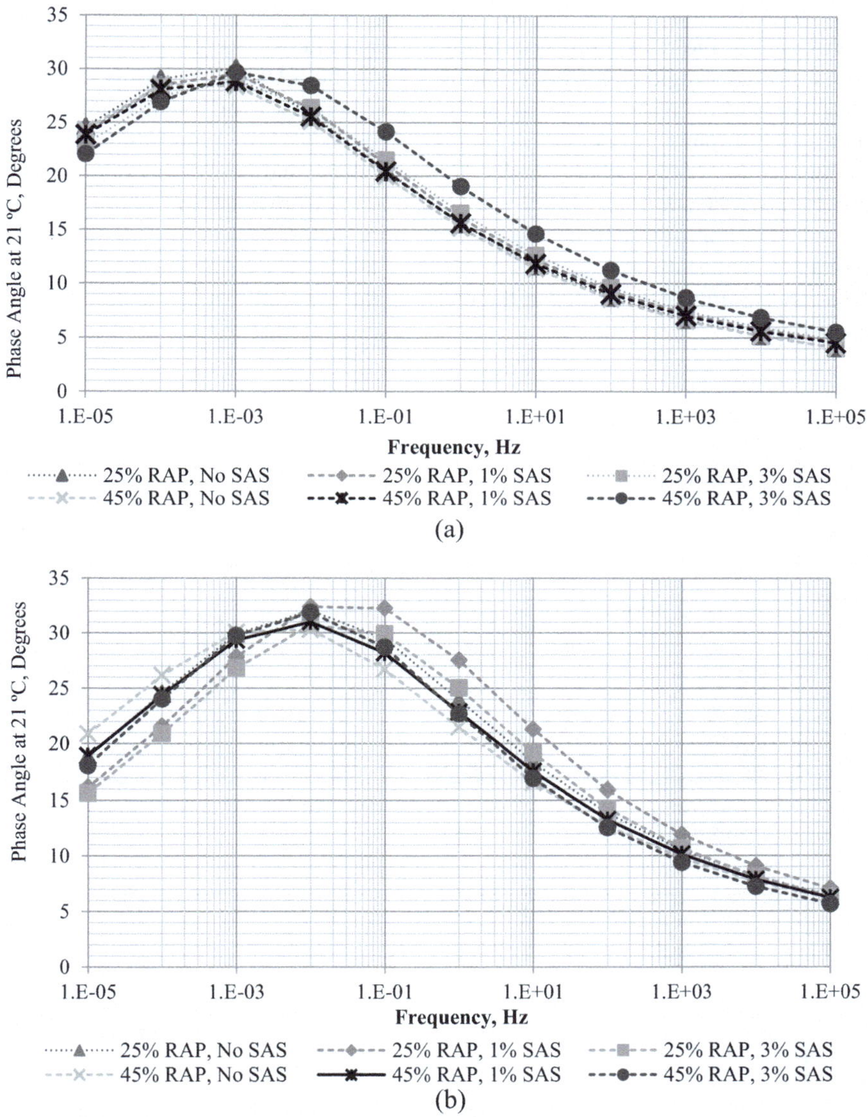

Fig. 3. Phase angle for mixtures containing RAP: (a) Type A; and (b) Type B.

Surprisingly, the dynamic modulus results show that adding SAS to the mixtures increases the stiffness. Addition of SAS produced a higher increase in dynamic modulus of the 25% RAP mixtures which were originally softer than the 45% RAP mixtures. Statistical analysis indicates that adding 1% SAS to the mixtures does not make a statistically significant difference in dynamic modulus, while there is a statistically significant difference between the stiffness of mixtures with 0% and 3% SAS. The trend is similar for the mixtures containing RAP A and B.

Although, inclusion of SAS in the trial asphalt mixtures resulted in higher dynamic modulus values, the relaxation capabilities of the mixtures also seem to improve. The analysis of phase angle data shows that increasing the SAS rejuvenator percentage in the test mixtures decreases the phase angle at low frequency (high temperature), while it increases the phase angle values at high frequency (low temperature). Both are desirable, as the higher phase angle at intermediate and low temperature help to improve the relaxation capability and cracking performance of mixtures against cracking; the improvement at high temperature helps protect the mixtures from deformation. The statistical analysis also shows adding 3% SAS rejuvenator causes a statistically significant difference in the phase angle, especially at low and intermediate temperatures.

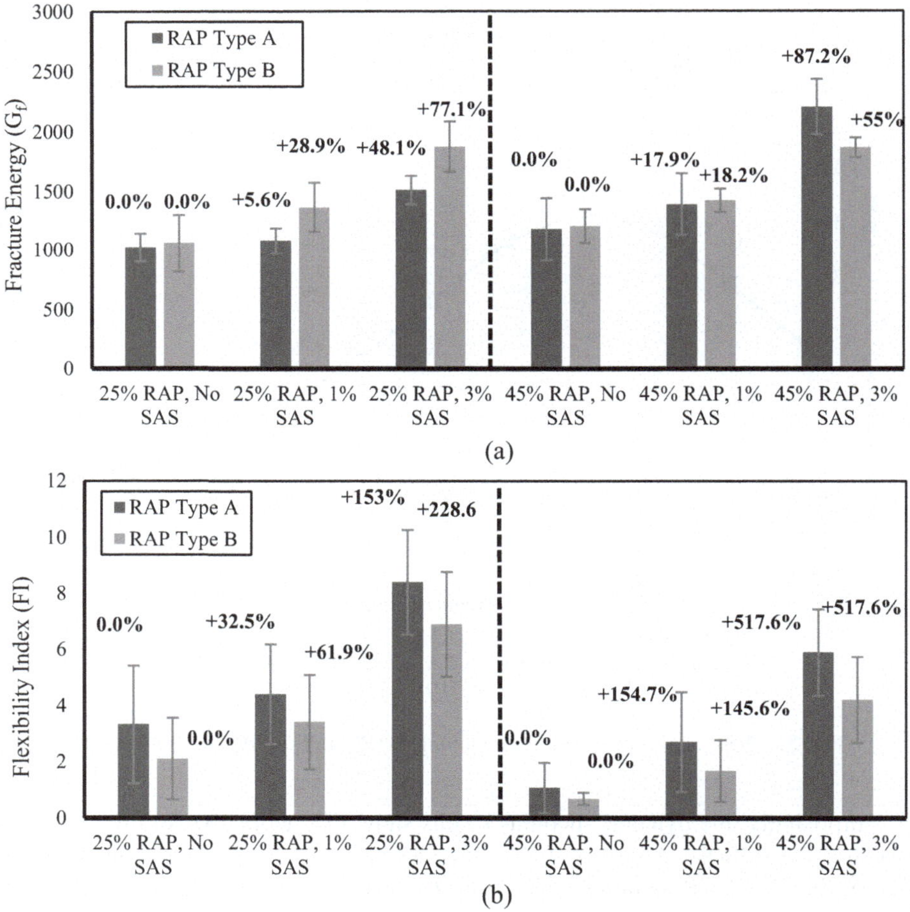

Fig. 4. Results of: (a) fracture energy and (b) flexibility index of the asphalt mixtures

The cracking indices (G_f and FI) of the mixtures are shown in Fig. 4, where the data labels shown on the bars indicate the percent change for each mixture compared to the similar, no SAS mixtures. The results indicate that increasing the percentage of SAS rejuvenator tends to improve the cracking resistance of asphalt mixtures containing RAP; the highest growth rate in the cracking parameters is observed when adding 3% SAS to the mixtures. The trend is similar for both types of RAP. Although the G_f values are very high for all the mixtures (between 1000 to 2500 J/m2), the FI values are low because of the high post-peak slopes in the load-displacement curves, especially for 45% RAP mixtures. Bahia et al. (2016) recommended that values of 12 and 5 be used as thresholds for FI of short term and long term aged mixtures. Statistical analysis of the results also indicates the improvement is statistically significant when the SAS percentage is increased from 1 to 3%. Although the fracture energy does not seem to be sensitive to the RAP content, the flexibility index shows a more desirable cracking behavior for 25% RAP mixtures than 45% RAP mixtures.

Figure 5 shows the flow number values of the asphalt mixtures. The rutting behavior of mixtures containing RAP A is better than the mixtures with RAP B, which is in a good agreement with the dynamic modulus results. As expected, the mixtures with 45% RAP show higher flow number values than do the 25% RAP mixtures. It is well known that increasing RAP contents in asphalt mixtures tends to lower deformation in the mixtures. Although increasing the amount of SAS rejuvenator resulted in a higher dynamic modulus and lower phase angle, it decreased rutting resistance for both RAP types and RAP percentages. This decrement is greater when the rejuvenator increases from 1 to 3 percent. The percentages on the bars show that adding 3% SAS rejuvenator may drop the FN values by about 50%. The NCHRP report 673 recommended the minimum acceptable flow number of hot mix asphalt based on the design traffic level (Jenks et al. 2011). For design traffic of 10 to 30 million ESAL, the minimum acceptable flow number is 190. In this study, the all measured flow number values met this criterion.

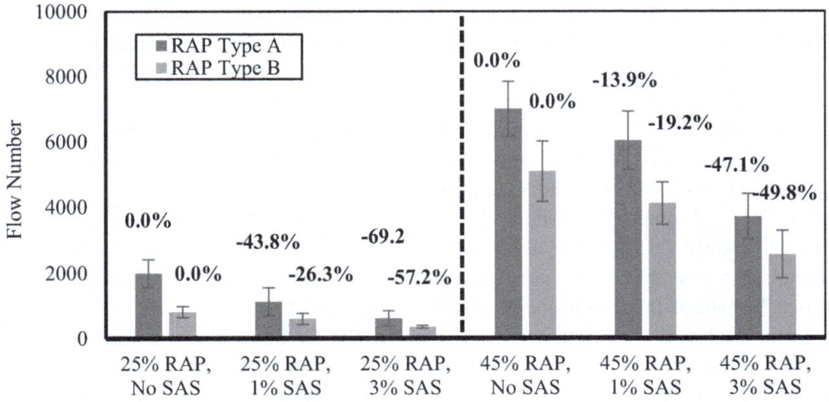

Fig. 5. Flow number results

5 Summary and Conclusion

In this study, the effect of different percentage of SAS rejuvenator on the performance of asphalt mixtures containing RAP was investigated using Complex modulus, SCB, and Flow Number testing methods. Given the findings, the following conclusions can be drawn:

- The SAS rejuvenator improves the viscoelastic behavior of asphalt mixtures by increasing stiffness (dynamic modulus) and relaxation capability (phase angle), especially at low and intermediate temperatures.
- The SAS rejuvenator increases the cracking resistance of asphalt mixtures containing RAP when compared to similar RAP mixtures without rejuvenator.
- In spite of increasing the mixtures' stiffness, the addition of SAS rejuvenator reduces the rutting resistance of asphalt mixtures. However, with a judicious choice of RAP and rejuvenator percentages, an asphalt mixture can be designed with acceptable deformation performance and increased cracking resistance.

Future work and analysis is planned to investigate the viscoelastic and fracture properties of extracted and recovered asphalt binders from the mixtures with SAS rejuvenator. Also, additional investigations will be performed to find an optimal percentage of SAS rejuvenator in order to obtain the best rutting and cracking performance. The pavement performance will also be modelled to evaluate the performance of mixtures with SAS rejuvenator.

Acknowledgments. The authors wish to express their gratitude to the Indiana Soybean Alliance for the financial support of this project.

References

AASHTO, T342-11 (2011) Standard method of test for determining dynamic modulus of hot mix asphalt (HMA)

AASHTO, TP124 (2016) Standard method of test for determining the fracture potential of asphalt mixtures using semicircular bend geometry (SCB) at intermediate temperature

Al-Qadi I, Ozer H, Lambros J, El Khatib A, Singhvi P, Khan T, Rivera J, Doll B (2015). Testing protocols to ensure performance of high asphalt binder replacement mixes using RAP and RAS. Illinois Center for Transportation/Illinois Department of Transportation

Bahia H et al (2016) Analysis and feasibility of asphalt pavement performance-based specification for WisDOT" WisDOT, Report No. 0092-15-04

Boyer RE (2000) Asphalt rejuvenators "Fact, or Fable". Transp Syst, 58:1

Elkashef M, Podolsky J, Williams RC, Cochran EW (2018) Introducing a soybean oil-derived material as a potential rejuvenator of asphalt through rheology, mix characterization and Fourier Transform Infrared analysis. RMPD 19(8):1750–1770

Epps Martin A, Arambula E, Kaseer F, Cucalon L, Yin F, Chowdhury A, Epps J, Glover C, Hajj E, Morian N, Pournoman S, Daniel J, Rahbar-Rastegar, R, King G (2017) The effects of recycling agents on asphalt mixtures with high RAS and RAP binder ratios, Phase II

Hansen KR, Copeland A (2017) Asphalt pavement industry survey on recycled materials and warm-mix asphalt usage: 2016 (No. IS 138(7e)). National Asphalt Pavement Association, Lanham, MD

Jenks CW, Jencks CF, Harrigan ET, Adcock M, Delaney EP, Freer H (2011) NCHRP Report 673: a manual for design of hot mix asphalt with commentary. Transportation Research Board, Washington, DC

Portugal ACX, Lucena LCDFL, Lucena AEDFL, Beserra da Costa D (2018) Rheological performance of soybean in asphalt binder modification. Road Mater Pavement Des 19 (4):768–782

Seidel JC, Haddock JE (2014) Rheological characterization of asphalt binders modified with soybean fatty acids. Constr Build Mater 53:324–332

Shen J, Amirkhanian S, Tang B (2007) Effects of rejuvenator on performance-based properties of rejuvenated asphalt binder and mixtures. Constr Build Mater 21(5):958–964

Zhang R, You Z, Wang H, Ye M, Yap YK, Si C (2019) The impact of bio-oil as rejuvenator for aged asphalt binder. Constr Build Mater 196:134–143

Effect of Water and Cement Content on the Mechanical Properties of Cold Recycled Mixtures (CRM) with Bitumen Emulsion

Simone Raschia[1](✉), Tushar Chauhan[2], Shalu Panwar[2], Alan Carter[1], Andrea Graziani[3], and Daniel Perraton[2]

[1] Construction Engineering Department, École de technologie supérieure (ÉTS), Montréal, Canada
simone.raschia.1@etsmtl.net

[2] Department of Civil Engineering, Indian Institute of Technology, Bombay, Mumbai, India

[3] Department ICEA of Civil and Building Engineering, and Architecture, Polytechnic University of Marche, Ancona, Italy

Abstract. Cold recycled mixtures (CRM) treated with bitumen emulsion are innovative materials in road pavement construction industry which gained visibility in the last decades. They are produced at ambient temperatures thanks to the employment of bitumen emulsion, water and additional hydraulic binders. In this research, four CRM mixes are studied analysing the effect of two water contents and two cement dosages. The mechanical properties of the mixes were studied in terms of Indirect Tensile Ratio (ITR), Indirect Tensile Stiffness Modulus (ITSM) test and dynamic modulus test, after a curing period of 14 days. Moreover, results analysis is improved by images obtained at the scanning electron microscope (SEM). From the results, the presence of more water decreased the indirect tensile strength (ITS) in both dry and wet conditions, even if the water susceptibility was generally low (ITR > 80%). The PUNDIT measurements along curing highlighted the progressive increase of the dynamic modulus, which after curing was higher in the mixtures with cement. At the end, SEM images highlighted the effect of the water content in the dispersion of the bituminous phase, as well as the presence of cement hydration products.

Keywords: Cold recycled mixtures · Bitumen emulsion · Indirect tensile ratio · Dynamic modulus · Scanning electron microscope

1 Introduction

Cold recycled mixtures (CRMs) are gaining interest in the road pavement construction industry thanks to the possibility to produce reliable materials for the pavement structure at ambient temperature (Davidson 2005; Tebaldi et al. 2014; Xiao et al. 2018). The bituminous phase is replaced by the use of bitumen emulsion, a two-phase suspension constituted by bitumen droplets suspended in an aqueous phase. In order to increase the short-term and long-term mechanical properties, some Ordinary Portland cement is normally added, which allows also a faster breaking process of the bitumen

M. Pasetto et al. (Eds.): ISAP APE 2019, LNCE 48, pp. 274–282, 2020.
https://doi.org/10.1007/978-3-030-29779-4_27

emulsion (ARRA 2001; Asphalt Academy 2009; Brown and Needham 2000). It is important to highlight that additional water is often considered in the production process to enhance the workability and compactability of the mix (Grilli et al. 2016; Raschia et al. 2019).

It is clearly demonstrated that the water content is of extreme importance for the reduction of the internal friction during compaction, as well as for the cement hydration. In fact, many studies demonstrated that with a higher amount of water in the mix, it is possible to reach higher density with less compaction effort (Cross 2003; Gaudefroy et al. 2008; Martínez et al. 2007). Once that the curing process starts, a small part of the water is employed by the cement for the formation of hydration products, whereas the major part is expelled by evaporation depending on the environmental conditions (temperature and humidity) (Cardone et al. 2014; Godenzoni et al. 2016; Kim et al. 2011). As curing proceeds, water partially evaporates, partially reacts with cement and partially remains in the solid structure of the mixture. In fact, it is often showed that it is difficult to reach the total loss of the water inside the mix (Cardone et al. 2014; García et al. 2013).

The objective of this study is to understand how the water content employed to satisfy compaction requirements affects the mechanical properties of the CRM with bitumen emulsion. To achieve this goal, mixtures with same volumetric properties are produced changing water content, and repeating the experimental plan with and without the addition of Ordinary Portland cement.

2 Materials and Methodology

2.1 Materials

The reclaimed asphalt pavement (RAP) used to produce the CRMs in this study was stockpiled in a production plant in Italy. The main physical properties of the RAP aggregate are collected in Table 1.

Table 1. RAP aggregate properties

Property	Standard	Unit	Value
Binder content	ASTM D6307	%	5.51
Nominal maximum particle dimension	ASTM D448-03	mm	16
Maximum specific gravity	ASTM C127-128	-	2.482

The bitumen emulsion used is a CSS-1 type (ASTM D2397) and its properties, as well as the ones of the residual bitumen, are listed in Table 2.

The cement used was a GU type (CSA A3000) with compressive strength at 28 days of 43.9 MPa (ASTM C109).

Table 2. Bitumen emulsion properties

	Standard	Unit	Value
Bitumen emulsion property			
Density	ASTM D6397-16	g/cm^3	1.0
Residue content	ASTM D6997-12	%	60.3
Storage stability @ 24 h	ASTM D6930-10	%	0.6
Residual bitumen property			
Needle penetration @ 25 °C	ASTM D5-13	mm	4.1
Softening point	ASTM D36-14	°C	48.6

2.2 Mixtures

In the present study, four mixtures were produced with the same gradation, which is represented in Fig. 1. The target distribution was the Fuller-Thompson maximum density curve, with exponent 0.45. In order to obtain the target gradation, the final aggregate blend was composed of 94.4% of RAP aggregate and 5.6% of limestone filler. The bitumen emulsion content was fixed at 5.0% by aggregate mass, which means a residual bitumen content of 3.0% by aggregate mass. Two cement contents were tested, as well as two different dosages of intergranular water: the mixtures 0C_2W and 0C_4W do not contain cement, and are produced with 2% and 4% of intergranular water, respectively; the mixtures 15C_2W and 15C_4W contain 1.5% of cement by aggregate mass and 2% and 4% of intergranular water, respectively. Water contents of 2% and 4% were chosen because the first is the amount of minimum water coming from the bitumen emulsion addition, whereas the second allows to employ half of the compaction energy compared to the first one to obtain the same amount of air voids.

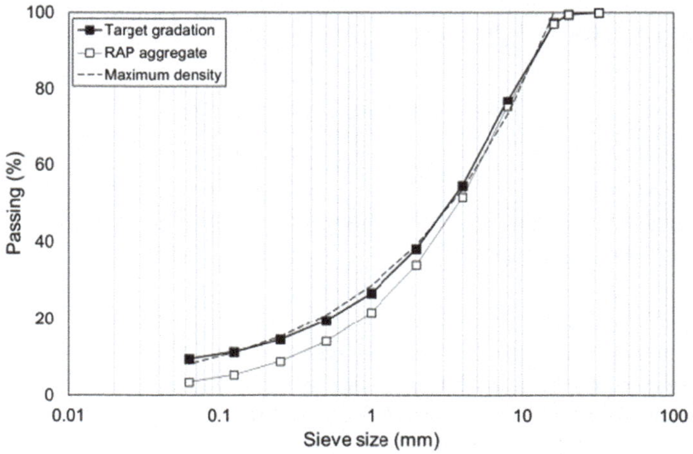

Fig. 1. Gradation of the studied granular material

The mixtures were produced initially blending the dry aggregates with the absorption water and letting the humid blend rest overnight. Afterwards, cement (when required), additional water (when required) and bitumen emulsion were added in this order, manually mixing the mixture at each addition. The specimens were compacted with a Superpave Gyratory Compactor (SGC) by using an undrained mould with D = 100 mm, 600 kPa of constant pressure, gyration rate of 30 rpm and external inclination angle of 1.25°.

The volumetric properties during and after compaction were analyzed through two parameters, the Voids in the mixture (V_m) and the Voids Filled with Liquids (VFL) (Grilli et al. 2016):

$$V_m = \frac{V_{V,A} + V_{W,I}}{V} \cdot 100 = \frac{V - (V_S + V_C + V_{B,R})}{V} \cdot 100 \quad (1)$$

$$VFL = \frac{V_{B,R} + V_{W,I}}{V_{B,R} + V_{W,I} + V_{V,A}} \cdot 100 = \frac{V_{B,R} + V_{W,I}}{V - V_S - V_C} \cdot 100 \quad (2)$$

where V is the total volume of the specimen, V_S is the bulk volume of aggregates, V_C is the volume of cement, $V_{B,R}$ is the volume of residual bitumen from emulsion, $V_{W,I}$ is the volume of intergranular water and $V_{V,A}$ is the volume of air.

In order to have specimens with same value of V_m, at first, mixes 0C_2W and 15C_2W were compacted at 200 gyrations and the final value of V_m was recorded. Afterwards, mixes 0C_4W and 15C_4W were compacted at fixed height to reach almost the same V_m value. The volumetric properties of the studied mixes are listed in Table 3.

Table 3. Volumetric composition of the studied mixtures

Components (%)	0C_2W	0C_4W	15C_2W	15C_4W
RAP aggregate	80.4	79.2	80.3	78.8
Filler	5.0	5.0	4.2	4.1
Unhydrated cement	0.0	0.0	1.0	1.0
Residual bitumen	6.1	6.0	6.0	5.9
Emulsion water	4.1	4.1	4.1	4.0
Additional water	0.0	4.1	0.0	4.0
Air	4.4	1.6	4.4	2.2
Total	**100**	**100**	**100**	**100**
Gyrations	200	95	200	100
V_m	8.0	9.7	8.5	10.2
VFL	72.8	89.9	69.3	87.2

After compaction, the specimens were stored in a chamber at fixed temperature of 40 ± 2 °C and relative humidity of 55 ± 5%.

The experimental program was developed performing Indirect Tensile Strength (ITS) test in dry and wet conditions, Indirect Tensile Stiffness Modulus (ITSM) test, dynamic modulus measurements with wave propagation at different curing ages, and image analysis through a Scanning Electron Microscope (SEM) (García et al. 2013; Godenzoni 2017; Richardson 1999; Rutherford et al. 2014).

The ITS test was performed on three dry specimens at 25 °C, in accordance to ASTM D6931. The ITS value was calculated as:

$$ITS = \frac{2000 \cdot P}{D \cdot t \cdot \pi} \tag{3}$$

where P is the maximum load (N), D is the specimen's diameter (mm) and t is the specimen's thickness (mm).

The ITS test in wet conditions was performed on three specimens which were previously submerged in water at 25 °C for 2 h. The Indirect Tensile Ratio was calculated:

$$ITR = \frac{ITS_{wet}}{ITS_{dry}} \tag{4}$$

The ITSM test was performed on one specimen each mixture at three temperatures: −10 °C, 0 °C and 10 °C, in accordance to UNI EN 12697-26 (Annex C). The test was repeated on both diameters and the average ITSM value was calculated as:

$$ITSM = \frac{F \cdot (\upsilon + 0.27)}{z \cdot h} \tag{5}$$

where F is the maximum load (N), υ is the Poisson ratio, z is the horizontal deformation (mm) and h is the specimen's thickness (mm).

Dynamic modulus measurements were performed by means of a Portable Ultrasonic Non-destructive Digital Indicating Tester (PUNDIT), using compression and shear waves. Shear waves transducers were attached on both faces of the specimen, using a coupling gel. The two probes were properly aligned in the same plane to achieve maximum transmission. The test was carried out on one specimen for each mixture, and the average value of three measurements was recorded at 1, 3, 5, 7 and 14 days of curing. The testing temperature was 40 °C.

3 Results Analysis

Figure 2 shows the results of ITS test in dry and wet conditions. It is highlighted that the ITR value is between 90% and 110%, indicating the low water susceptibility of all the studied mixtures. In terms of resistance, the mixtures with more intergranular water (0C_4W and 15C_4W) have in both cases less ITS than the respective mixtures with less water. Another important result is the slight effect of cement. In fact, mixtures 15C_2W and 15C_4W did not show a remarkable increase in the strength, as it was supposed to happen. However, mixes without cement showed very high ITS results, making less effective the addition of cement.

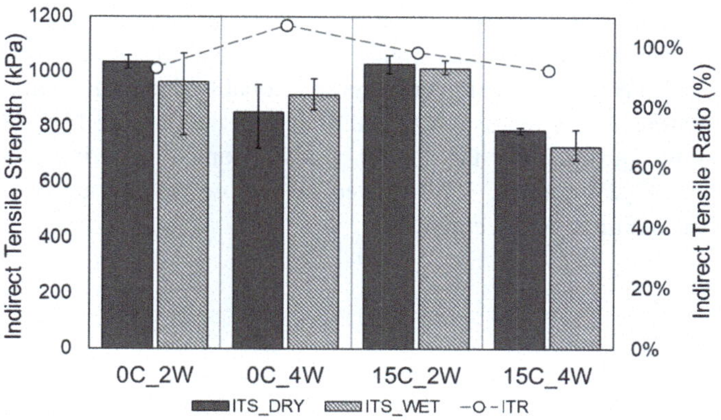

Fig. 2. Indirect tensile strength test results in dry and wet conditions

Figure 3 shows the results of the ITSM test. It is possible to observe that for all mixtures, stiffness is temperature sensitive. It is important to highlight that the mixture without cement 0C_2W showed the lower tensile stiffness, whereas the mix which was characterized by the highest stiffness is the 15C_2W. Mixtures 0C_4W and 15C_4W have similar ITSM values, which is unexpected since the presence of cement should improve stiffness. Hence, results for these two mixtures were probably influenced by the higher air voids. Furthermore, although the mixture 15C_4W contains cement and the mixture 0C_2W does not, their mechanical properties appear similar. Hence, the mixture 15C_4W showed ITSM values of a mixture without the effect of the emulsion, giving similar results to the mixture without cement (0C_4W). This evaluation can be confirmed also by the ITS results.

Figure 4 shows the dynamic modulus values at different curing ages. For all the studied mixes, it is possible to observe the effect of curing in the increase of the dynamic modulus. From those results, it can be highlighted that after 7 days of curing,

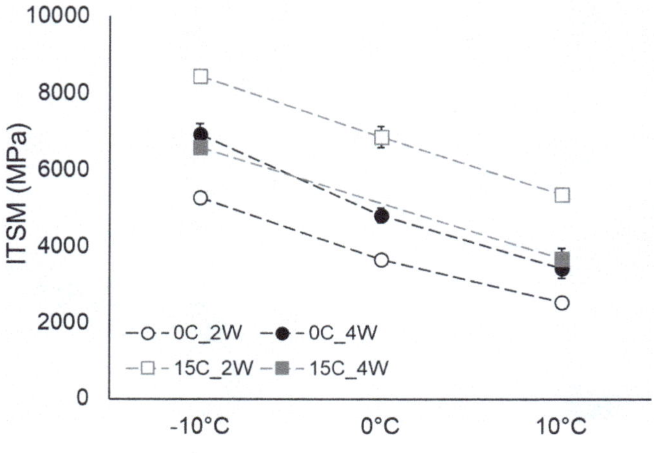

Fig. 3. Indirect tensile stiffness modulus test results

no significant increase in the modulus value occurred. In general, mixes with cement (15C_2W and 15C_4W) always had higher modulus than the mixes without cement. The mixture 15C_4W showed a slower increase, probably due to the higher amount of water in the mixture. Nevertheless, at the end of curing, the mixture 15C_4W has lower dynamic modulus, justified by the higher voids. Along the curing time, mixtures without cement 0C_2W and 0C_4W showed very close results, and values after 14 days lower than mixtures with cement.

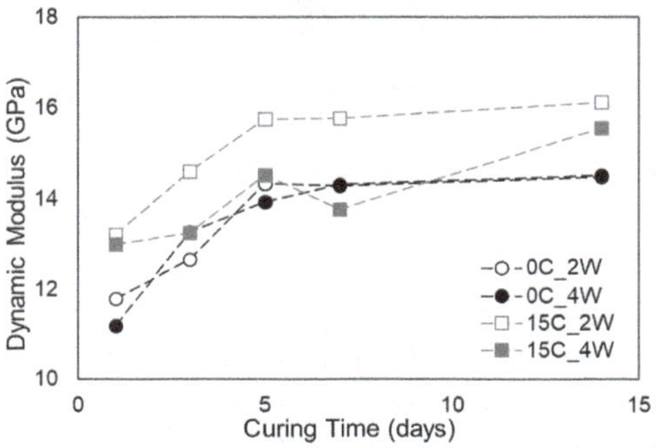

Fig. 4. Dynamic modulus test results (PUNDIT at 40 °C)

Figure 5 shows Scanning Electron Microscope images taken from broken samples after mechanical testing.

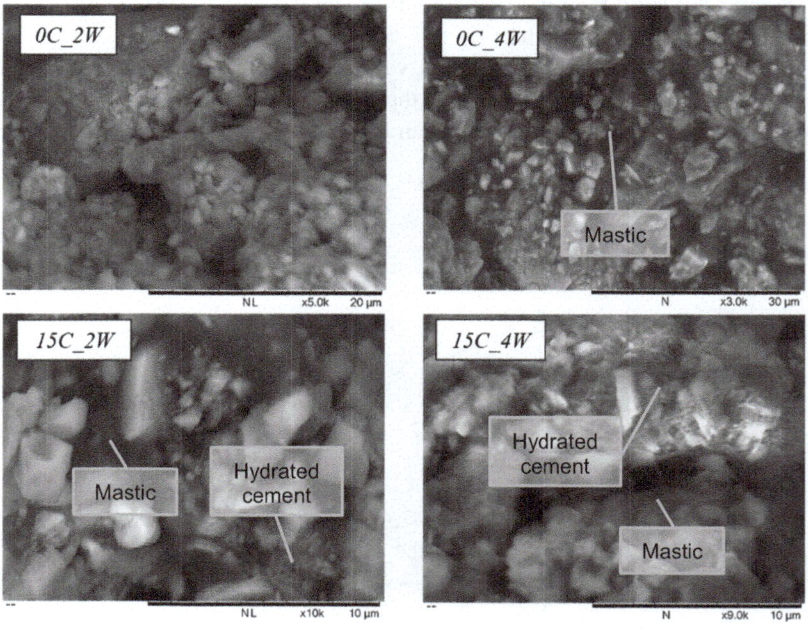

Fig. 5. Scanning electron microscope images

The test was performed on the material without any particular surface treatment prior testing. Comparing the pictures of mixes 0C_2W and 0C_4W, it is highlighted the contribution brought by the higher amount of water in the bitumen dispersion. Analyzing the mixes with cement it can be seen that the hydrates product formation is lower in the mixture 15C_2W, in which it is concentrated only in few spots. On the contrary, the mixture 15C_4W clearly shows the cement hydration process, which at the same time makes difficult to recognize the bituminous phase.

4 Conclusions

This research aimed to understand the effect of water content and cement content in CRM materials with bitumen emulsion. Four mixes, characterized by two cement dosages and two water contents, were analyzed in terms of ITS test, ITSM, dynamic modulus and SEM image analysis after a curing period of 14 days. The following conclusions can be drawn:

- From ITS results, the higher water content generally decreased the strength of the mix, whereas the addition of cement did not lead to a significant increase in resistance. It must be noted that mixes with more water required less compaction energy, which could have led to a lower aggregate packing;
- From ITSM results, the temperature susceptibility of the mixes is highlighted. Mixes with cement did not always show higher stiffness; in fact, the cement hydration seemed to have damaged the effect of the emulsion when more water was used. This led to lower stiffness values. In mixes with no cement, more water allowed a better dispersion of the bitumen phase and gave higher stiffness;
- Dynamic modulus measurements along curing showed an increase of the dynamic modulus significant in the first 7 days for all the four mixes. After 14 days, mixes with cement had higher modulus than the mixtures without cement, which were not different regarding the water content;
- From SEM image analysis, it was possible to capture important aspects of the mixes produced. In fact, the higher water content allowed a higher bitumen dispersion among the mixes 0C_2W and 0C_4W. In mixes with cement (15C_2W and 15C_4W), more water led to a higher cement hydration, which at the same time prevent a full contribution from the bitumen phase.

Further studies are needed in this perspective, especially at a microstructural level, to fully understand and estimate the collaboration between bitumen and hydraulic binders (such as cement). From the results, the contemporary presence of both did not lead to the most efficient mix.

References

ARRA (2001) Basic asphalt recycling manual

Asphalt Academy A (2009) Technical guideline (TG2): Bitumen stabilised materials

Brown S, Needham D (2000) A study of cement modified bitumen emulsion mixtures. Asph Paving Technol 69:92–121

Cardone F, Grilli A, Bocci M, Graziani A (2014) Curing and temperature sensitivity of cement–bitumen treated materials. Int J Pavement Eng 16(10):868–880. https://doi.org/10.1080/10298436.2014.966710

Cross S (2003) Determination of Superpave® gyratory compactor design compactive effort for cold in-place recycled mixtures. Transp Res Rec: J Transp Res Board 1819:152–160

Davidson JK (2005) Progress in cold mix processes in Canada. Canadian Technical Asphalt Association

García A, Lura P, Partl MN, Jerjen I (2013) Influence of cement content and environmental humidity on asphalt emulsion and cement composites performance. Mater Struct 46(8):1275–1289

Gaudefroy V, Wendling L, Odie L, Fabre J, De La Roche C, Hornych P, Dubois V (2008) Laboratory characterization of cold mix treated with bitumen emulsion. In: 4th eurosphalt and eurobitume congress presented at the meeting of 4th eurosphalt and eurobitume congress, France

Godenzoni C (2017) Multiscale rheological and mechanical charcaterization of cold mixtures (Doctoral dissertation). Università Politecnica delle Marche, Ancona

Godenzoni C, Cardone F, Graziani A, Bocci M (2016) The effect of curing on the mechanical behavior of cement-bitumen treated materials, vol 11, pp 879–890

Grilli A, Graziani A, Bocci E, Bocci M (2016) Volumetric properties and influence of water content on the compactability of cold recycled mixtures. Mater Struct 49(10):4349–4362. https://doi.org/10.1617/s11527-016-0792-x

Kim Y, Im S, Lee HD (2011) Impacts of curing time and moisture content on engineering properties of cold in-place recycling mixtures using foamed or emulsified asphalt. J Mater Civ Eng 23(5):542–553. https://doi.org/10.1061/(asce)mt.1943-5533.0000209

Martínez A, Miró R, Pérez-Jiménez F (2007) Spanish experience with gyratory compactor and indirect tensile test in design and control of cold recycled asphalt pavement. Transp Res Rec: J Transp Res Board 2001:163–168

Raschia S, Graziani A, Carter A, Perraton D (2019) Influence of RAP source and nominal maximum size on volumetric and physical properties of cement-bitumen treated materials. Paper presented at the transportation research board (TRB) 98th annual meeting, Washington D.C. USA

Richardson IG (1999) The nature of CSH in hardened cements. Cem Concr Res 29(8):1131–1147

Rutherford T, Wang Z, Shu X, Huang B, Clarke D (2014) Laboratory investigation into mechanical properties of cement emulsified asphalt mortar. Constr Build Mater 65:76–83

Tebaldi G, Dave EV, Marsac P, Muraya P, Hugener M, Pasetto M, Canestrari F (2014) Synthesis of standards and procedures for specimen preparation and in-field evaluation of cold-recycled asphalt mixtures. Road Mater Pavement Des 15(2):272–299. https://doi.org/10.1080/14680629.2013.866707

Xiao F, Yao S, Wang J, Li X, Amirkhanian SN (2018) A literature review on cold recycling technology of asphalt pavement. Constr Build Mater 180:579–604

Sustainable Warm In-plant SMA Mixtures with 80% Recycling and Produced at 115 °C

Jian Qiu(✉), Rien Huurman, Ernst Demmink, and Mark Frunt

BAM Infra Asphalt, Plantijnweg 32, 4104 BB Culemborg, The Netherlands
jian.qiu@bam.com

Abstract. Stone Mastic Asphalt, SMA, is widely used in the provincial roads in the Netherlands as surface layers due to its excellent performance and long lifespan of 15–20 years. However, this SMA mixture is made with 100% new materials and no recycling is permitted due to quality concerns. Together with provinces Gelderland, Noord-Brabant, Overijssel and Dutch Ministry of Transport (Rijkswaterstaat), sustainable SMA variants are developed in this paper with high quality, 80% of recycling and can be produced at low production temperature of 115 °C. The high quality SMA mixtures with up to 80% recycling are designed with reclaimed aggregate fractions (with bitumen content less than 1%) and bitumen-rich mortar fractions (with bitumen content 10–12%) obtained from an innovative decomposition technique with the support of the European Life+ program. To achieve the low production temperature of 115 °C, two foaming technologies have been applied, namely bitumen foaming and mortar foaming. With the positive results of the laboratory research, in July 2018, total 4 km demonstration sections, with sustainable SMA mixtures containing up to 80% recycling and produced at 115 °C, have been constructed successfully in two Dutch provincial roads N317 and N625 in Gelderland and Noord-Brabant.

Keywords: Stone Mastic Asphalt (SMA) · Recycling · Sustainability · Warm mix · Decomposition · LE2AP

1 Introduction

With the growing concerns of government bodies on climate change and sustainability, the historic Paris Agreement on Climate Change has been signed on 12 December 2015 by 195 nations. The Dutch Ministry of Transport (Rijkswaterstaat) states her ambition of 50% circular and 100% climate neutral in 2030 for her road network. To achieve these goals, developing of an innovative and revolutionary sustainable asphalt technology is of great importance. This technology should, on one hand, allow high percentage recycling with a lower production temperature aiming for sustainability; on the other hand, deliver high quality in which allows horizontal re-use of materials in the road constructions, for example, recycling of high-quality surface layer back to surface layer.

In 2013, with the support of the European Life+ program, the authors of this paper developed such abovementioned innovative technology under a project named LE2AP

M. Pasetto et al. (Eds.): ISAP APE 2019, LNCE 48, pp. 283–293, 2020.
https://doi.org/10.1007/978-3-030-29779-4_28

(Low Emission 2 Asphalt Pavement). The LE2AP technology allows producing high quality re-use of porous asphalt (PA) by the use of decomposition technology and mortar foaming technology (Huurman et al. 2018). In September and October 2016, a total of 2.3 km double layer noise reducing PA was installed in two Dutch provincial roads to demonstrate the feasibility of the LE2AP technology. The installation in both cases comprised above 80% re-use, and in both cases the temperature of the PA at the screed of the paver was about 105 °C. This technology reduces substantially the CO_2 emission up to 50%, toxicity (CO, C_xH_y, NO_x and SO_2) up to almost 80% and reduces energy usage up to 50% (Huurman et al. 2018).

Stone mastic asphalt, SMA is very popular in the Netherlands and all over the world as a high-performance surface layer. Due to quality concerns, the SMA mixtures in the Netherlands are always hot produced and using 100% virgin materials. In 2017, our company BAM, together with Dutch Ministry of Transport (Rijkswaterstaat) and three Dutch provinces (Gelderland, Noord-Brabant and Overijssel), started developing high-quality sustainable SMA mixture variants with the LE2AP technology, allowing a re-using percentage up to 80% and being produced at around 115 °C. This paper presents the multi-scale designing and evaluating of these sustainable SMA mixtures namely: (a) analysis of basic components PA-stone and mortar; (b) analysis of laboratory mixture performance; (c) analysis of field mixture performance.

2 LE2AP Mixture Designing Philosophy

2.1 LE2AP and LE2AP SMA

Figure 1a gives an overview of the LE2AP recycling process. Reclaimed PA is first decomposed into its components; mortar (<2 mm, bitumen content 10–15%) and stone aggregates (PA-stone, bitumen content less than 1%) as shown in Fig. 1b and c. The reclaimed aggregates are certified and can be directly reused replacing virgin aggregates (Huurman et al 2018). The mortar is then enriched, rejuvenated and homogenized to obtain LE2AP mortar with the same performance as its fresh equivalent. The LE2AP mortar is then foamed and mixed with PA-stone to produce a mixture at a temperature of 100–110 °C. This process gives control over ingredient quality and mix composition even at high rates of re-use and allows for production at lowered temperature.

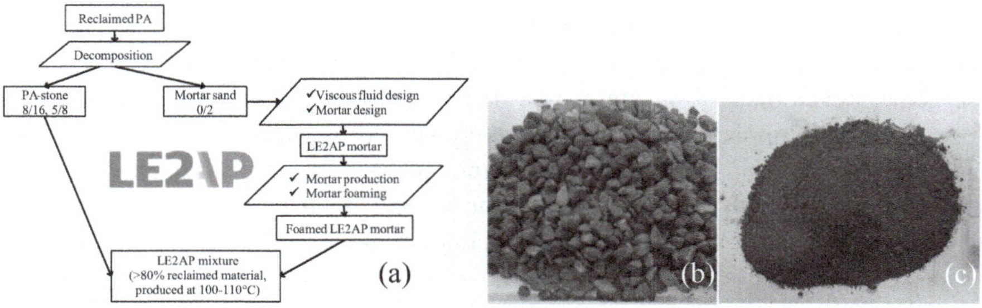

Fig. 1. (a) The LE2AP recycling process. (b) Reclaimed aggregates. (c) Reclaimed mortar sand

SMA mixtures, similar to that of the porous asphalt, can be defined as three components: stone, mortar and voids (Fig. 2). The stone fraction forms a skeleton of the mix, the mortar fraction forms as a binding medium between the stones. The only difference between SMA and PA is that the SMA mixture only has an air voids of 4–5%, depending on the traffic situations, compared to around 20% of the PA. The low air voids content requires that the variation of its components must be very limited, otherwise, one of the most common damage of the SMA mixture, bleeding or so-called fat spots, will occur. This damage is typically due to the incorrect volumetric combination of these three components. As a result, for designing a SMA, a volumetric critical mixture, using high percentage of PA-stone and the reclaimed mortar sand, special attentions have to be paid on quality of two basic components: the PA-stone and the mortar.

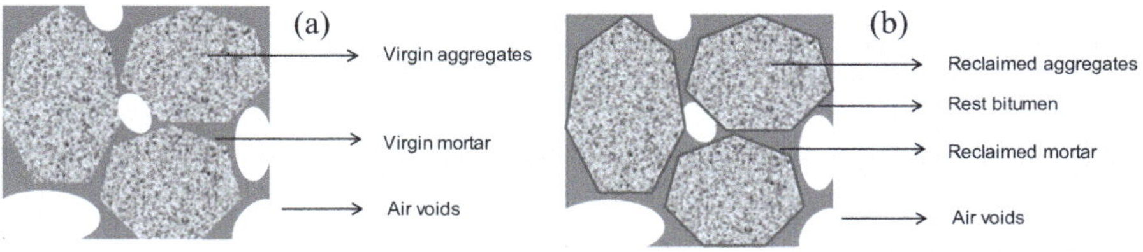

Fig. 2. SMA mixtures with (a) virgin stone/mortar and (b) reclaimed stone/mortar

2.2 PA-stone

The properties of PA-stone are similar to that of the virgin aggregates, and the PA-stone can be seen as a reclaimed aggregate, containing very limited sand and filler, with a bitumen content of around 0.7–1%. The PA-stone is obtained from reclaimed porous asphalt layers around 10 years old with a penetration of the bitumen around 5–10 (0.1 mm). Therefore, understanding the active percentage of the remaining bitumen on the PA-stone is important for a correct and successful volumetric design of the SMA mixture.

This can be achieved using the fracture energy concept. Research was conducted to compare the fracture energy of PA-stone mixture and Bestone mixture (virgin aggregates) with the same stone skeleton and mortar compositions (constant sand/filler/bitumen ratio). These tests were conducted at 15 °C with a loading speed of 50 mm/min and the results are shown in Fig. 3. The total fracture energy is calculated with the area under the full force-displacement curve. Through variating the virgin bitumen content of 3.5% to 5%, a linear relation between the total fracture energy in the mixture and the percentage of new bitumen can be observed. With the use of PA-stone, the total fracture energy increases constantly throughout all the tested scenarios to that of the Bestone mixture. This is an indication that the rest bitumen of the PA-stone is actively contributing to the mortar fraction. Therefore, for the 80% PA-stone (rest bitumen 0.7%) used in this mixture, a 0.35% active bitumen is calculated. This results in an active bitumen percentage of 0.35%/0.7% = 50%. With the results of the 50%

active bitumen, the stone skeleton of mixtures made with reclaimed aggregates is clearly defined.

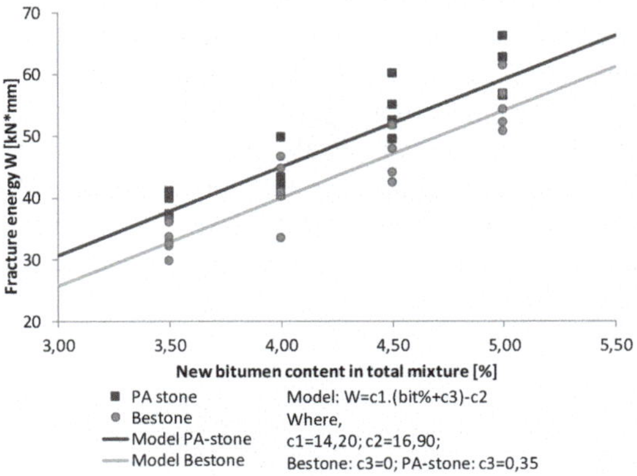

Fig. 3. Comparison of fracture energy of mixtures with Bestone and mixtures with PA-stone

2.3 LE2AP Mortar

Previous research of the authors indicates that the quality of the mortar, especially the flexibility of the mortar is responsible for the quality of the mixtures in resisting essential damages such as (micro-) ravelling in porous asphalt, SMA and thin surfacing layers (Huurman et al. 2010). As a result, designing of a good performed SMA mortar is important for high quality SMA mixture with high percentage recycling.

In the LE2AP process, the mortar sand, containing most of the bitumen, is separated from the mineral aggregates and treated separately. During treatment the mortar is heated, but no longer contacts with a flame or overheated air when compared with current day AC recycling. As a result, the reclaimed bitumen is no longer overheated and the lighter hydrocarbons in the reclaimed bitumen are preserved. In addition, treatment involves adding soft bitumen, rejuvenator and homogenization. This separate mortar treatment process ensures that issues of bitumen blending efficiency, homogeneity, oxidation or burning of bitumen, which are common in conventional AC recycling are overcome (Huurman et al. 2018).

Table 1 gives the information of the LE2AP mortar in comparison with the reference SMA mortar. It is known from previous research that the LE2AP mortar can be produced with identical bitumen penetration and bitumen content and without special gradation correction (Qiu et al. 2018). The LE2AP mortar is then designed by homogenizing and enriching the reclaimed mortar sand to a bitumen content of 25.9% and an expected penetration value of 89 (0.1 mm) with rejuvenators and soft bitumen (100/150 pen) by variating its bitumen properties and compositions. The total bitumen properties were designed by the log-pen relation of its components. The mortar performance was investigated using the Dynamic Shear Rheometer (DSR) as shown in Fig. 4a. The tests were conducted with a special mortar column setup with a height of 20 mm and a diameter of 6 mm and a height of 30 mm and a diameter of 9 mm Fig. 4b.

Table 1. Descriptions of mortar compositions

	Fresh crush sand + river sand [%]	Freshfiller [%]	Reclaimed mortar sand 0/2 [%]	Fresh bitumen [%]	Rejuvenator Sylvaroad RP1000 [%]	Total bitumen [%]	Expected penetration [0.1 mm]
Standard SMA mortar	22.9 + 22.9	24.4	–	25.9 (70/100)	–	25.9	89
LE2AP mortar	–	–	82.1	16.3 (100/150)	1.7	25.9	89

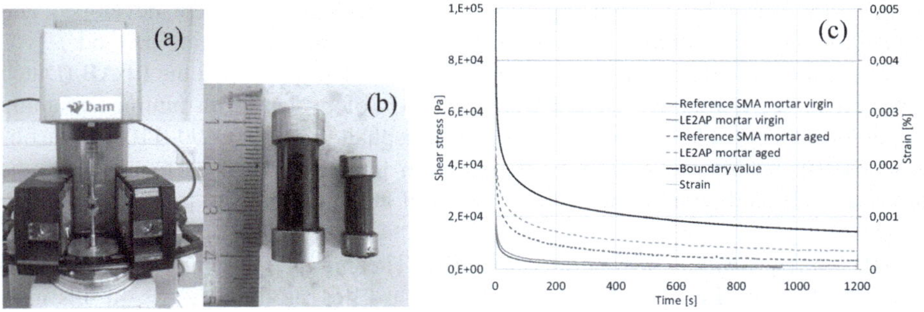

Fig. 4. (a) Dynamic Shear Rheometer. (b) Mortar column. (c) Relaxation of LE2AP SMA mortar and standard SMA mortar before and after ageing at −10 °C

Figure 4c shows the flexibility boundary value as a relaxation parameter at a very low temperature of −10 °C. The materials who are more flexible and under the boundary is safe for the winter whereas the materials who are not flexible enough and with their relaxation curve above the boundary are sensitive for winter damage (Huurman et al. 2010). It must be noted that the boundary value is obtained from indication of ravelling of porous asphalt, and this can only be indicative for the use of the SMA mortars. To investigate the flexibility performance of the mortars not only in their virgin state but also after ageing, a laboratory ageing procedure was also introduced. Aging was achieved by placing 2 mm thick mortar specimens in a stove at 135 °C for 44 h. It is expected that the aging hardening that is obtained in this manner is equivalent to 10 years of aging of porous asphalt in the Netherlands (Jemere 2010). The relaxation results of mortars before and after ageing show that both mortars, LE2AP mortar and the reference SMA mortar, are flexible enough, even after ageing to withstand the severe winter. It is also indicated that the LE2AP mortar, obtained from reclaimed mortar together with rejuvenator and soft bitumen, has identical performance than that of the reference SMA mortar.

3 Laboratory Production and Performance

After investigating of the performance of the individual components, PA-stone and the mortar, a laboratory mixture production and performance evaluation process was performed. Table 2 gives an overview of 5 mixtures investigated in this research. Mixture 1 is a reference SMA-NL 8B mixture with maximum aggregate size of 8 mm. Type "B" stands Dutch SMA mixtures with a designed air voids of 5% for high traffic category. Mixture 2 and 3 are mixtures with 20% PA-stone and 60% PA-stone, respectively. All these three mixtures were produced at 165–170 °C. Mixture 4 is a low temperature variant of mixture 3 with LEAB bitumen foaming technology. Mixture 5 is a LE2AP SMA mixture with 60% PA-stone and 18% reclaimed mortar sand, produced at 115 °C. In all these mixtures, the PA-stone is designed with 50% active bitumen in the new mixture (see Sect. 2.2).

To be able to produce the designed mixture 4 and 5 at a relative lower production temperature, e.g. 115 °C, two types of foaming technologies were developed and applied, namely the bitumen foaming method and the mortar foaming method.

(A) Bitumen foaming (Mixture 4, Fig. 5a): BAM has developed the LEAB (Dutch acronym for Low Energy Asphalt Concrete) technology based on bitumen foaming technology, being able to produce high quality asphalt mixtures at 90–110 °C (Jacobs et al. 2010). In this research, the required fresh bitumen content is foamed into the mixer and blended with pre-heated aggregates of around 120 °C to produce a mixture of 115 °C.

(B) Mortar foaming (Mixture 5, Fig. 5b): The mortar foaming technology was specially developed by BAM under the European Life+ project LE2AP. In this mixture, the reclaimed mortar sand is rejuvenated and enriched to a LE2AP mortar at a temperature of 170 °C with a bitumen content of around 25–30% by mass (see Sect. 2.3). This LE2AP mortar afterwards is foamed into the mixer with the with a special designed foaming unit (Fig. 5b) and mixed with the pre-heated aggregates of around 120 °C. This unit is specially designed considering both bitumen foaming principle and also the influence of the sand and filler particles (Huurman et al. 2018).

Table 2. Description of five SMA mixtures and their production methods

Mixture	1	2	3	4	5
	SMA	SMA 20%	SMA 60%	LEAB SMA 60% + 115C	LE2AP SMA 80% + 115C
Total recycling [%]	0	20	60	60	78
PA stone 5/8 [%]		20	60	60	60
LE2AP mortar, see Table 1 [%]					23 (18% reclaimed mortar sand)
Fresh Bestone 4/8 [%]	72.5	52.6	11.4	14.0	14.5
Fresh sand [%]	12.2	12.4	13.8	11.9	–
Fresh filler [%]	8.3	8.2	8.1	7.6	–
Fresh 70/100 bitumen [%]	6.7	6.6	6.3	6.4	–
Production temperature	165 °C	165 °C	165 °C	115 °C	115 °C
Production methods	Hot	Hot	Hot	Bitumen foaming	Mortar foaming
Air voids [%]	5.0	5.0	5.0	5.0	5.0

After mixing, these mixtures were compacted using the gyrator compactor with a targeting air voids of 5% and their performances were investigated using the indirect tension test according to NEN-EN 13108-20. Both the dry and the wet groups (specimens were retained around 72 h at 40 °C) were subjected to an indirect tension loading with a loading speed of 50 mm/min at a temperature of 15 °C. Due to limitation of the paper length, only strength and water sensitivity results are discussed in the following sections.

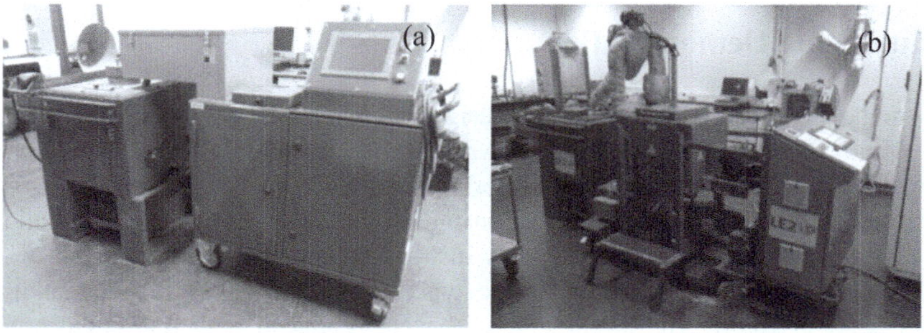

Fig. 5. Bitumen foaming apparatus (a) and mortar foaming apparatus (b)

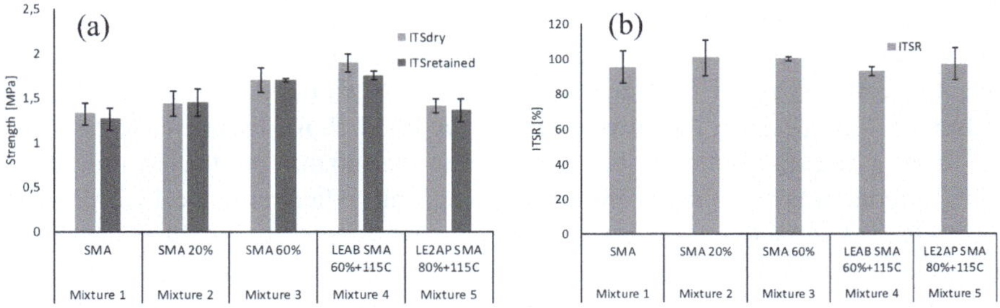

Fig. 6. Results of (a) strength and (b) water sensitivity ITSR for all the mixtures

Figure 6 gives the results of the mixture performance in terms of strength and water sensitivity. The strength of the mixture increases with the increasing PA-stone percentage with the contribution of remaining bitumen on the PA-stone. All the SMA mixtures show similar performance in terms of water sensitivity, even with mixtures containing 80% materials.

- By introducing PA-stone up to 60% in the SMA mixture in the hot production process, the strength and water sensitivity performances are similar to that of the reference mixture.
- By the use of LEAB bitumen foaming technology, the mixture with 60% PA-stone can be produced at a 115 °C. The performance of this mixture is similar to its hot equivalent and also the reference mixture.

- By the use of LE2AP mortar foaming technology, the reuse percentage of the SMA mixture can be further increased to 80% with a production temperature of 115 °C. The strength and water sensitivity of this mixture is identical to that of the reference mixture. It is expected that, with the use of rejuvenating and enriching technology, the mortar quality in the mixture is controlled and homogenized, which is very positive for mixture with high percentage recycling and produced at low temperature.

4 Full Scale Demonstration

With the positive results of the mixture design and laboratory performance testing, full scale demonstration of these mixtures was carried out in July 2018 on two Dutch provincial roads N317 of Province Gelderland and N625 of Province Noord-Brabant, respectively. The total length of the demonstration sections is about 4 km, and all these above mentioned 5 mixtures were evenly distributed in these two sections.

The production process of these five types of mixtures was carried out in one asphalt plant of BAM. The mixtures 1 to 3 were produced with traditional hot production method. The mixture 4 was produced with the LEAB foaming technology, see Fig. 7a for an impression of the equipped LEAB foaming located at the side of the mixer.

For the production of the LE2AP mixture, mixture 5, a semi-full scale production method was developed with additional mortar production line on the existing asphalt plant. First, 50 ton of LE2AP mortar was produced with a stirring kettle at a temperature of 165 °C. Then the stirring mixer was lifted above the mixer to be able to doss in the production process (Fig. 7b). During the mixture production process, the LE2AP mortar was first dossed in the weighing unit. Afterwards the LE2AP mortar was discharged into a foaming pipe connecting to the mixer (Fig. 7c). With the effect of the injected water in the foaming pipe, the LE2AP mortar was foamed to the mixer and mixed together with PA-stone aggregates to produce an SMA mixture with 80% recycling at 115 °C. The laying and compaction process of these mixtures was carried out with standard procedures which are applied for conventional SMA mixtures.

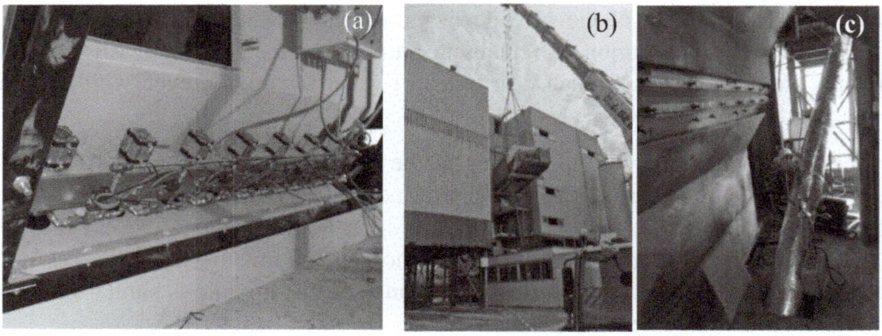

Fig. 7. (a) LEAB foaming unit; (b) Lifting stirring kettle; (c) mortar foaming pipe

Special attentions were given to two important critical elements during the whole process, segregation and skid resistance.

Segregation of the mixtures can cause premature problems for SMA mixtures like fat spots, especially for mixtures with high percentage recycling. First, visual inspections have been carried out during all the process. No segregation phenomenon was observed for all five mixtures. Furthermore, the bitumen content of the mixtures was investigated during three phases, namely mixing (from mixing plant), laying (from the hopper) and compaction (from core specimens) as shown in Fig. 8. The bitumen content from these three phases in the process do not differ from each other, indicating no segregation happening during the process.

Due to the use of high percentage reclaimed materials, the skid resistance performance is another important parameter to consider. The skid resistance was measured with a Dutch standard RAW 2015/72, in the longitudinal direction with a slip ratio of 86% and with a standardized non-profiled PIARC test tyre. Figure 9 gives the results of the skid resistance properties. The measurements were carried out direct after construction (year 0) and after 6-month. It must be noted that the year 0 measurements were measured before opening to traffic, which expresses the value with full bitumen film and mostly with the influence of sand spreading. After usage of 6 month, due to remove of the sand and wear out of the bitumen film by traffic, the skid resistance performance is more representative for the mixtures. All the skid resistance values (year 0 and 6 month) shown in Fig. 9 are higher than the requirement, indicating good initial skid resistance performance. Further monitoring, including visual inspections and skid resistance, will be carried out to monitor the performance of mixtures for after 2 and 5 years.

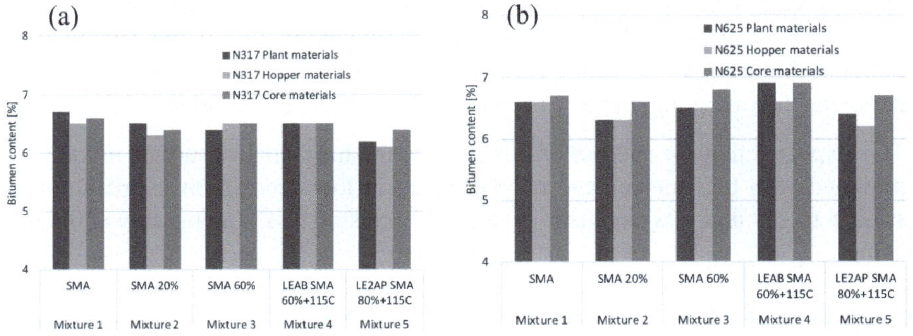

Fig. 8. Bitumen content results of five mixtures from demonstration sections N317 and N625

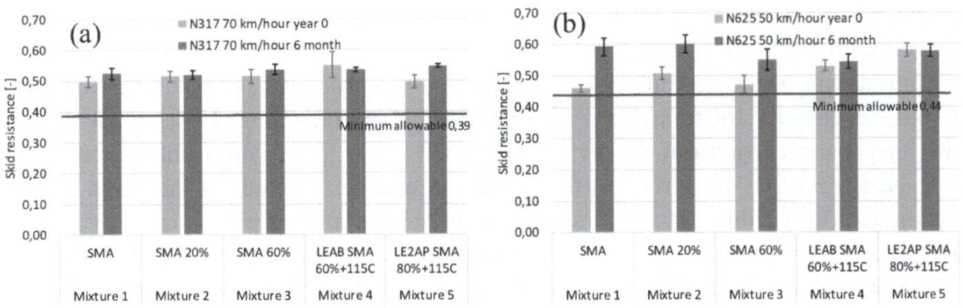

Fig. 9 Skid resistance results 0 and 6 months for demonstration sections N317 and N625

5 Conclusions

In this research sustainable SMA mixtures were developed with up to 80% of recycling and can be produced at 115 °C.

- The reclaimed porous asphalt mixtures can be decomposed into bitumen-poor reclaimed stones (less than 1% bitumen content) and bitumen-rich reclaimed mortar sand (10–12% bitumen content). The reclaimed mortar can then be rejuvenated and enriched to a LE2AP mortar which has identical performance than that of the virgin mortar, together with the reclaimed stones, making a designing of high-quality SMA mixtures with up to 80% recycling possible.
- Through LEAB bitumen foaming and LE2AP mortar foaming technology, these mixtures can be produced at 115 °C and have good mechanical performance in terms of strength and water sensitivity than that of the reference mixtures.
- A 4 km demonstration sections of sustainable SMA mixtures were constructed on Dutch provincial road N317 and N625. These asphalt mixtures were constructed for a monitoring period of 5 years. Based on the results obtained so far, the in-service performances are comparable with that of the reference mixtures.

It is expected that the success of this project, the high-quality surface mixtures can be designed with high percentage recycling and at low production temperature, will contribute to circular development of asphalt industry and slowing-down of global warming.

Acknowledgements. Province Gelderland and Province Noord-Brabant are thankful for their financial support to make the sustainable SMA demonstration sections possible.

References

Huurman M, Mo LT, Woldekidan MF (2010) Porous asphalt ravelling in cold weather conditions. Int J Pavement Res Technol 3(3):110–118

Huurman M, Demmink E, Qiu J, de Bruin B, Kuijper C, Kersten H (2018) Low Emission2 asphalt pavement. In: 93rd annual meeting of association of asphalt paving technologists (AAPT 2018), Jacksonville, USA, 18–21 March 2018

Jacobs MMJ, van den Beemt CMA, Sluer BW (2010) Successful Dutch experiences with low energy asphalt concrete. In: 11th international conference on asphalt pavements, Nagoya, Japan

Jemere Y (2010) Development of a laboratory ageing method for bitumen in porous asphalt. MSc thesis, Delft University of Technology, The Netherlands

Qiu J, Huurman M, de Bruin B, Demmink E, Frunt M (2018) Towards 90% warm re-use of porous asphalt using foaming technology. J Clean Prod 190:251–260

Reclaimed Asphalt Usage: Handling, Processing, Management and Future Trends in Lithuania

Mindaugas Martišius(✉)

Department of Transport Engineering, Vilnius Gediminas Technical University, Saulėtekio al. 11, 10223 Vilnius, Lithuania
mindaugas.martisius@vgtu.lt

Abstract. Predictably asphalt pavements are 100% recyclable objects. Reclaimed asphalt pavements (RAP) could be recycled hot, warm and cold. RAP recycling should be focused not only on costly bitumen re-use and reduction of virgin bitumen usage in hot and warm mix asphalt but also in natural resources such as aggregates and filler savings. In Lithuania 100% of RAP is recycled, but in average only 10–30% RAP is added producing hot mixed asphalt (HMA). The usage of RAP could be increased creating RAP processing and handling applications with accompanying modern in-plant production technologies, premium quality dosing and asphalt mixing plants (AMP) and expert knowledge in mix designs. This paper describes and summarizes various possible methods to process RAP recycling including its milling, handling, sampling, testing, preparation and storage before warm and hot recycling. As well recommendations for the specific climate countries as Lithuania are also summarized.

Keywords: Asphalt recycling · Recycled asphalt pavement · Bitumen · Aggregates

1 Introduction

Pavement maintenance is needed when its surface and structural characteristics do not meet the requirements due to deterioration caused by traffic loading, climatic influence and/or bad material quality (Mollenhauer and Gaspar 2012).

Lithuania as being a high road transit country where the roads suffer from dramatically increased heavy traffic load. There are 14 357 km of roads with asphalt pavements (stand 01.01.2018, www.lakd.lrv.lt, 2019). Old roads are wearing out faster and even breaking down starting from the top layers until subbase. At the other hand it is needed to be considered that appearing in a specific climate zone where the daily mean temperature asphalt layers varies from −18 °C during winter to 47 °C during summer or even higher under more severe climatic conditions (Vaitkus et al. 2017). Lithuanian road asphalt pavements are suffering from many freezing-melting cycles (temperature change from positive to negative and back during the night) in autumn to spring, low temperatures influence during winter, high temperatures influence in summer, sunlight ultraviolet (UV) rays, rain and snow, moisture and everything

M. Pasetto et al. (Eds.): ISAP APE 2019, LNCE 48, pp. 294–302, 2020.
https://doi.org/10.1007/978-3-030-29779-4_29

combined with possible heavy transport overload. Lithuania as well as other Baltic countries has poor natural foundation for the roads. Most of Baltic states territories has swampy grounds, poor soil bearing capacity and altitude mark do not cross more than 300 m above sea level.

There are still left many bitumen road surfaces paved 30–40 and more years ago in Lithuania and Baltic countries. Due to poor bitumen quality with high sulphur content road surfaces were (and still are) affected more by environmental factors. Also needed to be mentioned that bitumen itself naturally ages slowly over the years and is hardening (becomes more brittle) and could not withstand usual or increased stresses. Most of the European countries use RAP generally for unbound layers without taking the best advantage of the valuable components in the RAP (Zaumanis and Mallick 2015).

According European Asphalt Pavement Association's (EAPA) report for 2016, 100% of available RAP is all recycled in Lithuania (EAPA 2017) and for 2017 there is no official data provided. Apart of that, there is no exact official data which part of RAP was used for the HMA, cold recycling, unbound road layers or other civil engineering applications.

According gathered data by author Lithuanian estimated yearly average RAP quantity is approx. 80 000–100 000 tons. This is approx. 5% of total HMA produced yearly in Lithuania.

2 RAP Handling and Processing

Since 1970's after Wirtgen GmbH had introduced rotary cold asphalt milling technology in the road repair, RAP and quantities of RAP starting to grow each year. RAP is widely used and recycled using cold and hot recycling technologies straight on the road (in-situ) or in plants. The second phase of the RAP usage increase was oil embargo at the beginning of 1980's when companies find out huge potential of bitumen saving using rest bitumen found in RAP. Next step was approx. 10–15 years later when the road surface repair was increasing tremendously, and more and more cold and hot asphalt recycling technologies were introduced. In Lithuania the first cold milling machines were delivered in 1983. Some years later Wirtgen Remix technology was introduced.

The most used method of rehabilitation of worn out road asphalt surface is milling and replacing. According intensiveness of bitumen ageing and loosing the bearing capacity asphalt pavement wearing course layers must be replaced each 7 to 15 years (Wirtgen GmbH 2012).

Usually information about the asphalt layers thickness, types, used recipes is found in road construction projects. Clearly planning of road rehabilitation projects and milling selectively layer by layer in all its thickness leads for great recycling results saving bitumen, virgin materials and overall costs.

To ensure homogeneity of RAP milling procedures should be followed-up. Milling layer by layer ensures that old asphalt from top, bottom and subbase road layers with different properties would be not mixed during milling process. This could ensure homogeneity of RAP. Method is reasonable when milling volumes are high enough. At least top and bottom layers should be milled separately from subbase layer as in most cases different types (granite, dolomite, others) of aggregates, as well as their screening curves, and binders were used during production. When developing asphalt pavement recycling methods and technologies, the most important principles of this relatively new and promising method of recycling road pavement by reusing its asphalt materials shall be known and applied (Sivilevicius et al. 2017).

Road rehabilitation sites in Lithuania mainly are small-scaled. Therefore, milled asphalt quantities from each site are not enough to have big volumes of homogenous RAP storages. That means reclaimed asphalt comes from very different sites, in small quantities and is mixed at stockpiles (Fig. 1).

Fig. 1. Various milled RAP layers in a stockpile

Before mixing in AMP, RAP should be prepared, e.g. crushed or granulated and classified. Crushing is required to reduce grains of RAP bigger than 22 mm into smaller sizes. RAP crushing with impact crushers is combined with screening (classifying).

After the crushing process RAP is separated to 0/16 mm, 16/22 mm and 22/x mm sizes (Fig. 2). 0/16 mm is used to add RAP directly to the hot aggregate bin or mixer. This avoids fine particles sticking on the lifting mantels of drying or parallel drum at AMP. 16/22 mm is used to heat in parallel drum. 22/x mm is returned for the further size reducing in the crusher. At this stage various objects such as geotextile, steel (milling picks, picks holders, etc.), wood or at least safety wests and helmets are eliminated from delivered RAP.

Crushing process allows to dry RAP as in crusher chamber due to high energy dissipation humidity of RAP is reduced. At the other hand crushing leads higher portion of fine particles to be produced in impactor crusher. Figure 3 indicates that increase of 0/16 mm portion is 7.5%.

In this case, recommendation is to pre-screen raw RAP from stockpile out 0/16 and 16/22 m and left 22/x mm fractions could be processed by impactor crusher later.

After crushing and screening at least 0/16 mm RAP is recommended be placed under the roof as it is containing fine particles which use to intake moisture. To stock RAP under the roofed areas extremely important gets when RAP was prepared in

Fig. 2. Crushing and classifying (0/16, 16/22 and 22/x) of RAP in Lithuania

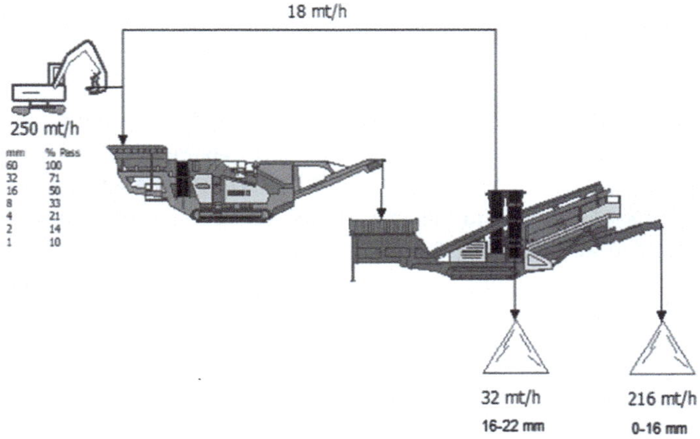

Fig. 3. RAP crushing and classifying process (metric tons per hour)

autumn. During winter season snow and melting snow penetrates to the deepest stockpile layers where it freezes.

According 48 years weather statistics (www.weatherbase.com, 2018) in Lithuanian capital Vilnius in average there was 175 days with precipitation, yearly average precipitation is 653.5 mm and average weather humidity 78.5%.

Absolute RAP quantity in Lithuania is stored in open areas without roofing. As there is rain every second day, huge quantities of water stay in RAP stockpiles which could increase RAP humidity up to 8–9%. In such cases more fuel is required for the AMP burner asphalt mixture to dry RAP in drying drum and AMP productivity drops down.

3 RAP Quality Determination

Author's research showed that RAP in Lithuania mainly used to produce HMA for base layers. This is influenced by inhomogeneous RAP distribution in stockpiles as many different types of used asphalt are mixed. The properties of such RAP should be evaluated during laboratory tests. These cases require properly sampling and properly characterized to find out which requirements RAP meets (Table 1). As a base for Lithuanian asphalt technical guidelines paper German TL Asphalt-StB 07 is taken.

Table 1. Total tolerances for properties depending on the type of asphalt mixtures.

Properties	$N_{allowed,\ i}$	
	Top and bottom layers	Base layers
Softening point °C	8	8
Binder content mass %	1,0	1,2
Filler content mass %	6,0	10,0
Fine aggregates mass %	16,0	16,0
Coarse aggregates mass %	16,0	18,0

The amount of possible RAP quantity K_i, respectively, depending on the homogeneity, is determined by evaluating the ranges of characteristics a_i and the total tolerances for $N_{allowed\ i}$ in (AST ASFALT 08 2008). The quantity K_i is calculated according to "Eqs. (1) or (2)". In the case of HMA should be used for the base layers "Eq. (1)" applies to all properties. In the case of HMA should be used for the top or bottom layers "Eq. (1)" applies to the softening temperature characteristic and "Eq. (2)" to all other properties.

$$K_i = \frac{100 \cdot (0.5 \cdot N_{allowed\ i})}{a_i} \tag{1}$$

$$K_i = \frac{100 \cdot (0.33 \cdot N_{allowed\ i})}{a_i} \tag{2}$$

where K_i - is the possible percentage of used asphalt granules to be added; a_i is the range of characteristics (difference between the highest and lowest value of the results) and $N_{allowed,i}$ is the total tolerance.

According author's experience and market research "AMPs in Lithuania (JAN 2019)" (Table 2) is provided.

Table 2. AMPs fleet in Lithuania (JAN 2019)

AMP total number	AMP with RA addition	AMP without RA addition
47	21	26

In Lithuania there are 47 AMPs and in 21 AMPs there is possibility to add up to 10–40% RAP into the HMA. Only 1 plant in Lithuania can able to add up to 70% RAP into the mix (plant is equipped with parallel drum).

Research (Zaumanis et al. 2018) in Latvia showed that best results are after RAP is properly mixed in a stockpile. Needed to say that there is no evaluated particles > 0,063 mm content sampling from road and binder content is less after mixing (Fig. 4).

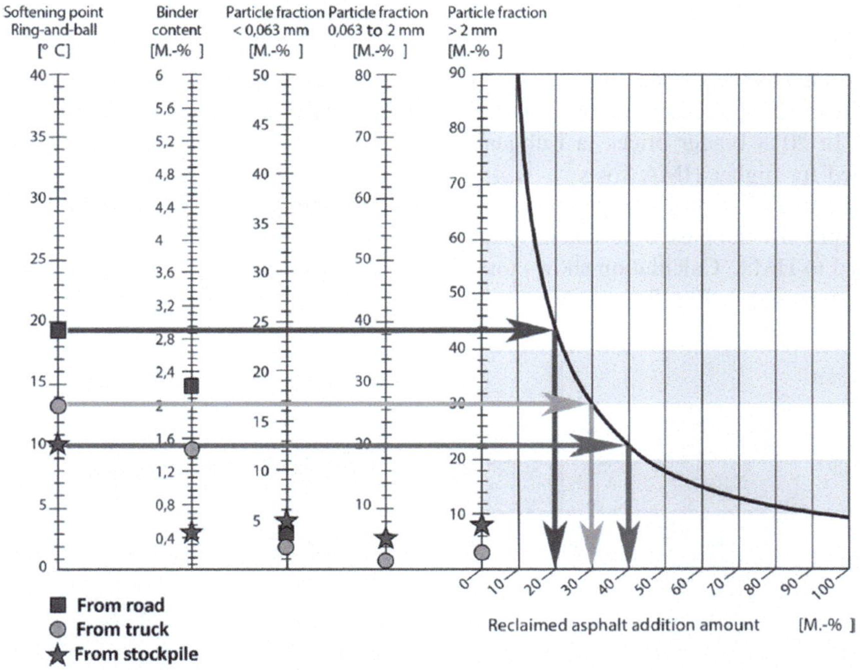

Fig. 4. Determination of the maximum RA addition amount at each stage (road, truck, stockpile). (Zaumanis et al. 2018)

4 RAP Management

Estimated 1 t HMA price excluding AMP costs (depreciation, financial costs, fuel, electricity, maintenance, labor, etc.) without 0% of RAP addition are given in (Table 3). This calculation is based on AC 16 AN type asphalt recipe where average values of binder, filler, fine and coarse particles are given. As well as average Lithuanian market prices are taken as a basis for the calculation.

Table 3. Estimated 1 t HMA price excluding AMP costs (AMP costs: depreciation, fuel, electricity, maintenance, labor, etc.)

Name	Content, %	Price, €/t	Total, €
Binder	4	330,00	13,20
Filler	7	30,00	2,10
Fine aggregates	25	3,50	0,88
Coarse aggregates	64	19,00	12,16
			28,34

As it was mentioned above due to low quantity and quality of RAP portion in new HMA is 10–30%. This amount could be the base for the calculations in savings using RA. In 2018 binder prices in Lithuania raised up to 40% comparing to 2017 which leaded for higher HMA costs.

In June of 2018 and experiment was made in Eastern part of Lithuania. From RAP stockpile 15 samples were taken, tested and evaluated as up to 30% of RA could be added to HMA. Calculation shows total 21.5% savings adding 30% of RA and results are expressed in Fig. 5.

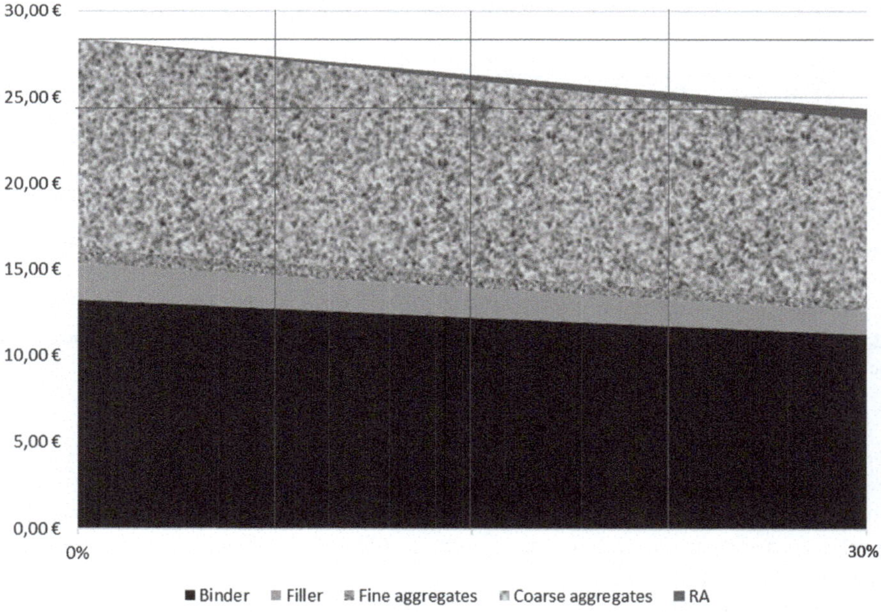

Fig. 5. Cost analysis of 1 t HMA with 0 to 30% RAP content in HMA

Results from calculations comparisons show greater savings using (producing) hot RAP versus new mixed hot asphalt mixes from the virgin materials. This study showed that using 30% RA binder savings are 22.5%, then aggregates 28.7%.

This could be critically taken as Lithuania produces binder only from imported oil. Secondary, mostly virgin imported aggregates (granite) are used for HMA production.

As for the future research it is planned to make tests when RAP containing granite (imported aggregates) would be used mixing HMA with local aggregates (river stones and dolomite).

5 Conclusions

1. RAP quantities will significantly grow in Lithuania in nearest future as more roads with asphalt pavement are built and reconstructed. Usage of RAP should be considered as there are no satisfactory equipment to recycle high quantities of RAP producing HMA.
2. The data of RAP usage in Lithuania needed to be systemized as there is no exact RAP accounting. EAPA data indicate that 100% RAP in Lithuania is used, but it is not detailed where: new hot or cold asphalt production, road base mixtures, etc.
3. Further steps should be to developed Lithuanian road database which contains the construction and/or rehabilitation history with dates, road layers, mix design data of the existing road network. This database would help to select the possible options for maintenance and rehabilitation applications economically in the best cost-benefit ratio.
4. Database of long-term performance of former asphalt recycling test roads should be created. Road pavement design and construction, traffic loads and ambient weather conditions are relatively reliably known. Data collection and investigation of experimental road testing sections would open more possibilities to evaluate long-term perspectives.
5. Preferably to plan big-scaled road rehabilitation projects. Having insufficient amounts of different composition and quality parameters in RAP stockpiles influence inhomogeneous RAP, therefore only small portions could be used in production process. There is no method for preparing RAP (i.e. milled asphalt is sieved into fractions, but is not crushed or crushed properly, as the primary granulometry changes after the rotary crusher and a larger fraction of the fines up to 4 mm is formed).

References

Aurangzeb Q, Al-Qadi IL, Ozer H, Yang R (2014) Hybrid life cycle assessment for asphalt mixtures with high RAP content. Resour Conserv Recycl 83:77–86

Automobilių kelių asfalto mišinių techninių reikalavimų aprašas TRA ASFALTAS 08 (2008) Technical requirements of asphalt – Lithuanian standards

Deutscher Asphaltverband (German Asphalt Pavement Association) (2011) Recycling of asphalt

EAPA paper (2017) Asphalt in figures 2016

EAPA position paper (2012) Asphalt the 100% recyclable construction product

Karlsson R, Isaksson U (2006) Material-related aspects of asphalt recycling-state-of-the-art. J Mater Civ Eng 18(1):81–92

LST EN 13108-20. Bituminiai mišiniai. Medžiagos ir techniniai reikalavimai. 8 dalis. Naudotas asfaltas. (Bituminous mixtures. Materials and specifications. Part 8 Recycled asphalt.)

Mollenhauer K, Gaspar L (2012) Synthesis of European knowledge on asphalt recycling: options, best practices and research needs. In: 5th Eurasphalt & Eurobitume Congress, Istanbul, Turkey, 13–15 June

Sivilevičius H, Bražiūnas J, Prentkovskis O (2017) Techologies and principles of hot recycling and investigation of preheated reclaimed asphalt pavement batching process in an asphalt mixing plant. Appl Sci 7(11):1–20

Technische Lieferbedingungem fuer Asphaltgranulat TL AG-StB 09 (2009) (Technical delivery conditions for recycled asphalt – German Standards)

Technische Lieferbedingungem fuer Asphaltgranulat TL Asphalt-StB 07 (2007) (Technical delivery conditions for asphalt – German Stadards)

Vaitkus A, Vorobjovas V, Kleizienė R, Šernas O, Gražulytė J (2017) Modified asphalt mixtures for heavy duty pavement wearing layers. Constr Build Mater 131:503–511

Wirtgen GmbH (2012) Wirtgen cold recycling technology, 1st edn. Wirtgen GmbH

Zaumanis M, Mallick RB (2015) Review of very high-content reclaimed asphalt use in plant-produced pavements: state of the art. Int J Pavement Eng 16(1):39–55

Zaumanis M, Oga J, Haritonovs V (2018) How to reduce reclaimed asphalt variability: a full-scale study. Constr Build Mater 188:546–554

Experimental Study to Re-refine Aged Binder Using Water

Kengo Akatsu(✉), Yousuke Kanou, and Shouichi Akiba

College of Industrial Technology, Nihon University, 1-2-1 Izumi-Cho, Narashino, Chiba 275-8875, Japan
k.akatsu0620@gmail.com

Abstract. In Japan, asphalt pavement waste is recycled about 99%, of which about 60% is utilized as a hot recycled mixture. However, conventional technology does not guarantee the quality of repeatedly recycled pavement. In the future, it is essential to establish sustainable recycling technology for asphalt pavement waste. Based on above, the authors are working on developing reconstruction technology that can separate aggregate and aged binder in waste materials and restore the separated material to its original state. We confirmed that aggregate separated and recycled by the hot water rubbing method can be maintained and used in the same way as with virgin aggregate, in previous study. This study attempted to recycle asphalt using water with a focus on solvent characteristics of high-temperature and high-pressure water. As a result, hydrothermal decomposition technology utilizing high-temperature and high-pressure water reduced the molecular weight and oxidation degree in aged binder and confirmed to re-refined to near chemical properties and physical properties of virgin asphalt.

Keywords: Recycling technology · Aqueous pyrolysis technology · Sub-critical water · Hydrolysis · Rheological properties · Aged binder

1 Introduction

In Japan, the development of technologies for the utilization of asphalt pavement waste materials has a history of nearly half a century. In recent years, about 99% of asphalt pavement waste is mechanically crushed for recycling, of which about 60% is utilized as a recycled aggregate in reclaimed asphalt mixture. Thus, a quantitatively advanced recycling system has been established.

Meanwhile, in addition to the current situation where from the standpoint of materials supply, there is not easy to secure good quality aggregate on the background of preservation and development of coarse and fine aggregate, there is a possibility that the supply of asphalt will be narrowed of oil refining technology that has been encouraged by the increased proportion of heavier crude oils and the higher demand for lighter oils. Thus, in the future, it will still be imperative to establish sustainable recycling technologies by investigating a means of repeat recycling based on conventional methods and by developing new reconstruction technologies in order to maintain a stable supply of materials based on the utilization of waste materials. Also,

M. Pasetto et al. (Eds.): ISAP APE 2019, LNCE 48, pp. 303–312, 2020.
https://doi.org/10.1007/978-3-030-29779-4_30

a qualitatively more sophisticated recycling system is desired in order to develop a technological foundation to support the maximization and diversification of pavements.

The authors engaged in the development of a reconstruction technology that can separate aggregate and aged binder in waste materials and restore the separated materials into their original state, and we confirmed that aggregate separated and recycled by the hot water rubbing method can be maintained and used in the same way as with virgin aggregate, as reported in our previous study (Kano et al. 2016). This study attempted to recycle asphalt using water with a focus on two solvent characteristics of high-temperature and high-pressure water: (1) A decrease in the dielectric constant along with an increase in temperature causes dissolution or high-dispersion of organic substances; and (2) An increase in ion product along with an increase in temperature promotes hydrolysis. As a result, this confirms the possibility that an aqueous pyrolysis technology based on the application of the solvent characteristics of high-temperature and high-pressure water can cause a reduction of the molecular weight and oxidation degree in aged binder, the aged binder is thereby re-refined to near the chemical and physical properties of virgin asphalt.

2 Solvent Characteristics of High-Temperature and High-Pressure Water and the Aqueous Pyrolysis Method

Water changes its state depending on temperature and pressure, as illustrated in Fig. 1 (a). Solvent characteristics of high-temperature and high-pressure water are roughly understood as outlined in Fig. 1(b) based on indexes of the dielectric constant (i.e. polarity) and ion product (i.e. hydrolysis resolution).

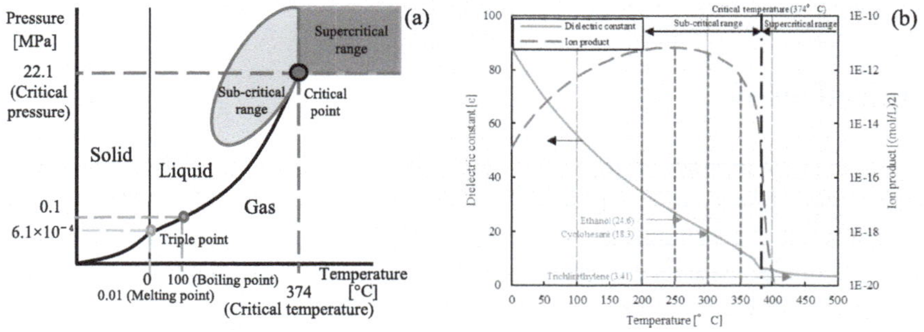

Fig. 1. P-T diagram (a) and dielectric constant and product (b) of *water*

As is generally known, water at ambient temperature barely dissolves most organic substances of low polarity, which exhibits the so-called “water-and-oil state.” However, as shown in Fig. 1(b), the dielectric constant of water decreases as temperature increases, and it reaches an equivalent level to that of organic solvent near the critical temperature, and various organic substances can be dissolved.

On the other hand, the ion product of water can incrementally increase up to around 250 °C as illustrated in the figure. Accordingly, the ion product of water, which is 1.0×10^{-14} at ambient temperature, can rise up to 1.0×10^{-11} in the sub-critical range, in which water itself has strong acid and alkaline properties and hydrolysis progresses with no catalyst.

Recently this aqueous pyrolysis method that utilizes the properties of high-temperature and high-pressure water has been applied to reformation processes of oil sand bitumen, etc. The authors have confirmed in the course of development of an asphalt extraction test using sub-critical water as solvent that properties of aged binder can be re-refined with the solvent (Akatsu et al. 2016).

Based on the findings described above, this study focused on the aqueous pyrolysis method using hot water at 200–350 °C, at which the dielectric constant can be as low as possible and hydrolysis is expected to occur, and examined the effects of reaction temperature and reaction time on restoration property and compared the restoration property effects with the conventional method. The methods used for experimentation and assessment are detailed in the next section.

3 Experimental Overview

3.1 Effects of Reaction Temperature and Reaction Time on Restoration Property

The experimental procedure of aqueous pyrolysis and an overview of equipment are shown in Figs. 2 and 3, respectively. Both are roughly the same as those of the sub-critical water extraction method except that the asphalt under test was left standing to cool so that it could be collected under water because cooling at the top may cause the asphalt to cure on top and promotes deterioration. As shown in Fig. 1(b), solvent characteristics of high-temperature and high-pressure water are primarily dependent on reaction temperature. In order to observe the effects of reaction temperature and reaction time on asphalt restoration property, the chemical and physical properties of aged binder as described in (a) and (b) below were assessed after 15 min of reaction at 200 °C, 250 °C, 300 °C, and 350 °C, and after 0, 15, and 45 min at 250 °C and 350 ° C. For the purpose of this article, "after 0 min" means that asphalt was cooled immediately once the reaction temperature was reached. The amount of water is the amount required for the vapor pressure, which rises with heating, to maintain the liquid phase.

To circumvent the effect of extraction and collection of asphalt, virgin asphalt were used that had undergone accelerated deterioration by means of a rolling thin film oven test (RTFOT) at 163 °C for 5 h and a pressure aging vessel (PAV) at 100 °C · 2.1 MPa for 23 h so that the penetration was 20 before the experiment. Table 1 shows the characteristics of virgin asphalt before and after accelerated deterioration.

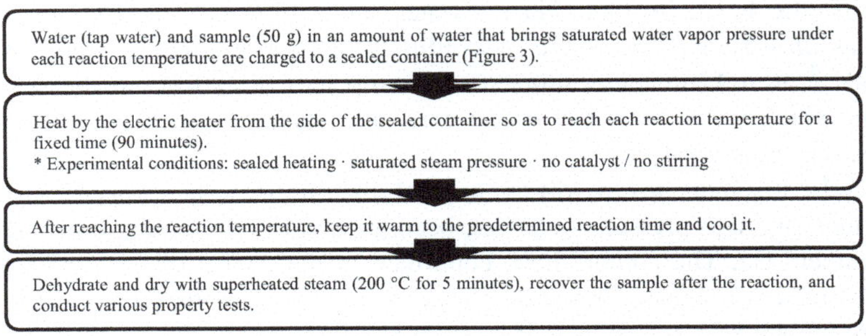

Fig. 2. Experimental procedure

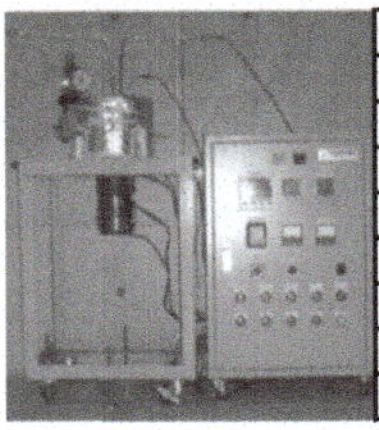

Manufacturing company	Taiatsu Techno® corporation
Brand name	Type TAS-1-19 reaction container
Inner volume	1000mla
Highest pressure range	19MPa
Highest temperature range	350°C
Heater	Microring heater
Specification voltage	200V single phase, 1.5KW
Specification current	20A
Law applicable	Small pressure container

Fig. 3. Overview of equipment

Table 1. Properties of virgin asphalt

Penetration (25 °C)	[0.1 mm]	68
Softening point	[°C]	47.0
Elongation	[cm]	100+
Density	[g/cm^3]	1.035

(a) Chemical properties

The effects of reaction conditions on the chemical properties of re-refined asphalt were observed by infrared absorbance and composition analysis.

Infrared absorbance was measured by the attenuated total reflection method by using a Fourier transform infrared spectrophotometer. For this analysis, the carbonyl index was obtained and assessed based on Formula (1) with a focus on an increase in oxygen-containing functional groups along with oxidation degradation.

$$CI = \frac{\log(I_{o1}/I_1)}{\log(I_{o2}/I_2)} \quad (1)$$

Where,

I_1, I_2: Peak spectrum transmittance near 1700 cm^{-1} and 1600 cm^{-1}, respectively

I_{o1}, I_{o2}: Background transmittance of respective wavenumbers above

CI represents the ratio between the peak height near 1600 cm^{-1} (C-C bond) not impacted by deterioration and the peak height near 1700 cm^{-1} (C-O bond) that was raised by deterioration, and can be used as an index of the degree of oxidation degradation.

For composition analysis, component ratio was analyzed by thin layer chromatography with flame ionization detection (TLC/FID).

(b) Physical properties
The effects of reaction conditions on the physical properties of re-refined asphalt were observed by penetration, softening point, penetration index, and ductility (15 °C). Each test conformed to the Pavement Design and Implementation Guideline and PI was calculated using Formula (2) (Japan Road Association 2007).

$$PI = \frac{30}{1 + 50A} - 10,\ A = \frac{\log 800 - \log P}{\mathrm{SP} - 25} \quad (2)$$

Where,
P: Penetration, SP: Softening point, A: Thermosensitivity

3.2 Characteristics of Restoration Property Effects Compared with the Conventional Method

To comprehend the characteristics of restoration property effects of aqueous pyrolysis, asphalt treated for accelerated deterioration as described in the previous sub-section was used and the properties of asphalt re-refined by the aqueous pyrolysis method and re-refined asphalt with penetration adjusted with additives were compared by component ratio and rheological properties. The aqueous pyrolysis method was implemented by following the procedure shown in Fig. 3, reaction temperature and time conditions were set the most effective conditions for restoring asphalt characteristics based on the findings mentioned above. The properties of the additives are listed in Table 2. For this purpose, the amount of additives applied to the aged binder was 10.03% in line with the penetration of re-refined asphalt.

Table 2. Properties of additives

Density		[g/cm^3]	1.0
Composition	Asphaltene compounds	[%]	0.0
	Resin compounds	[%]	5.2
	Aromatic compounds	[%]	48.2
	Satiation compounds	[%]	46.6

Analyses of component ratio was conducted as outlined in the previous sub-section.

For rheological properties, the dynamic share rheometer (DSR) method was used to measure the phase angle and complex elastic modulus. The test temperature was maintained in the range of 10–30 °C for ambient temperature using a φ8 mm parallel plate, and 50–70 °C for high temperature using a φ25 mm parallel plate. Angular velocity was 0.1–100 rad/s, and distortion was 5%.

4 Experimental Results

4.1 Effects of Reaction Temperature on Restoration Property

(a) Chemical properties
A comparison of CI, which represents the degree of oxidation, is shown in Fig. 4. Cl of deteriorated asphalt is higher than that of virgin asphalt, which represents advanced oxidation degradation. Asphalt re-refined by aqueous pyrolysis method presented slightly lower Cl at 200–300 °C than deteriorated asphalt, but was halved lower Cl at 350 °C (15 min).

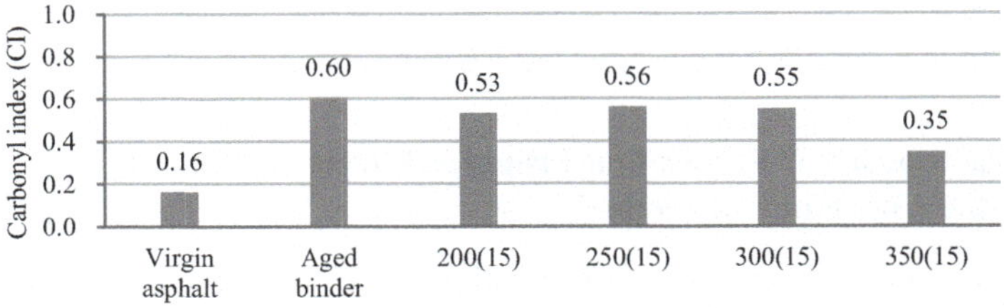

Fig. 4. Carbonyl index (200–350 °C, 15 min)

Figure 5 shows the component ratio figures. Compared with virgin asphalt, deteriorated asphalt exhibited a general deterioration trend where asphaltene and resin compounds are increased while aromatic compounds is decreased. On the other hand, asphalt re-refined by the aqueous pyrolysis method presented that asphaltene and resin compounds decreased, and aromatic compounds increased at higher reaction temperatures; at 350 °C (15 min), in particular, approached virgin asphalt.

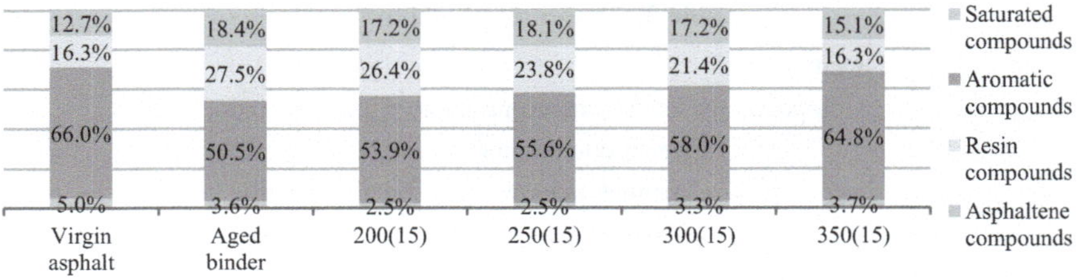

Fig. 5. Asphalt composition (200–350 °C, 15 min)

(b) Physical properties
Results of the physical property testing are listed in Table 3. Penetration was generally increased from 200 °C to 300 °C and substantially increased up to the standard value of virgin asphalt at 350 °C (15 min). Restoration property along with increase of reaction temperature was also observed at the softening point, PI, and ductility. In

particular, softening point and ductility figures at 350 °C (15 min) satisfied the standard values of virgin asphalt and PI was also re-refined to a similar degree.

Table 3. Physical properties (200–350 °C, 15 min)

	Virgin asphalt	Aged binder	200 (15)	250 (15)	300 (15)	350 (15)
Penetration [0.1 mm]	68	20.3	26.0	28.7	33.7	60.3
Softening point [° C]	47.0	64.9	60.1	59.0	55.8	48.7
PI	−1.20	0.00	−0.38	−0.40	−0.72	−1.09
Elongation [cm]	100+	5.7	8.7	9.0	25.3	100+

From these findings, it was confirmed that the restoration property effects from the aqueous pyrolysis method are influenced by reaction temperature and are substantially increased in the temperature range of 300–350 °C. This tendency was correlated with the dielectric constant indicated in Fig. 1(b), which suggests a possibility that hydrolysis was promoted due to a higher dispersion of asphalt.

4.2 Effects of Reaction Time on Restoration Property

(a) Chemical properties
Changes in Cl by temperature are shown in Fig. 6(a) and (b). At 250 °C, CI, which represents the degree of oxidation, decreased between 0 min and 15 min, but increased between 15 min and 45 min. At 350°C, CI, decreased between 0 min and 15 min, but no changes were observed between 15 min and 45 min.

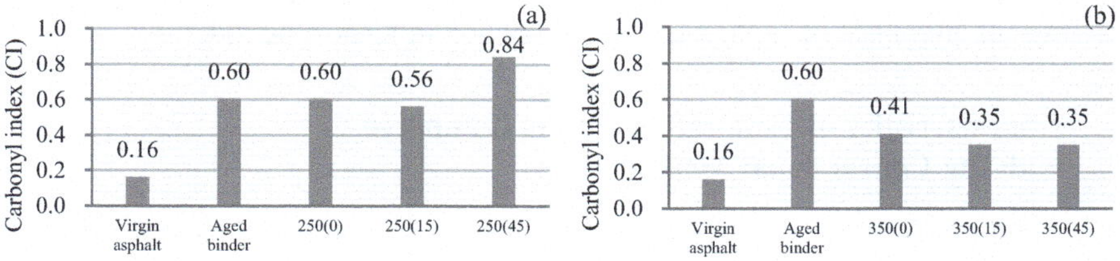

Fig. 6. 250 °C, 0–45 min (a) and 350 °C, 0–45 min (b) of *carbonyl index*

Changes in component ratio by temperature are shown in Fig. 7(a) and (b). With re-refined asphalt, asphaltene and resin compounds were decreased respectively, and aromatic compounds was increased, in the period between 0 min and 15 min at both 250 °C and 350 °C, however, between 15 min and 45 min, conversely, the same tendency as the composition change of the deteriorated asphalt was observed.

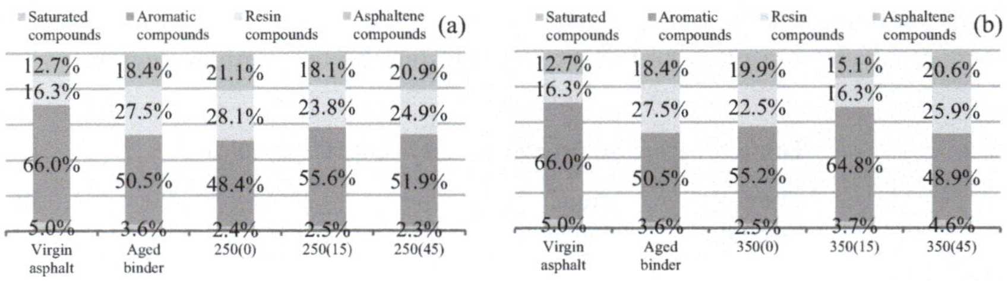

Fig. 7. 250 °C, 0–45 min (a) and 350 °C, 0–45 min (b) of *asphalt composition*

(b) Physical properties
The results of the physical property testing by temperature are listed in Table 4. Restoration of penetration, softening point, and ductility were all the most significant at a reaction time of 15 min and deterioration advanced between 15 min and 45 min, as with chemical properties.

Table 4. Physical properties (250 °C, 0–45 min and 350 °C, 0–45 min)

	Virgin asphalt	Aged binder	250 (0)	250 (15)	250 (45)	350 (0)	350 (15)	350 (45)
Penetration [0.1 mm]	68	20.3	24.3	28.7	25.3	42.3	60.3	49.3
Softening point [° C]	47.0	64.9	60.3	59.0	60.7	51.0	48.7	50.4
PI	−1.20	0.00	−0.47	−0.40	−0.33	−1.32	−1.09	−1.13
Elongation [cm]	100+	5.7	8.3	9.0	9.0	62.7	100+	69.3

From these findings, it was confirmed that the restoration property effects on aged binder by the aqueous pyrolysis method are influenced by reaction time and are substantially increased in the time range of 0–15 min. This finding suggests a possibility that aqueous pyrolysis of asphalt can be completed in a short period.

4.3 Characteristics of Restoration Effects Compared with the Conventional Method

Component ratio re-refined by the aqueous pyrolysis method and asphalt re-refined by the conventional method are shown in Fig. 8. In composition ratio, with re-refined asphalt at 350 °C (15 min), than those asphalt rejuvenated by additives, asphaltene and resin compounds were decreased, and aromatic compounds was increased.

For rheological properties, the correlation between phase angle and complex elastic modulus in different temperature ranges is shown in Fig. 9(a) and (b). In the ambient temperature range, re-refined asphalt exhibited a little restoration of phase angle at 350 °C (15 min) in comparison to rejuvenated binder. Complex elastic modulus was lower in rejuvenated binder, whereas re-refined asphalt exhibited a complex elastic modulus figure equivalent to that of virgin asphalt. The same tendency as in the ambient temperature range was observed in the high temperature range.

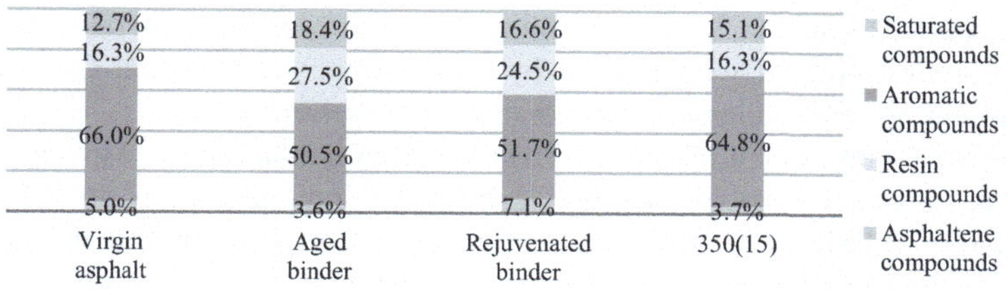

Fig. 8. Asphalt compositional (rejuvenated binder and re-refined asphalt)

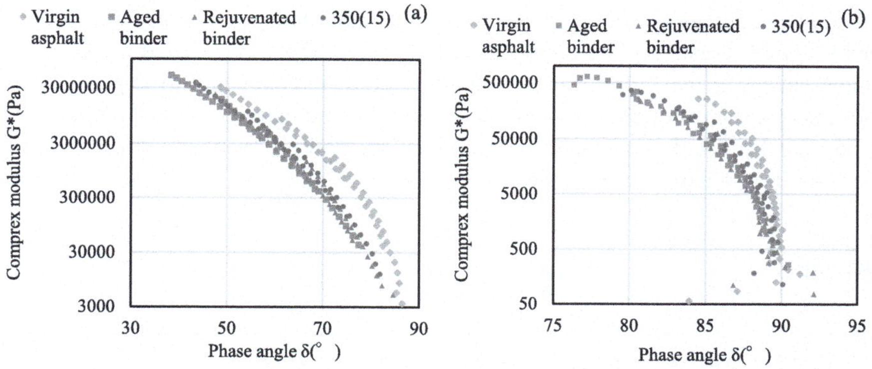

Fig. 9. Ambient temperature range (a) and high temperature range (b) of *rheological properties*

The findings discussed above demonstrate that asphalt re-refined by the aqueous pyrolysis method characteristically has component ratio and rheological properties closer to that of virgin asphalt than rejuvenated binder. The aqueous pyrolysis method to dissolve and reduce higher molecular components suggested the possibility of contributing to the recovery of substantial properties of aged binder than the conventional method of adds low-molecular component.

5 Conclusions

The findings of this study are summarized below.

(a) Restoration property effects of the aqueous pyrolysis method substantially vary depending on reaction temperature and reaction time. In particular, re-refined asphalt at 350 °C (15 min) exhibited lower asphaltene content and resin compounds decrease. Aromatic compounds are substantially increased compared to the aged binder.

(b) Re-refined asphalt at 350 °C (15 min) satisfied standard of virgin asphalt for penetration, softening point, and ductility and its PI was also re-refined to near the pre-deterioration level.

(c) In re-refined asphalt at 350 °C (15 min), the carbonyl group generated through the deterioration process was reduced by its reductive reaction and Cl, which was increased due to accelerated deterioration, was halved.
(d) Compared with asphalt rejuvenated by the addition of additives, asphalt re-refined by the aqueous pyrolysis method exhibited component ratio and rheological properties a little close to that of virgin asphalt, suggesting that the process of dissolving and reducing high-molecular components can more effectively contribute to the restoration of properties in aged binder than the conventional method.

For these findings, it was confirmed that it is possible that the aqueous pyrolysis method can cause a lowering of molecular weight and oxidation degree in aged binder, thereby partially restoring the chemical and physical properties of aged binder to their original states. In the future, further studies should be conducted to explore optimal conditions for the restoration property technology including environmental effects, etc., and to reveal a detailed mechanism behind the method by analyzing reacted solvents and gases in order to improve efficiency.

References

Kano Y, Akiba S, Akatsu K (2016) A basic study the separation recycling technology of reclaimed asphalt pavement. J Jpn Soc Civil Eng 72(3):I_61–I_68

Akatsu K, Kano Y, Akiba S (2016) A proposal of the extraction test of asphalt using sub-critical water. J Jpn Soc Civil Eng 72(1):31–41

Japan Road Association (2007) Pavement design and implementation guideline. Japan Road Association, [2]281–292

Cold In-place Recycling for a Base Layer of an Italian High-Traffic Highway

Edoardo Bocci[1(✉)], Andrea Graziani[2], and Maurizio Bocci[2]

[1] Faculty of Engineering, eCampus University, Via Isimbardi 10, 22060 Novedrate, Italy
edoardo.bocci@uniecampus.it
[2] Department of Civil, Environmental and Architecture Engineering (DICEA), Università Politecnica delle Marche, Via Brecce Bianche, 60131 Ancona, Italy
{a.graziani,m.bocci}@univpm.it

Abstract. Cold in-place recycling (CIR) has increasing interest worldwide due to the large economic and environmental benefits deriving from reduction of atmospheric emissions, preservation of natural resources, no hauling of the material to the recycling plant and back to the jobsite. However, a unique protocol for the structural rehabilitation of highly trafficked highways based on CIR has not been still developed, discouraging their application in the upper layers of the pavement (base layer). This paper describes a case study where a cold recycled asphalt base (CRAB), containing 100% of reclaimed asphalt treated with polymer-modified bitumen emulsion and cement, has been applied in the reconstruction of the SS268 divided highway in southern Italy. The technical approach followed in this project included the preliminary execution of falling weight deflectometer tests to estimate the subgrade stiffness, the design of the pavement structure through a mechanistic-empirical procedure, the mix design of the CRAB and the validation of the construction procedure and mix properties in a trial section. This paper aims to provide a protocol for the design and construction of CIR mixtures, allowing their use as base layer for the highways interested by a high traffic.

Keywords: Cold in-place recycling · Bitumen emulsion · Reclaimed asphalt · Cold recycled asphalt base · Construction protocol

1 Introduction

In recent years, the interest of researchers, practitioners and road agencies in cold in-place recycling (CIR) techniques has increased worldwide. This is due to the large economic and environmental benefits related to the reduction of atmospheric emissions, the preservation of natural resources and the elimination of the milled material hauling to the recycling plant and back to the jobsite (Grilli et al. 2016). According to AASHTO (1998), a CIR mixture is the result of the stabilization of the reclaimed asphalt (RA) from distressed bituminous layers. This operation is directly carried out in site without heating the components.

CIR mixtures include RA (integrated with virgin aggregate in the case of incorrect gradation), water, bitumen emulsion (or foamed bitumen) and/or Portland cement.

M. Pasetto et al. (Eds.): ISAP APE 2019, LNCE 48, pp. 313–322, 2020.
https://doi.org/10.1007/978-3-030-29779-4_31

The nature and dosage of the binding agents allow regulating the mix behavior from cement-like (stiff and brittle behavior) to asphalt-like (cumulative damage and time-temperature sensitivity) (Pérez et al. 2013; Godenzoni et al. 2018).

Within the pavement structure, CIR mixtures can be used as subbase layer or base layer. In the latter case, cold recycled asphalt base (CRAB) layer can replace the traditional hot-mix asphalt (HMA) base, even in high-trafficked roads (Bocci et al. 2011). In some countries, this CRAB application is discouraged because a reliable pavement rehabilitation protocol based on CIR has not been still developed. In Italy, CIR mixtures have been used in highway rehabilitation since the 90s and CRAB layers have been constructed since 2000 (Bocci et al. 2002). With years, the technique has improved and an application protocol has been defined, even if it has to be adapted for the specific situations. This protocol provides the characterization of the distressed pavement, the design of the new pavement structure, the optimization of each paving mix (including the CIR materials) composition, the construction of a trial section to set the production and laying phases, and the validation of material quality and construction operations through laboratory and field tests.

The paper describes a case study where a CIR mixture has been used. The novelty of the application deals with the content of RA (pushed to 100%), the importance of the road (a high-traffic highway) and the use of the CIR mixture in place of HMA as base layer. The objective of the paper is to present a procedure for the design, construction and validation of CIR mixtures, contributing to the spread of this technique and helping to define, in the future, guidelines for application.

2 Project Description

The project deals with the reconstruction of the four-lane divided highway SS268, close to city of Naples (southern Italy) from km 0+000 to km 19+554. The traffic volume, measured in 2011 using pneumatic tubes, was estimated in 17 million 80 kN ESALs for twenty years of service.

The existing pavement consisted of 18 cm of distressed HMA and 35 cm of granular foundation on the natural soil subgrade. The new pavement, designed through a mechanistic-empirical method, was built in 2017, and consisted of 35 cm of in-place cement-stabilization of the foundation material, 20 cm of CRAB containing 100% RA and treated with polymer-modified bitumen emulsion and cement, 6 cm of HMA binder course and 5 cm of porous asphalt surfacing.

3 FWD Tests on the Subgrade

In 2016, a falling weight deflectometer (FWD) test campaign was carried out on the old pavement. This analysis allowed assessing the structural condition of the unbound foundation, to evaluate the necessity of stabilizing it with hydraulic binders. Moreover, it allowed determining the modulus of the subgrade, to be used in the mechanistic-empirical design.

The FWD campaign provided over 250 measurement points on the four lanes. The 15th percentile of the modulus data was assumed as design value for each layer. The results showed that the foundation had a poor bearing capacity (E_f = 64 MPa), despite the effect of the heavy traffic in the many years of service. The subgrade showed a 15th percentile modulus E_{sub} = 77 MPa.

4 Design of the Pavement Structure

The structural design of the pavement was carried out using a mechanistic-empirical approach. In particular, the pavement was modelled with a multi-layer linear elastic system, where each layer was characterized by thickness, Young's modulus and Poisson's ratio.

To estimate the Young's modulus of the HMA layers, the seasonal average temperatures, registered by the weather station of Naples, were used in the Witczak equation (1972) to calculate the HMA temperatures at specific depths. Hypothesizing the penetration and softening point of the bitumen typically used in southern Italy and fixing the loading frequency, the stiffness modulus of the bitumen was determined through Van Der Poel nomograph (Van der Poel 1954). Then, HMA elastic moduli for the different seasons were estimated by means of the Francken equation (1996), assuming specific volumetric properties for each layer. The Poisson's ratio was assumed equal to 0.30 for winter, 0.40 for summer and 0.35 for spring and autumn, according to the results by Graziani et al. (2014).

The service life of CRAB and cement-stabilized foundation (CSF) layers was considered to have two phases: a pre-cracked and a post-cracked phase. The Young's modulus and Poisson's ratio of CRAB were assumed as indicated by Liebenberg and Visser (2004), while for CSF the parameters suggested by Theyse et al. (1996) were adopted.

The service life of each material was determined according to the empirical laws reported in Table 1. In particular, these laws were defined by different authors through laboratory and field testing considering the specific phenomena that bring each layer to collapse (fatigue for HMA, CRAB Ph1 and CFS Ph1, permanent deformation for CRAB Ph2, CSF Ph2 and subgrade). The results obtained for each season were combined using Miner's law (Miner 1945).

The described design method provided a service life of the pavement of about 24 million 80-kN ESALs, approximately corresponding to 27 years of service.

5 Mix Design of the CRAB Mixture

Before the construction, the CRAB mixture was designed in laboratory to determine the optimum content of bitumen emulsion, cement and water. The RA, that represents the aggregate component of the CRAB mixture, was milled from SS268 and characterized in terms of gradation (black curve). The particle size distribution was suitable for the use in CRAB mixture, without the addition of any virgin aggregate.

Table 1. Damage laws used in the mechanistic-empirical design of the pavement

Material	Damage law	Reference
HMA	$N_t = 10^{-3.083} \cdot \varepsilon_t^{-3.291} \cdot E^{-0.854}$	Finn et al. 1977
CRAB, Ph1	$N_f = 10^{\left(8.0331-1.2775\frac{\varepsilon_t}{\varepsilon_b}\right)}$	Liebenberg and Visser 2004
CRAB, Ph2	$N_{PDb} = \left(\frac{54,005}{t} + 4,5389\right), (SR + 0,0664)^{-0,2313}$	Liebenberg and Visser 2004
CSF, Ph1	$N_f = 10^{6.84\left(1-\frac{\varepsilon_t}{7.63\varepsilon_b}\right)} \cdot SD$	Theyse et al. 1996
CSF, Ph2	$N_{Ca} = 10^{8.184\left(1-\frac{\sigma_v}{1.2UCS}\right)}$	Theyse et al. 1996
Subgrade	$\log N_z = -7.21 - 3.95 \cdot \log \varepsilon_z$	Powell et al. 1984

The CRAB mixture was produced in laboratory using pozzolanic cement (classified as CEM IV/B 32.5 R according to EN 197-1) and slow-setting cationic polymer-modified bitumen emulsion (classified as C60BP10 according to EN 13808). Six water contents (3–8% by aggregate mass), three cement contents (1.5, 2.0 and 2.5% by aggregate mass) and three bitumen emulsion contents (3.0, 3.5 and 4.0% by aggregate mass) were considered. Cylindrical specimens with 150 mm diameter were prepared by means of a gyratory compactor (GC) and cured at 40 °C for 72 h (accelerated curing). Then, dry bulk density and indirect tensile strength at 25 °C (EN 12697-23) were determined.

The job mix formula defined through the mix design study provided 4.0% of bitumen emulsion (2.4% of residual bitumen), 2.0% of cement and 4.0% of total water. The optimized CRAB mixture showed a dry bulk density of 2.147 Mg/m^3 after 100 gyrations and an indirect tensile strength at 25 °C of 0.40 MPa. Moreover, the indirect tensile stiffness modulus at 20 °C, measured according to EN 12697-26-Annex C, was equal to 3242 MPa. These values were used as reference for the validation of the CRAB mixture constructed in site.

6 Construction of the Trial Section

Before the application on the whole length of the highway, the construction procedures were set on a 1.0 km trial section, involving two lanes. This operation aimed at verifying the mechanical properties of the designed materials, adjusting the construction protocol, assessing the effectiveness of the construction equipment and measuring the material properties to be used as reference for the control tests.

The trial section was built on October 2016 according to the following program.

Phase 1: cement stabilization on the right lane.

- Milling of the HMA to a depth of 18 cm for the width of the right lane and the shoulder (5.0 m) and storage of the RA on the left lane.
- Milling of the unbound foundation to a depth of 6 cm and hauling of the material to a landfill site.

- Spreading of cement using a vehicle equipped with a volumetric batcher.
- Passing of the recycling machine to mix the cement with the existing granular foundation to a depth of 37 cm (Fig. 1a). The mixed material was sampled to check the moisture content, then the layer was compacted with a 14 tons single drum vibrating roller and a 25 tons pneumatic roller (Fig. 1b).
- Moving of the RA from the left to right lane, on top of the compacted CSF layer.

Fig. 1. Photos taken during the construction of the experimental pavement section: (a) first recycler pass for mixing CSF layer; (b) compaction of CSF layer; (c) levelling of the RA; (d) second recycler pass for mixing CRAB layer; (e) compaction of CRAB layer; (f) compacted CRAB layer before sealing.

Phase 2: cement stabilization on the left lane.

- Same operations described for Phase 1. During the mixing with the recycling machine, about 40 cm of the new CSF layer of the slow lane were re-processed to improve the longitudinal joint.

Phase 3: in-place recycling.

- Levelling of the RA on the entire width of the carriageway, on top of the new CSF layer (Fig. 1c), and light compaction with a single drum vibrating roller to obtain an adequate support for the wheels of the recycling machine.

- Spreading of the cement and passing of the recycling train (Fig. 1d) including the recycling machine and the tank truck with the bitumen emulsion (the tank truck containing water was not used as additional moisture was unnecessary). The treatment interested a thickness of 20 cm, involving 18 cm of lightly compacted RA and 2 cm of the underlying CSF material.
- First compaction with two passes of a single drum vibrating roller.
- Regularization of the surface with a grader (creation of the target longitudinal and transversal slopes).
- Final compaction with a single drum vibrating roller and pneumatic-tired roller (Fig. 1e).

Phase 4: construction of asphalt surfacing.
- Sealing of the CRAB surface with bitumen emulsion.
- Laying of the binder and porous asphalt layers (a curing period of few days before CRAB sealing was programmed but finally deleted because of a rainfall).

7 Validation of the CRAB Mixture

The volumetric and mechanical properties of the CRAB mixture were verified through laboratory tests on the material sampled and compacted directly in site with a GC and on cores taken from the trial section. FWD tests were also carried out.

7.1 Laboratory Tests on GC Specimens

The specimens compacted with the GC were cured for 72 h at 40 °C and then tested to measure the indirect tensile strength (ITS) and unconfined compressive strength (UCS). The average ITS and UCS values were 0.47 MPa and 2.9 MPa, respectively. The ITS was comparable to the value obtained during mix design and complied with the construction specifications (acceptable range between 0.32 and 0.55 MPa). Differently, the UCS was higher than the maximum limit (acceptable range between 1.20 and 2.50 MPa), probably due to an excess of cement in the mixture.

7.2 Laboratory Tests on Cores

Ten cores were extracted from the trial section 45 days after construction. The visual analysis of the cores allowed noting the good quality of the construction operations, as no irregularities or voids were present on the lateral surface (Fig. 2).

The laboratory tests on the cores included determination of the air voids content, ITS tests and complex modulus tests.

The average dry bulk density of the CRAB cores was equal to 2.160 Mg/m^3, thus comparable to the value determined in the mix design. The corresponding air voids content of the cores was 10.6%. This result confirmed the high degree compaction of the CRAB layer, because the volume of water in the fresh mix was about 8.5%.

Fig. 2. Core taken from the trial section including CRAB and binder layers.

ITS tests were carried out on 50 mm height specimens obtained by cutting the central part of the CRAB cores. The average ITS from 4 specimens was 0.71 MPa. This value, higher than the ITS determined in the mix design phase or on the GC specimens, suggests that the in-place curing of the CRAB mixture was more advanced than the laboratory-simulated curing (72 h at 40 °C).

The complex modulus E^* was measured by means of cyclic compression tests, at 4 temperatures (10, 20, 30 and 40 °C) and 5 frequencies (10, 3, 1, 0.3 and 0.1 Hz), with a target strain amplitude of 25 microstrain. The complex modulus ($E^* = E_1 + jE_2 = E_0\, e^{j\phi}$) was calculated from the steady-state sinusoidal component of stress and strain.

Figure 3a shows the results from the two tested cores in the Black diagram (E_0 as a function of ϕ). It can be observed that E_0 varied between 2852 MPa and 15058 MPa while the phase angle ϕ varied from 4.4 to 13.0°. Some variability in the results from the two cores was noticed, suggesting that the construction was not perfectly homogeneous.

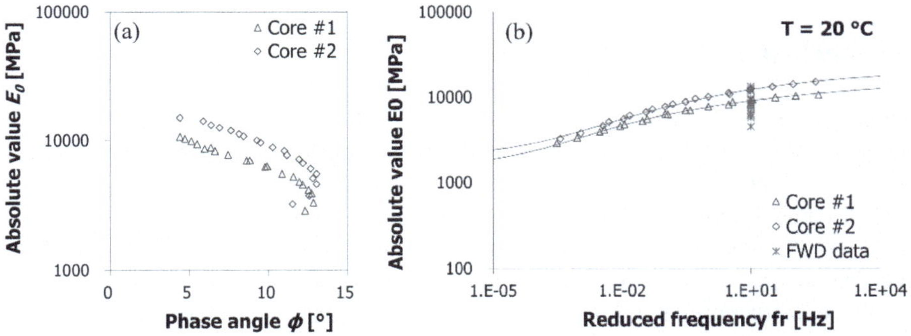

Fig. 3. The measured values of E^*: (a) Black diagram; (b) Master curves of the stiffness modulus E_0 at 20 °C and back-calculated FWD moduli.

The results show that CRAB mixture had a thermo- and frequency-dependent behavior, due to the presence of the fresh bitumen from emulsion and the aged bitumen in the RA. However, compared to HMA (Graziani et al. 2014; Godenzoni et al. 2018),

CRAB mixture showed a lower variability of E_0, lower values of ϕ and higher modulus values at high temperature/low frequencies.

As the experimental data were represented by a single smooth curve in the Black diagram, the time-temperature superposition principle was considered valid. So, temperature shift factors were applied to plot the master curves of E_0 (Fig. 3b) and the Huet-Sayegh model (Sayegh 1967) was used to simulate the rheological behaviour of CRAB.

7.3 In Situ FWD Tests

FWD tests were carried out the same day of the core extraction. At that time, the pavement did not include the porous asphalt surfacing layer, thus the FWD blows were applied on the top of the binder layer. The pavement temperature registered during FWD tests was 22.6 °C. So, the moduli of HMA and CRAB layers were corrected to the reference temperature of 20 °C using the law indicated by Asphalt Institute (1982).

The stiffness moduli determined through back-calculation (15th percentile of the data population) were 6278 MPa, 6385 MPa, 2510 MPa and 46 MPa for binder layer, CRAB layer, CSF layer and subgrade, respectively. This denoted the good quality of the CRAB layer, which showed a stiffness modulus comparable to that of the upper HMA layer.

The individual values of the back-calculated moduli of the CRAB layer, plotted in Fig. 3b, are generally lower than the stiffness modulus E_0 measured on the cores. This fact had been noted in another study (Godenzoni et al. 2018). The main reason is that CRAB cores probably experienced a further curing after extraction, because of the direct exposition to air. Moreover, it should be considered that FWD test investigates the behavior of a large volume of pavement that can include local micro-damages, determining a scale effect.

8 Conclusion

The present paper describes a case study where a CIR mixture, including 100% of RA aggregate, polymer-modified bitumen emulsion and cement, was used to build the base layer of the high-trafficked highway SS268, in southern Italy.

The rehabilitation of SS268 highway provided a FWD test campaign on the old pavement, the mechanistic-empirical design of the new pavement including a CRAB layer, the laboratory study of all the mixtures to define the optimum job mix formulas and the construction of a 1-km trial section to validate procedures, materials and equipment.

The construction protocol provided:

- the milling of the HMA layers and the RA storage on the jobsite;
- the cement-stabilization of the old foundation;
- the RA replacing and recycling with polymer-modified bitumen emulsion and cement (CRAB layer);
- the laying of the binder and porous asphalt layers.

The validation of the CRAB mixtures allowed observing that:

- the laboratory tests on the GC specimens provided ITS and UCS values higher than those determined in the mix design phase;
- the cores taken from the trial section had a low air voids content and high ITS;
- the CRAB mixtures had a thermo- and frequency-dependent behavior, due to the presence of the fresh bitumen from emulsion and the aged bitumen in the RA;
- the stiffness moduli determined through back-calculation from FWD tests denoted the good quality of the CRAB layer, which showed a stiffness modulus comparable to that of the upper HMA layer.

In conclusion, the analysis allowed assessing the good quality of the construction operation and the high performance of the CRAB mixture. In the future, the structural properties of the pavement, with particular attention to the CRAB layer, will be periodically measured to analyze the evolution of the mechanical behavior during the road service life.

At the light of the success in the described case study, the application protocol that was followed can represent a reference for future rehabilitation projects involving CIR and encourage its use also in the construction of CRAB layers.

References

AASHTO-AGC-ARTBA Joint Committee Task Force 38 Report (1998) Report on cold recycling of asphalt pavements. American Association of State Highway and Transportation Officials, Washington, D.C

Asphalt Institute (1982) Research and development of the asphalt institute's thickness design manual (MS-1), 9th edn. Research Report No. 82-2, College Park, Maryland

Bocci M, Grilli A, Cardone F, Graziani A (2011) A study on the mechanical behaviour of cement-bitumen treated materials. Constr Build Mater 25(2):773–778. https://doi.org/10.1016/j.conbuildmat.2010.07.007

Bocci M, Virgili A, Colgrande S (2002) A study of the mechanical characteristics of cold recycled bituminous concretes. In: 4th European symposium on performance of bituminous and hydraulic materials in pavement, Nottingham, 11–12 April

Finn F, Saraf C, Kulkarni R, Nair K, Smith W, Abdullah A (1977) The use of distress prediction subsystems for the design of pavement structures. In: 4th international conference on structural design of asphalt pavements, Ann Arbor, Michigan, 22–26 August

Francken L, Partl MN (1996) Complex modulus testing of asphaltic concrete: RILEM interlaboratory test program. Transp Res Rec 1545:133–142

Godenzoni C, Graziani A, Bocci E, Bocci M (2018) The evolution of the mechanical behavior of cold recycled mixtures stabilised with cement and asphalt: field and laboratory study. Road Mater Pavement Des 19(4):856–877. https://doi.org/10.1080/14680629.2017.1279073

Graziani A, Bocci E, Canestrari F (2014) Bulk and shear characterization of bituminous mixtures in the linear viscoelastic domain. Mech Time-Dependent Mater 18(3):527–554. https://doi.org/10.1007/s11043-014-9240-x

Grilli A, Graziani A, Bocci E, Bocci M (2016) Volumetric properties and influence of water content on the compactability of cold recycled mixtures. Mater Struct 49(10):4349–4362. https://doi.org/10.1617/s11527-016-0792-x

Liebenberg JJE, Visser AT (2004) Towards a mechanistic structural design procedure for emulsion-treated base layers. J S Afr Inst Civil Eng 46(3):2–8

Miner MA (1945) Cumulative damage in fatigue. J Appl Mech 3:159–164

Pérez I, Medina L, Del Val MA (2013) Mechanical properties and behavior of in situ materials which are stabilized with asphalt emulsion. Road Mater Pavement Des 14(2):221–238. https://doi.org/10.1080/14680629.2013.779301

Powell WD, Potter JF, Mayhew HC, Nunn ME (1984) The structural design of bituminous roads. Transport and Road Research Laboratory, Report 1132

Sayegh G (1967) Viscoelastic properties of bituminous mixtures. In: International conference on structural design asphalt pavements, Washington, D.C.

Theyse HL, de Beer M, Rust FC (1996) Overview of South African mechanistic pavement design method. Transp Res Rec 1539:6–17

Van der Poel C (1954) Representation of rheological properties of bitumens over a wide range of temperature and loading times. In: 1st International congress of rheology, vol. 2. Butterworth's Scientific Publications, London, pp. 331–337

Witczak M (1972) Design of full depth air field pavement. In: 3rd international conference on the structural design of asphalt pavements, London, 11–15 September

Test Methods and Performance

Effect of Nano SiO_2, TiO_2 and ZnO Modification to Rheological Properties of Neat and Polymer Modified Bitumen

Rita Kleizienė(✉), Miglė Paliukaitė, and Audrius Vaitkus

Road Research Institute of Faculty of Environmental Engineering of Vilnius Gediminas Technical University, Linkmenu 28, 08217 Vilnius, Lithuania
rita.kleiziene@vgtu.lt

Abstract. With development of nanotechnology, the application of nano materials to improve the bitumen became a focus of many researchers. The manipulation of material composition in nanoscale has led to creation of macro-materials with better mechanical performance and durability. However, the influence of nano materials to bitumen behaviour is still unclear. The objective of this research was to determine the influence of nano materials to rheological properties of bitumen. Based on literature analysis the most promising nano-materials and their composition were selected for experimental research. The rheological properties were determined for 29 specimens produced from two bitumen (non-aged 70/100 and PMB 45/80-55) and three nano-materials (nano SiO_2, nano TiO_2 and nano ZnO). The dynamic shear modulus and phase angle, fatigue performance and non-recoverable creep compliance parameters were determined with Dynamic Shear Rheometer. Statistical analysis was conducted to relate the bitumen modification variables with the rheological properties. The research results showed that nano materials improve bitumen resistance to permanent deformation, however, the excessive amount of nano materials in bitumen may result lower bitumen resistant to fatigue.

Keywords: Nano-materials · Fatigue resistance · Rutting resistance · Rheological properties · Modified bitumen

1 Introduction

Sustainable and durable transport infrastructure is a persistent goal of every highway and airport agency. From the first asphalt pavement, build in 19th century, the asphalt pavement technology evolved rapidly and today is the leading industry in infrastructure of transportation sector. Nowadays almost 95% of bitumen produced worldwide is applied to asphalt pavement industry (Lesueur 2009), which also corresponds to the total road length constructed with asphalt pavement. As traffic load and climate changes are increasing, the preservation of pavement surface condition is becoming more and more challenging and leading to many severe distresses of road surfaces. The development of bitumen modification with polymers opened the new page for improved durability asphalt pavements (Zhu et al. 2014). Asphalt mixtures with polymer modified bitumen characterized by resistance to fatigue cracking at low temperatures

M. Pasetto et al. (Eds.): ISAP APE 2019, LNCE 48, pp. 325–336, 2020.
https://doi.org/10.1007/978-3-030-29779-4_32

improved elastic recovery, substantial resistance to rutting (Dreessen et al. 2010; Santagata et al. 2013). However, bitumen modification with plastomer (polypropylene PE, polypropylene PP, ethylene–vinyl acetate EVA, ethylene–butyl acrylate EBA) or thermoplastic elastomer (styrene-butadiene-styrene SBS, styrene–isoprene–styrene SIS, styrene–ethylene/butylene– styrene SEBS) polymers has disadvantages (Zhu et al. 2014): signally raised construction costs (SIS, SEBS); complicated compatibility (SBS); lower resistance to heat and oxidation (SIS) and instability storage and transportation (SEBS).

Thus, the need for new, improved bitumen (asphalt binder) modification methods is still relevant, especially by improving adhesion between bitumen and aggregates and resistance to aging, assuring recyclability, and lowering modification costs. In recent years, the development of technologies allowed to increase the production of nanomaterials. Nano material is a pure simple chemical composition material composed of smaller than 100 nm size particles. The physical properties of nano-materials differ from conventional materials because of surface area-to-volume ratio of particles, shape, chemical composition (Teizer et al. 2011). The manipulation of material composition at nanoscale allows creating novel macro-materials with not conventional properties. In civil engineering, nano-materials mostly used for Portland cement concrete and cement mortar modification (Teizer et al. 2011). The application of nano-materials in order to improve the bitumen became a focus of many researchers. However, many factors influence the physical and mechanical properties of nano-modified bitumen, such as bitumen type, nano-material type, nano-material content, particle size, modification method (Li et al. 2017).

The objective of this research was to determine the influence of nano materials to rheological properties of bitumen. The research scope of the work included:

- Analysis of bitumen nano-material modification methods
- Based on literature review selection of the most promising nano materials for bitumen modification
- Experimental test on rheological properties of nano-modified bitumen
- Analysis of nano materials influence to bitumen rheological properties and performance

2 Background of Bitumen Modification with Nano Additives

Nanotechnology has gradually penetrated into the field of bitumen modification. Currently, nanomaterials such as, nano-zinc oxide (nano-ZnO), nano titanium dioxide (TiO_2) and nano silica dioxide (SiO_2) have improved bitumen properties and performance of asphalt pavement. Researches have conducted lots of work in order to prove a positive effect of nano material to bitumen properties and asphalt pavement performance.

Zhang et al. (2018) determined the effect of nano ZnO particle size on properties of asphalt and asphalt mixture. Results showed that the decrease of ZnO size in bitumen weakened its creep stiffness, but increased its softening point, ductility, viscosity, anti-rutting index, creep rate and anti-aging ability. In addition, with the decrease of ZnO size, the high temperature stability, water stability and low temperature crack resistance of asphalt mixture were improved using the determined optimum size of ZnO particle 80 nm. Azarhoosh et al. (2016) analysed hot mix (HMA) asphalt containing 1%, 3%, 5%, and 7% of nano ZnO mixed with bitumen 85/100. The results showed that fatigue life of mixtures containing nano ZnO were higher than those of the control mixtures due to improved cohesion energy and higher resistance to fatigue cracking in asphalt.

Yao et al. (2013) evaluated the rheological properties and chemical bonding of nano-SiO_2 modified bitumen. The findings from this study showed that the antiaging property, rutting and fatigue cracking performance of nanosilica modified bitumen was enhanced. Ganjei and Aflaki (2016) analysed the effect of combination of nanosilica, also SBS in the self-healing of HMA was investigated by applying Indirect Tensile Test (IDT) Superpave set-up at a temperature of 25 °C, using the Taguchi Design of Experiments (DOE). The obtained results of Dissipated Creep Strain Energy (DCSE) recovery facilitated the measurement of High Index (HI). Experimental approaches under laboratory conditions with mentioned materials and methods showed that the highest HI corresponds to the asphalt mixture of 7% bitumen, 2% nanosilica with 4% SBS (optimum condition).

Yang et al. (2018) analysed nano TiO_2 influence to asphalt binder properties under the long-term aging UV (Ultraviolet) and PAV (Pressure Aging Vessel). Since titanium dioxide (TiO_2) is known as a photocatalytic material, which can decompose the nitrogen oxides and sulphur oxides under the condition of ultraviolet radiation, therefore it can have a positive effect on bitumen properties. Buhari et al. (2018) analyzed the physical and rheological properties of bitumen modified with TiO_2 at different concentrations: 2%, 4%, 6% 8% and 10%. According to the results, higher concentrations of TiO_2 have a significant effect on the physical properties and rheological properties of bitumen. As a result, the susceptibility to temptation decreased and the bitumen stiffness and elasticity increased compared to unmodified bitumen. In addition, Gong et al. (2018) concluded that the high- and low-temperature performance and water stability of nano-TiO_2/$CaCO_3$ and basalt fibre composite modified asphalt mixture were better than those of an ordinary asphalt mixture before and after freeze-thaw cycles. Finally, Yang et al. (2018) analysed the rheological properties of SBS and Sasobit modified bitumen blended with various contents (0%, 1%, 3% and 5%) of nano TiO_2 after Ultraviolet (UV) aging and PAV aging procedures. Test results showed that nano materials did not affect complex modulus and phase angle of bitumen. UV aged binders showed better fatigue and cracking resistance than the PAV aged binders.

Based on a comprehensive literature review, three nano materials and their impact on bitumen properties were analysed:

- Nano Zinc Oxide (nZnO) - increases bitumen resistance to aging, permanent deformation and fatigue. The optimum amount of nano zinc oxide to increase resistance to aging is 3%, long-term deformation and fatigue is 7%.
- Nano-Silica (nSiO_2) - increases bitumen resistance to aging, permanent deformation, fatigue and thermal cracking. The optimum amount of nano SiO_2 is 3–4%.
- Nano Titanium Dioxide (nTiO_2) - improves bitumen resistance to UV aging. Reduces temperature sensitivity, increases the stiffness of the asphalt mixture during freeze-thaw cycles. Based on research conducted by different scientists, the

influence of Nano-TiO_2 on bitumen properties is different and not fully defined. The optimum amount is about 5–10%.

3 Materials and Methods

3.1 Materials

The original (unaged) base bitumen used was neat 70/100 and polymer modified PMB 45/80-55 (SBS elastomer polymer). The conventional properties of the bitumen evaluated with penetration, softening point, resistance to hardening and flash point tests. Characteristics of base bitumen are presented in Table 1. The original bitumen is modified adding 1% and 3% nano titanium dioxide (nTiO_2), nano silica dioxide (nSiO_2) and nano zinc oxide (nZnO). The physical properties of nano materials are presented in Table 2.

Table 1. Characteristics of base bitumen (LST EN 12591 2009)

Characteristics	Bitumen type	
	70/100	PMB 45/80-55
Penetration at 25 °C	70–100	45–80
Softening point	43-51 °C	≤ 55 °C
Resistance to hardening at 163 °C:		
Residual penetration at 25 °C	≥ 46%	≥ 50%
Increase in softening temperature	≤ 9 °C	≤ 2 °C
Change in mass	≤ 0.8%	≤ 0.5%
Flash point	≥ 230 °C	≥ 235 °C

Table 2. Physical properties of nano additives

Properties	Nano additive type		
	nZnO	nSiO_2	Rutile nTiO_2
Appearance	White powder	White powder	White powder
Molecular weight	81.37	60.80	79.87
Melting point	1975 °C	>1700 °C	1830 °C
Purity	99.5%	99.5%	99.5%
Characteristic particle size	30 nm	15 nm	25 nm
Solubility	Insoluble	Insoluble	Insoluble

3.2 Experimental Program Ant Methods

3.2.1 Bitumen Nano-Modification and Sample Preparation Method

Modified bitumen samples were produced by mixing base bitumen 70/100 with three nano additives: titanium dioxide (nTiO_2), silica (nSiO_2) and zinc oxide (nZnO).

Modification of bitumen was performed on 50 ± 2 g of bitumen adding 3% and 1% of nano additive. The coding system of bitumen samples is shown in Table 3.

First of all, bitumen was heated up to 150 °C for 30 min in order to pour 50 ± 2 g of bitumen into the containers, then the required amount of nano additive was recalculated and mixed with the heated bitumen. Bitumen was mixed with a glass rod to keep the nanoparticles inside the bitumen (not to spread into the air) and reheated up to 150 °C. Mixing was performed at 5000 rpm for 30 min, maintaining a bitumen temperature of 150 °C. After modification, bitumen was poured into 8 and 25 mm moulds for further testing with metal cans with clamping caps.

Table 3. Coding system of bitumen samples

Bitumen type	Nano additive type	Content of nano additive, %	Code
70/100	–	0	1-OR
	$nSiO_2$	1.0	1-OR-nS.1
		3.0	1-OR-nS.3
	$nTiO_2$	1.0	1-OR-nT.1
		3.0	1-OR-nT.3
	nZnO	1.0	1-OR-nZ.1
		3.0	1-OR-nZ.3
PMB 45/80-55	–	0.0	4-OR
	$nSiO_2$	1.0	4-OR-nS.1
		3.0	4-OR-nS.3
	$nTiO_2$	1.0	4-OR-nT.1
		3.0	4-OR-nT.3
	nZnO	1.0	4-OR-nZ.1
		3.0	4-OR-nZ.3

3.2.2 Linear Amplitude Sweep Test (LAS) Method

Accumulated strain in asphalt layers influenced by traffic repetitive loading at low and moderate temperatures results fatigue cracking. In order to evaluate the influence of nano-material and the resistance of nano-modified bitumen to fatigue, the LAS test was carried out. Linear Amplitude Sweep (LAS) test is based on viscoelastic continuum damage theory to predict bitumen fatigue life as a function of strain (Hintz et al. 2011a, b). This test method includes the bitumen resistance to damage using a cyclic load applied to a linearly increasing strain amplitude.

LAS test was conducted with DSR MCR 302 regarding standard AASHTO TP 101-12 (AASHTO TP 2012). Test was performed in two steps: 1st – constant strain of 0.1% frequency sweep (FS) test, 2nd – constant frequency of 10 Hz LAS test. All types of bitumen were tested at 20 °C temperature with 8 mm parallel plate and 2 mm gap, before each test a 10 min temperature equilibrium was used. At the beginning, the FS test was conducted applying load of 0.1% strain over a range of frequencies from 0.25 to 40 Hz. LAS test was conducted at a constant 10 Hz frequency consisting of cycling loading at linearly increasing strain from 0.1% to 30%. Test data was analysed using Excel spreadsheet provided by Modified Asphalt Research Center of Civil Engineering

Department of University of Wisconsin-Madison (Hintz et al. 2011a, b). The FS test was conduct to determine the shear modulus (G*) of nano-modified bitumen and to obtain α parameter (of storage modulus over frequency) of undamaged bitumen behaviour. This test is used to determine parameters 'a' and 'b' for bitumen fatigue function:

$N_f = A_{35}(\gamma_{\max})^{-B}$, where N_f – load cycles to failure, A_{35} – regression parameter determined from LAS tests results at damage corresponding to 35% decrease from initial $|G^*| \cdot sin\delta$, B – regression parameter determined as -2α from FS test results, and γ_{max} – maximum expected binder strain in %. Moreover, the amplitude sweep to determine damage accumulation and the 35% damage level regarding to determine maximum shear stress value. To compare nano-modified bitumen resistance to fatigue performance the parameter N_f was estimated at 2.5% and 5% strain.

3.2.3 Multi Stress Creep Recovery Test (MSCR) Method

Accumulated strain in asphalt layers influenced by traffic repetitive loading at high temperatures results rutting. The asphalt mixture resistance to rutting (or permanent deformation) is determined performing cyclic load test under high temperatures for asphalt concrete and for bitumen (D'Angelo 2009; Wasage et al. 2011). In order to evaluate the influence of nano-material and the resistance of nano-modified bitumen to rutting, the MSCR test was used. MSCR test was conducted with dynamic shear rheometer MCR 302 regarding standard EN 16659 (2015). The test was performed in four steps: 1st – constant strain of 1% frequency sweep test, 2nd – constant stress of 0.1 kPa MSCR test, 3rd – constant stress of 1.6 kPa MSCR test, 4th – constant stress of 3.2 kPa MSCR test. All types of bitumen were tested at 60 °C temperature with 25 mm parallel plate and 1 mm gap, before each test a 10 min temperature equilibrium was used. At the beginning, the FS test was conducted applying load of 1% strain over a range of frequencies from 0.25 to 40 Hz. FS test was conduct to determine the shear modulus (G*) of nano-modified bitumen. MSCR test was conducted by applying load for 1 s duration and following measure of recovery for 9 s, this repeated for 10 cycles. MSCR test is a combination of shear creep and recovery experiments, which allows determining the bitumen resistance to accumulate strain by two parameters: the non-recoverable compliance J_{nr} and the $R_{\%}$ recovery. Requirements for traffic grade according to non-recoverable compliance J_{nr} and traffic level and load rate (AASHTO MP 19-10) are presented in Table 4.

Table 4. Requirements for traffic grade according non-recoverable compliance J_{nr} and traffic level and load rate (AASHTO MP 19-10)

Characterization of traffic load rate				Non-recoverable compliance		Traffic grade
Design 8 t ESALs (mill.)	Design 10 t ESALs (mill.)	Relation with traffic speed	Speed (km/h)	J_{nr} at 3.2 kPa (1/kPa)	J_{nrdiff} at 0.1–3.2 kPa (%)	
<10	<4.1	and	>70	<4.0	75	S
10–30	4.1–12.3	or	20–70	<2.0	75	H
>30	>12.3	or	<20[a]	<1.0	75	V
>30	>12.3	and	<20[a]	<0.5	75	E

[a] Standing traffic with speed less than 20 km/h.

[b] S, H, V, and E represent the Standard, High, Very High, and Extreme high of traffic level and loading rate grade.

4 Result Analysis

For each test, at least two replicates were used and averaged values were reported. Based on the test results, nano-modified bitumen is ranked from 1 to 6, with referring 1 to the highest and 6 to the lowest. Due to different rheological behaviour, nano-materials affect neat bitumen 70/100 and polymer modified bitumen 45/80-55 was analysed separately. Summary of experimental results of nano-modified 70/100 and PMB 45/80-55 bitumen binders are documented in Tables 5 and 6 respectively.

Table 5. Summary of experimental results of nano-modified 70/100 bitumen

Test	Parameter	Units	Measured values						
			Original bitumen	$nSiO_2$		$nTiO_2$		nZnO	
				1%	3%	1%	3%	1%	3%
FS	G* at 20 °C, 0.25 Hz	kPa	224.0	286.6	342.8	235.8	265.8	235.8	261.4
	G* at 20 °C, 10 Hz	kPa	3588.3	4234.5	4688.6	3751.8	4169.8	3619.1	4081.8
	G* at 60 °C, 0.25 Hz	kPa	0.22	0.27	0.32	0.23	0.24	0.21	0.23
	G* at 60 °C, 10 Hz	kPa	7.90	9.57	12.21	8.41	8.54	7.74	8.32
			–	2.0	1.0	5.0	3.0	6.0	4.0
LAS	Parameter A_{35}	–	70469	77977	88884	66621	62287	76760	67011
	Parameter B	–	−2.335	−2.431	−2.540	−2.344	−2.354	−2.375	−2.358
	Nf at 2.5%	Cycles	8295	8406	8667	7776	7205	8556	7721
	Nf at 5.0%	Cycles	1644	1559	1490	1531	1409	1645	1506
			–	2.0	3.0	4.0	6.0	1.0	5.0
MSCR	%R at 0.1 kPa	%	−1.0	−0.6	1.1	−1.1	−1.0	−1.3	0.2
	%R at 1.6 kPa	%	−2.2	−1.7	−1.2	−1.8	−1.6	−2.1	8.3
	%R at 3.2 kPa	%	−2.2	−1.9	−1.6	−2.4	−2.6	−2.6	−2.4
	J_{nr} at 3.2 kPa	1/kPa	8.0	6.5	5.0	7.4	7.4	8.4	7.6
	J_{nrdiff} ant 0.1–1.6 kPa	%	7.0	6.1	9.8	5.8	6.4	6.6	6.6
	J_{nrdiff} ant 0.1–3.2 kPa	%	11.3	10.8	15.9	10.1	11.2	11.2	11.3
			–	2	1	4	3	6	5

Table 6. Summary of experimental results of nano-modified PMB 45/80-55 bitumen

Test	Parameter	Units	Measured values						
			Original bitumen	$nSiO_2$		$nTiO_2$		nZnO	
				1%	3%	1%	3%	1%	3%
FS	G* at 20 °C, 0.25 Hz	kPa	379.3	335.3	419.3	332.1	239.2	293.4	313.7
	G* at 20 °C, 10 Hz	kPa	4855.2	4237.1	4882.3	4172.6	3177.5	3761.7	4012.9
	G* at 60 °C, 0.25 Hz	kPa	1.05	1.12	1.53	0.99	0.99	1.03	1.02
	G* at 60 °C, 10 Hz	kPa	16.35	17.18	20.33	15.49	15.33	16.21	15.93
			–	2	1	3	6	5	4
LAS	Parameter A_{35}	–	170750	196296	194942	207403	186263	164196	185940
	Parameter B	–	−2.657	−2.696	−2.818	−2.715	−2.653	−2.688	−2.678
	Nf at 2.5%	Cycles	14966	16597	14737	17052	16389	14052	15992
	Nf at 5.0%	Cycles	2373	2561	2090	2592	2607	2189	2500
			–	3	5	1	2	6	4

(continued)

Table 6. (*continued*)

Test	Parameter	Units	Measured values						
			Original bitumen	$nSiO_2$		$nTiO_2$		nZnO	
				1%	3%	1%	3%	1%	3%
MSCR	%R at 0.1 kPa	%	72.0	78.3	91.1	74.1	75.8	73.2	73.8
	%R at 1.6 kPa	%	73.2	79.0	89.8	75.6	77.3	74.3	75.1
	%R at 3.2 kPa	%	67.6	72.1	87.1	69.7	70.3	69.5	69.3
	J_{nr} at 3.2 kPa	1/kPa	0.5	0.349	0.108	0.451	0.435	0.439	0.451
	J_{nrdiff} ant 0.1–1.6 kPa	%	−2.2	−4.0	12.6	−4.5	−5.3	−2.7	−3.5
	J_{nrdiff} ant 0.1–3.2 kPa	%	15.6	21.5	28.5	15.5	20.1	13.3	16.8
			–	2	1	6	3	4	5

The dynamic shear modulus G^* at 20 °C and 60 °C is used to determine the effect of nano-material to bitumen visco-elastic behaviour. Comparison of dynamic shear modulus is presented in Fig. 1. The results showed that the dynamic shear modulus of bitumen 70/100 was enhanced by adding 3% and 1% of $nSiO_2$, also 3% of $nTiO_2$ influenced higher G^* at 20 °C. In case of bitumen 70/100, the nano materials have positive influence to dynamic shear modulus. However, nano material effect to G^* of polymer modified bitumen PMB 45/80-55 was positive only with $nSiO_2$, $nTiO_2$ while nZnO has minimal or negative effect on rheological properties.

According to the results of LAS test, the effect of nano materials on bitumen resistance to fatigue by load cycles to failure (Nf) are evaluated. Comparison of number of cycles to failure is presented in Fig. 2. At lower strain level (2.5%) the bitumen 70/100 Nf increased by 4.5% and 3.2% adding 3% of $nSiO_2$ and 1% of nZnO respectively. But bitumen 70/100 Nf at higher strain level (5%) decreased significantly and all nano materials showed negative effect to the resistance to fatigue. The nano material influence to neat and polymer-modified bitumen differs. The best effect on Nf for PMB 45/80-55 bitumen was determined adding 1% and 3% of $nTiO_2$, which increased Nf by 23% and 19% comparing to base bitumen. The negative effect on Nf was determined after adding 1% of nZnO and 3% of SiO_2, which decreased cycles by 14% and 13% respectively.

The non-recoverable creep compliance (J_{nr}) of MSCR tests determined at 60 °C and different stress levels: 0.1 kPa, 1.6 kPa and 3.2 kPa 70/100 and PMB 45/80-55 bitumen were returned back to their original shape and vice versa. Finally, the higher non-recoverable creep compliance (J_{nr}), asphalt mixtures are less resistant to rutting. Figure 3 presents a plot of the influence of stress levels on J_{nr} of 70/100 bitumen modified with different nano materials. Bitumen 70/100 modified with 3% of $nSiO_2$ showed the lowest values of J_{nr} comparing with other bitumen samples, whereas bitumen modified with 1% of nano ZnO showed the highest J_{nr} values. Figure 4 presents a plot of the influence of stress levels on J_{nr} of PMB 45/80-55 bitumen additionally modified with different nano materials. Bitumen PMB 45/80-55 modified with 1% and 3% of $nSiO_2$ showed significantly lower values of J_{nr} comparing with other bitumen samples and the original PMB 45/80-55 (without nano additives). Distribution of results is different from the 70/100 bitumen since J_{nr} depends not only on the stress level but also on the polymer used in bitumen.

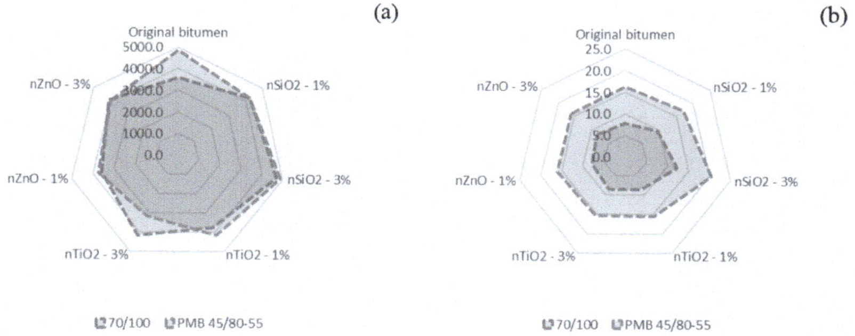

Fig. 1. Comparison of dynamic shear modulus measured at 20 °C, 10 Hz (a) and at 60 °C, 10 Hz (b)

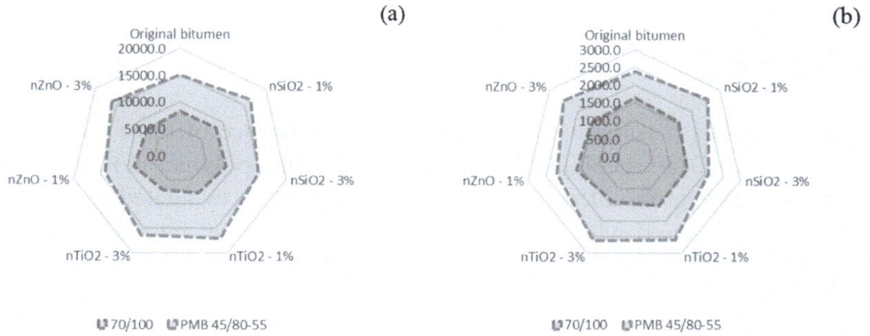

Fig. 2. Comparison of number of cycles to failure measured at 20 °C and 10 Hz (a) for 2.5% strain level and (b) for 5.0% strain level

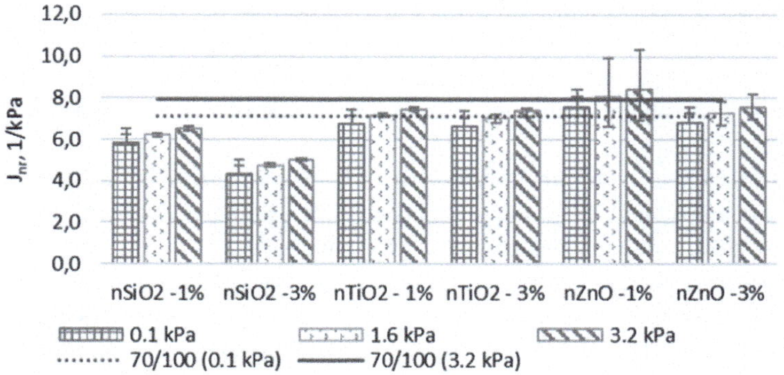

Fig. 3. Non-recoverable creep compliance J_{nr} of 70/100 bitumen with nano-materials tested at 60 °C

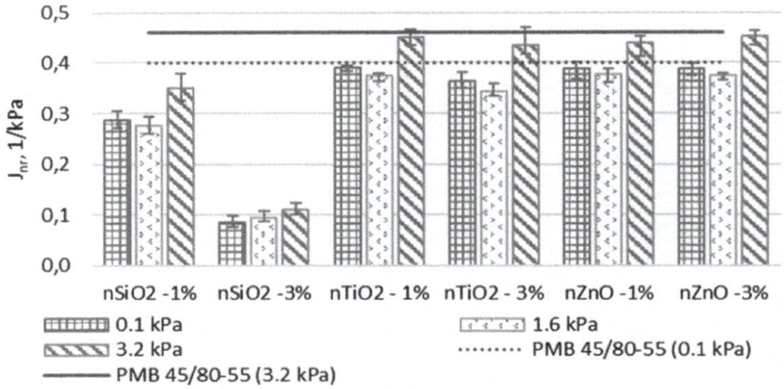

Fig. 4. Non-recoverable creep compliance J_{nr} of PMB 45/80-55 bitumen with nano-materials tested at 60 °C

5 Conclusions

The objective of this research was to determine the influence of three nano materials (SiO_2, TiO_2 and ZnO) to rheological properties of neat and polymer modified bitumen. The rheological properties of nano-modified bitumen were evaluated using two test procedures and compared with the properties of original (base) bitumen (neat 70/100 and SBS polymer modified PMB 45/80-55). Based on the results, the following findings are drawn:

1. This study confirms SiO_2 positive effect on bitumen resistance to rutting. The 3% of nano SiO_2 reduced the non-recoverable creep compliance (J_{nr}) by 36.9% for 70/100 and 76.6% for PMB 45/80-55 comparing with the bitumen without nano SiO_2. The positive effect to J_{nr} also determined adding 1% of nano SiO_2, which decreased J_{nr} by 18.4% for 70/100 and 24.3% for PMB 45/80-55.
2. The positive effect on increasing the resistance to fatigue determined adding 1–3% of nano TiO_2 to polymer modified bitumen PMB 45/80-55. However, the influence of nano TiO_2 on neat bitumen rheological properties is different and not fully defined.
3. The lowest effect to bitumen binder rheological properties is determined to nano ZnO. It can be stated, that bitumen modification with nano ZnO with 3% and lower does not affect bitumen mechanical performance.
4. Nano-particle as modification additive has the potential to enhance the complex shear modulus, reduce the phase angle and finally improve rheological performance of base binder. However, the excessive amount of nano materials may result the lower asphalt binder resistant to fatigue.
5. Using nano SiO_2 or nano TiO_2 as a second additive in polymer modified bitumen improves the properties of bitumen. The nano additives influence the stability and resistance to aging of the polymer itself has to be evaluated in the future research.

6. The determined rheological properties of nano-modified bitumen have high results variation between the samples. The biggest results variation was determined for nano ZnO, the smallest – for nano SiO_2. The evaluation of nano particles dispersion and uniformity was still unsolved yet, so the quantitative experimental tests should be carried out in order to eliminate the risk of the nano-particle dispersion influence.

Acknowledgements. This project has received funding from European Social Fund (project No 09.3.3-LMT-K-712-02-0111) under grant agreement with the Research Council of Lithuania (LMTLT).

References

AASHTO TP 101-12 (2012) Estimating fatigue resistance of asphalt binders using the linear amplitude sweep

Azarhoosh AR, Nejad FM, Khodaii A (2016) Nanomaterial and fatigue cracking of hot mix asphalt. Road Mater Pavement Des 629:1–14

Buhari R, Abdullah ME, Ahmad MK, Ching AL, Haini R, Bakar SKA (2018) Physical and rheological properties of Titanium Dioxide modified asphalt. In: International conference on civil & environmental engineering (CENVIRON 2017), Penang, Malaysia, 28–29 November

D'Angelo JA (2009) The relationship of the MSCR test to rutting. Road Mater Pavement Des 10:61–80

Dreessen S, Planche JP, Ponsardin M, Pittet M, Dumont AG (2010) Durability study: field aging of conventional and polymer-modified binders. In: Transportation research board 89th annual meeting, Washington, DC, 10–14 January

EN 16659 (2015) Bitumen and bituminous binders - multiple stress creep and recovery test, Brussels

Ganjei MA, Aflaki E (2016) Application of nano-silica and styrene-butadiene-styrene to improve asphalt mixture self-healing. Int J Pavement Eng 8436:1–11

Gong Y, Bi T, Tian Z, Tan G (2018) Pavement performance investigation of modified asphalt mixture under freeze-thaw cycles. Appl Sci 8(12):2581

Hintz C, Velasquez R, Li Z, Bahia H (2011a) Effect of oxidative aging on binder fatigue performance. J Assoc Asphalt Paving Technol 80:527–547

Hintz C, Velasquez R, Johnson C, Bahia H (2011b) Modification and validation of linear amplitude sweep test for binder fatigue specification. Transp Res Rec: J Transp Res Board 2207:99–106

Yang S, Yan K, He B, He W, Wang D, Wang H (2018) Ultraviolet and PAV aging procedures influence on rheological characteristics of Sasobit/SBS modified binder containing titanium dioxide nanoparticles. Petrol Sci Technol 36(19):1524–1530

Yao H, You Z, Li Z, Lee CH, Wingard D, Yap YK, She X, Goh SW (2013) Rheological properties and chemical bonding of asphalt modified with nanosilica. J Mater Civil Eng 25 (11):1619–1630

Lesueur D (2009) The colloidal structure of bitumen: consequences on the rheology and on the mechanisms of bitumen modification. Adv Colloid Interface Sci 145(1–2):42–82

Li R, Xiao F, Amirkhanian S, You Z, Jia H (2017) Developments of nano materials and technologies on asphalt materials – a review. Constr Build Mater 143:633–648

Liu HY, Zhang HL, Hao PW, Zhu CZ (2015) The effect of surface modifiers on ultraviolet aging properties of nano-zinc oxide modified bitumen. Pet Sci Technol 33(1):72–78

Santagata E, Baglieri O, Dalmazzo D, Tsantilis L (2013) Evaluation of the anti-rutting potential of polymer-modified binders by means of creep-recovery shear tests. Mater Struct/Materiaux et Constructions 46(10):1673–1682

Teizer J, Venugopal M, Teizer W, Felki J (2011) Nanotechnology and its impact on construction: bridging the gap between researchers and industry professionals. J Constr Eng Manag 138 (5):594–604

Wasage TLJ, Stastna J, Zanzotto L (2011) Rheological analysis of multi-stress creep recovery (MSCR) test. Int J Pavement Eng 12(6):561–568

Zhang H, Gao Y, Guo G, Zhao B, Yu J (2018) Effects of ZnO particle size on properties of asphalt and asphalt mixture. Constr Build Mater 159:578–586

Zhang HB, Zhang HL, Ke NX, Huang JH, Zhu CZ (2015) The effect of different nanomaterials on the long-term aging properties of bitumen. Pet Sci Technol 33(4):388–396

Zhu J, Birgisson B, Kringos N (2014) Polymer modification of bitumen: advances and challenges. Eur Polymer J 54(1):18–38

Impregnation of Lightweight Aggregate Particles with Phase Change Material for Its Use in Asphalt Mixtures

Muhammad Rafiq Kakar[1(✉)], Zakariaa Refaa[1,2], Jörg Worlitschek[2], Anastasia Stamatiou[2], Manfred N. Partl[1], and Moises Bueno[1]

[1] Empa, Swiss Federal Laboratories for Material Science and Technology, 8600 Dübendorf, Switzerland
muhammad.kakar@empa.ch
[2] Lucerne University of Applied Sciences and Arts, 6048 Horw, Switzerland

Abstract. Phase change materials (PCM) are widely investigated nowadays for building applications due to their ability of passive heating and cooling, thus regulating indoor temperature. Recently, PCMs have been studied also as additives of asphalt binders for road applications. However, PCM modifications of asphalt binder limits the effect due to the low amount of binder (5–6%wt.) used in asphalt mixtures. In order to increase the PCM content in a mixture an effective option may consist of modifying aggregate particles as main mixture component (94–95%wt.). In this research, foam glass and burnt expanded clay (agriculture beads) particles are impregnated with PCM (Tetradecane, T_{melt} = 6 °C) and investigated for relatively low temperature applications. PCM impregnation together with the protective coating and sealing of these PCM soaked lightweight aggregate (LWA) particles are performed in the lab. The particles are coated using epoxy adhesives combined with Ordinary Portland Cement (OPC). The results reveal that porous lightweight particles are indeed promising PCM carriers and comparatively easy to impregnate with liquid PCM. However, the choice of coating technique depends on the final application as hot, warm or cold mix asphalt.

Keywords: Tetradecane · Low temperature distresses · Specific heat capacity · Asphalt pavement

1 Introduction

The endothermic and exothermic cycle of phase change materials (PCM) may help to stabilize the interior room temperature of buildings. It decreases heating and cooling, not by affecting the thermal resistance of the building envelope but by rinfluencing the surface temperatures (Cabeza et al. 2007; Tyagi et al. 2011). In this way, PCM implemented in wallboards, shutters, under-floor heating systems and ceiling boards can be used as part of the building for heating and cooling applications (Kissock et al. 1998). The use of PCM in building materials and building elements has been widely investigated within the past decades. Their capability of storing heat energy in a latent form through a change in phase, leads to a heat storage capacity per unit volume greater

M. Pasetto et al. (Eds.): ISAP APE 2019, LNCE 48, pp. 337–345, 2020.
https://doi.org/10.1007/978-3-030-29779-4_33

than of conventional building materials (Sharma et al. 2009; Baetens et al. 2010; Refaa et al. 2018; Kakar et al. 2019a).

The addition of modifiers to asphalt binder is a common practice for improving pavement properties. Bituminous materials are viscoelastic and therefore, their performance is directly influenced by their temperature. In general, asphalt mixtures become softer presenting more viscous at high temperatures which may lead to ruts and permanent deformation (e.g. Wang et al. 2018). However, when the temperature changes from high to low, the bituminous binder slowly transforms from a ductile viscoelastic into a brittle elastic material eventually experiencing low temperature cracks during cooling (e.g. Hamzah et al. 2014; Liu et al. 2017). These temperature dependent properties of asphalt binder become more critical upon aging under environmental conditions along with applied traffic (Kakar et al. 2015). Numerous studies have attempted to improve the thermal properties of bituminous binder and asphalt mixtures (Ryms et al. 2015; Kheradmand et al. 2015; Manning et al. 2015; Ryms et al. 2017).

However, although the benefits of PCM for modifying asphalt roads appear very promising, its practical feasibility has only been studied recently (Kakar et al. 2018; Wei et al. 2019). Using PCM as an asphalt binder modifier may be one way for avoiding or delaying extreme low or high temperatures in asphalt mixtures. However, the lower amount of asphalt binder (5–6%) in the whole mixture limits the use of significant quantities of PCM in asphalt mixtures. Therefore, considering the large amount of aggregates (94–95%) in asphalt mixtures, it appears attractive to study ways of using aggregates as carrier of PCM, e.g. by impregnating lightweight aggregates (LWA), such as expanded clay. LWA has been widely used in cement concrete (Zhang and Gjvorv 1991), however has found limited applications in asphalt concrete, where its possible usage is primarily for surface treatments on low volume roads and for reducing dead weight of pavements for special applications, like runways on weak soil. Wycoff (1959) studied the application of LWA in asphalt concrete and reported acceptable performance in highway pavements (Mallick et al. 2004). Qian et al. (2011) investigated the use of LWA in epoxy asphalt concrete for bridge paving with focus on the performance of the material.

This research aims at exploring the possibility of impregnating different porous materials with PCM as core materials and its subsequent coating with the intention of protecting it against possible leakage (Kakar et al. 2019b). This was done particularly regarding their potential of minimizing low temperature-related problems. Computed tomography (CT) was used to acquire the images manifesting the internal voids structure in porous material used. This analysis was performed on both impregnated and non-impregnated different porous mediums, foam glass and expanded burnt clay, in order to study the possibility of incorporating tetradecane as PCM.

2 Materials and Methods

The foam glass porous aggregate particles shown in Fig. 1(a) were supplied by Misapor® (Switzerland). Particles of expanded burnt clay were commercially available products used for gardening purpose and are shown in Fig. 1(b). The liquid tetradecane

(PCM) with 99% purity was supplied by SIGMA-ALDRICH® (Switzerland). Since tetradecane as PCM in direct contact with bitumen results in diluting and eventually softening the bitumen rather than acting as short term thermal energy storage (Kakar et al. 2019a; Kariznovi et al. 2013). Therefore, in order to prevent the softening effect of the bitumen due to the PCM leakage, proper coating for sealing the PCM impregnated aggregates must be achieved. In this study, the process followed for the impregnation of the different LWA with tetradecane and its coating is described next. The process for coating was similar as described by Garcia et al. (2016). A controlled amount of porous particles was submerged in liquid tetradecane as shown in Fig. 2 for 15 min at room temperature. Then the submerged particles were placed in an oven at 40 °C under continuous vacuum for another 30 min or until no more air bubbles (tetradecane replacing air) emerging from the particles were visible. Later on, the single particles were taken out from the tetradecane liquid, allowing excess tetradecane draining from the surface of the particles during 1 min. Immediately after, the specimens were placed in a freezer (below −10 °C) in order to crystalize the tetradecane inside the porous particles. After 20 min, the particles were taken out and coated with epoxy glue supplied by Aldried® (Switzerland) as shown in Fig. 3(a). The gluing material was prepared in a separate container and the frozen particles were introduced while gently revolving the container horizontally by hand. Afterward, Ordinary Portland Cement (OPC) shown in Fig. 3(b) was used to cover the glue on the surface of particles and after assuring the complete coating of particles visually, these coated particles were left for 10 h in the air to dry at room temperature. Next, a final second layer of epoxy coating done was applied and the particles were left at room temperature for another 10 h. The final coated foam glass and expanded burnt clay beads filled with PCM are shown in Fig. 4(a) and (b). As control particles, the same coating procedure was carried out for particles without impregnation of tetradecane.

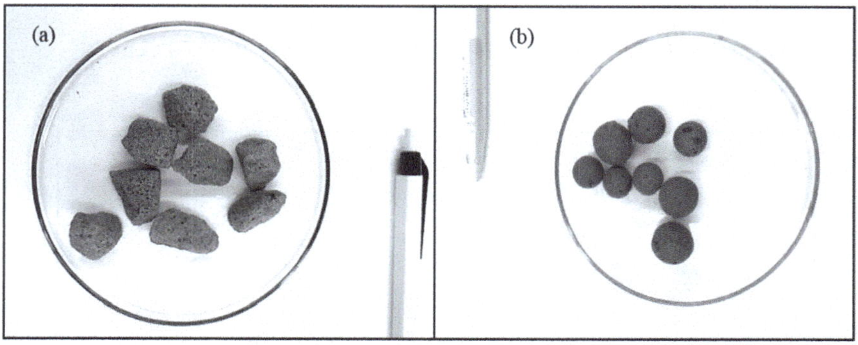

Fig. 1. (a) Foam glass aggregate (b) Burnt clay beads

In order to investigate the thermal transitions of the PCM (tetradecane), a Differential Scanning Calorimeter (DSC) 8000 from PerkinElmer® (USA) was used. A sample of 10 ± 1 mg was prepared for the DSC analysis. Firstly, the sample was cooled down to −20 °C and maintained at this temperature for 5 min; then the sample

Fig. 2. Foam glass and burnt clay submerged in PCM

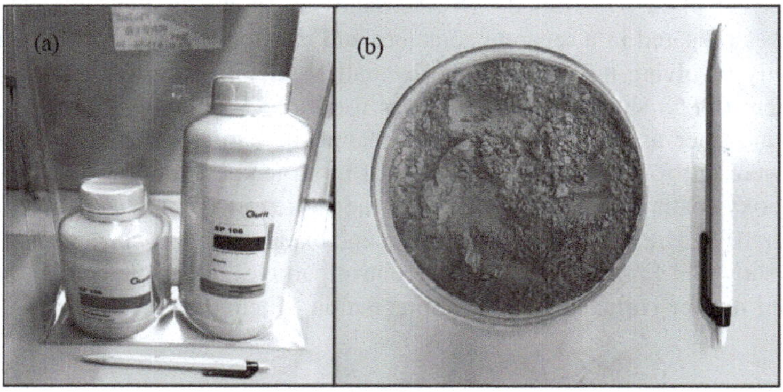

Fig. 3. (a) Liquid Epoxy Resin (SP 106) (b) Ordinary Portland Cement

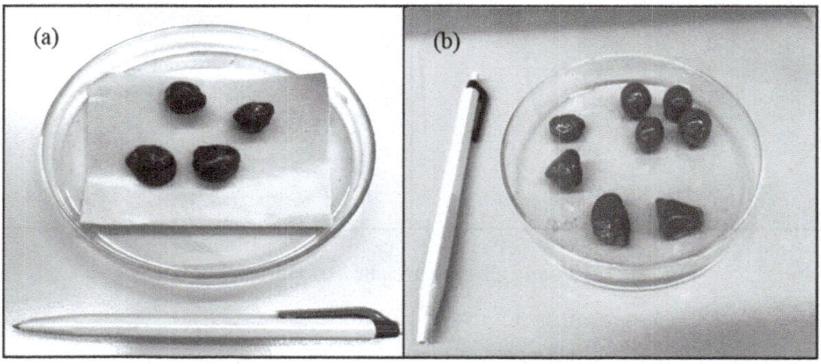

Fig. 4. (a) Coated specimens with impregnated PCM (b) Coated specimens without impregnated PCM

was heated up to 20 °C and kept at this temperature for 5 min before cooling again to −20 °C, where it remained for 5 min until being finally heated up to 20 °C. The cooling and heating ramps were conducted with a rate of 10 °C/min. Each formulation was analyzed at least three times for reproducibility purposes and for standard deviation calculation.

X-ray Computer Tomography CT-scan imaging technique was applied to determine the distribution of impregnated tetradecane inside the coated porous particles. The X-Ray MicroCT (Easy Tom XL Ultra 230-160) by RX Solutions® (France) was used to scan the different particles. In this investigation, an 85 kV CT was used; the scanning nominal current (270 μA); the number of projections was 1440; the source-to-detector-distance (sdd, 516.4 mm) and the source-to-object-distance (sod, 48.8 mm) resulting in a magnification of 10.58.

3 Results and Discussion

Melting and crystallization behavior of tetradecane (PCM) was observed during the heating and cooling process. The heat flow versus temperature for tetradecane (PCM) is presented in Fig. 5. During the heating cycle the heat flow curve shows a melting peak at 8.71 °C and a cooling peak at −1.03 °C, which indicates the ability of PCM to store and release thermal energy during the heating and cooling cycles respectively.

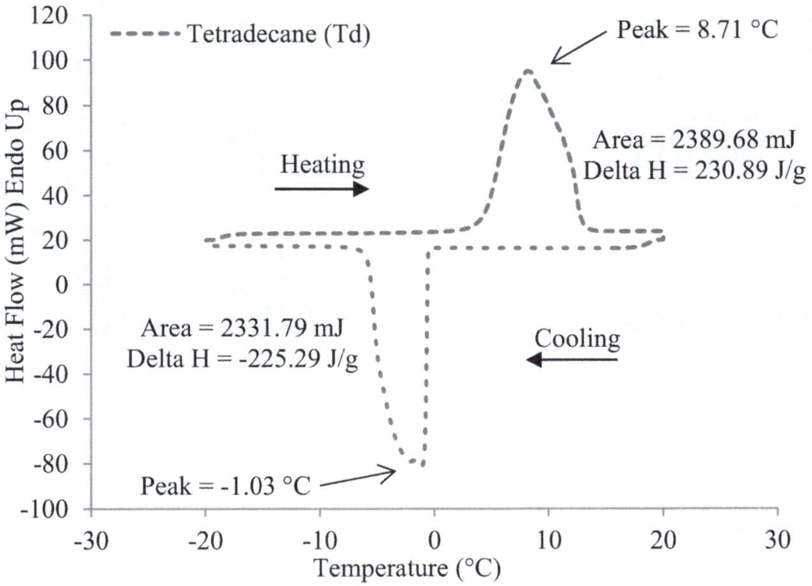

Fig. 5. Heat flow curves versus temperature for tetradecane (PCM), scanning rate of 10 °C/min

The thermal energy generated by the crystallization process can be estimated by integrating the crystallization peak at −1.03 °C. The crystallization enthalpy is 225.3 J/g, which reflects a high thermal energy compared to other comparable PCMs. The difference between the melting temperature and the crystallization temperature is known as the supercooling degree, which depends on the PCM nature and shape (raw or microencapsulated). Moreover, it is also dependent on cooling conditions. For instance a slow scanning rate (cooling/heating rate) would lead to a small supercooling degree.

CT-scan images from foam glass particles with and without PCM for identifying the filled and unfilled voids with PCM are shown in Fig. 6. The gray scale image in Fig. 6(a) presents the coating layers and the void structure which is highly porous. The gray scale image clearly distinguishes the voids filled with PCM from the unfilled voids. A large portion of the porous foam glass particle is saturated with PCM, which demonstrates its suitability as a PCM carrier for applications such as asphalt pavement materials. As a reference, Fig. 6(b) shows the internal skeleton of a foam glass particle before coating and PCM filling. It can be easily seen from that the thin pore walls promote the formation of continuous pores and thus the impregnation with PCM liquid.

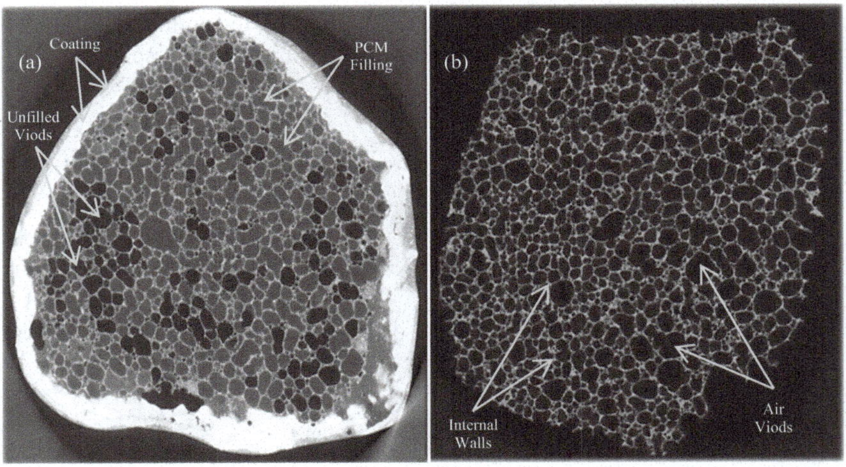

Fig. 6. CT-Scan Image slice, foam glass (a) coated and filled with PCM (b) Uncoated without PCM

On the other hand, Fig. 7(a), illustrates that the expanded burnt clay beads are less filled with PCM than foam glass. Furthermore, according to Fig. 7(b), the internal voids in the expanded burnt clay beads appear more irregular in shape and less connected as compared with foam glass.

It was calculated by mass differences that, on average, the foam glass particles were saturated up to 98%. However, the expanded burnt clay beads were able to adsorb less PCM, i.e. only 56% by mass. The results clearly imply that with respect to impregnation properties, foam glass particles are more feasible PCM carriers than burnt clay

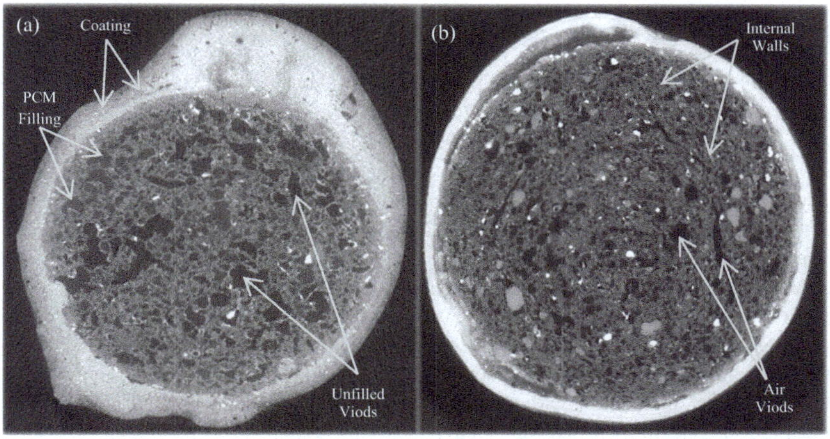

Fig. 7. CT-Scan Image slice, coated burnt clay (a) with PCM (b) without PCM

beads. Furthermore, as shown in Fig. 7(b) the non-impregnated expanded burnt clay beads consist of an irregular narrow void skeleton that is mainly responsible for the higher resistance against PCM saturation compared to foam glass.

4 Conclusions and Recommendations

In the present study, two different types of porous lightweight aggregate (LWA) particles, foam glass and expanded burnt clay beads were investigated for their suitability as phase change material (PCM) carrier aggregates for asphalt mixtures. The impregnation with tetradecane as PCM was performed in the laboratory using the vacuum saturation method. A coating treatment with double layer epoxy glue combined with ordinary Portland cement was used after applying PCM impregnation of the particles. The results based on mass loss calculation before and after saturation show that foam glass has an almost double saturation rate than expanded burnt clay beads, thus elucidating the high propensity of PCM soaked foam glass over expanded burnt clay beads. The CT-scan results confirm that the internal structure of foam glass contains well established interconnected voids compared with expanded burnt clay beads. This highly porous nature of foam glass allowed impregnations with a comparatively high amount of PCM, whereas the irregular and less connected nature of pores in expanded burnt clay beads resisted the accommodation of PCM and, hence, resulted in a lower amount of PCM inside. However, further studies are recommended to investigate the thermal effect of coated LWA particles filled with PCM. In parallel, an evaluation of the suitability of these modified aggregates for its final application in asphalt mixtures must be conducted.

Acknowledgements. The authors would like to acknowledge the Swiss National Science Foundation (SNSF) for the financial support of the project number 200021_169396 and Dr. Thomas Lüthi at Center for X-ray Analytics for conducting CT-Scan test.

References

Baetens R, Jelle BP, Gustavsen A (2010) Phase change materials for building applications: a state-of-the-art review. Energy Build 42(9):1361–1368

Cabeza LF, Castellon C, Nogues M, Medrano M, Leppers R, Zubillaga O (2007) Use of microencapsulated PCM in concrete walls for energy savings. Energy Build 39(2):113–119

Garcia A, Austin CJ, Jelfs J (2016) Mechanical properties of asphalt mixture containing sunflower oil capsules. J Clean Prod 118:124–132

Hamzah MO, Kakar MR, Quadri SA, Valentin J (2014) Quantification of moisture sensitivity of warm mix asphalt using image analysis technique. J Clean Prod 68:200–208

Kakar MR, Hamzah MO, Valentin J (2015) A review on moisture damages of hot and warm mix asphalt and related investigations. J Clean Prod 99:39–58

Kakar MR, Refaa Z, Bueno M, Worlitschek J, Stamatiou A, Partl MN (2019a) Investigating bitumen's direct interaction with Tetradecane as potential phase change material for low temperature applications. Road Mater Pavement Des, 1–8

Kakar MR, Refaa Z, Worlitschek J, Stamatiou A, Partl MN, Bueno M (2018) Use of microencapsulated phase change materials in bitumen to mitigate the thermal distresses in asphalt pavements. In: RILEM 252-CMB-Symposium on Chemo Mechanical Characterization of Bituminous Materials. Springer, Cham, pp 129–135

Kakar MR, Refaa Z, Worlitschek J, Stamatiou A, Partl MN, Bueno M (2019b) Thermal and rheological characterization of bitumen modified with microencapsulated phase change materials. Constr Build Mater 215:171–179

Kariznovi M, Nourozieh H, Guan JGJ, Abedi J (2013) Measurement and modeling of density and viscosity for mixtures of Athabasca bitumen and heavy n-alkane. Fuel 112:83–95

Kheradmand M, Castro-Gomes J, Azenha M, Silva PD, de Aguiar JL, Zoorob SE (2015) Assessing the feasibility of impregnating phase change materials in lightweight aggregate for development of thermal energy storage systems. Constr Build Mater 89:48–59

Kissock JK, Hannig JM, Whitney TI, Drake ML (1998) Testing and simulation of phase change wallboard for thermal storage in buildings. In: Proceedings of the International Solar Energy Conference, New York, USA, pp 45–52

Liu J, Zhao S, Li L, Li P, Saboundjian S (2017) Low temperature cracking analysis of asphalt binders and mixtures. Cold Reg Sci Technol 141:78–85

Mallick R, Hooper F, O'Brien S, Kashi M (2004) Evaluation of use of synthetic lightweight aggregate in hot-mix asphalt. Transp Res Rec: J Transp Res Board 1891:1–7

Manning BJ, Bender PR, Cote SA, Lewis RA, Sakulich AR, Mallick RB (2015) Assessing the feasibility of incorporating phase change material in hot mix asphalt. Sustain Cities Soc 19:11–16

Qian Z, Chen L, Jiang C, Luo S (2011) Performance evaluation of a lightweight epoxy asphalt mixture for bascule bridge pavements. Constr Build Mater 25(7):3117–3122

Refaa Z, Kakar MR, Stamatiou A, Worlitschek J, Partl MN, Bueno M (2018) Numerical study on the effect of phase change materials on heat transfer in asphalt concrete. Int J Therm Sci 133:140–150

Ryms M, Denda H, Jaskuła P (2017) Thermal stabilization and permanent deformation resistance of LWA/PCM-modified asphalt road surfaces. Constr Build Mater 142:328–341

Ryms M, Lewandowski WM, Klugmann-Radziemska E, Denda H, Wcisło P (2015) The use of lightweight aggregate saturated with PCM as a temperature stabilizing material for road surfaces. Appl Therm Eng 81:313–324

Sharma A, Tyagi VV, Chen CR, Buddhi D (2009) Review on thermal energy storage with phase change materials and applications. Renew Sustain Energy Rev 13(2):318–345

Tyagi VV, Kaushik SC, Tyagi SK, Akiyama T (2011) Development of phase change materials based microencapsulated technology for buildings: a review. Renew Sustain Energy Rev 15 (2):1373–1391

Wang J, Molenaar AA, van de Ven MF, Wu S (2018) Influence of internal structure on the permanent deformation behavior of a dense asphalt mixture. Constr Build Mater 171:850–857

Wei K, Wang Y, Ma B (2019) Effects of microencapsulated phase change materials on the performance of asphalt binders. Renew Energy 132:931–940

Wycoff JC (1959) Suitability of lightweight aggregate for bituminous plant mix. ASTM Bull 235:33–36

Zhang MH, Gjvorv OE (1991) Mechanical properties of high-strength lightweight concrete. Mater J 88(3):240–247

The Use of a Polyethylene-Based Modifier to Produce Modified Asphalt Binders on Site

Antonio Roberto(✉), Antonio Montepara, Elena Romeo, and Saŝa Tatalović

Department of Engineering and Architecture, University of Parma, Parco Area delle Scienze 181/A, Parma, Italy
antonio.roberto@studenti.unipr.it

Abstract. Modified asphalt binders are commonly employed to reduce severity of flexible pavement distresses and increases service life. In some countries, modified binders need to be imported from abroad. Further problems may arise due to the geographical location of the plants (e.g. desert areas or areas with low industrial development), as well as to the economic inconvenience in the supply of raw materials necessary for small processes. This study focuses on the investigation of a technology that allows the production of modified binder directly on site by adding a polyethylene-based-modifier during the mixing stage. Eight different mixtures were prepared combining two different asphalt binders, neat and SBS modified, together with the modifier and RAP material. The mechanical and energy properties of the mixes were investigated performing the SuperPave IDT test. The results showed significant mechanical improvements for neat asphalt mixtures combined with the modifier, showing a decrease of the rate of damage accumulation and an increase in energy thresholds. A more critical interpretation was obtained when the modifier was added to SBS modified mixtures.

Keywords: Modified asphalt binders · Polyethylene · HMA mixtures · Superpave indirect tension test · Reclaimed asphalt pavement

1 Introduction

It is widely recognized that the use of modified asphalt binders reduces the amount and severity of flexible pavement distresses and increases service life. The primary benefit of using these high-performance binders is improved rutting and fatigue resistance, with overall improved mixture durability (FHWA-HIF-11-038 2011; Mehta et al. 2013). Indeed, they are usually employed to contrast ageing effects (Airey 2003). In some applications, the use of modified asphalt binders can be complicated due to the need of specific mills for their production and tanks dedicated to the storage at high temperatures. Road construction companies or operators of asphalt plants that are unable to produce modified binders, often have difficulties in supplying these materials. In some countries, modified binders need to be imported from abroad. Further problems may arise due to the geographical location of the plants (e.g. desert areas or areas

M. Pasetto et al. (Eds.): ISAP APE 2019, LNCE 48, pp. 346–355, 2020.
https://doi.org/10.1007/978-3-030-29779-4_34

with low industrial development), as well as to the economic inconvenience in the supply of raw materials required for small processes.

This study focuses on the evaluation of a new production technology for asphalt mixtures modified directly on site employing an innovative modifier. This material consists in a polymer-compound (P) made with both low-density polyethylene (LDPE) and high-density polyethylene (HDPE), ethylene vinyl acetate (EVA) and polypropylene (PP). It is characterized by semi-soft and flexible granules with dimensions varying between 1 and 4.5 mm. Previous analyses showed that PE-modified asphalt mixtures exhibit good performance when compared to traditional mixtures (Punith and Veeraragavan 2007). Indeed, the inclusion of PE in the asphalt mixtures reduces rutting and temperature susceptibility. Al-Dubabe et al. (1998) found that the softening point tend to increase with the addition of PE, which indicates an improvement in fracture resistance. Celauro et al. (2019) studied a similar polymer compound correlating laboratory analyses with on-site applications. They showed an increasing in mechanical properties when the traditional Hot Mix Asphalt (HMA) is modified with P.

In this experimental study, the main objective is to verify the performance of the on-site modification methodology by performing a laboratory analysis. Therefore, P was evaluated like an on-site additive and mixed with two different asphalt binders, one neat, PG 64-22, and one SBS modified, PG 70-22. Eight different asphalt mixtures were prepared combining the two different asphalt binders with the modifier P (6% by the weight of the optimum bitumen content) and 20% of Reclaimed Asphalt Pavement (RAP). The evaluation of mechanical and facture behaviors was evaluated according to visco-elastic HMA Fracture Mechanics framework (Roque et al. 2002 and Zhang et al. 2001). This model is based on Superpave IDT procedure, which uses three different tests performed at 10 °C, Resilient Modulus (Roque and Buttlar 1992), Creep Compliance (Buttlar and Roque 1994) and Tensile strength (Roque et al. 1999). This HMA Fracture Mechanics framework introduced an innovative concept, the failure of HMA is governed by Fracture Energy density threshold (FE) and Dissipated Creep Strain Energy threshold ($DCSE_f$). The introduction of these two energy thresholds indicates that there is a damage limit before the propagation of the permanent fracture being, and $DCSE_f$ represents it. Before that limit, the micro-cracks are absorbed by the materials, when the energy exceed $DSCE_f$ the micro-cracks are not healable and coalesce in a macro-crack.

2 Asphalt Mixture Cracking Behavior: HMA Fracture Mechanics

The cracking behaviour of the asphalt mixtures was evaluated according to the visco-elastic model “HMA Fracture Mechanics” (Roque et al. 2002; Zhang et al. 2001) which identifies five properties of the material to define its performance in terms of cracking resistance and permanent deformation accumulation. These parameters are obtained from three tests in indirect tensile configuration performed at 10 °C, according to the Superpave IDT procedure. Failure of asphalt mixtures is governed by two fundamental properties: Fracture Energy density threshold (FE) and Dissipated Creep Strain Energy threshold ($DCSE_f$). The FE limit is determined as the area under the stress–strain curve at fracture point, while the $DCSE_f$ is the fracture energy minus the elastic energy, both obtainable from an indirect tensile-strength-test. The $DCSE_f$ represents the measure of how much the micro-damage mixture can absorb before it becomes in a macrocrack initiation. To predict top-down cracking performance of the mixture in the field, an Energy Ratio criterion was proposed (Roque et al. 2004). The Energy Ratio is defined as follows:

$$ER = \frac{DCSE_F}{DCSE_{min}} \tag{1}$$

where $DCSE_f$ is the dissipated creep strain energy threshold of the mixture and $DCSE_{min}$ is the minimum dissipated creep strain energy required, a function of the creep compliance power low parameters. Essentially, $DCSE_{min}$ describes how much damage, during service life, the material will accumulate in the field. For a good performance level of the mixture $ER > 1$ is required. More details on the model and the parameters employed can be found in previous publications (Zhang et al. 2001; Roque et al. 2002, 2004).

3 Materials

Two different asphalt binders were employed: N is a PG 64-22 unmodified binder and H is a PG 70-22 modified binder. The mixtures were composed by the same aggregates and gradation (12.5 Nominal Maximum Size). For each type of asphalt binder, mixtures containing 20% RAP on weight of fine aggregates (up to 2 mm), were also prepared. The filler employed, with a density of 2.71 g/cm^3, was obtained from the grinding of coal rocks. To label the mixtures, an acronym was associated to each material; a detailed description of the mixtures is given in Table 1.

Table 1. Asphalt mixtures definition.

Mixture label	Binder	Filler	Aggregates	Polymer compound
N	Natural	Coal rock	Virgin	No
NP	Natural	Coal rock	Virgin	Yes
NR	Natural	Coal rock	Virgin + RAP	No
NRP	Natural	Coal rock	Virgin + RAP	Yes
H	SBS modified	Coal rock	Virgin	No
HP	SBS modified	Coal rock	Virgin	Yes
HR	SBS modified	Coal rock	Virgin + RAP	No
HRP	SBS modified	Coal rock	Virgin + RAP	Yes

3.1 Polymer Compound

This material consists in a polymer-compound (P) made with both low-density polyethylene (LDPE) and high-density polyethylene (HDPE), ethylene vinyl acetate (EVA) and polypropylene (PP). It is characterized by semi-soft and flexible granules with dimensions varying between 1 and 4.5 mm and a density of 0.55 to 0.65 g/cm^3.

The plastic polymers may have three kind of behaviors, linear elastic, rigid and elasto-plastic. The P mechanical behaviour is elasto-plastic, which means that when the applied effort is not very high, the mechanical behavior is linear elastic; while when the applied effort goes over the linear-elastic stress limit, the behavior changes and the material shows plastic deformation. The polymer compound percentage used in the mixture is within the range of 4–8%. This study aimed at verifying the performance level obtained on site using 6% of P by the binder weight. Table 2 shows all the physical parameters that characterise P.

Table 2. Polymer compound parameters.

Softening point [°C]	160–180
Fusion point [°C]	180
Melt index	1–5
Bulk density [g/cm^3]	0.55–0.65
Bulk density at 25 °C [g/cm^3]	0.40–0.60
Specific gravity [g/cm^3]	0.96–0.98
Density at 20 °C [g/cm^3]	0.90–0.98
Grains dimension [mm]	1.00–4.50

3.2 HMA Specimen Preparation

Eight different mixtures were prepared using 4500 g aggregate batches for a total of 32–152 mm diameter cylindrical specimens.

The batches were mixed with the asphalt binder at 160 °C for unmodified mixes and 170 °C for modified mixes. The design asphalt content is 5.2%, while modifier P content is 6% by the weight of total amount of asphalt binder. The mixing procedure was divided into three different phases as follow:

1. Mixing aggregates, pre-heated at 170 °C for 4 h, and P for 1 min;
2. Addition of optimum bitumen content and mixing for 1 min at asphalt binders equiviscous temperatures (Yildirim et al. 2000);
3. Addition of filler, pre-heated at 170 °C for 4 h, and mixing until obtaining a uniform materials-blending.

These three steps were followed to simulate the P-additive introduction during the mixing on-site. After the mixing procedure, the material obtained was heated for two hours at 135 °C for short-term aging, as indicated into the SuperPave-protocol. The mixtures were then compacted using the Pine Gyratory Compactor at N = 126 gyrations to achieve 6% (±0.5%) air voids into 150 mm diameter specimens. Each cylindrical specimen was sawn to obtain two effective plates, each 30 mm thick discarding the top and the bottom plates for reducing density gradient effects. For each mixture, four circular-shaped specimens were used to perform the Superpave IDT test, at 10 °C, according to the procedures developed by Roque and Buttlar (1992) and Buttlar and Roque (1994).

4 Results and Analysis

All the tests were performed using an MTS closed-loop servo-hydraulic loading system. All the results obtained from the Superpave IDT test are listed in Table 3.

Table 3. Results for all parameter investigated during this study.

Asphalt mixtures	Creep compliance [MPa]	Resilient modulus [GPa]	Tensile strength [MPa]	Fracture energy [kJ/m^3]	Elastic energy [kJ/m^3]	$DCSE_f$ [kJ/m^3]	Energy ratio
H	1.09	18.08	3.87	2.43	0.433	1.20	2.24
HP	0.68	19.84	3.99	2.39	0.40	1.20	1.67
HR	0.53	17.10	4.01	3.15	0.47	2.68	4.31
HRP	0.56	14.01	3.42	3.02	0.44	2.58	4.92
N	1.63	23.50	3.73	1.13	0.30	0.83	0.34
NP	1.07	19.91	3.96	1.73	0.35	1.39	0.84
NR	0.81	20.92	4.08	2.30	0.43	1.87	1.11
NRP	0.54	16.68	4.42	3.51	0.61	2.89	2.29

4.1 Resilient Modulus Test

The resilient modulus is a measure of the material's elastic stiffness since it corresponds to the ratio of the applied stress to the recoverable strain with repeated loading. P-containing mixtures show a decrease in the resilient modulus when neat and modified asphalt binders have been mixed with it. This trend has been confirmed with and without RAP. The H combination did not confirm this trend. Probably, this effect is due to the presence of a high content of polymers in this type of mixtures.

4.2 Creep Compliance Test

The Creep-Compliance curve was used to describe and evaluate the HMA damage-rate. Therefore, it represents the asphalt mixtures-aptitude at permanent damage accumulation in function of the time. However, considering the HMA Fracture Mechanics Framework, the greater the asphalt mixture permanent deformation the greater the crack growth process. Therefore, when the mixtures exhibit high creep rates, they show higher crack growth rates. The creep test was conducted in a load control mode by performing a static load keeping the horizontal strain in the linear viscoelastic range. The creep-curves for both neat and modified asphalt binders are shown in Figs. 1 and 2, respectively. For both the asphalt binders, P plays an important role, in fact, the creep compliance parameter decreases when the polymer compound is added in the mixtures.

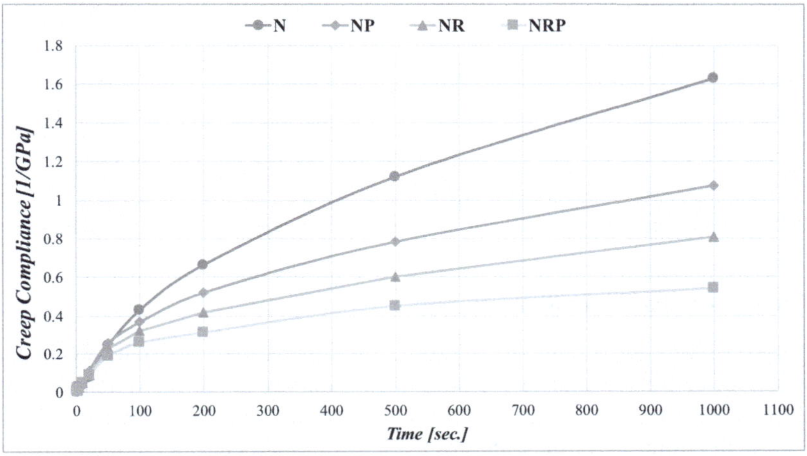

Fig. 1. Creep compliance in function of time of mixtures made with neat asphalt bitumen.

Another important parameter is the m-value which describes the damage accumulation speed. The greater is the m-value the greater is the speed of the damage accumulation rate.

Analysis of the results shows as follow:

- The mixtures made with neat asphalt binder exhibit an increase of performance when P has been blended in the mixtures. This trend is not confirmed when the RAP was used in mixtures although the difference between the mixtures made with and without P is small.
- The mixture made with modified asphalt binder show an increase in m-value when the RAP was blended in the mixtures. In this case, probably, the aged bitumen inside the RAP helps to increase the stiffness of the final bending asphalt binder and reduce the damage accumulation speed. Indeed, the presence of different kind of polymer does not help understanding correctly the behaviour of this mixture's types.

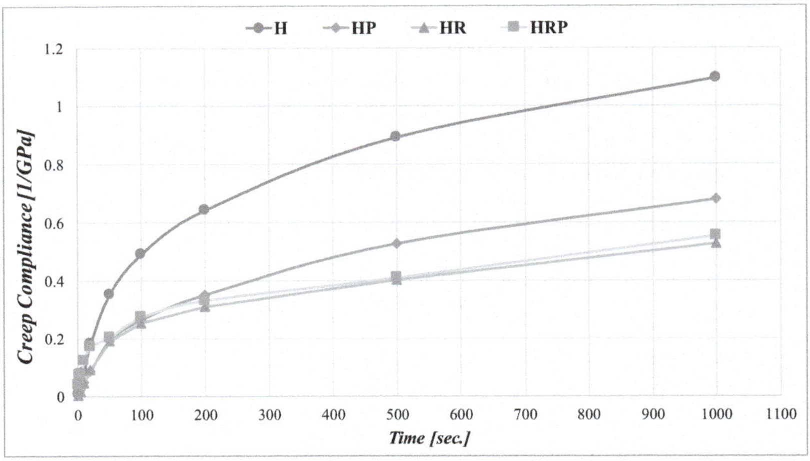

Fig. 2. Creep compliance in function of time of mixtures made with modified asphalt bitumen.

4.3 Energy Limits Evaluation

Zhang et al. (2001) and Roque et al. (2002) introduced the possibility evaluation of energy parameter by the analysis of stress-strain curves obtained by tensile strength test. All the parameters investigated are summarised in Table 3. The Results show that tensile strength is similar for the mixtures made with modified asphalt binder, this trend is not confirmed only for HRP, showing the lower tensile strength. For the mixtures made with neat asphalt binder this trend is not confirmed. NP shows an increase in performance regarding the base-mixture N, the same behaviour has been confirmed when the RAP material is inside the mixtures. For this mixture the difference between the mixture with and without RAP is higher. This difference could be due to the aged-asphalt binder which is inside the recycling material. This analysis suggests that the polymer compound influences the tensile strength when it is mixed with neat asphalt binder. However, the high tensile strength may be a problem for the mixture NRP. The energy-analysis of stress-strain curves is shown in Figs. 3 and 4. Therefore, the energy parameters evaluation showed an important increasing in performance for the mixtures made with neat asphalt binder, in fact FE and $DCSE_f$ showed an increasing trend for these mixtures. However, this trend is not confirmed for the mixtures containing modified asphalt binder which showed a decrease performance when P was added in the mixtures. This is probably due to a higher percentage of polymers inside the mixtures. Finally, the energy analysis gives the possibility of understanding if the mixtures are susceptible or not to top down cracking. To obtain a good top-down cracking performance the concept suggests that the ER should be over 1.

The results obtained are shown in Fig. 5. The analysis of these results indicates that the mixtures made with modified asphalt binder have a high top-down cracking performance. This trend has been confirmed with and without the presence of P into the mixtures. The mixtures prepared with neat asphalt binder have not showed an optimal top-down cracking performance. In fact, only the mixtures containing RAP has shown an energy-ratio over one. Considering the results showed in Fig. 5 and Table 3, we could observe an increase in performance over 50% for the mixtures without RAP, and over 100% for those with RAP.

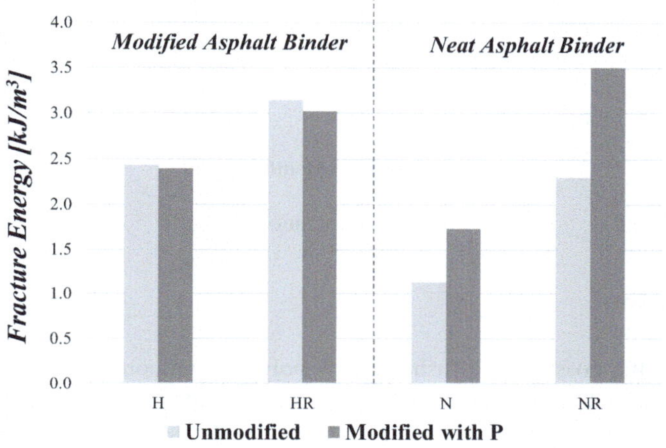

Fig. 3. Fracture energy in function of the base asphalt binder.

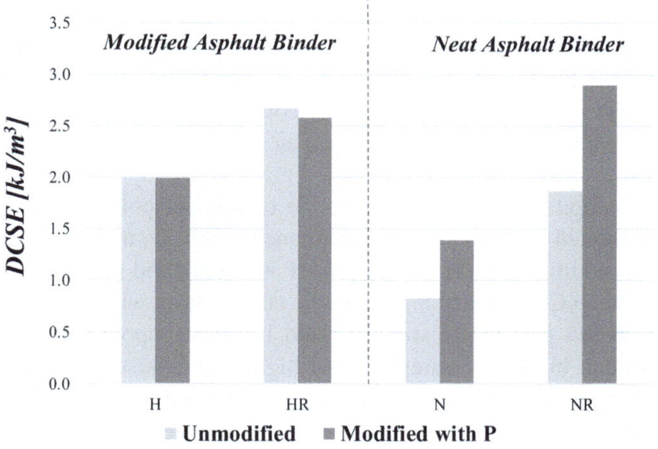

Fig. 4. Dissipated creep strain energy in function of the base-asphalt binder.

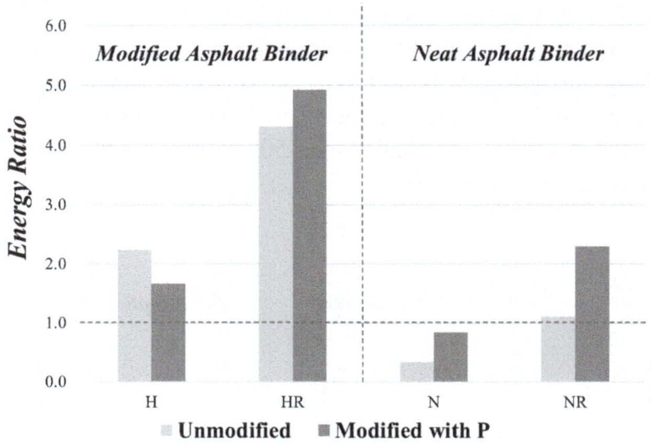

Fig. 5. Energy ratio results

5 Conclusion

In this study, the possibility of using a new polymer compound that allows the production of modified binder directly on site was evaluated. The analysis was conducted under the condition proposed by HMA Fracture Mechanics visco-elastic cracking model, for the energy parameter, and SuperPave protocol, for the mechanical behaviors. Neat and SBS modified asphalt binders and new compound were mixed with virgin aggregates and RAP materials. The following findings on mixtures mechanical and energy cracking behaviours were obtained:

- The presence of new compound affects Resilient Modulus, in fact the presence of polymer reduces the elastic-stiffness for the mixtures made with neat asphalt binder. This effect was exhibited by the mixtures containing RAP material. The mixtures made with modified asphalt binder showed an increase of elastic stiffness when Rap is not added.
- The Creep Compliance was affected by the compound polymer presence. All the mixtures exhibited a decreasing in permanent damage accumulation. Only the combination modified asphalt binder + RAP + Compound polymer showed a little increase of Creep Compliance regarding the combination without additive polymer.
- The energy-based parameters were affected by the compound polymer presence. The neat asphalt binder mixtures showed higher aptitude to increase their energy parameters when the compound polymer is contained. Differently, the modified asphalt binder mixtures have showed a more critical interpretation with and without the RAP material inside the gradation curves. The highest energy-parameter increase was obtained for top-down cracking performance. The mixtures made with neat asphalt binder showed an increase in performance over 50% with respect to the mixtures without RAP, and over 100% with respect to those containing RAP. This means that, when the compound polymer is mixed in this kind of mixtures, the aptitude to top-down cracking decreases.

References

Office of Pavement Technology - FHWA-HIF-11-038 (2011) The multiple stress creep recovery (MSCR) procedure

Mehta Y, Nolan A, DuBois E, Zorn S, Shirodkar P (2013) Correlation between Multiple Stress Creep Recovery (MSCR) Results and Polymer Modification of Binder. Report No. FHWA-NJ-2014-002

Airey GD (2003) State of the art report on ageing test methods for bituminous pavement materials. Taylor & Francis Ltd. https://doi.org/10.10850/1029843042000198568

Roque R, Birgisson B, Sangpetgnam B, Zhang Z (2002) Hot mix asphalt fracture mechanics: a fundamental crack growth law for asphalt mixtures. J Assoc Asphalt Paving Technol 71:816–827

Zhang Z, Roque R, Birgisson B, Sangpetgnam B (2001) Identification and verification of a suitablecrack growth law. J Assoc Asphalt Paving Technol 70:206–241

Punith VS, Veeraragavan A (2007) Behavior of asphalt concrete mixtures with reclaimed polyethylene as additive. J Mater Civ Eng 19(6):500–507

Al-Dubabe IA, Al-Abdul Wahhab HI, Asi IM, Mohammed FA (1998) Polymer modification of Arab asphalt. ASCE-J Mater Civil Eng 10(3):161–167

Celauro C, Bosurgi G, Sollazzo G, Ranieri M (2019) Laboratory and in-situ tests for estimating improvements in asphalt concrete with the addition of an LDPE and EVA polymeric compound. Constr Build Mater 196:714–726

Roque R, Buttlar WG (1992) The development of a measurement and analysis system to accurately determine asphalt concrete properties using the indirect tensile mode. J Assoc Asphalt Paving Technol 61:304–332

Buttlar WG, Roque R (1994) Development and evaluation of the strategic highway research program measurement and analysis system for indirect tensile testing at low temperatures. Transp Res Rec 1454:163–171

Yildirim Y, Solaimanian M, Kennedy TW (2000) Mixing and compaction temperatures for hot mix asphalt concrete. South Central Superpave Center Research Project 0-1250

Roque R, Zhang Z, Sankar B (1999) Determination of crack growth rate parameters of asphalt mixtures using the superpave indirect tension test (IDT). J Assoc Asphalt Paving Technol 68:404–433

Roque R, Birgisson B, Drakos C, Dietrich B (2004) Development and field evaluation of energy-based criteria for top-down cracking performance of hot mix asphalt. J Assoc Asphalt Paving Technol 73:229–260

Effect of Moisture on Fatigue Characteristics of Asphalt Concrete Mixtures

Mohit Chauhan(✉) and Atul Narayan

Department of Civil Engineering, Indian Institute of Technology Madras, Chennai 600036, India
mohitchauhankunj@gmail.com, atulnryn@iitm.ac.in

Abstract. Moisture in asphalt concrete (AC) pavements does not only cause distresses like potholes, ravelling, etc. but also exacerbates distresses like fatigue and rutting. Moisture susceptibility of AC mixtures is usually characterized using the Modified Lottman Test (AASHTO T283) adopted in Superpave mix design. This test has been observed to provide a good prediction of moisture susceptibility of AC mixtures. However, it does not provide insight on how moisture affects the fatigue characteristics of AC mixtures. Therefore, there is a need for a test procedure that can characterize the effect of moisture on the fatigue behaviour of AC mixtures. In this study, changes in fatigue characteristics after moisture conditioning were evaluating by conducting four-point beam (4 PB) fatigue tests on dry and moisture conditioned beam specimens. For this purpose, specimens of AC mixed with two different types of binders with target air voids of $4 \pm 0.5\%$ were prepared. These specimens were subjected to partial vacuum saturation by submerging completely in water and applying vacuum pressure to the system. The fatigue tests were conducted at four different strain amplitudes of 200, 400, 600 and 800 micro-strains. The results were then compared with the fatigue tests results obtained with dry beam specimens. Test results show that the conditioning reduces both the fatigue life and the initial flexural stiffness of the beam specimens. Moreover, it was also observed that the binder type plays a significant role in the degree of saturation.

Keywords: Asphalt concrete · Fatigue life · Four-point beam fatigue test · Moisture conditioning

1 Introduction

Moisture damage causes the asphalt concrete (AC) pavement to undergo several types of distress, such as stripping, potholes, etc. But, in addition to them, its presence may increase the extent and severity of other distresses, such as fatigue cracking, top-down cracking, rutting, ravelling, and bleeding (Lu and Harvey 2008). Amongst these distresses, fatigue cracking is a predominant form of distress in AC pavements and one of the key failure criteria used for AC pavement designs. It is thus pertinent to study the effect of moisture on fatigue resistance of AC mixtures.

The presence of moisture in the AC layer of pavements reduces the cohesion in the asphalt mastic, and the adhesion at the asphalt-aggregate interface (Kringos et al. 2008). Moreover, moisture increases pore pressure within the AC layer during vehicle

M. Pasetto et al. (Eds.): ISAP APE 2019, LNCE 48, pp. 356–366, 2020.
https://doi.org/10.1007/978-3-030-29779-4_35

loading, which in turn causes excessive stresses/strains in the pavement structure (Lu and Harvey 2008). Moisture in this way weakens the AC mixture and makes it more susceptible to fatigue cracking in pavements.

When characterizing the influence of moisture on the fatigue behaviour of AC mixtures through experiments, the observed behaviour may change markedly with the type of test used to quantify fatigue resistance and the type of moisture conditioning used to simulate moisture induced damage. Typically, the effect of moisture on fatigue resistance of AC mixtures has been quantified in the literature through three different types of tests. This includes cyclic indirect tension test (IDT), dynamic mechanical analysis test (DMA) and flexural beam test (Mehrara and Khodaii 2013).

The cyclic indirect test has been used in a number of researches. Gilmore et al. (1984) studied the effect of lime in controlling the detrimental effects of moisture on asphalt concrete. Birggisson et al. (2003) have also extracted a well-developed energy base parameter named DCSE (dissipated creep strain energy) using IDT. In the DMA test method, a cylindrical fine aggregate mixture is subjected to cyclic torsion. Kim et al. (2004) first used this test to evaluate the stripping potential of sand asphalt mixtures.

Among them, the flexural beam test with strain-controlled sinusoidal loading is possibly the suitable test, being used for characterization of the effect of moisture on the fatigue behaviour of AC mixtures (Shatnawi et al. 1995; Lu and Harvey 2008). Using this test, one can predict the performance life an AC pavement in a saturated condition. Table 1 summarizes the relevant literature related to fatigue characteristics of moisture conditioned mixtures.

Table 1. Literature on fatigue characteristics of moisture conditioned mixtures

Author	Shatnawi et al. 1995	Lu and Harvey 2008
Mode of test	3-point beam test with sinusoidal loading	4-point beam test with Sinusoidal loading
Controlling variable	Displacement	Displacement
Test details	200 and 300 με, 10 Hz, 20 °C	200 and 400 με, 10 Hz, 20 °C
Fatigue life criteria	50% reduction in initial stiffness	50% reduction in initial stiffness
Fatigue life	$N_f^{dry} > N_f^{wet}$	$N_f^{dry} > N_f^{wet}$

To introduce the effect of moisture, a moisture conditioning process needs to be adopted prior to the fatigue testing. The nature of moisture conditioning can also significantly affect the fatigue resistance of specimens with moisture. Three parameters were determined in literature for the conditioning process: saturation level, conditioning temperature, and conditioning duration. Table 2 summarizes the relevant literature related to the conditioning of AC mixture.

Table 2. Literature on preconditioning parameters

Author	Saturation level	Preconditioning temperature	Preconditioning duration
Shatnawi et al. 1995	60–80%	60 °C, 25 °C, −18 °C	5, 4, 5 h respectively
Lu and Harvey 2008	50–80%	60 °C	24 h

The effect of moisture on the fatigue life of mixtures and its dependence on the type of binder, mixture volumetric, etc. has been extensively investigated in the literature (Shatnawi et al. 1995; Lu and Harvey 2008). However, the effect of moisture on the variation of fatigue life with amplitude of applied loading has received relatively less attention from researchers. In the present study, fatigue behaviour was observed at different strain levels at both dry and moisture conditioned states to evaluate the effect of moisture on fatigue life at different strain levels. A particular vacuum saturation technique and four-point beam fatigue tests were employed for this purpose, the details of which are presented in the following sections.

2 Experimental Investigation

Two different AC mixtures were prepared for the study, each with a different binder, but with the same aggregate matrix. The aggregate gradation chosen for the study conforms to Bituminous Concrete Grade II (BC-II) mid-gradation with nominal maximum aggregate size of 13.2 mm as per MoRTH (2013) specification. The binders used for the study are unmodified binders conforming to VG10 and VG30 viscosity grading as per IS 73-2013 (2013). A binder content of 5% by mass of the total mix was selected. Beams of size 450 × 150 × 160 mm were fabricated using a shear compactor following ASTM D7981-15 (2015) specification. These beams were further sliced to the size of 380 ± 6 (length) × 63 ± 2 (width) × 50 ± 2 (height) mm for 4-point beam fatigue test with target air voids of 4 ± 0.5%. Tables 3 and 4 show the aggregate gradation for BC-II and binders respectively used in tests. All tests were conducted at a temperature of 20 °C with a frequency of 10 Hz. Fatigue life was calculated using two post processing methods: AASHTO T321-07 (2007) and ASTM D7460-10 (2010). The various combinations of testing conditions used in the current study is given in Table 5.

Prior to the testing, half of the beam samples were moisture conditioned in a vacuum saturation fixture. A vacuum saturation apparatus was specifically designed and fabricated for this purpose. Figure 1a shows the apparatus fabricated for moisture conditioning of the samples. The apparatus consists of an airtight cylindrical glass container with an outlet valve connected to a vacuum pump. The dimensions of the glass container were made sufficient to keep a beam of standard size for four-point beam fatigue tests.

Table 3. Aggregate gradation (BC – II) used in tests

IS Sieve (mm)	Mid-gradation	Cumulative weight retained (%)	Cumulative weight passing (g)	Individual weight (g)
19	100	0	0	0
13.2	95	5	335	335
9.5	79	21	1405	1070
4.75	62	38	2542	1137
2.36	50	50	3344	802
1.18	41	59	3946	602
0.600	32	68	4548	602
0.300	23	77	5150	602
0.150	16	84	5618	468
0.075	7	93	6220	602
Passing 0.075	0	100	6688	468
Binder (5%)				352

Table 4. Properties of different binders as per IS:73-IS 2013

Characteristics	VG10		VG30	
	Results obtained	IS 73 (2013) recommendation	Results obtained	IS 73 (2013) recommendation
Absolute viscosity at 60 °C (Poises)	1432	800–1200	3335	2400–3600
Kinematic viscosity at 135 °C (cSt), Min.	412	250	534	350
Penetration at 25 °C (0.1 mm), Min.	104	80	43	45
Softening point, R&B (°C), Min.	46	40	52	47
Tests on residue from rolling thin film oven tests/RTFOT				
(a) Viscosity ratio at 60 °C, Max.	3.5	4.0	2.93	4.0
(b) Ductility at 25 °C after rolling thin film oven test, Min. (cm)	100+	75	100+	40

Table 5. Test parameters

Mode of testing	Displacement controlled
Amplitude (με)	200, 400, 600 and 800
Waveform	Sinusoidal
Termination criteria	1×10^6 cycles or 80% reduction in stiffness
Saturation level	55–80%
Conditioning temperature	60 °C
Conditioning duration	24 h

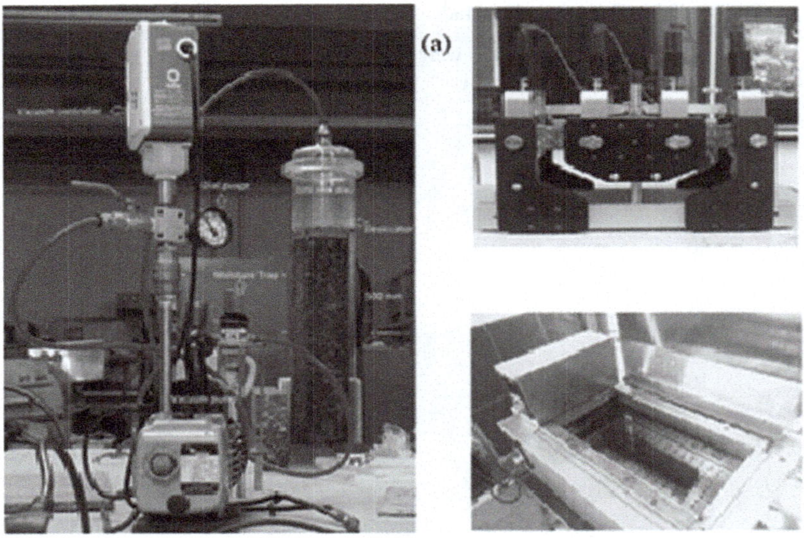

Fig. 1. Saturation equipment (a), 4-point beam jig (b) and Preconditioning at 60 °C water bath (c)

Sufficient amount of water was added such that there was at least 25 mm of water above the top surface of the specimen. A vacuum of 740 mm-Hg partial pressure was applied for a duration of 3 h. Following partial vacuum application, the specimen was then allowed to remain submerged in water for 10 min after the release of vacuum, to allow the water to penetrate the voids in the specimen under atmospheric pressure. The degree of saturation is then calculated using Eq. (1) as the difference in air weight of beam specimen before and after saturation and the volume of air voids is computed using Eq. (2). The beam sample was then kept for conditioning by placing in a 60 ± 1° C water bath for 24 ± 1 h. After preconditioning, the specimen was cooled to room temperature by placing the beam specimen in a water bath for 30 ± 10 min then wrapped with an impermeable plastic sheet, to retain its internal moisture. Then, beam samples were placed inside an environmental chamber at 20 °C for 2 h prior to testing as per AASHTO T321-14 (2014).

$$S = \frac{100 \times W_{sat} - W_{dry}}{V_{air}} \tag{1}$$

$$V_{air} = \frac{V_a - V_{beam}}{100} \tag{2}$$

where, S is the degree of saturation (%), W_{sat} is the saturated surface dry weight of beam specimen after saturation (g), W_{dry} is the weight of dry beam in air (g), V_{air} is the volume of air voids (cm^3), V_a is the air voids (%), and V_{beam} is the volume of beam (cm^3). Figure 2 shows the saturation results. It was interesting to note that samples prepared with VG30 binder attained the degree of saturation between 70–80% after 3 h of saturation. However, specimens with VG10 binder achieved only 55–65% degree of saturation after 3 h at similar air voids and vacuum pressure. Therefore, it could be inferred from the above findings that the reason could be the different internal micro-structure specifically inter-connectivity between the voids for both these mixtures. It is therefore not necessary that degree of saturation shall be identical for the mixtures prepared with two different unmodified binders at 4 ± 0.5% air voids, subject to similar vacuum and time. Thus, it could be inferred that higher is the connectivity of voids, more would be the degree of saturation.

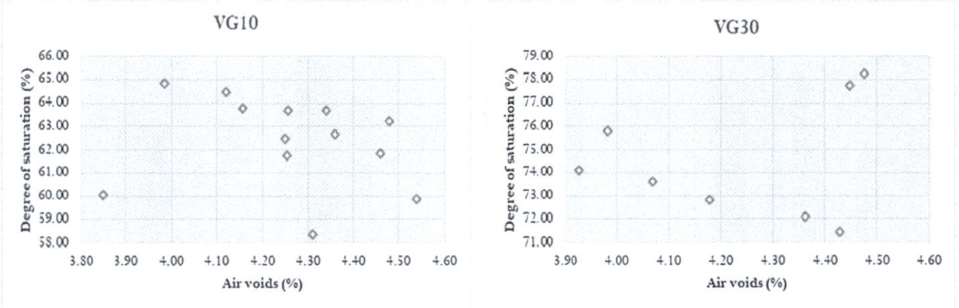

Fig. 2. Degree of saturation of samples

3 Results and Discussions

Figure 3 shows the variation of average initial stiffness modulus at different strain levels as a function of conditioning for both VG10 and VG30 samples. Mixtures containing the VG10 binder have stiffness much lower than that of mixtures containing the VG30 binder. It is interesting to note that the initial stiffness modulus is significantly affected by the binder type and moisture condition. This is because of the moisture conditioning and the associated damage. It is not necessary that flexural stiffness would decrease when the beam is saturated. It usually increases. Figure 4a and b show moisture conditioning significantly reduces the number of cycles to reach 50% of initial stiffness modulus and increases the rate of reduction in stiffness respectively. Hence the fatigue life of the saturated samples was observed to be lesser based on the fatigue criterion defined as per AASHTO T321-07 (2007) i.e. 50% reduction of initial stiffness. Similar results were also observed for all other strains levels. For the lower strain levels, the beam stiffness did not reduce to 50% of the initial value even after 1 million cycles. In such cases, the variation of stiffness with the number of cycles was fit with the function, $S = ae^{bx} + ce^{dx}$, where S represents the initial stiffness, x represents the number of cycles and *a, b, c and d* are model constant. The number of load cycles corresponding to the fatigue life was obtained from the exponential fit. The sample fit is shown in Fig. 5a and b and the average fatigue life calculated from the exponential fit is tabulated in Table 6.

Fatigue life was also determined as per ASTM D7460-10 (2010). As per ASTM D7460-10 (2010), the peak of normalized modulus versus number of cycles is considered as the fatigue failure initiation point. The normalized modulus can be calculated using below Equation,

$$NM_i = \frac{S_i}{S_o} \times \frac{N_i}{N_o}$$

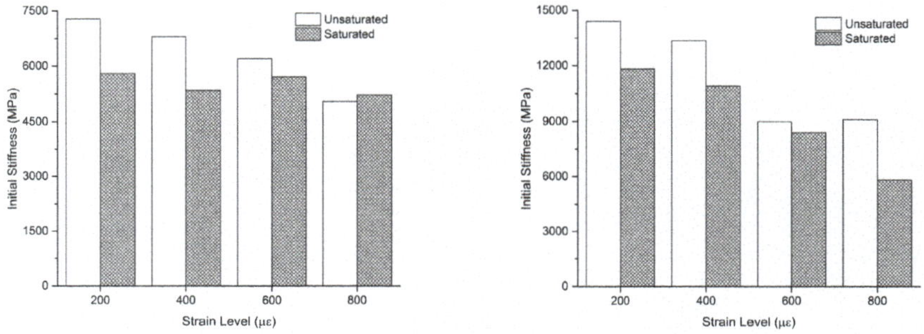

(a) Specimens with VG10 (b) Specimens with VG30

Fig. 3. Variation of initial flexural stiffness at different strain amplitudes

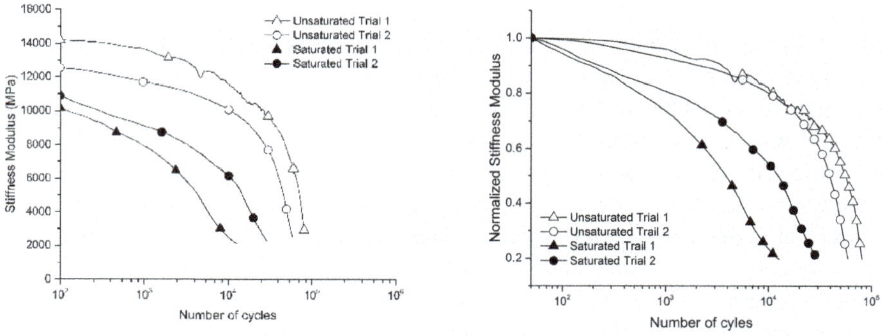

(a) Reduction in flexural stiffness (b) Rate of reduction of flexural stiffness

Fig. 4. Variation of flexural stiffness at 400 με after moisture conditioning

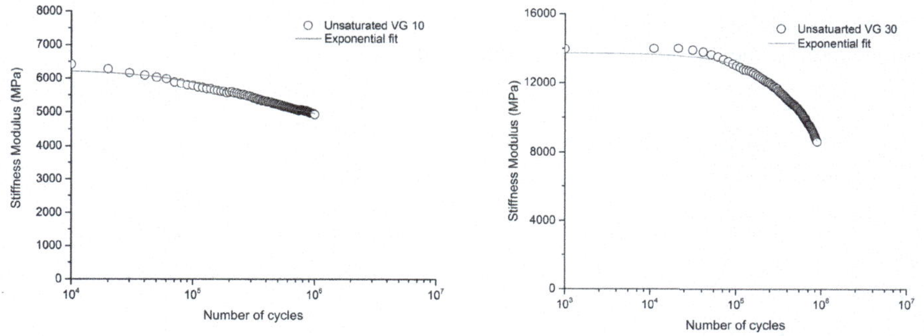

(a) Specimen with VG10 (b) Specimen with VG30

Fig. 5. Exponential fit for VG10 and VG30 at 200 με

where NM_i = normalized modulus at the i cycle, S_i = flexural beam stiffness at cycle i (Pa), $N_i = i$ cycle, S_o = initial flexural beam stiffness at 50 cycle (Pa) and N_o = number of cycles where the initial stiffness is estimated. Figure 6a shows the normalised modulus vs. number of cycles for VG30 at 200 με. At low strain level (200 με), the peak value for one VG30 sample was not attained and hence the peak value for the normalized modulus was determined by assuming Weibull distribution. In the Weibull plot, the horizontal axis is the number of cycles to failure and the vertical axis is the cumulative distribution function, which describes the percentage of failure at a given number of cycles. In ASTM D7460-10 (2010), Weibull distribution is used to extrapolate the fatigue failure point especially at low strain levels, where the peak value of the 'Normalized Modulus' cannot be quantified within the test duration. In this procedure, the number of cycles to failure is calculated using the equation, $\ln(-\ln(SR)) = \gamma \times \ln(N) + \ln(A)$, where SR = flexural beam stiffness ratio, N = number of cycles, γ and A are regression constants. Figure 6b shows the Weibull curve for VG30 at 200 με. The failure point was estimated for the value of N where SR is equal to 0.5 for 50% reduction in stiffness criteria. It was observed from the Fig. 6a that the normalized modulus curve for saturated samples attained the peak value in lesser number of cycles as compared to unsaturated samples and the number of the cycle corresponding to the peak value was considered as fatigue life. Similar results were also observed for all strain levels. This implies that lower fatigue life for saturated samples. In this analysis, for all the tests at 200 με, Weibull distribution method was used to determine the fatigue life and for higher strain levels, normalized modulus method was used. The fatigue life obtained as per ASTM D7460-10 were tabulated in Table 6. The fatigue responses of mixtures containing two different binders are quite distinct from each other.

Table 7 shows the percentage reduction in the fatigue life after moisture conditioning at different strain level for both VG10 and VG30 samples. It can be observed that the percentage reduction in fatigue life for VG30 samples was found to be more than that of samples with VG10. It is hypothesized that the difference in the percentage reduction in fatigue life for the two mixtures could mainly be due to the difference in the degree of saturation. Furthermore, it was also observed from Table 7 that percentage reduction in stiffness after moisture conditioning is significantly affected by post processing method used for calculating fatigue life.

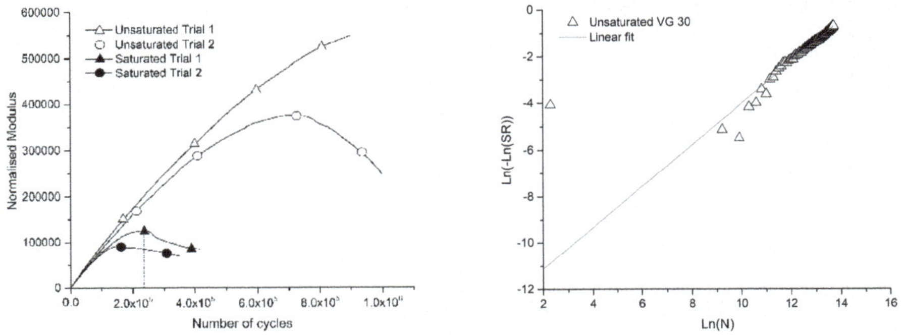

(a) Normalized modulus (b) Weibull curve

Fig. 6. Normalized modulus and Weibull curve for VG30 at 200 με

Table 6. Fatigue life

Strain level (με)	Fatigue life as per AASHTO T321-07 (2007)			
	VG10		VG30	
	Unsaturated	Saturated	Unsaturated	Saturated
200	6.81E+06*	2.81E+06*	1.05+06*	2.09E+05
400	5.82E+05	2.04E+05	4.75E+04	8.32E+04
600	4.95E+04	1.82E+04	1.17E+04	2.74E+03
800	1.20E+04	8.20E+03	2.46E+03	5.66E+02
Strain level (με)	Fatigue life as per ASTM D7460-10 (2010)			
	VG10		VG30	
	Unsaturated	Saturated	Unsaturated	Saturated
200	1.07E+07#	4.76E+06#	1.03E+06#	1.92E+05
400	6.60E+05	2.93E+05	4.73E+04	9.77E+03
600	1.04E+04	2.61E+03	1.06E+04	3.71E+03
800	2.92E+04	1.27E+04	2.71E+03	8.92E+02

*Values obtained after fitting Exponential fit #Values obtained after fitting Weibull fit

Table 7. Effect of conditioning on fatigue life

Strain level (με)	% Reduction in fatigue life after conditioning			
	VG10		VG30	
	AASHTO	ASTM	AASHTO	ASTM
200	59%	56%	80%	81%
400	65%	56%	82%	79%
600	63%	75%	77%	65%
800	32%	56%	77%	60%

4 Conclusion

A new methodology for preconditioning of AC beam samples particularly at lower air voids (4 ± 0.5%) has been proposed in this study. The study shows that VG30 samples achieved 70–80% degree of saturation in 3 h, whereas VG10 samples were saturated to 55–65% in the same duration of time. It is hypothesized that the degree of saturation is dependent on the type of binder governing variability in the internal micro-structure. Moisture conditioning causes a reduction in the initial stiffness modulus and increases the rate of reduction of stiffness. From the two post processing methods adopted for this study, it was observed that the percentage reduction in fatigue life after moisture conditioning depends upon the adopted post processing method. Moreover, it can also be observed that the percentage reduction in fatigue life for VG30 samples was found to be more than that of samples with VG10. It is hypothesized that the percentage reduction in fatigue life for samples with VG10 and VG30 samples could mainly be due to the difference in the degree of saturation. From the overall analysis, it was understood that moisture reduces fatigue life, this may be due to the premature failure by the loss of cohesion in the asphalt mastic and adhesion in the asphalt aggregate interface.

5 Limitations and Scope for Future Study

- This study was limited to two different binders and one aggregate matrix. Hence, further research on internal micro-structure of the mixture with different binders and aggregates matrix and its relation with the degree of saturation needs to be conducted.
- Further studies can be carried out to quantify the effect of different post processing methods in the estimation of fatigue life of moisture conditioned samples.

References

AASHTO T321-07 (2007) Determining the fatigue life of compacted hot mix asphalt subjected to repeated flexural bending. American Association of State Highways and Transportation Officials, Washington, D.C., USA

ASTM D7981-15 (2015) Standard Practice for Compacting of Prismatic Asphalt Specimens by Means of Means of the Shear Box Compactor. ASTM International, West Conshohocken, Pennsylvania, USA

ASTM D7460-10 (2010) Standard test method for determining fatigue failure of compacted asphalt concrete subjected to repeated flexural bending. ASTM International, West Conshohocken, Pennsylvania, USA

Birgisson B, Roque R, Page GC (2003) Evaluation of water damage using hot mix asphalt fracture mechanics. J Assoc Asphalt Pav Technol (AAPT) 72:516–562

Gilmore DW, Lottman RP, Scherocman A (1984) Use of indirect tension measurements to examine the effect of additives on asphalt concrete durability. J Assoc Asphalt Pav Technol (AAPT) 53:495–525

IS 73-2013 (2013) Paving bitumen – specification. Bureau of Indian Standards, New Delhi, India
Kim YR, Little DN, Lytton RL (2004) Effect of moisture damage on material properties and fatigue resistance of asphalt mixtures. J Transport Res Rec (TRB) 1891:48–54
Kringos N, Scarpas T, Kasbergen C, Selvadurai P (2008) Modelling of combined physical-mechanical moisture-induced damage in asphaltic mixes, Part 1: governing processes and formulations. Int J Pavement Eng 9(2):115–128
Mehrara A, Khodaii A (2013) A review of state of the art on stripping phenomenon in asphalt concrete. Constr Build Mater 38:423–442
MoRTH (2013) Specification for road & bridge works, fifth revision. IRC publication, Ministry of Road Transport & Highways, New Delhi, India
Lu Q, Harvey JT (2008) Inclusion of moisture effect in fatigue test for asphalt pavements. In: Transportation and development innovative best practices, pp 498–504
Shatnawi S, Nagarajaiah M, Harvey J (1995) Moisture sensitivity evaluation of binder-aggregate mixtures. Transp Res Rec 1492:71–84. TRB, National Highways Research Council, Washington, D.C.

A New Approach to Determine Absorption Water of Reclaimed Asphalt Pavement Aggregate (RAP) for the Production of Cold Recycled Mixtures (CRM)

Simone Raschia[1(✉)], Shalu Panwar[2], Tushar Chauhan[2], Alan Carter[1], Andrea Graziani[3], and Daniel Perraton[2]

[1] Construction Engineering Department, École de Technologie Supérieure (ÉTS), Montréal, Canada
simone.raschia.1@etsmtl.net
[2] Department of Civil Engineering, Indian Institute of Technology, Bombay, India
[3] Department ICEA of Civil and Building Engineering, and Architecture, Polytechnic University of Marche, Ancona, Italy

Abstract. Absorption ability of aggregates is one of the most important material characteristics in the perspective of a bituminous mixture mix design. In particular, this aspect gains more importance in the framework of cold recycled mixtures (CRM). Since such materials are produced at ambient temperature, the workability is ensured by the employment of bitumen emulsion and water. As a consequence, the water absorption of the aggregate phase needs to be clearly stated, in order to know the amount of effective water which will affect workability and mechanical properties of the mixtures. The determination of absorption water is clearly described in international standards (ASTM and European Standard). Nevertheless, both practices leave some aspects unclear, leading to misleading results if applied to the same material. In this paper, a new approach to determine the water absorption of reclaimed asphalt pavement (RAP) used to produce CRM is proposed. Results showed that both standards used in this work gave different absorption water values, making difficult to assess which one is reliable. The new approach highlighted the possibility to evaluate in a more precise and scientific way the absorption water of RAP aggregate, studying the volumetric changes at different water content under compaction.

Keywords: Cold recycled mixtures · Bitumen emulsion · Absorption water · Volumetric properties

1 Introduction

In the production of hot mix asphalts (HMA), the porous nature of the mineral aggregates plays an important role on the definition of the amount of bituminous binder to be used. Whereas on one side, the bitumen generally enhances workability and laydown, on the other hand, a certain percentage is absorbed by the aggregates surface

M. Pasetto et al. (Eds.): ISAP APE 2019, LNCE 48, pp. 367–375, 2020.
https://doi.org/10.1007/978-3-030-29779-4_36

and it does not result as effective in the mixture produced. The definition of the real bitumen absorption by the aggregates is not simple, and many researchers and standards correlated this value to the water absorption of the aggregates (Cominsky et al. 1994; Kandhal and Khatri 1991). This parameter is indeed easier to calculate, thanks to the presence of various standards (ASTM C127, ASTM C128, EN 1097-06, LC 21-065, LC 21-066 and LC 21-067).

The issue with the procedures proposed by the standards is the high impact of the operator experience. The major part of the process for the determination of the absorption water is extremely subjective, and the given indications are purely based on visual analysis (Kasemchaisiri and Tangtermsirikul 2007; Rodrigues et al. 2013). This aspect gets importance in the framework of cold recycled mixtures (CRMs) mix design (Grilli et al. 2016). Such materials need high amount of water to be workable and compactable at ambient temperature, to allow a good dispersion of bitumen and to react with hydraulic binders, if present (Cardone et al. 2014; Grilli et al. 2012). This motivates the need for a clear procedure, with physical meaning, for the determination of the water absorption by the aggregate phase. In this way, especially in terms of CRM, a better estimation of the effective water in the mix can be obtained.

The objective of this study is to propose a different approach to measure the water absorption of a typical CRM granular blend, and to compare the results with the ones obtained following two reference standards.

2 Materials and Methodology

2.1 Materials

The aggregate blend studied in this work is composed by 94.4% of reclaimed asphalt pavement (RAP) and 5.6% of limestone filler. Gradations of RAP aggregate and the resulting mixture are illustrated in Fig. 1. As it can be seen, the target gradation is close to the maximum density curve with exponent 0.45.

The main characteristics of the RAP aggregate used are listed in Table 1.

2.2 Methodology

The experimental program was divided in three steps. In the first two steps, the water absorption was measured only on the RAP aggregates following two standards, the Quebec Standards and the European standard, respectively. In the third phase, the absorption water of the aggregate blend (RAP and filler) was stated through a volumetric analysis on compacted specimens.

The Quebec standard requires to calculate the absorption water on three different aggregate fractions: 0/2.5 (LC 21-065), 2.5/5 (LC 21-066) and $d \geq 5$ mm (LC 21-067). Initially, the three samples are immersed in water for at least 24 h to allow saturation of the pores. After this step, the procedure to determine the three absorption values is different.

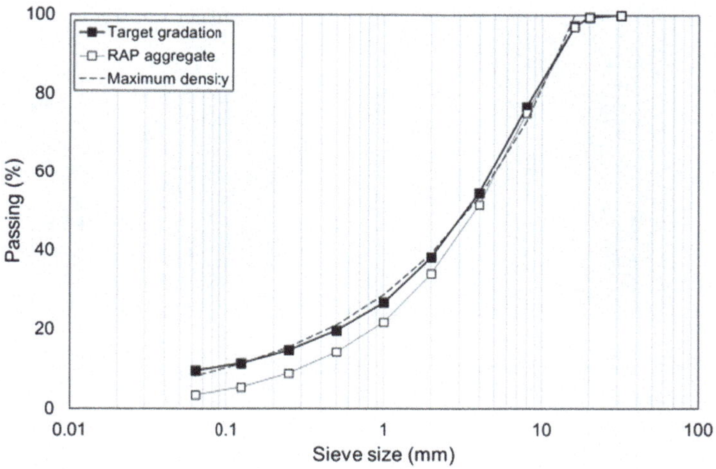

Fig. 1. Gradation of the studied granular material

Table 1. RAP aggregate properties

Property	Standard	Unit	Value
Binder content	ASTM D6307	%	5.51
Nominal maximum particle dimension	ASTM D448-03	mm	16
Maximum specific gravity	ASTM C127-128	–	2.482

Regarding the fraction 0/2.5, as much water as possible is removed manually from the aggregates, paying attention to not lose any particle. The sample is then spread in a uniform layer and undergoes a moderate hot air system while it is continuously mixed, to accelerate the drying process. To establish the saturated surface dried condition (for which the absorption water is calculated), the cone test is performed during the drying process. The test consists in pouring some material inside a standard cone placed on a clean surface. Afterwards, a light compaction is applied by means of a small hammer, applying 25 blows. The cone is raised and the material is visually examined. The standard establishes that "if there is still surface moisture, the fine aggregate will keep the cone shape. On the contrary, when the fine aggregate collapses slightly, it means that it has reached the desired saturated surface dried condition". Figure 2a and b show the test setup.

The saturated surface dried condition on the fraction 2.5/5 is obtained initially draining a sample of the material, and then wrapping it in a dry cloth to have the shape of a ball. It must be shaken until the cloth is completely damp. This step must be repeated with other dry cloths until after shaking they do not appear quite dry. At this point, the saturated surface dried condition is reached.

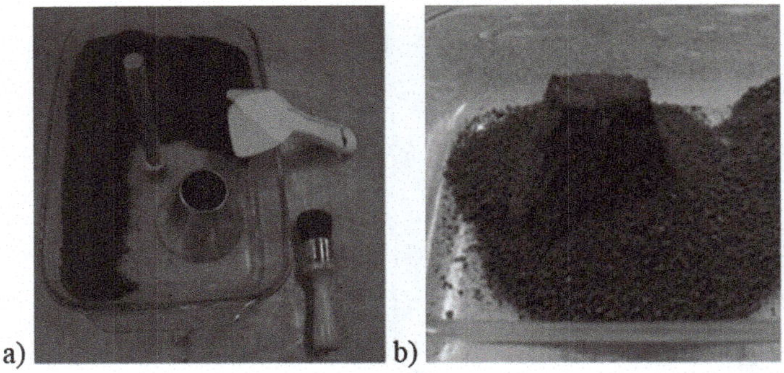

Fig. 2. Test performed on 0/2.5 fraction in accordance to LC 21-065: (a) Cone test; (b) Saturated surface dried condition

For the fraction with d $\geq$ 5 mm, the sample needs to be drained and placed on a dry cloth. The aggregate surface is slightly wiped, and when the cloth is not able to dry anymore, the material is moved on another dry absorbent cloth. The aggregates are spread in a layer with thickness equal to one grain and exposed to air. The saturated surface dried condition is reached when every visible water film is disappeared, but the aggregates still have a humid appearance.

The European standard EN 1097-06 requires to calculate the absorption water on two fractions of the aggregate: 0/4 and d $\geq$ 4 mm.

Regarding the fraction 0/4, the procedure is very similar to the one described in the LC 21-065. The same cone test is performed with 25 blows. The only difference is in the visual analysis of the material once the cone is lifted. In fact, the standard gives a figure as a reference for the saturated surface dried condition (Fig. 3a and b).

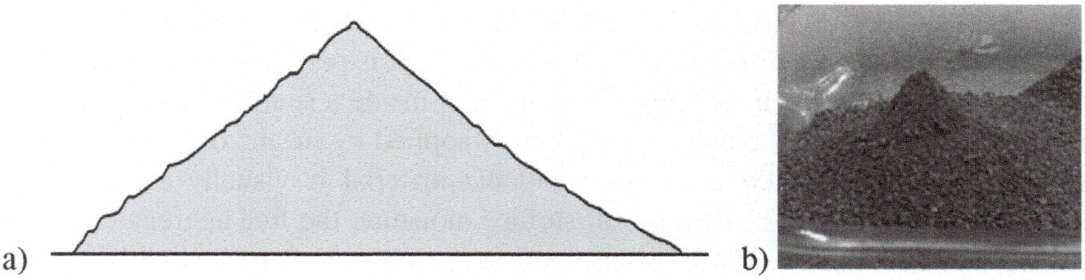

Fig. 3. Saturated surface dried condition in accordance to EN 1097-06: (a) Reference; (b) Sample after the cone test

The procedure to determine the absorption water on the fraction d = 4 mm follows exactly the same steps as the LC 21-067.

It is possible to characterize the aggregate components of a granular blend considering the volumetric proportioning of the different elements. The volume of the aggregates calculated refers to the oven-dried aggregates, hence to their apparent density, which volume considers the aggregate phase and the impermeable voids. The permeable voids on the surface of the aggregates are responsible of the water absorption, which will be included in the aggregate particle bulk volume.

Figure 4a shows the volumetric representation of virgin or reclaimed aggregates, whereas Fig. 4b shows the volumetric representation of reclaimed asphalt aggregates.

As long as the water fills the permeable voids on the surface of RAP aggregates, it does not contribute to decreasing the internal friction among aggregate particles when mixed, hence, if compaction is performed, no densification occurs. In other terms, the packing rate of the granular blend when no water is added and when the absorption water is added should be the same. On the contrary, when the water is higher than the absorption water, the lubricant effect takes place, leading to an increase of the density. So:

$$\text{If } V_{air,DRY} = V_{air,Wi} - \text{no densification (saturation of permeable voids)} \quad (1)$$

$$\text{If } V_{air,DRY} > V_{air,Wi} - \text{densification} \quad (2)$$

where $V_{air,DRY}$ is the volume of air in the dry aggregate blend and $V_{air,Wi}$ is the volume of air in the aggregate blend when the ith amount of water is added.

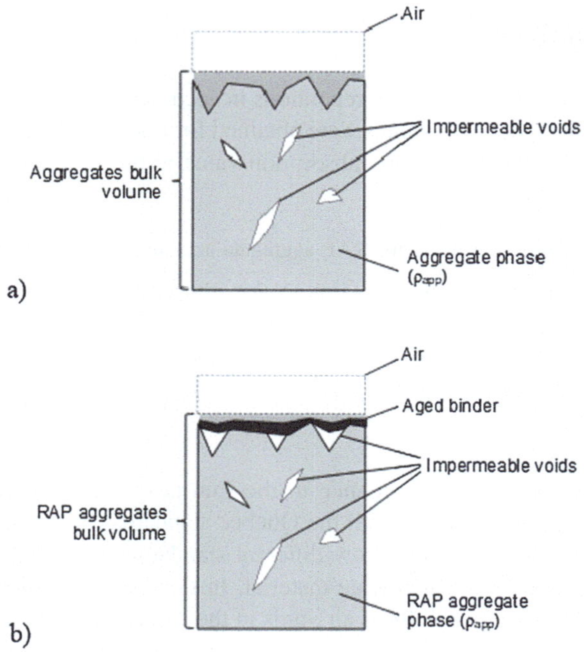

Fig. 4. Water absorption and volumetric representation for: (a) virgin aggregates, (b) RAP aggregate

The absorption water W_{abs} is defined as the ith amount of water that gives:

$$V_{air,Wi+1} > V_{air,Wi} \tag{3}$$

In this study, $i = \{0\%, 0.5\%, 1\%, 1.5\%, 2\%, 2.5\%, 3\%, 4\%\}$, referring to the dry RAP aggregate weight. The filler is considered as not absorbent.

The dry aggregates were initially mixed with the ith amount of water and sealed in a plastic bag overnight, to allow a good humidification. For each i value, two specimens were compacted with a Superpave Gyratory Compactor (SGC) by using an undrained mould with D = 100 mm, 600 kPa of constant pressure, gyration rate of 30 rpm and external inclination angle of 1.25°. Compaction was performed at 150 gyrations.

During and after compaction, the air volume V_{air} inside the mixture was calculated as:

$$V_{air} = \frac{V - (V_{,RAP} + V_{,FILL} + V_{,W})}{V} \cdot 100 \tag{4}$$

where V is the specimen's volume, $V_{,RAP}$ is the dry volume of the RAP aggregates (referred to apparent density), $V_{,FILL}$ is the volume of filler, $V_{,W}$ is the volume of water.

3 Results Analysis

Table 2 shows average values of two repetitions in accordance to the Quebec standards. It is highlighted that quite high values were obtained for fractions 0/2.5 and 2.5/5, which at the end affected the average water absorption value of the RAP aggregate.

Table 2. Absorption water for RAP aggregate according to Quebec standards

Fraction	Standard	Unit	Value	Proportion	Average
0/2.5	LC 21-065	%	3.60	37.7	2.85
2.5/5	LC 21-066	%	3.56	21.7	
d > 5 mm	LC 21-067	%	1.79	40.6	

Table 3 shows results in accordance to the European standard (two repetitions). Compared to the results obtained with the Quebec standards, the average value of the mix is much lower. This indicates how different standardized procedures give totally different results for the same aggregate material. In this case, the difference is 250%.

Figure 5 shows average results of air voids in the aggregate blend after compaction with different water contents (2 specimens each amount), whereas Table 4 shows the average air voids loss at each water addition (0.5%-step addition) and at 1%-step water addition.

Table 3. Absorption water for RAP aggregate according to European standards

Fraction	Standard	Unit	Value	Proportion	Average
0/4	UNI EN 1097-06	%	1.05	49.6	1.14
d > 4 mm		%	1.22	50.4	

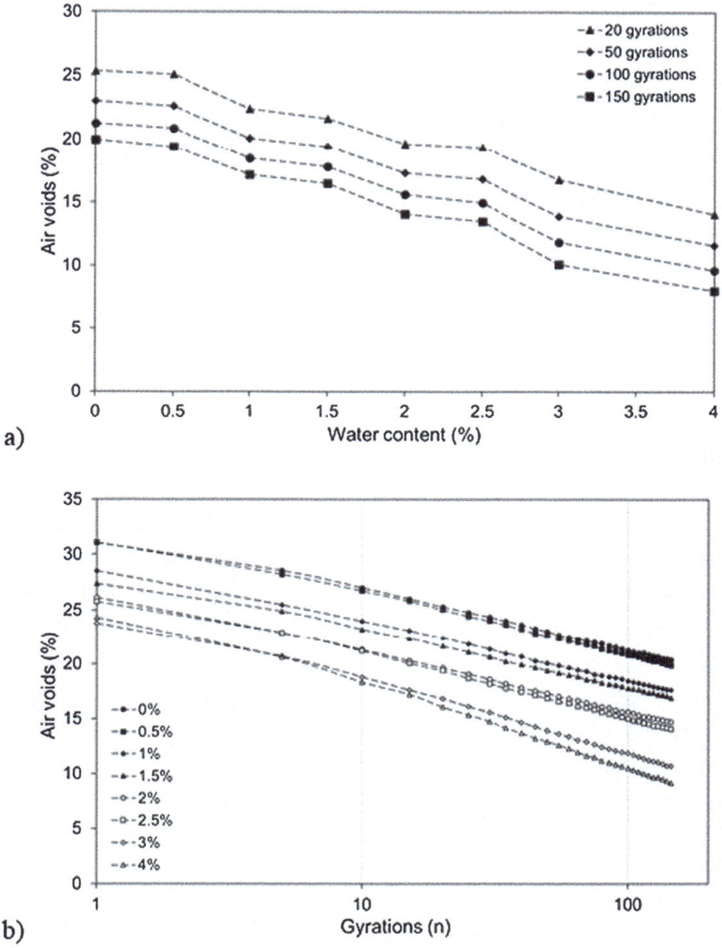

Fig. 5. SGC compaction results: (a) effect of water content on air voids, (b) effect of number of gyrations on air voids

In general, it can be observed that adding water, the aggregate blend gets compacted in any case, regardless the compaction energy considered. Adding 1% of water by mass of RAP aggregate to the mix, an air voids decrease between 2.78 and 2.92 is already visible, at all the compaction energies (Table 4). This range of loss in the air voids is fairly repeated at each addition of water, until 4%. Hence, these results show that densification of the granular mix occurs with only 1% of water by RAP aggregate mass, whereas with 0.5% almost no loss in the air voids was observed. As a consequence, it is reasonable to assume that the amount of water absorption is between 0.5% and 1% by mass of RAP aggregate.

Table 4. Air voids loss at different compaction energies and at different water contents

	Water contents (%)						
	0–0.5	0.5–1	1–1.5	1.5–2	2–2.5	2.5–3	3–4
20 gyr	0.26	2.66	0.76	1.99	0.25	2.56	2.73
50 gyr	0.40	2.55	0.65	2.09	0.47	2.94	2.29
100 gyr	0.41	2.37	0.67	2.17	0.63	3.14	2.20
150 gyr	0.55	2.23	0.68	2.39	0.56	3.41	2.09
	0–1		1–2		2–3		3–4
20 gyr	2.92		2.75		2.81		2.73
50 gyr	2.95		2.74		3.41		2.29
100 gyr	2.78		2.84		3.77		2.20
150 gyr	2.78		3.07		3.97		2.09

4 Conclusions

This study proposes a new methodology to evaluate the absorption water of a granular blend. Standards in Quebec and Europe evaluate such parameter by visual analysis or empirical tests. The approach proposed studies the phenomenon from a physical point of view, being at the same time rather easy to perform. The following conclusions can be drawn:

- Quebec standards require to calculate the water absorption on three different fractions of the material. The RAP aggregate considered in this study is characterized by an average water absorption of 2.85%;
- European standards establish to measure the water absorption on two different fractions of the material. For the RAP aggregate studied, the average water absorption was 1.14%;
- The volumetric analysis compared the air void content only in the granular mix under compaction with the air voids content obtained with gradual addition of water. In this manner, a critical water content that led to a significant decrease in the air voids was estimated to be between 0.5% and 1% by mass of RAP aggregate, regardless the compaction energy applied.

From these results, the European standard and the volumetric analysis are able to predict closer values than the Quebec standards. However, additional studies are recommended to assess the repeatability of the test, as well as a validation on standard virgin aggregates for which the absorption water is well-established.

References

ASTM (2015a) Standard test method for relative density (specific gravity) and absorption of coarse aggregate

ASTM (2015b) Standard test method for relative density (specific gravity) and absorption of fine aggregate

Cardone F, Grilli A, Bocci M, Graziani A (2014) Curing and temperature sensitivity of cement–bitumen treated materials. Int J Pavement Eng 16(10):868–880. https://doi.org/10.1080/10298436.2014.966710

Cominsky RJ, Huber GA, Kennedy TW, Anderson M (1994) The superpave mix design manual for new construction and overlays. Strategic Highway Research Program Washington, DC

EN 1097-6: Test for mechanical and physical properties of aggregates - Part 6: Determination of particle density and water absorption (2013)

Grilli A, Graziani A, Bocci M (2012) Compactability and thermal sensitivity of cement–bitumen-treated materials. Road Mater Pavement Des 13(4):599–617. https://doi.org/10.1080/14680629.2012.742624

Grilli A, Graziani A, Bocci E, Bocci M (2016) Volumetric properties and influence of water content on the compactability of cold recycled mixtures. Mater Struct 49(10):4349–4362. https://doi.org/10.1617/s11527-016-0792-x

Kandhal PS, Khatri MA (1991) Evaluation of asphalt absorption by mineral aggregates

Kasemchaisiri R, Tangtermsirikul S (2007) A method to determine water retainability of porous fine aggregate for design and quality control of fresh concrete. Constr Build Mater 21 (6):1322–1334. https://doi.org/10.1016/j.conbuildmat.2006.01.009

LC 21-066: Détermination de la densité et de l'absorption du granulat fin de classe granulaire d/D Québec, M. d. T.-G. d. (2012a)

LC 21-067: Détermination de la densité et de l'absorption du gros granulat Québec, M. d. T.-G. d. (2012b)

LC 21-065: Détermination de la densité et de l'absorption du granulat fin Québec, M. d. T.-G. d. (2012c)

Rodrigues F, Evangelista L, de Brito J (2013) A new method to determine the density and water absorption of fine recycled aggregates. Mater Res 16(5):1045–1051. https://doi.org/10.1590/s1516-14392013005000074

Effect of Air Void Topology on the Hydraulic Conductivity and Clogging Properties of Pervious Asphalt Roads

Alavaro Garcia(✉), Mustafa Aboufoul, Kassra Gerami, and Frank Asamoah

Department of Civil Engineering, Nottingham Transportation Engineering Centre (NTEC), University of Nottingham, Nottingham NG7 2RD, UK
alvaro.garcia@nottingham.ac.uk

Abstract. This paper explains the effect of air void topology on the hydraulic conductivity and clogging properties of pervious asphalt. Cylinders of asphalt mixtures with a single type of aggregates and five gradations have been manufactured, Computed Tomography (CT) scanned and, transparent resin blocks with equivalent pore structure have been 3D printed to allow the visual inspection and quantification of the clogs. Geometrical properties, such as the macroporosity, air void diameter, Euler number and tortuosity, have been recorded from the CT scans. Furthermore, the hydraulic conductivity and clogging susceptibility of the asphalt cylinders and 3D-printed resin blocks have been measured.

It has been found that the air void diameter and porosity are the parameters that most affect the movement of water and solids through the porous asphalt. In the mixtures studied in this paper, these two parameters can be directly correlated to the tortuosity and Euler number of the asphalt mixtures. Higher macroporosity implies higher hydraulic conductivity so as lower clogging susceptibility.

In order to use the results of this paper to design new pervious pavements, further research is required where the clogging properties of mixtures with a wider range of aggregate gradations, maximum sizes and geometries is studied.

Keywords: Pervious asphalt · Porosity · Pore geometry · Hydraulic conductivity · Clogging · 3D printing

1 Introduction

Porous asphalt's primary use is to increase the drainage capacity and storm water retention of asphalt pavements (Fwa et al. 2014). The Interconnecting air void structure acts as a drainage channel for storm water, serving to reduce surface runoff and ponding. Benefits include improved driver safety, less surface spray from vehicles and improve visibility of road markings (Taylor et al. 2009). A further feature of porous asphalt is the infiltration reservoir that sits beneath the porous surface layer, providing a storm water retention system. The overall result is that porous asphalt serves as a flood alleviation measure (Ferguson 2005). In addition, vehicle noise reduction is the other main use of porous asphalt in many urban areas (Nicholls 1997), as the interconnecting

M. Pasetto et al. (Eds.): ISAP APE 2019, LNCE 48, pp. 376–384, 2020.
https://doi.org/10.1007/978-3-030-29779-4_37

air void structure acts to dampen both the pitch and amplitude of the tyre/road noise, that is generated from traffic moving at more than 30 km/h (Sandberg 1987).

One of the main problems preventing the wide-spread use of porous asphalt is that it tends to clog. Physical clogging occurs when fine particles such as soil, tire dust, or stripped fine bitumen coated aggregate (Hassan et al. 2015) block the surface texture and internal pore passages of the porous asphalt, and in doing so, the permeability and noise reduction properties of the pavement are reduced (Fwa et al. 2014).

A porous asphalt pavement can become fully-clogged over a period of 12 years, however, hydraulic conductivity is seen to decrease long before this timeframe (Yong et al. 2013).

Although effective cleaning techniques have been devised, see Matsuda et al. (1998), for an example, this represents an ongoing maintenance cost, on top of the higher capital construction costs of porous asphalt compared to dense asphalt mix. As a result, there has been much research into the causes of clogging and the effect it has on the hydraulic conductivity of porous pavements.

Previous research has concluded that the size and composition of the clogging particles has very high importance to determine the rate of clogging of an asphalt mixture. For example, it is possible that fines and clay minerals become attached to the aggregates and mastic particles of the porous asphalt, which build up to clog the air voids over time (Hassan et al. 2015). On the other hand, it has been found that particles between 600 μm and 1.18 mm are the most effective clogging the porous asphalt (Guwe et al. 1999). The reason for this can simply be that this is the one of the most common air pore sizes in porous asphalt mixture. Furthermore, it has been previously concluded that higher concentration of clogging particles in the runoff water will increase the susceptibility of pervious asphalt to clogging (Hassan et al. 2015).

In the porous asphalt the air voids are not normally homogeneously distributed within the structure; at the surface of the asphalt, the size of the air voids tends to be smaller than at higher depths. As a result, the air void size and connectivity increase with depth into the road, and the surface of porous asphalt clogs first (Hassan et al. 2015).

Fwa et al. (2014), who conducted a laboratory experiment to examine the drainage and clogging characteristics of porous asphalt and pervious concrete with porosity ranging from 10% to 25%, along several cycles of clogging, concluded that materials with higher porosity maintained a higher permeability through the clogging cycles. Furthermore, these authors proposed a deterioration trend to describe the evolution of hydraulic conductivity (k) over a period of clogging cycles. The exponential relationship found is as follows:

$$k = a\,e^{-bN}. \tag{1}$$

Where, N is the number of clogging cycles and, a and b regression coefficients that depend of the air void properties, such as air void content and diameter, and on the concentration of clogging material in the runoff water.

Aboufoul and Garcia (2016), concluded that the governing parameters affecting the asphalt's hydraulic conductivity are the macroporosity, and the average air void diameter. On the other hand, Aboufoul et al. (2018), concluded that the hydraulic conductivity, average air void diameter and macroporosity are also controlled by the aggregate maximum size and geometry. For this reason, it is easy to imagine that from two asphalt mixtures with equal initial hydraulic conductivity, the one with the smallest air pores will show the highest clogging susceptibility.

The aim of this paper is to give an introduction of the influence of air void properties, such as air void content, and diameter on the clogging susceptibility of porous asphalt mixture. In addition, we will introduce the use of 3D printing to reproduce the porosity and visualize the clogged air voids without the need for CT-Scanning.

2 Materials and Methods

2.1 Clogging Material Gradation

The size of the clogging material was selected based on research from Hassan et al. (2015), who measured it from a highway location, as it can be seen in Table 1.

Table 1. Clogging material gradation

Sieve size (mm)	Cumulative percentage passing (%)
2.36	89.76
1.18	68.62
0.6	46.52
0.3	30.44
0.15	15.28
0.075	3.9

2.2 Asphalt Composition

Asphalt with target air void content 13%, 17%, 21%, 26% and 29% was manufactured by using bitumen 60/40 pen and crushed limestone aggregates with maximum size 14 mm. 30 cm × 30 cm × 5 cm slabs were built, following the standard BS EN 12697-33. Then, 100 mm cores were cut from the slabs. The internal air void geometry of the cores was characterized by using X-ray Computed Tomography scans (Table 2).

Table 2. Asphalt composition

Sieve size (mm)	Target air void content				
	13%	17%	21%	26%	29%
10–20 mm	63	67	75	81	86
4–10 mm	17	16	13	8	8
Dust: 4–0.063 mm	20	17	12	11	6
Filler < 0.063 mm	4	4	3	3	2
Binder content (%)	5	4.7	4.5	4.3	4

2.3 Impression of Resin Blocks with Realistic Pore Geometries

The air void topology of the test samples was captured by means of CT-Scanning the asphalt cores, using the procedure described in Aboufoul and Garcia (2016). By using this procedure, 3D models of the pores were generated. ImageJ was used to quantify the topological properties of the air voids. The dimensions of the blocks is 8 cm × 2.5 cm × 5 cm.

After, resin blocks with imprinted air voids were 3D printed using the methodology described in Aboufoul et al. (2018). See Fig. 1 for a view of the printed test samples. This resin had a wetting angle of 68.7° ± 7.67, which will be equivalent to the combined wetting angle of bitumen, 93.04°, (Ahmed 2011) and limestone, 59.0°, (Ishutov et al. 2017), in mastic.

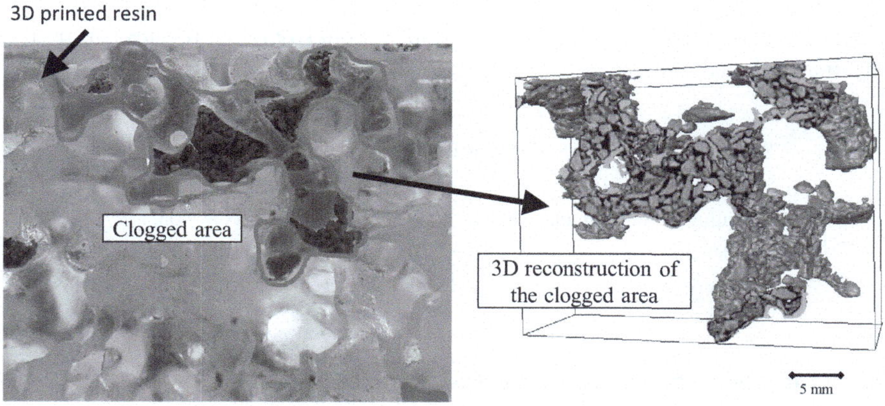

Fig. 1. 3D printed resin test sample and detail of a clog.

2.4 Hydraulic Conductivity Measurements

Hydraulic conductivity of the asphalt was measured using the Florida falling head method and the hydraulic conductivity was calculated based on the Darcy law as the average of 3 test samples. The methodology has been defined in Aboufoul et al. (2018).

2.5 Clogging Measurements

Clogging was measured in the asphalt test samples. The first step is to measure the hydraulic conductivity. Then, 1 g or 4 g of clogging material were sprinkled on the surface of the 3D printed resin blocks and real asphalt test samples, respectively, and the hydraulic conductivity was measured after. This process was repeated 8 times. Each value was obtained from the average of three measurements.

After each clogging cycle, the transparent samples were shone through the samples to be able to see the clogs, see Fig. 1. These were recorded using the software ImageJ and the total area of clogs was determined.

2.6 Measurement of Macroporosity, Pore Diameter, Euler Number and Tortuosity

The macroporosity or air void content, pore diameter, Euler number, and tortuosity were measured using the BoneJ and AnalyzeSkeleton plugins, from ImageJ, as it is described in Doube et al. (2010) and Arganda-Carreras et al. (2010).

3 Results and Discussion

3.1 Geometrical Properties of Air Voids

In Table 3 the geometrical properties of the air voids, obtained through the analysis of the asphalt's CT-Scans, are represented. These properties are highly interrelated and show linear relationships among them (see Table 4). The reason for the uniformity of this set of data is that the same geometry and maximum size of aggregates was used for every material analysed.

Table 3. Geometrical properties of the air voids

Macroporosity (%)	13.6	15.8	20.9	21.7	28.6
Average pore diameter (mm)	1.95	2.21	2.59	2.68	3.03
Euler number	−37	−140	−254	−361	−499
Tortuosity	1.39	1.34	1.25	1.15	1.06

3.2 Hydraulic Conductivity of the Test Samples

Figure 2 shows hydraulic conductivity data for unclogged asphalt materials, versus the macroporosity for the 5 materials analysed. These results are in the range of those reported previously by Aboufoul and Garcia (2016), who developed a function based on a statistical model that the hydraulic conductivity to the macroporosity. Because these results have been obtained from aggregates with the same maximum size and similar geometry, the dispersion of results is minimal.

3.3 Clogging of Porous Asphalt

Figure 3(a) shows how the hydraulic conductivity of the test samples analysed deteriorates with the clogging cycles. Similar deterioration curves have been observed previously by Fwa et al. (2014) and Hassan et al. (2015), with the former researcher concluding that the hydraulic conductivity decays with time following Eq. (1). In the figure it can be observed that the clogging of mixtures with higher initial hydraulic conductivity is slower than that of mixtures with lower initial hydraulic conductivity.

To be able to compare the decay of the hydraulic conductivity in the different materials analysed, the authors have developed the concept of Clogging Level (*CL*), which is defined as the relationship between the logarithm of hydraulic conductivity after each clogging cycle and the logarithm of the initial hydraulic conductivity.

Table 4. R^2 coefficients of the linear relationships between the topological properties of asphalt and the clogging ratio.

	Clogging ratio	Macroporosity	Pore diameter	Euler number
Clogging ratio				
Macroporosity	0.81			
Pore diameter	0.83	0.97		
Euler number	0.81	0.96	0.98	
Tortuosity	0.84	0.95	0.96	0.99

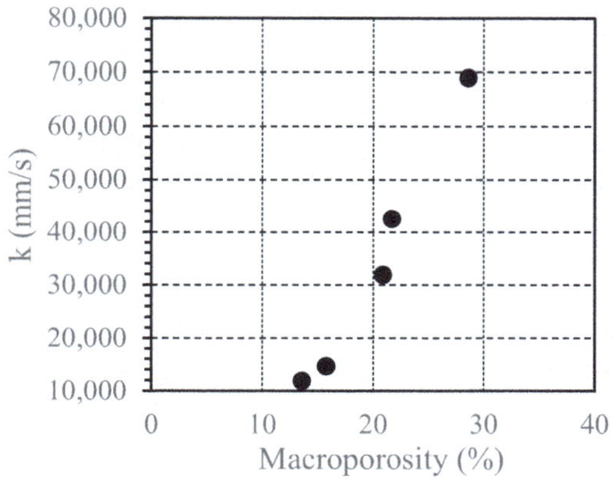

Fig. 2. Macroporosity vs hydraulic conductivity for unclogged samples.

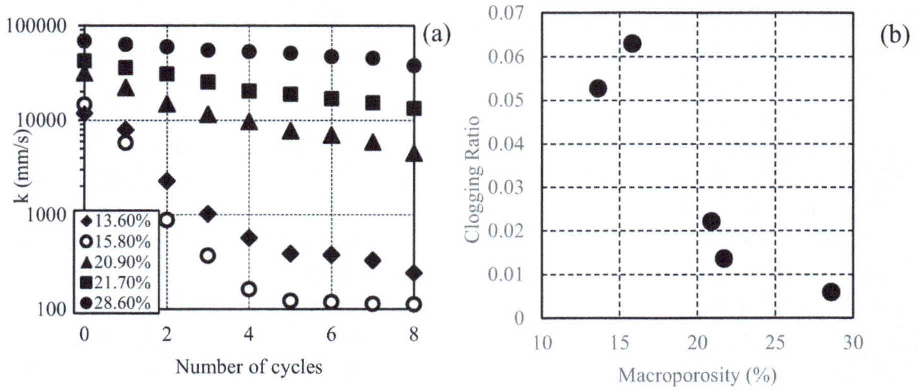

Fig. 3. (a) Macroporosity vs hydraulic conductivity during the clogging cycles. (b) Clogging ratio vs macroporosity.

Finally, the Clogging Ratio (*CR*) of each of the asphalt mixtures analysed has been defined as the absolute value of slope of the fitting line that relates the clogging levels to the number of clogging cycles. If the clogging ratio is higher, the material clogs faster, while if the clogging ratio is smaller, the materials does not clog so fast.

From Fig. 3(b), it can be concluded that the clogging ratio decreases with increases in the air void content. The reason for this is that the size of the air voids in asphalt increases with the macroporosity. As a result, the pores in the asphalt mixture are not as effective retaining the particles in the runoff water, which are smaller than the pores. The clogging ratio could eventually become zero when the size of all the air voids is higher than the size of every clogging particle.

3.4 Effect of the Air Void Topology on the Clogging Ratio

To understand the influence of the air void topology, such as macroporosity, pore diameter, and Euler number on the clogging ratio, the values for all the variables have been represented against each other and fitted using a linear approach. For example, all the clogging ratios have been represented against their respective macroporosities. The coefficient of determination, R^2, of the linear relationships can be found in Table 4.

It can be observed that for the materials studied in this paper there is an equally high correlation between all the parameters. For this reason, it is not possible to conclude which is the most influential value for the Clogging Ratio, although the authors believe that the air void diameter will be the most influential parameter, because clogging particles will be retained in air voids with similar diameter. In order to answer this question, it will be necessary to obtain additional results, for example, of porous asphalt made with a wider range of maximum aggregate sizes, gradations, and shapes.

3.5 Distribution of the Clogs in the Asphalt Mixture

Figure 4(a) shows the percentage of clogged air voids after 8 clogging cycles. It can be observed that the percentage of clogged voids increases with macroporosity up to the 20.9% sample. After, the percentage of clogging reduces. The reason is that when the macroporosity increases above a certain size, the filter efficiency of asphalt mixture reduces, because the air void diameter will be bigger than the clogging particles size. This experiment was done using the transparent resin blocks. The data shown is an average taken from two repeats of the clogging experiment.

To prove this, Fig. 4(b) shows that in the mixtures with less than 21% air void content, most of the clogging occurs in the upper 10 mm of the test samples. Thus, the voids become so clogged that any further transport of material through the sample is prevented. As porosity increases, the clogging material can pass with greater ease through the sample and therefore total clogging of pores increases up to a maximum, which corresponds to approximately 21% of air voids content. For this size of clogging material, this is the mixture with highest clogging efficiency. After, increasing porosity enables the clogging material to wash through the deeper sections of the sample and so the overall clogging percentage reduces.

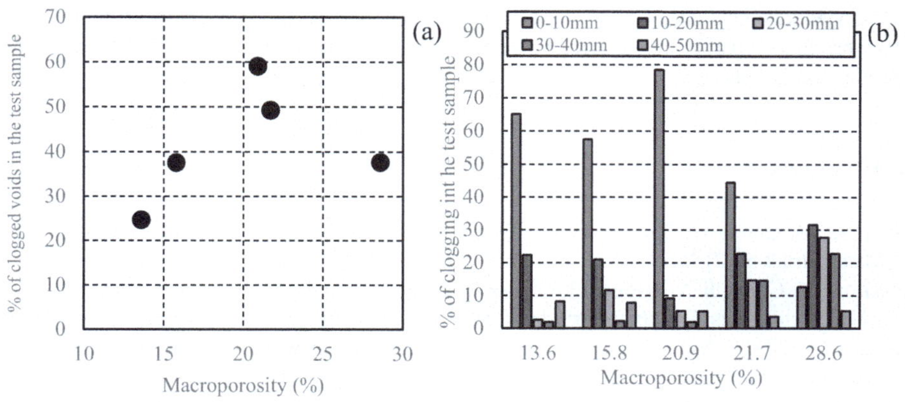

Fig. 4. (a) Macroporosity vs percentage of clogged voids in the test samples. (b) Location of the clogs in the test samples analysed.

4 Conclusions

In the present study, the clogging properties of porous asphalt made with constant aggregate geometry and maximum aggregate size, and a range of aggregate gradations have been studied. Furthermore, to analyse the percentage of clogged pores and location of the pores within the asphalt, the air void geometries have been captured by means of CT scanning and image analysis and reproduced in a transparent resin material by means of a 3D printer. Based on the research discussed in this paper, the following conclusions can be drawn:

The pores in asphalt mixture work as a sieve for the particles on the road surface. They retain the particles of approximately the same size, while smaller particles may flow through the pores.

From the information provided in this paper, it is not possible to determine the topological property of air voids, such as tortuosity, pore size diameter, macroporosity, or Euler number that influence clogging the most, although the authors theorize that the air void diameter is the most influential factor for clogging. In order to determine the most influential factors that control clogging of asphalt mixture, a more extensive study is required, which will involve the study of asphalt with a wide range of aggregate properties and gradation.

For the most commonly used porous asphalt, which has 21% air void content, the clogging materials tend to accumulate in the first 10 mm from the road's surface.

The main recommendation of this article is that before any pervious asphalt pavement is built, the size of the clogging particles is analysed, and the asphalt mixture is designed to avoid that the particles get retained in the pores. In addition, the development of pervious pavements with anticlogging properties may require that the porosity changes with the distance to the surface, for example, to maximise the retention properties of asphalt; therefore, the development of more precise design and manufacturing techniques for pavements may be required in the future.

Acknowledgements. The authors would like to acknowledge the University of Nottingham, which has funded part of this research through the Dean of Engineering Research Scholarship for International Excellence.

References

Aboufoul M, Garcia A (2016) Factors affecting the hydraulic conductivity of asphalt mixture. Mater Struct 50(2):116

Aboufoul M, Chiarelli A, Triguero I, Garcia A (2018) Virtual asphalt to predict road's air voids and hydraulic conductivity. https://doi.org/10.20944/preprints201810.0710.v1. ppr: ppr61080

Ahmad N (2011) Asphalt mixture moisture sensitivity evaluation using surface energy parameters. Ph.D. thesis in Civil Engineering, University of Nottingham

Arganda-Carreras I, Fernandez-Gonzalez R, Munoz-Barrutia A, Ortiz-De-Solorzano C (2010) 3D reconstruction of histological sections: application to mammary gland tissue. Microsc Res Tech 73(11):1019–1029

Bacher M (2013) 3D-printing of undisturbed soil imaged by X-ray. Master's thesis in Environmental Science, Upsala University

BS EN 12697-33, B.S.I.B. (2003) Bituminous Mixtures, Test Methods for Hot Mix Asphalt: Specimen prepared by roller compactor

Doube M, Kłosowski MM, Arganda-Carreras I, Cordeliéres F, Dougherty RP, Jackson J, Schmid B, Hutchinson JR, Shefelbine SJ (2010) BoneJ: free and extensible bone image analysis in ImageJ. Bone 47:1076–1079

Fwa T, Lim E, Tan S (2014) Comparison of permeability and clogging characteristics of porous asphalt and pervious concrete pavement materials. Transp Res Rec J Transp Res Board 2511:72–80

Guwe Y, Fwa T, Tan S (1999) Laboratory evaluation of clogging potential of porous asphalt mixtures. Transp Res Rec J Transp Res Board 1681:43–49

Hassan N, Abdullah N, Shukry N, Mahmud M, Yunus N, Putrajaya R, Hainin M, Yaacob H (2015) Laboratory evaluation of clogging behaviour of porous asphalt pavements. J Teknologi 76(14)

Ishutov S et al (2017) Using resin-based 3D printing to build geometrically accurate proxies of porous sedimentary rocks. Groundwater 56(3):482–490

Matsuda T, Inagaki T, Masuyama Y (1998) Investigation and examination concerning function recovery of drainage asphalt pavement. In: 9th Road Engineering Association of Asia and Austalasia (REAAA) Conference, Wellington, New Zealand

Nicholls JC (1997) Review of UK Porous Asphalt Trials, Thomas Telford

Sandberg U (1987) Tire/road noise – studies of the mechanisms of noise generation, methods of measurement and road surface characterization. Linköping Studies in Science and Technology, Dissertations 166, Department of Physics and Measurement Technology, Linköping University, Linköping

Taylor M, Boudreau H III, Huang T, Tatel S, Najafi F (2009) The real world advantages and disadvantages of permeable pavements as roadways. Int J Interdiscip Soc Sci Annu Rev 4 (1):341–354

Yong C, McCarthy D, Deletic A (2013) Predicting physical clogging of porous and permeable pavements. J Hydrol 481:48–55

Fatigue Performance of Bituminous Binders Tested by Linear Amplitude Sweep Test

Krzysztof Błażejowski[1(✉)], Marta Wójcik-Wiśniewska[1], Wiktoria Baranowska[1], Przemysław Ostrowski[1], Radek Černý[2], and Petr Jisa[2]

[1] ORLEN Asfalt sp. z o.o., ul. Łukasiewicza 39, 09-400 Płock, Poland
krzysztof.blazejowski@orlen.pl
[2] Unipetrol výzkumně vzdělávací centrum, a.s., Areál Chempark, 436 70 Litvínov-Záluží, Czech Republic

Abstract. Fatigue cracking is one of the most important phenomena connected to durability of asphalt pavements. The fatigue resistance of asphalt pavements depends on many factors such as asphalt mixture composition and binder properties. Also temperature of asphalt pavement is important, what is directly related to viscoelastic nature of the bituminous binder. The fatigue cracks may occur within the bitumen (cohesive cracking) or at the bitumen-aggregate interface (adhesion cracking). The cohesive cracking resistance can be evaluated by bitumen testing. The Linear Amplitude Sweep (LAS) test has recently been proposed for fatigue characterization of bituminous binders. The LAS test is performed acc. to AASHTO TP 101 standard. This procedure is based on viscoelastic continuum damage approach to predict binder fatigue life as a function of strain level in the pavement. In the research fatigue performance of three types of binders: paving grade, typically polymer modified and highly polymer modified were evaluated. Tests were performed using the LAS method at temperatures: 5 °C, 10 °C, 20 °C, on samples after RTFOT+PAV aging. Obtained results have shown that each group of binders behave in different way which is connected to its internal network or lack of it. Finally, the best fatigue properties have been presented by highly modified bitumen (HiMA). The paper presents description of the LAS method, test results containing comparison of fatigue characteristics of used binders and conclusions.

Keywords: Bitumen · Fatigue · LAS test · VECD · Highly modified bitumen

1 Introduction

The phenomenon of fatigue is one of the most important factors influencing the durability of asphalt pavements. It depends on many factors, e.g. the composition and design of the asphalt mixture as well as the properties of the used bituminous binder. Fatigue cracking may occur within the binder (cohesive cracking) or at the bitumen-aggregate interface (adhesion cracking) (Pereira et al. 2016). Literature describes several methods using DSR for the testing of the fatigue life of bituminous binders. Currently, the most popular one is method developed in 1987 in the USA as a part of the Superpave system. It uses the $|G^*|\cdot\sin\delta$ parameter for the quantitative determination of the bitumen fatigue resistance. The testing is performed in the intermediate

M. Pasetto et al. (Eds.): ISAP APE 2019, LNCE 48, pp. 385–394, 2020.
https://doi.org/10.1007/978-3-030-29779-4_38

temperature which depends on the PG grade of a particular binder. The requirements limit the stiffness of the bitumen to the maximum of 5000 [kPa] (for the S grade) and 6000 [kPa] (for the H, V, E grades). The main disadvantage of this specification is that it does not describe in any way the real fatigue phenomena occurring in the binder, such as: higher strains or different level of the frequency of the applied load (Bahia et al. 2001, 2002).

Another widely applied method allowing to specify the fatigue performance of bitumen is time sweep test. The time sweep test consists of the use of a cyclic load application to a binder sample at a constant frequency and a variable strain amplitude. The time sweep test was developed as part of the NCHRP Project 9-10 (Bahia et al. 2001), in order to improve test methodology described in the AASHTO T315 standard. An advantage of the time sweep test is the possibility to evaluate the influence of bituminous binders self-healing phenomena on the fatigue life of asphalt pavement (Pereira et al. 2016), as well as the high correlation with the fatigue tests results of asphalt mixtures (Bahia et al. 2001).

In view of the on-going discussion in the scientific community about the adequacy of using above described methods, in the USA were conducted works in order to develop a new method for the testing of the fatigue life of bitumens. The result of these endeavors is the method described in the AASHTO TP 101-14 standard - *Estimating Damage Tolerance of Asphalt Binders Using the Linear Amplitude Sweep* (abbreviation: LAS), executed with using DSR. It consists a cyclic loading of the tested sample with a constant frequency and a gradually increasing strain amplitude in order to cause accelerated fatigue damage. The test begins with a frequency sweep performed to specify the initial properties of not damaged material. Subsequently, on the same sample the amplitude sweep test is conducted. The frequency sweep is executed at constant shear strain amplitude of 0.1% during the entire test and at a variable frequency, which increases in the range of 0.2–30.0 [Hz]. The amplitude sweep phase begins with 100 initial cycles of sinusoidal loading at the 0.1% strain and at the frequency of 10 [Hz]. The frequency remains unchanged during the entire test. The amplitude increased along with the step every 1% until it reaches the level of 30%. Each level of the amplitude increase includes 100 cycles applying load onto the sample whereby the cumulatively tested material is subjected to 3100 cycles of loading. As the results of the amplitude sweep test complex shear modulus G*, phase angle δ, a shear stress and the accompanying shear strain in the sample are registered. Fatigue characterization – N_f in the LAS test is calculated at the damage level corresponding to the peak stress response.

The AASHTO TP 101-14 standard does not specify the referential temperature at which the testing should be conducted. It merely states that the LAS test should be performed at an intermediate temperature specified in accordance with the PG functional type of particular binder. However, correct choice of the test temperature is extremely important because it influences the sample damage mechanism during test (Safaei and Castorena 2016). Researchers Soenen and Eckmann (2000), Soenen et al. (2004) conducted tests of bitumen at different temperatures and noted that when the samples were subjected to cyclic loading in DSR at too high temperature, the binder exhibited mainly viscous properties - edge flow was observed. In turn, during testing at too low temperature, adhesion cracking was occurred. Anderson et al. (2001) made similar observations and defined the transition between the real fatigue cracking and an instability flow of the sample, based on the dependence of the fatigue life with the temperature. Safaei and Hintz (2014) evaluated the influence of temperature on the bituminous binders cyclically loaded at a constant strain amplitude during testing in DSR. The main conclusion from their research was that the test temperatures should be chosen in such way that the dynamic shear modulus corresponding to the test temperature and frequency was in the range from 10 to 60 [MPa] to avoid effects of flow and adhesion loss. Safaei and Castorena (2016) distinguished three mechanisms of sample damage, depending on the test temperature: too low temperature - adhesion loss; too high temperature - bulging of specimens; intermediate temperature - cohesive cracking. The test temperatures for fatigue tests, according to the LAS methodology, should be selected in such way that the tested samples fail by only as a result of cohesive cracking.

2 VECD - Viscoelastic Continuum Damage Theory

The LAS test is based on the VECD - *Viscoelastic Continuum Damage* theory. The main advantage of VECD is possibility to prediction of fatigue life of tested materials at any loading amplitude. Asphalt mixtures and bituminous binders present a well-defined relationship between the amplitude of the applied load and the fatigue life (N_f). This correlation may be described as follows (Monismith et al. 1970):

$$Nf = A \cdot (\gamma)^B \tag{1}$$

where *A* and *B* are materials dependent parameters, and γ is strain amplitude.

The VECD theory is based on the Schapery works (Schapery 1975, 1984, 1990). According to that theory for a viscoelastic material, the correlation between the propagation of cracks - *D* - and the work performed - *W* - necessary for their occurrence may be described as follows (Kim et al. 2006):

$$\frac{dD}{dt} = \left(-\frac{\partial W}{\partial D}\right)^{\alpha} \tag{2}$$

where: *W* - work performed (the pseudo strain energy density function); *D* - damage intensity; *t* - time; α - materials dependent constant.

In accordance with the AASHTO TP 101-14 standard, the α parameter is calculated according to the formula: $\alpha = 1/m$, where *m* is the slope of the frequency-shear modulus graph.

In order to estimate the work performed during the occurrence of damages in the sample, the equation developed by Kim et al. (2006) is used, based on the dissipated energy theory:

$$W = \pi \cdot I_D \cdot \gamma^2 \cdot G^* \cdot \sin\delta \tag{3}$$

where: W - dissipated energy; I_D - initial undamaged value of the dynamic shear modulus |G*|; γ - applied shear strain amplitude; G^* - complex shear modulus; δ - phase angle.

Using Eqs. (2) and (3), damage level at time - $D(t)$ (damage accumulation), is calculated as follows (Kim et al. 2006; Johnson 2010; Hintz et al. 2011):

$$D(t) = \sum_{i=1}^{N} \left[\pi I_D \gamma^2 (C_{i-1} - C_i)\right]^{\alpha/1+\alpha} (t_i - t_{i-1})^{1/1+\alpha} \tag{4}$$

where: $D(t)$ - damage accumulation, N - load cycle for which damage accumulation D is calculated, $C(t)$ - material integrity coefficient, described by the equation:

$$C(t) = \frac{G^*(t)}{I_D} \tag{5}$$

The maximum stress - τ_{max} reached by the sample during LAS test is specified as the criterion of the fatigue life. Failure damage level in the sample at this time - $D(f)$ - is described by the correlation:

$$D_f = \left(\frac{C_0 - C_{atpeakstress}}{C_1}\right)^{1/C_2} \tag{6}$$

where: $C_0 = 1$ - initial value of the material integrity; C_1 *and* C_2 - calculation coefficients of the curve fitting (y axis – $log(C_0 - C(t))$, x axis – $logD(t)$).

By solving the system of equations consisting of the expressions: (2), (3) and (4) the final equation describing the fatigue performance can be described as follows:

$$N_f = \frac{f(D_f)^k}{k(\pi C_1 C_2)^\alpha} (\gamma_{max})^{-2\alpha} \tag{7}$$

where: f - frequency [Hz]; k - calculation coefficient, described with the equation: $k = 1 + (1 - C_2)\alpha$, D_f – failure damage level, γ_{max} – maximum set strain.

Applying the VECD theory, it is possible to calculate the number of cycles to failure of the sample for any selected strain amplitude. In this way, it is possible to predict the fatigue life of the tested bitumens at any level of pavement loading.

3 Highly Polymer Modified Binders - HiMA

The HiMA binders are the new type of bitumen modified by more than 7% m/m of SBS block polymers. Such a high quantity of SBS causes that the volume proportions between bitumen and polymer after the modification process are reversed and the final binder is characterized by the reversed bitumen-polymer phase. The volume advantage of the polymer network and its physical continuity gives the binder its unique properties, more similar to the properties of an elastomer than bitumen. One of the main characteristics of the new binder is the significant improvement of the flexibility and high tolerance to increasing tensile strains, as well as the other properties which result from it - fatigue performance, resistance to cracking etc. (Błażejowski et al. 2016). The research and the implementation works of new bituminous binders showed that they are products with above standard functional properties (Błażejowski et al. 2016). Full-scale testing conducted since 2009 on the experimental track in the US (NCAT Pavement Test Track) indicated that the pavement designed with the use of highly modified bitumen is extraordinarily resistant to rutting and fatigue cracking (Timm et al. 2013; West et al. 2012). Due to its properties, highly modified bitumen are particularly suitable for use in situations requiring very high durability, e.g.: asphalt pavements subject to high stresses and strains, courses requiring high resistance to low temperatures, asphalt base courses with very high fatigue durability, e.g. for perpetual pavements.

4 Experimental

4.1 Materials and Test Methods

The goal of this studies was to checked and compared fatigue properties of three different type of bitumens. Fatigue characterization of the binders was conducted using the LAS test in accordance with the AASHTO TP 101-14. All binders were aged in the standard RTFOT and PAV prior to testing. Tests were performed at three temperatures: 5 °C, 10 °C, and 20 °C to allow for assessment of the effect of temperature on fatigue resistance.

The following types of binders were used in the test program:

- Paving grade bitumen 50/70
- Polymer modified bitumen PMB 45/80-55 (SBS polymer content 3–4% m/m)
- Highly polymer modified bitumen PMB 45/80-80 HiMA (SBS polymer content ~7.5% m/m) –polymer - bitumen reversed phase binders.

Table 1 provides a summary of the bituminous binders used.

Table 1. Summary of the bituminous binders used

Properties	Unit	Bitumen type		
		Paving grade 50/70	PMB 45/80-55	PMB 45/80-80 HiMA
Penetration at 25 °C	[0.1 mm]	56	70	59
Softening point	[°C]	50.0	65.9	94.4
Elastic recovery at 25 °C	[%]	–	89	94
Performance grade	[–]	64–22	70–28	94–28
FCCT, $G^*\sin\delta = 5000$ kPa	[°C]	21.3	14.1	11.7
FCCT, $G^*\sin\delta = 6000$ kPa	[°C]	19.4	12.5	10.1
Jnr 3.2 kPa^{-1} MSCR test, 64 °C (samples after RTFOT)	[kPa^{-1}]	2.49	0.63	0.02

4.2 Influence of the Test Temperature on the Sample Damage Mechanism During LAS Test

Safaei and Castorena (2016) described three mechanisms of sample damage depending on the temperature at which the measurements are taken: adhesion loss, bulging of specimens and cohesive cracking. The temperature in the LAS test should be selected so that the product of the initial value of the complex stiffness modulus and the phase angle specified at the frequency of 10 [Hz] is in the linear viscoelasticity range of the tested binder. Acc. to Safaei and Castorena works, the $|G^*|_{LVE}$ range for cohesive cracking should be between $10 < |G^*|_{LVE} < 60$ [MPa]. However, it should be noted that this range does not have to correct for all type of binder, especially for PMB or HiMA. The HiMA binders are characterized by a significantly lower stiffness, what can be observed of the FCCT comparison in Table 1 or the |G*| comparison in Table 2.

Table 2 shows the results of $|G^*|_{LVE}$ taking into account the mechanism of sample damage, depending on the test temperature for three tested bitumens.

Table 2. Relationship between $|G^*|_{LVE}$, test temperature and failure mechanism

Test temperature [˚C]	5˚C	10˚C	20˚C
		|G*|LVE [MPa]	
Paving grade50/70	**69.2**	**40.6**	**13.4**
PMB 45/80-55	**50.7**	**27.9**	10.2
PMB 45/80-80 HiMA	**38.9**	**23.2**	4.9

Adhesion loss	Cohesive cracking	Instability flow

According to data presented in Table 2, the way in which the sample is damaged depends on the test temperature. Only at the temperature of 10 °C the damage of all of the tested samples occurred as the result of cohesive cracking.

4.3 Obtained Parameters and Discussion

The LAS test begins with a frequency sweep test conducted to specify the initial properties of the tested material. The result of this measurement is the α parameter which describing the initial characteristics of the tested material. In Table 3 values of the α parameters which were obtained at three test temperatures for the tested bituminous binders are presented.

Table 3. α parameters obtained from frequency sweep test at three temperatures

Test temperature [°C]	5 °C	10 °C	20 °C
	α parameters		
Paving grade 50/70	2.779	2.489	1.999
PMB 45/80-55	2.377	2.139	1.730
PMB 45/80-80 HiMA	2.511	2.286	1.917

The value of the α parameter decreases along with the increase of the test temperature. This relationship is present in case of all tested bitumens, regardless of the degree of SBS polymer modification.

The AASHTO TP 101-14 standard as the fatigue criterion for the tested materials specifies the moment during the amplitude sweep test when the maximum shear stress occurs in the sample is registered - τ_{max}. Figure 1 presents changes of shear stress occurring in the tested samples, depending on the test temperature.

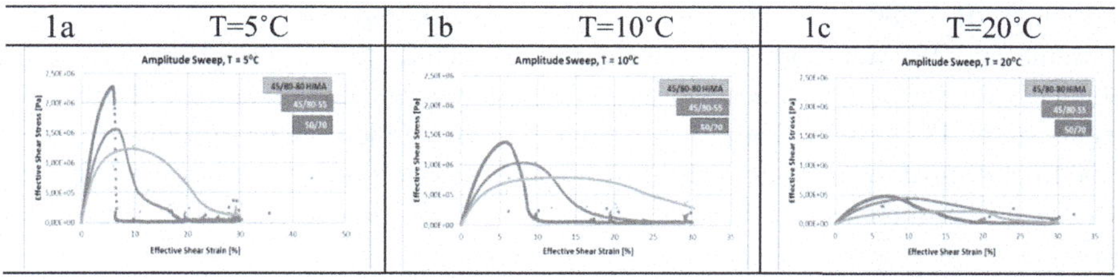

Fig. 1. Diagrams of shear stress-shear strain dependence of test temperatures for three bitumens

The lower the test temperature, the higher values of the shear stress in the samples. On the presented graphs three types of binders can easily be recognized. In the case of the stress-strain curve for the paving-grade bitumen 50/70 and PMB 45/80-55 it is possible to easily indicate the moment of testing when the level of stress occurring in the sample reaches the maximum value. In case of PMB 45/80-80 HiMA, that point is not clearly visible. In case of that bitumen, also no sudden decrease in the stress can be seen after the maximum value is reached, as it happens in case of the paving-grade bitumen 50/70 and PMB 45/80-55. Such behaviour of PMB HiMA means that they are not damaged in the used test conditions and that it is able to transfer the applied load continuously. A similar correlation was observed by other researchers (Wesołowska and Ryś 2018).

The fatigue life criterion – N_f of the tested materials can be specified as sample damage level corresponding to the peak stress response - τ_{max}. Masad et al. (2001) shown that the strain level in the bituminous binder working in the pavement is about 50 times higher than the strain level in the asphalt mixture. Because of that, it is recommended to calculate the fatigue parameter for the two strain levels: γ = 2.5% and γ = 5.0%. Figure 2 shows a comparison of fatigue characteristics of the tested binders, conducted at three test temperatures.

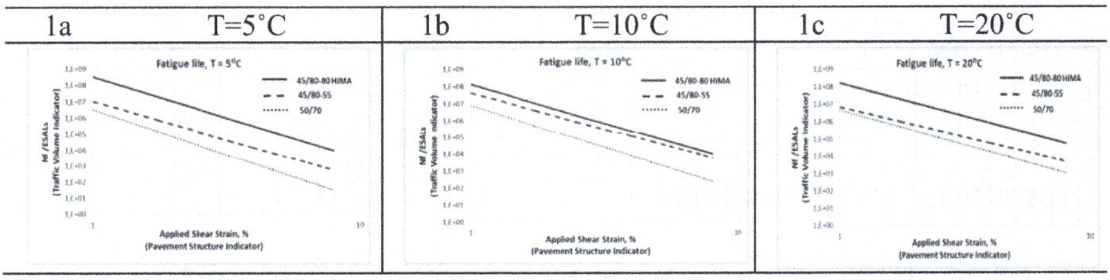

Fig. 2. Variation of fatigue life for three bitumens at different test temperatures

Comparison of the fatigue curves of the tested bitumens, shown that the highly modified bitumen PMB 45/80-80 HiMA is characterized by the best fatigue properties for both applied strain: $\gamma = 2.5\%$ as well as $\gamma = 5\%$. Regardless of the tested temperature, this bitumen withstands the largest number of load cycles to destruction. A fundamental role in this behaviour of HiMA binders is the predominance of the polymeric phase over the bitumen phase as well as the continuity of the elastomer network, which substantially enhances the tensile strength of the binder. As expected, the weakest fatigue properties was obtained for the paving-grade bitumen. Modified bitumen PMB 45/80-55 is characterized by intermediate properties.

Bahia and Teymourpour (2014) proposed to include the results of the fatigue life obtained with the LAS test as an additional property in the classification of bituminous binders in the PG system. They suggested a division of the obtained results in relation to the thickness of the asphalt layer in which a particular binder has to be used. Table 4 presents a classification of the tested bitumens in relation to their usefulness for a particular traffic grades, for each test temperature.

Table 4. Bitumens classification based on LAS test results, acc. to additional PG classification

Test temperature [°C]	Nf at 2.5% for asphalt layer > 4″			Nf at 5.0% for asphalt layer < 4″		
	5 °C	10 °C	20 °C	5 °C	10 °C	20 °C
Paving grade 50/70	H	V, E	V, E	*	*	*
PMB 45/80-55	V, E	V, E	V, E	*	V, E	H
PMB 45/80-80 HiMA	V, E	V, E	V, E	V, E	V, E	V, E

* *result out of classification*

Analyzing the data from the Table 4, it can be stated that in case of pavements in which the thickness of the asphalt layer exceeds 4″ (the so-called strong pavements), the tested binders meet the requirements for Very Heavy/Extreme grades. In the case of pavements in which the thickness of the asphalt layer will not exceed 4″ (the so-called weak pavement), only the PMB 45/80-80 HiMA meets the requirements for the Very Heavy/Extreme grades at all test temperatures. This confirms the results obtained at the NCAT Pavement Test Track which proved that the pavement of a reduced thickness designed with using HiMA is resistant to fatigue cracking (Timm et al. 2013; West et al. 2012). The paving-grade bitumen 50/70 shows too weak fatigue properties for the strain $\gamma = 5.0\%$ and does not meet the fatigue criteria in any of the set temperature values. PMB 45/80-55 tested at 10 °C shows properties appropriate for the Very Heavy/Extreme grades. However, at other temperatures it shows significantly weaker fatigue performance.

5 Summary and Conclusions

- The presented results clearly show that the PMB 45/80-80 HiMA is characterized by the best fatigue properties. As expected, the weakest fatigue properties is obtained for the paving-grade bitumen 50/70. The PMB 45/80-55 has shown intermediate properties.

- The LAS test results confirm the previous research conducted by ORLEN Asfalt (Błażejowski et al. 2016) regarding the functional properties of HiMA binders. It has been found that PMB 45/80-80 HiMA may successfully be used in pavement construction which require very good fatigue properties.
- The test temperature is significant importance for the obtained results of the fatigue life and for the mechanism of sample damage during the LAS test. For that reason the way in which the temperature is chosen should be clearly specified.

Acknowledgements. The tests were performed within as part of the research program conducted in 2017 by ORLEN Asfalt company, Poland: "*Studies on functional properties on bituminous binders based on modern test methods*". The experimental activities presented in the paper have also been included in the project financially supported by the Ministry of Industry and Trade of the Czech Republic which has been providing institutional support for long-term conceptual development of research organization. The project has been integrated into the National Sustainability Programme I of the Ministry of Education, Youth and Sports of the Czech Republic through the project Development of the UniCRE Centre (LO1606).

References

Anderson DA, Le Hir YM, Marasteanu M, Planche J-P, Martin D, Gaauthier G (2001) Evaluation of fatigue criteria for asphalt binder. Transp Res Rec 1766:48–56

Bahia HU, Hanson DI, Zeng M, Zhai H, Khatri MA, Anderson RM (2001) Characterization of modified asphalt binders in Superpave Mix Design. NCHRP Report 459. TRB, National Research Council, Washington, D.C., USA

Bahia HU, Teymourpour P (2014) Linear amplitude sweep test: binder grading specification and field validation. In: Presentation on Binder Expert Task Group Meeting, Baton Rouge, LA, USA

Bahia HU, Zhai H, Zeng M, Hu Y, Turner P (2002) Development of binder specification parameters based on characterization of damage behavior. J Assoc Asphalt Paving Technol 70:442–470

Błażejowski K, Wójcik-Wiśniewska M, Peciakowski H, Olszacki J (2016) The performance of a highly modified asphalt for long-lasting asphalt pavements. Transp Res Procedia 14:679–684

Hintz C, Velasquez R, Johnson CM, Bahia HU (2011) Modification and validation of linear amplitude sweep test for binder fatigue specification. Transp Res Rec 2207:99–106

Johnson CM (2010) Estimating asphalt binder fatigue resistance using an accelerated test method. Ph.D. thesis, University of Wisconsin – Madison, USA

Kim YR, Little DN, Lytton RL (2006) A simple testing method to evaluate fatigue fracture and damage performance of asphalt mixtures. J Assoc Asphalt Paving Technol 71:176–206

Masad E, Somadevan N, Bahia HU, Kose S (2001) Modelling and experimental measurements of strain distribution in asphalt mixes. J Transp Eng 127:477–485

Monismith CM, Epps JA, Kasianchuk DA, McLean DB (1970) Asphalt mixture behavior in repeated flexure. Institute of Transportation and Traffic Engineering, University of California, Berkeley, USA

Pereira A, Micaelo R, Cidade MT, Quaresma L (2016) Evaluation of different methods for the estimation of the bitumen fatigue life with DSR testing. RILEM Bookseries. https://doi.org/10.1007/978-94-017-7342-3_81

Safaei F, Castorena C (2016) Temperature effects of linear amplitude sweep testing and analysis. Transp Res Rec 2574(1):92–100

Safaei F, Hintz C (2014) Investigation of the effect of temperature on asphalt binder fatigue. In: International Society for Asphalt Pavement Conference, pp 1491–1500

Schapery RA (1975) A theory of crack initiation and growth in viscoelastic media. Analysis of continuous growth. Int J Fract 11:549–562

Schapery RA (1984) Correspondence principle and generalized integral for large deformation and fracture analysis of viscoelastic media. Int J Fract 25:195–223

Schapery RA (1990) A theory of mechanical behavior of elastic media with growing damage and other changes in structure. J Mech Phys Solids 2:215–253

Soenen H, De La Roche C, Redelius P (2004) Predict mix fatigue tests from binder fatigue properties measured with a DSR. In: 3rd Eurasphalt and Eurobitume Congress, Vienna, Austria

Soenen H, Eckmann B (2000) Fatigue testing of bituminous binders with a dynamic shear rheometer. In: Proceedings of the 2nd Eurasphalt and Eurobitume Congress, Barcelona, Spain

Timm DH, Robbins MM, Willis JR, Tran N, Taylor AJ (2013) Field and laboratory study of high-polymer mixtures at the NCAT test track. Draft Report, National Center for Asphalt Technology, Auburn University, USA

West R, Timm DH, Willis JR, Powell B, Tran N, Watson D, Brown R, Robbins M, Vargas-Nordcbeck A, Nelson J (2012) Phase IV NCAT pavement test track findings. Draft Report. National Center for Asphalt Technology, Auburn University, USA

Wesołowska M, Ryś D (2018) Analysis of the fatigue life of neat and modified bitumens using linear amplitude sweep test. Roads Bridges Drogi I Mosty 17:317–336

Oxidative Aging Effects on Damage-Healing Performance of Unmodified and Polymer Modified Asphalt Binders

Yifang Chen and Chao Wang(✉)

Department of Road and Urban Railway Engineering, Beijing University of Technology, Beijing 100124, People's Republic of China
wangchao@bjut.edu.cn

Abstract. This paper investigates the damage-healing properties of unmodified and polymer modified asphalt binders under different aging levels using newly developed linear amplitude sweep based healing (LASH) test. One neat asphalt 70# and one Styrene-Butadiene-Styrene (SBS) modified asphalt are selected in this study. The original binders are subjected to the Rolling Thin Film Oven (RTFO) and Pressurized Aging Vessel (PAV) tests for stimulating the short- and long-term aging of asphalt binders during the pavement construction and service phases. Through the LASH tests under different aging levels of the two binders, it is found that the damage-healing performance of both neat and modified asphalt binders gradually decrease with the increasing aging levels. It is also observed that the healing performance of 70# neat asphalt is higher than that of SBS binder in pre-failure conditions, however, the healing behaviors of SBS binder is becoming obviously better than that of 70# binder after fatigue failure. Furthermore, a promising relationship is found between the macroscale healing properties of asphalt binders and their microscale chemical composition under various aging levels.

Keywords: Asphalt · Healing · Aging · Rheology · Chemical composition

1 Introduction

Due to the repeated traffic loading of the vehicles, the fatigue cracking is becoming one of the main diseases of asphalt pavements. The continuous fatigue loading mode is utilized in traditional laboratory based fatigue evaluation. However, the traffic loading in field is actually always discontinuous rather than continuous mode and thus, during the traffic loading gap, the asphalt pavement materials can show a healing recovery. Bazin and Saunier (1967) firstly found the healing phenomenon of asphalt concrete, in which the tension strength of the damaged concrete could recover to its 90% initial level after 3 days' rest period. Kim et al. (1995) utilized the spectral analysis of surface waves testing to verify the healing occurrence of asphalt concrete layers in pavements (Kim and Kim 1997). Jose and Irene (2017) studied the healing behaviour of asphalt mixture in microwave heating process. Meanwhile, it should be also noted that the

M. Pasetto et al. (Eds.): ISAP APE 2019, LNCE 48, pp. 395–403, 2020.
https://doi.org/10.1007/978-3-030-29779-4_39

healing performance of asphalt concrete and pavement is strongly affected by the healing properties of asphalt binders (Bhasin and Motamed 2011; Mannan and Tarefder 2018). Therefore, characterization of asphalt binder healing behaviour is of great significance to accurately predict the fatigue and healing performance of asphalt pavements. The effects of different modifiers on binder healing were widely studied in recent years (Pan et al. 2018; Al-Mansoori et al. 2018; Chen et al. 2015). The molecular dynamics approach can also be applied to simulate the binder healing behaviors and further investigate the effect of temperature on healing (Sun et al. 2015, 2018). Recently, a linear amplitude sweep based healing (LASH) protocol was proposed to quantify the healing potential of neat and modified asphalt binders, in which the damage level and rest period effects on healing behaviors can be unified through a healing mastercurve approach (Xie et al. 2017).

Up to date, the oxidative aging impact on healing performance on asphalt materials are relatively limited from the literature. During the construction and service of asphalt pavement, the chemical composition of asphalt undergoes many complicated physical and chemical changes due to the contact with oxygen, atmospheric ozone and ultraviolet light in the air and thus, the asphalt aging is growing which is manifested by the fact that the light components gradually harden and the molecular weights increase. Petersen summarized the oxidative aging mechanism and modeling of asphalt binder (Petersen 2009). The Strategic Highway Research Plan (SHRP) program proposes the use of a Rolling Thin Film Oven (RTFO) test to simulate the short-term aging that occurs in pavement construction, and uses Pressurized Aging Vessel (PAV) test to simulate long-term aging (Peterson and Anderson 1994a; Anderson et al. 1994; Peterson and Anderson 1994b). The current RTFO (163 °C, 85 min) and PAV (100 °C, 20 h) tests are still main test methods to simulate asphalt aging. It is important to study the aging mechanism of asphalt, but it is more meaningful to study the effect of aging on asphalt pavement performance.

The macroscale engineering performance of asphalt binder is generally determined by their microscale chemical properties. The difference in chemical composition, molecular weight and structures in asphalt will cause corresponding distinguished rheological performance. The chemical composition analysis of asphalt usually separates petroleum asphalt into saturate, aromatic, resin, and asphaltene (SARA), which is a very classic method. The properties of each part in SARA fractions are independent and they contribute different performance of asphalt respectively (Shi et al. 2017). A point of view was proposed that the lower branched-chain and higher long and thin molecule content can accelerate healing in asphalt (Sun et al. 2017) and they also used molecular dynamics simulation to dish some indices of asphalt healing (Sun et al. 2016). There are some studies focused on the effects of microscale chemical components on asphalt healing properties, but the effect of aging is still relatively limited in these works. Thus, it would be meaningful to quantify the effect of aging on healing from the microscale perspective.

2 Objectives

The objectives of this paper is to:

(1) Quantify the effect of aging on asphalt healing performance using LASH test and related performance index.
(2) Compare the macroscale healing properties and microscale chemical composition of asphalt under various aging levels.

3 Materials and Testing

3.1 Materials

Two types of asphalt binders commonly used by the paving industry in Beijing, P. R. China were evaluated in this study: one neat asphalt with penetration grade of 70 and one SBS polymer modified asphalt binder. Both the unaged original binders (OB) and aged binders were tested. The RTFO test was utilized to simulate the short-term aging of asphalt binder during pavement construction followed by the PAV test to simulate the long-term aging that happen during pavement service.

3.2 Testing Methods

The rheological LAS and LASH tests were conducted using an Anton Paar MCR 302 dynamic shear rheometer (DSR) with the 8-mm parallel plate geometry and 2-mm gap setting. The SARA chemical composition of the asphalt binders was also conducted. At least two replicates were run for all rheological and chemical tests to ensure the coefficient of test variation was within 10%.

3.2.1 LAS Test

The frequency sweep of 0.1–100 rad/s was firstly completed respectively at the temperature of 35 °C, 20 °C, 5 °C. Then the LAS test was performed at 20 °C and 10 Hz with the strain amplitude increasing from 0.1% to 30% in 5 min. The typical stress-strain response measured from LAS test is shown in Fig. 1(a) and the corresponding pseudo strain energy (PSE) evolution during the LAS test is shown in Fig. 1(b). The maximum stored PSE was proposed to define the cohesive failure in LAS test (Wang et al. 2015). The damage characteristic curve (DCC) of asphalt binder can be represented from the relationship between the pseudo stiffness (C) and the damage intensity (S) that derived from the simplified-viscoelastic damage (S-VECD) model, details of which can be found elsewhere (Wang et al. 2015). Then the strain amplitude and damage intensity corresponding to the cohesive failure (γ_f and S_f) can be obtained for a given asphalt binder.

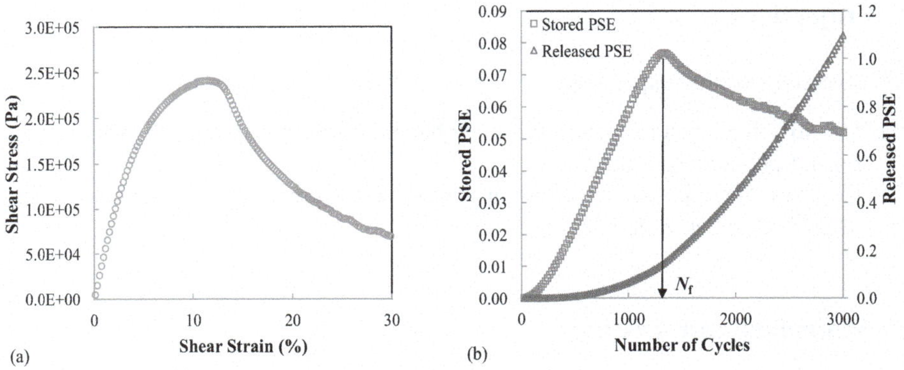

Fig. 1. Typical LAS test results (a) stress-strain curve (b) PSE-based failure definition

3.2.2 LASH Test

The LASH test was established based on LAS test and proposed to quantify the healing potential of asphalt binder (Xie et al. 2017). A rest period is applied at various damage levels of 25%S_f, 50%S_f, 75%S_f and 125%S_f. The rest periods normally include 60s, 300s, 900s and 1800s, respectively. The DCC curves of typical LASH test is shown in Fig. 2, where the percent healing (%H_S) is defined by the change of the damage intensity (S) before and after the rest period.

The "rest-damage superposition principle" can be further employed to develop the healing mastercurve of asphalt binder, which eliminates the effect of damage level and rest period on %H_S results (Xie et al. 2017). Four performance indexes, which consist of instantaneous %H_S (%H_{S0}), minimum rest period (RP_{min}), rate of healing (H^R), and maximum rest period (RP_{max}), can be obtained from the healing mastercurve and indicate the healing characteristics for a given asphalt binder. The %H_{S0} represents the instantaneous healing recovery ability of asphalt, which should be a constant number that independent of rest period duration; RP_{min} is the shortest rest period that required to activate the healing potential and achieve a rapid growth from %H_{S0}; The H^R

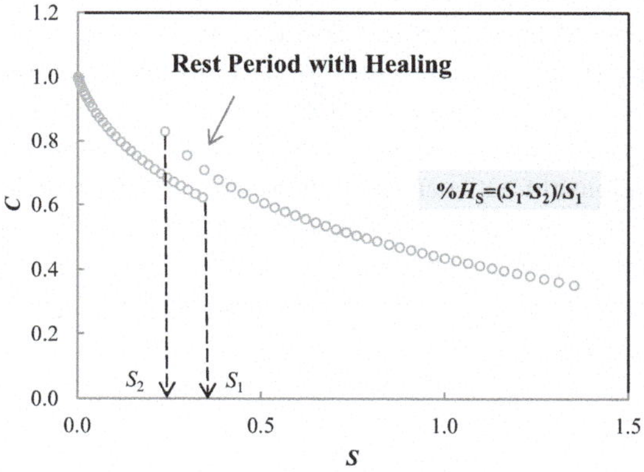

Fig. 2. Schematic illustration for %H_S calculation

characterizes the linear slope from steady growth of $\%H_S$; The RP_{max} represents the rest period duration that required for a 100% healing recovery.

3.2.3 Chemical SARA Composition Test

The chemical compositions of OB, RTFO- and PAV-aged samples of the neat 70# and SBS binders were also tested with the thin-layer chromatography following the ASTM D4124 (ASTM 2009). The relative fractions of saturates, aromatics, resins and asphaltenes (SARA) can be measured basically from their solubility difference in organic solvents.

4 Results and Discussions

4.1 LASH-Based Healing Performance

In this study, the OB, RTFO and PAV aging levels of neat70# and SBS binders were firstly subjected to the standard LAS test to quantify the cohesive failure. Then the strain values corresponding to the damage levels of $25\%S_f$, $50\%S_f$, $75\%S_f$ and $125\%S_f$ were calculated respectively. The LASH test was conducted at each damage level with various rest period duration. Figure 3(a) to (d) respectively show the measured $\%H_S$ results of OB, RTFO- and PAV-aged neat70# binders under different damage conditions. Firstly, it can be seen that the healing ability increases simultaneously with the longer rest period durations under the same damage level. At the same rest period duration, the increased damage level generally decreases the binder healing performance. More importantly, the healing potential of unaged OB binder is much better than RTFO- and PAV-aged binders, indicating a negative aging effects on binder healing behaviours. The healing behaviour of OB, RTFO- and PAV-aged SBS modified binders during LASH tests with various damage levels and rest period are also given in Fig. 4, in which similar findings to neat 70# binder can be found for SBS binder. The only remarkable difference for SBS binder is that its RTFO binder also shows obvious better healing recovery than PAV binder. Additionally, it is also observed that the healing performance of neat 70# binder is better than that of SBS binder in pre-failure conditions; however, the healing recovery of SBS binder is becoming obviously higher than that of 70# binder after the cohesive failure.

Figure 5(a) and (b) respectively show the healing mastercurves of neat 70# and SBS modified binder under different aging levels. It is observed that the aging progress negatively impact the healing potential in different degrees for the two tested binders. Table 1 summarizes the four determined healing indexes of $\%H_{S0}$, RP_{min}, H^R and RP_{max} of neat 70# and SBS binders under different aging conditions. The rate of healing (H^R) is generally decreased with the elevated aging degree for both two binders. Therefore, it can be concluded that the RTFO and PAV aging reduce the healing potential of both unmodified and modified asphalt binders based on LASH protocol and analysis.

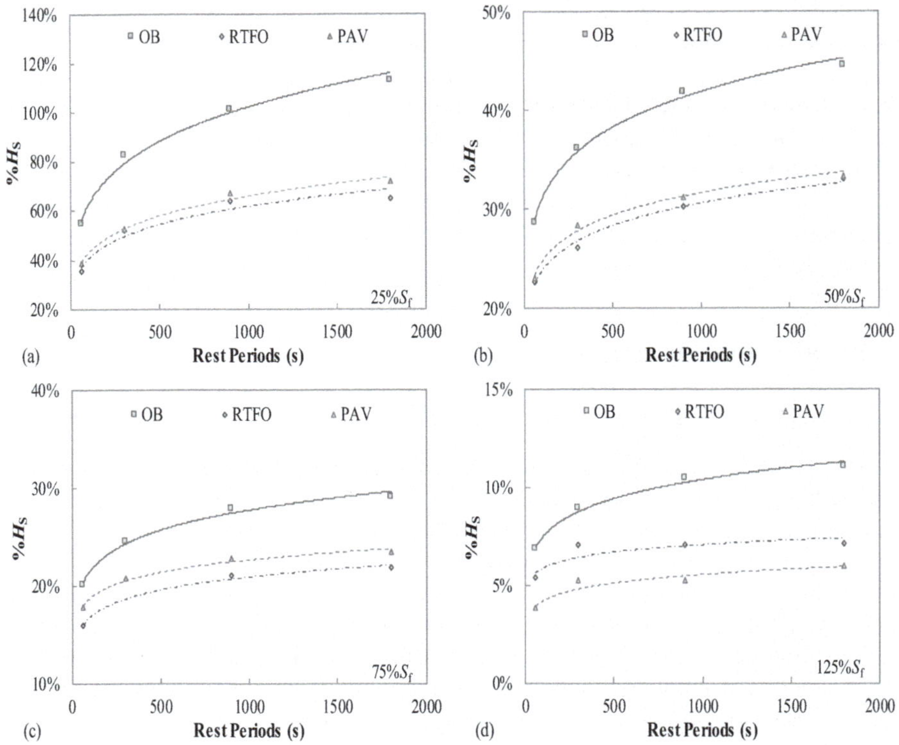

Fig. 3. Percent healing (%H_S) results of neat 70# binder

4.2 Comparison Between Rheological Healing Properties and Chemical Composition

The macroscale engineering properties of asphalt binder are normally related to the microscale chemical characteristics. In this study, the SARA composition results were preliminary determined for the neat 70# and SBS binders with various aging levels, as shown in Table 2. It can be seen from the Table 2 that the weight percent of saturates and aromatic components decrease during the RTFO and PAV aging process, whereas the contents of resins and asphaltenes are increasing. It is well accepted that the characteristics of saturates and aromatic components are to provide the fluidity within asphalt binder and thus, the reduced fluidity from elevated aging would negatively impact the cracking healing behaviour around the crack interface, indicating a slower rate of healing (H^R). Therefore, the previous quantified macroscale healing performance of two binders under oxidative aging can be preliminary explained from SARA analysis and also will be further addressed in the future study with more chemical characterization.

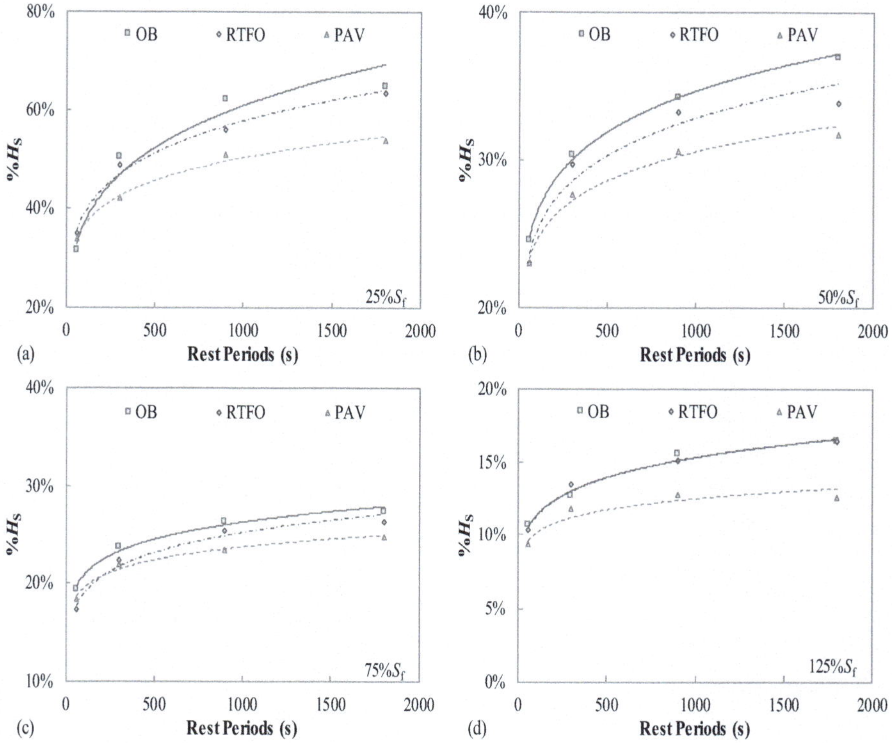

Fig. 4. Percent healing (%H_S) results of SBS modified binder

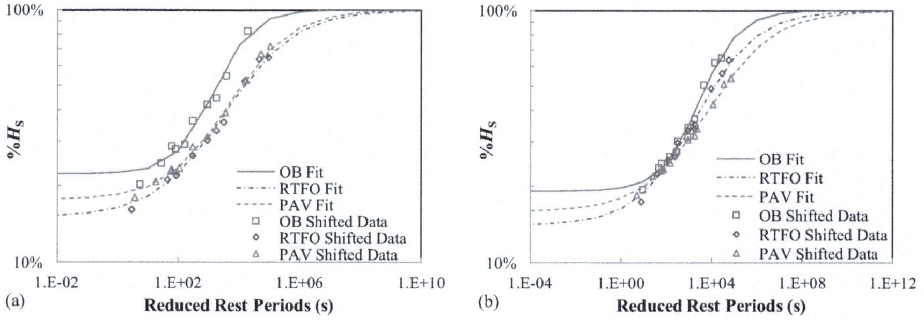

Fig. 5. Healing mastercurves of neat 70# and SBS binders (a) 70# neat binder (b) SBS binder

Table 1. Healing indexes of neat 70# and SBS binders

Sample ID	$\%H_{S0}$	RP_{min} (s)	H^R	RP_{max} (s)
70#-OB	22%	1	0.211	1.E+07
70#-RTFO	15%	0.1	0.161	1.E+11
70#-PAV	18%	10	0.158	1.E+10
SBS-OB	19%	1	0.179	1.E+09
SBS-RTFO	14%	0.01	0.152	1.E+12
SBS-PAV	16%	0.1	0.123	1.E+13

Table 2. SARA results of neat 70# and SBS binders

Sample ID	Saturate (%)	Aromatic (%)	Resin (%)	Asphaltene (%)
70#-OB	14.07	42.26	25.53	15.59
70#-RTFO	12.02	39.23	26.17	19.33
70#-PAV	11.67	36.33	26.53	22.16
SBS-OB	12.29	39.31	22.76	24.18
SBS-RTFO	11.72	38.35	23.93	23.72
SBS-PAV	10.60	34.79	24.57	26.75

5 Conclusion

This paper investigated the aging effects on damage-healing potential of unmodified and polymer modified asphalt binders by rheological testing and chemical composition analysis. Specific findings are summarized as follows:

(1) The percent healing ($\%H_S$) results measured from LASH test under various damage levels and rest periods demonstrated that RTFO and PAV aging negatively decreased the healing potential of neat 70# and SBS binders.
(2) The rate of healing (H^R) that defined from healing mastercurve provided a clearer comparison for the two binders subjected to oxidative aging. The unaged original binder showed advanced healing performance.
(3) The macroscale healing properties of two binders under RTFO and PAV aging can be explained by the SARA composition analysis. The saturates and the aromatic were reduced during the RTFO and PAV aging process, which resulted a reduced fluidity within asphalt and a slower rate of healing (H^R).

Additionally, the findings in this paper are summarized only from two typical asphalt binders in the Beijing area. Therefore, more unmodified and modified binders will be further validated in the future study.

Acknowledgements. The authors would like to gratefully acknowledge the sponsorship from National Natural Science Foundation of China (51608018), Beijing Municipal Education Commission (KM201810005020) and Beijing Natural Science Foundation (8174059).

References

Al-Mansoori T, Norambuena-Contrer J, Micarlo R, Garcia A (2018) Self-healing of asphalt mastic by the action of polymeric capsules containing rejuvenators. Constr Build Mater 161:330–339

Anderson DA, Christensen DW, Bahia HU, Dongre R (1994) Binder characterization and evaluation. Physical Characterization SHRP-A-369, vol 3. National Research Council

ASTM (2009) Standard test method for separation of asphalt into four fractions. ASTM D4124

Bazin P, Saunier JB (1967) Deformability, fatigue and healing properties of asphalt mixes. In: Proceeding of 2nd international conference on the structural design of asphalt pavement, pp 553–569

Bhasin A, Motamed A (2011) Analytical models to characterise crack growth in asphaltic materials and healing in asphalt binders. Int J Pavement Eng 12(4):371–383

Chen YL, Gong MH, Yao ZH, Dong XX, Song KM (2015) Investigation on the self-healing abilities of base and ionomer-modified asphalt binder with a – peel test. In: New frontiers in road and airport engineering, pp 64–73

Jose NC, Irene GT (2017) Influence of the microwave heating time on the self-healing properties of asphalt mixtures. Appl Sci 7(10):1076

Kim Y, Kim YR (1997) In situ evaluation of fatigue damage growth and healing of asphalt concrete pavements using stress wave method. Transp Res Rec 1568:106–113

Kim YR, Whitmoyer SL, Little DN (1995) Healing in asphalt concrete pavements: is it real? In: Transportation research record TRB, National Research Council, Washington, D.C., no 1454, pp 89–96

Mannan UA, Tarefder RA (2018) Investigating different fatigue failure criteria of asphalt binder with the consideration of healing. Int J Fatigue 114:198–205

Pan CL, Tang P, Riara M, Mo LT, Li ML, Guo M (2018) Effect of healing agents on crack healing of asphalt and asphalt mortar. Materials 11(8):1373

Peterson JC (2009) A review of the fundamentals of asphalt oxidation: chemical, physicochemical, physical property, and durability relationships. Transportation Research Circular E-C140, Transportation Research Board of the National Academies, Washington, D.C

Peterson JC, Anderson DA (1994a) Binder characterization and evaluation. SHRP-A-367, vol 1. National Research Council

Peterson JC, Anderson DA (1994b) Binder characterization and evaluation. Test Methods SHRP-A-370, vol 4. National Research Council

Shi HQ, Tao T, Pei Z, Jiang RL (2017) Combustion properties of saturates, aromatics, resins and asphaltenes in asphalt binder. Constr Build Mater 136:515–523

Sun DQ, Lin TB, Zhu XY, Cao LH (2015) Calculation and evaluation of activation energy as a self-healing indication of asphalt mastic. Constr Build Mater 95:431–436

Sun DQ, Lin TB, Zhu XY, Tian Y, Lin FL (2016) Indices for self-healing performance assessments based on molecular dynamics simulation of asphalt binders. Comput Mater Sci 114:86–93

Sun DQ, Sun GQ, Zhu XY, Ye FY, Xu JY (2018) Intrinsic temperature sensitive self-healing character of asphalt binders based on molecular dynamics simulations. Fuel 211:609–620

Sun DQ, Yu F, Li LH, Lin TB, Zhu XY (2017) Effect of chemical composition and structure of asphalt binders on self-healing. Constr Build Mater 133:495–501

Wang C, Castorena C, Zhang JX, Kim YR (2015) Unified failure criterion for asphalt binder under cyclic fatigue loading. Road Mater Pavement Des 84:269–299

Xie W, Castorena C, Wang C, Kim YR (2017) A framework to characterize the healing potential of asphalt binder using the linear amplitude sweep test. Constr Build Mater 154:771–779

Pavement Structures, Maintenance and Management

Performance Evaluation of Innovative and Sustainable Pavement Solutions for Road Tunnels

Pier Paolo Riviera(✉), Eldho Choorackal, and Ezio Santagata

Department of Environment, Land and Infrastructure Engineering, Politecnico di Torino, Turin, Italy
pierpaolo.riviera@polito.it

Abstract. This paper summarizes the studies which were carried out to support pavement design and construction operations in a major motorway tunnel. Candidate pavement cross sections subjected to analysis included a subgrade and foundation layer constituted by self-compacting cement-bound granular mixtures. The potential use of standard cement-stabilized foundations was also considered. The self-compacting mixtures were designed to incorporate significant amounts of recycled materials including Reclaimed Asphalt Pavement and mineral sludge retrieved from aggregate washing operations. Expected performance of the innovative pavement solutions and of a reference standard cross section were assessed by means of a mechanistic-empirical approach which in most part relied on the outcomes of laboratory and field tests. In particular, a full-scale test section was constructed with the use of self-compacting mixtures in the subgrade and foundation, overlaid by two asphalt layers. The subsequent investigation included Falling Weight Deflectometer measurements and laboratory tests carried out for the assessment of volumetric and mechanical properties of laid mixtures. Obtained results led to the identification of an optimal pavement cross section and highlighted the potential performance achieved by the proposed innovative solutions. The additional advantages related to the greater efficiency of construction operations in tunnels and to the lower consumption of virgin aggregates were also discussed.

Keywords: Self-compacting cement-bound mixtures · Recycling · Road tunnels · Mechanistic-empirical pavement design

1 Introduction

Selection of a pavement solution for road tunnels depends upon several factors ranging from estimated traffic, safety, cost, availability of materials, presence of buried utility lines, logistics of construction operations, and expertise of involved contractors. Various pavement types can be selected, including flexible, rigid and semi-rigid, but designers may also need to deviate from standard solutions, identifying innovative materials and technologies which are compatible with site-specific requirements.

Pavements in road tunnels are subjected to a temperature regime which is different from that of open roadways. In particular, temperatures tend to be significantly higher

M. Pasetto et al. (Eds.): ISAP APE 2019, LNCE 48, pp. 407–416, 2020.
https://doi.org/10.1007/978-3-030-29779-4_40

and characterized by a lower daily and seasonal variability, thus leading to lower stiffness values of the bitumen-bound upper layers. Hence, in order to reduce deflections and strains under loading, pavements of the semi-rigid type can be selected, in which asphalt concrete layers are placed over a relatively stiff cement-stabilized foundation (Solanki and Zamam 2017). Semi-rigid pavements constitute a significant portion of the European road network (Ferne 2006) and have proven to exhibit satisfactory performance in service (Merrill et al. 2006). Furthermore, they may easily incorporate waste materials in the foundation and may lead to a reduction of bitumen consumption as a result of the use of thinner asphalt layers (Zheng 2012).

In the specific case of a newly constructed major motorway tunnel (13 km in length), a standard semi-rigid pavement cross section was chosen in the preliminary design phase, which focused on its load bearing capacity. However, the Authors of this paper were involved in its optimization which was deemed necessary in order to fulfill several additional needs. These were related to the potential constraints encountered in the compaction of subgrade and foundation materials, to the presence of underground utilities, including a high-voltage transmission line, and to the desire of reducing the consumption of virgin aggregates. Consequently, innovative pavement cross sections were identified, in which the use of self-levelling and self-compacting cement-bound mixtures was considered for the formation of the subgrade and foundation. These mixtures were designed in order to incorporate significant amounts of recycled materials such as Reclaimed Asphalt Pavement (RAP) and mineral sludge retrieved from aggregate washing operations. Laboratory and field studies included the assessment of flow, mechanical and thermal properties. Results obtained during these investigations are fully described elsewhere (Riviera 2018; Choorackal et al. 2019a, b).

This paper summarizes the studies which were carried out as part of pavement optimization. Expected performance of the innovative solutions and of a reference standard cross section were assessed by means of a mechanistic-empirical approach in which predicted traffic and environmental conditions were taken into account. Mechanical properties of materials were mainly derived from the outcomes of field investigations carried out on a full-scale test section and of laboratory tests performed on materials sampled on site.

2 Pavement Cross Sections

The standard semi-rigid pavement cross section considered in the preliminary design phase (indicated as "CS") included a compacted soil subgrade, a cement-stabilized foundation (thickness 20 cm) and three layers of dense-graded asphalt concrete (total thickness 19 cm).

The innovative pavement solutions included self-levelling and self-compacting cement-bound mixtures (SCMs) for the formation of both the subgrade and foundation (cross section "GG") or of the subgrade only (cross section "CG"). In the latter case, it was hypothesized that the foundation would be constructed with the same type of standard cement-stabilized mixture (CBM) considered as part of cross section CS. For both pavements, as in the case of cross section CS, the foundation thickness was maintained equal to 20 cm and total thickness of the asphalt layers was fixed at 19 cm since such a value was considered adequate for the prevention of reflective cracking (Austroads 2017). It was hypothesized that the 19 cm would be constituted by 10 cm base course, 5 cm binder course and 4 cm wearing course.

For all pavement solutions the total thickness of subgrade considered in calculations was equal to 1 m. Such a value results from the geometry of the tunnel invert, which is filled in its lower portion by lean Portland cement concrete that acts as a rigid base.

It should be mentioned that after completion of preliminary design, the tunnel administration accepted to allow the installation of a high-voltage transmission line within the pavement subgrade. The corresponding design led to the identification of an appropriate layout of conduits and cables to be buried at a depth below the subgrade surface equal to approximately 80 cm.

3 Performance Evaluation of Pavement Solutions

Performance evaluation of the considered pavement cross sections was carried out by means of a mechanistic-empirical approach which required the prediction of traffic volumes, environmental conditions and mechanical properties of materials.

The motorway tunnel for which the study was carried out is a new infrastructure, recently excavated in parallel to an existing tunnel, open to traffic since 1980. Thus, data available for the existing tunnel were employed for the identification of traffic and temperature conditions relevant for the design of the new pavement (see Sects. 3.1 and 3.2). Mechanical properties of materials constituting the various pavement layers were derived from field and laboratory tests, from available literature, and from technical specifications (see Sect. 3.3). Transfer functions employed for the calculation of design lives, briefly discussed in Sect. 3.4, were those provided by the SAPEM, South African Pavement Design Manual (SANRA 2014).

3.1 Design Traffic

A synthesis of the two-way traffic data collected for the existing tunnel for a period of five years is shown in Table 1, in which the number of passes are given for light (motorcycles, cars and vans) and heavy vehicles (buses, trucks and trailers). The percentage of heavy vehicles, of the order of 40%, was approximately constant in time. It was observed that annual traffic growth values were quite variable, ranging between 0.1% and 7.0%. Thus, the average value, equal to 3.0% was assumed for traffic predictions.

Table 1. Traffic data collected for the existing tunnel

Year	2014	2015	2016	2017	2018
Light vehicles	998,342	1,115,511	1,129,051	1,100,382	1,099,845
Heavy vehicles	701,728	703,653	736,248	766,561	813,233
Total	1,700,070	1,819,164	1,865,299	1,866,943	1,913,078

In subsequent calculations, light vehicles were neglected, whereas the heavy vehicles were converted into 80 kN equivalent single axle loadings (ESALs) by taking into account the vehicle sub-categories and relative percentages recorded at the tunnel entrance for toll collection. Individual truck factors were derived from the axle loads indicated in the Italian pavement catalogue (CNR 1995) and by referring to the equivalent axle load factors (EALFs) reported in the 1993 AASHTO Guide for a Structural Number (SN) equal to 5 and a final Present Serviceability Index (PSI) equal to 3.0. The average truck factor obtained by means of such an approach was found to be equal to 2.0. A lane factor equal to 0.8 was assumed for design purposes.

Monthly traffic distribution was also made available by the tunnel administration and it was found that no significant changes occurred in the considered years (with the only exception of the year 2015, which was treated as an outlier). These data were later used in damage calculations for the considered pavement cross sections (see Sect. 3.5).

It was planned that pavement construction would occur in two stages. In particular, it was envisioned that after laying of the binder course, the tunnel would be accessed for 2 years exclusively for the instalment of all accessory utilities, including ventilation, lighting and safety systems. The pavement would then be completed with the laying of the 4 cm wearing course. Construction traffic anticipated in the 2 years of finishing operations was estimated based on the planned daily activities and was found to correspond to 7,508 passes of three- and four-axle trucks on each lane. For the considered vehicles the average truck factor was equal to 2.14.

Upon request of the tunnel administration, pavement design life, inclusive of the 2 years of tunnel completion, was set equal to 20 years. Based on the data and assumptions illustrated above, corresponding total design traffic expressed in ESALs was found to be equal to 16.2 million.

3.2 Design Temperatures

Available data consisted in hourly air temperatures recorded for one year along the existing tunnel in 4 fixed stations. As expected, lower values were reached near the tunnel entrance. Thus, as part of a conservative approach, average values derived only from the 3 stations closer to the tunnel center were considered in subsequent analyses. Hourly data were converted into average daily and monthly temperatures. As a result, four different periods characterized by a similar average monthly temperature were identified and were then associated to average air temperature values ($T_{air,p}$).

In order to calculate pavement design temperatures, the presence of the buried high-voltage line was considered. Thus, its operating temperature was calculated by taking into account the requirements fixed by line designers and by making use of the model proposed by Neher and McGrath (1957). In each of the abovementioned periods, air temperature was considered representative of far field conditions, while the resistivity of the materials surrounding the cables and conduits varied from one cross section to the other. In the case of cross section CS, calculations were carried out by considering the presence of a lean concrete duct-bank and of a soil backfill, with thermal resistivity values equal to 1.368 K m/W and 1.2 K m/W, respectively. For the GG and CG pavements, no duct-bank was planned to be constructed, so modelling was based on a single resistivity value, equal to 0.860 K m/W, representative of the SCM intended for use for the formation of the subgrade. Thermal resistivity values indicated above for lean concrete and for the SCM were measured by the Authors by means of the thermal needle probe technique indicated by ASTM D 5334-14 (Choorackal et al. 2019a). In the case of the backfill soil, the considered value was retrieved from literature.

In order to derive pavement temperature profiles, surface temperature was considered equal to air temperature, whereas at 120 cm below the surface, temperature was assumed to be equal to the average of the temperatures reached inside the conduits of the 4 cables constituting the high-voltage line. Temperature gradients within the pavement layers and subgrade were assumed to be inversely proportional to the corresponding thermal resistivity values. Representative pavement design temperatures ($T_{pav,p}$) were thereafter calculated for each period at a depth from the surface equal to one-third of the total thickness of the asphalt layers.

Results obtained by means of the modelling approach synthesized above are provided in Table 2.

Table 2. Air and pavement design temperatures

Period	Months	$T_{air,p}$ (°C)	$T_{pav,p}$ (°C)			
			Final phase of construction	Service life		
			CS-GG-CG	CS	GG	CG
1	January–March	20.8	20.8	22.7	22.2	22.2
2	April–May	24.6	24.6	26.3	25.9	25.9
3	June–November	28.4	28.4	30.0	29.6	29.5
4	December	20.8	20.8	22.7	22.2	22.2

3.3 Mechanical Properties of Materials

Assessment of the mechanical properties of materials constituting the various pavement layers was carried out by focusing on elastic moduli. Values of Poisson's ratio were derived from relevant literature.

A full-scale pavement section was constructed by replicating cross section GG with a reduced asphalt thickness of 15 cm (10 cm binder course, 5 cm wearing course). The test section had a length of 14 m and width of 4 m. The SCM subgrade and foundation had a thickness of 100 cm and 20 cm, respectively. Both asphalt mixtures, which were intended for use in the final construction of the pavement in the tunnel, contained polymeric additives for the improvement of rutting and fatigue resistance.

Following previous investigations, the SCMs were designed to achieve a satisfactory flowability, while guaranteeing adequate short-term and long-term mechanical properties. In particular, a 170–300 mm acceptance range was fixed for spread diameter (ASTM D 6103) and threshold values were identified for compressive strength (EN 12390-3), which was required to be greater than 0.5 MPa at 3 days of curing and less than 2.0 MPa at 28 days of curing.

The SCMs employed in the test section and proposed for use in the tunnel contained 20% gravel, 36% fine sand, 20% RAP and 24% aggregate sludge (percentages by weight). Cement content was set equal to 100 kg/m^3 and 60 kg/m^3 in the subgrade and foundation, respectively. Water content was that corresponding to a water-to-powder ratio equal to 0.8 ("powder" being the sum of cement and aggregate sludge).

During the construction of the test section, samples of the two SCMs were taken for the immediate assessment of spread diameter which in both cases was found to be of the order of 250 mm. Specimens were also prepared for the laboratory evaluation of compressive strength. Corresponding results obtained after 3 and 28 days of curing, respectively equal to 0.84 MPa and 1.67 MPa for the mixture with 100 kg/m^3 cement dosage and to 0.50 MPa and 0.9 MPa for the mixture with 60 kg/m^3 cement dosage, satisfied the previously mentioned acceptance requirements.

Falling Weight Deflectometer (FWD) tests were performed over the finished pavement surface at three different loading levels (47, 63 and 83 kN). At the time of testing, a temperature of 16 °C was recorded at mid-depth in the asphalt layers. Back-calculation was carried out with an iterative procedure by employing the BISAR software (Shell 1998) and by hypothesizing full slip between the bottom asphalt course and the extremely smooth underlying foundation. Computed elastic modulus values were equal to 3,500 MPa, 350 MPa and 1,000 MPa for the asphalt layers, foundation and subgrade, respectively. For the subgrade soil and for the CBM foundation included in cross section CS, values of 200 MPa and 800 MPa, respectively, were assumed consistently with preliminary design hypotheses.

In the case of the asphalt layers, cores were taken in order to assess their volumetric characteristics. Unfortunately, due to the physical constraints imposed by the test section, which was quite narrow and with a working surface located 1 m above the ground, achieved compaction level was lower than expected. In particular, measured air void content, equal to 9.7%, was definitely higher than the target value of 7.0%, compatible with mix design studies and technical specifications.

For the purpose of pavement evaluation, the elastic modulus of the asphalt layers was adjusted for temperature (WSDOT 2005). Moreover, it was assumed that during construction operations in the tunnel, target void content would be reached. Thus, the modulus was adjusted further by making use of the same functional dependency proposed by Bonnaure et al. (1977) for stiffness values measured in the bending mode.

Results obtained by means of the procedure outlined above are given in Table 3, where they are presented for each temperature period, pavement cross section and phase.

3.4 Transfer Functions

Transfer functions used for performance evaluation were drawn from the SAPEM (SANRA 2014). Minor adjustments were needed in order to take into account the specific properties of the employed materials.

In the case of the asphalt layers, analyses focused exclusively on fatigue cracking, the transfer function of which depends upon elastic modulus and thickness. However, an additional shift factor of 1.10 was introduced in calculations in order to account for the performance-related benefits deriving from the use of polymeric additives.

Performance of the CBM and SCM foundation layers was analyzed in two phases. In phase 1, the layers were assumed to be intact and were consequently modelled in

Table 3. Elastic modulus of asphalt layers

Period	Months	Elastic modulus (MPa)			
		Final phase of construction	Service life		
		CS-GG-CG	CS	GG	CG
1	January–March	3,185	2,742	2,845	2,855
2	April–May	2,334	2,007	2,083	2,090
3	June–November	1,660	1,428	1,481	1,486
4	December	3,185	2,742	2,845	2,855

terms of their fatigue resistance under bending. The corresponding transfer functions were selected by fitting the considered materials into the categories indicated by the SAPEM (C3 for the CBM, C4 for the SCM). In phase 2, reached as a result of cracking, they were considered as unbound granular materials with a reduced modulus (of classes EG4 and EG5, respectively, for the CBM and for the SCM). In such a state, their performance was assessed in terms of their resistance to shear failure.

For all pavement cross sections, the same transfer function was employed for the performance assessment of the subgrade. In particular, a reliability of 95% was selected, which corresponds to a terminal rut depth of 10 mm.

Design calculations were performed by combining the results obtained in the various temperature periods and by introducing the concept of cumulated damage. The response under loading of the pavement cross sections in each period was assessed by means of the BISAR software by assuming full adhesion between the layers and by considering a standard dual-wheel single axle with a contact pressure of 577.4 kPa.

3.5 Design Life

The design life of each pavement solution, expressed in terms of allowable ESALs and corresponding years of traffic, was computed as the sum of three terms associated to the final stage of construction and to the previously mentioned phases 1 and 2.

Obtained results are synthesized in Table 4. In all cases final conditions were reached as a result of fatigue cracking in the asphalt layers, thus indicating that their performance potential was fully exploited. Furthermore, accumulation of permanent deformation in the subgrade was not of concern for any of the considered solutions.

From the data listed in Table 4 it can be observed that the CG pavement cross section provided the highest design life, equal to 26 years, followed in ranking by

Table 4. Design life of considered pavement solutions

	ESALs			Years		
	CS	GG	CG	CS	GG	CG
Construction + Phase 1	9.19E+06	9.83E+06	9.71E+06	13.3	14.0	13.8
Phase 2	7.25E+06	1.78E+06	1.42E+07	6.9	1.8	12.2
Total	1.64E+07	1.16E+07	2.39E+07	20.2	15.8	26.0

solutions CS and GG. Both cross sections with a CBM foundation satisfied the requirement of 20 years design life. On the contrary, the cross section with the two-layer SCM supporting system was found to have a shorter design life, close to 16 years, which in any case corresponds to a remarkably high traffic (equal to 11.6 million ESALs).

It can be observed that the three solutions exhibited a similar design life associated to the sum of the construction completion phase and of the phase 1 pre-cracking stage. However, the highest number of allowable loadings were found in the case of cross section GG, followed in ranking by CG and CS. This is due to the fact that phase 1 transfer functions employed for calculations yield numbers of loadings to cracking of the cement-stabilized layers which depend not only upon load-induced tensile strains but also on the ductility of the mixtures (expressed in terms of the so-called strain-at-break). Thus, less stiff materials such as the SCMs (associated to category C4), may yield higher fatigue lives in comparison to stiffer materials such as the C3-type CBMs.

Despite their similar early behavior, the three cross sections showed a totally different response in phase 2, the highest and lowest lives being associated to solutions CG and GG, respectively. As highlighted in Fig. 1, this is essentially due to the fact that asphalt damage in both phases progressed with completely different rates, which were affected by the stiffness of both the foundation and the subgrade. Figure 1 also shows that a similar dependency upon stiffness support was exhibited by the cement-stabilized foundations in phase 2, the lowest damage rates being associated to the cross sections containing the stiffer SCM subgrade (GG and CG).

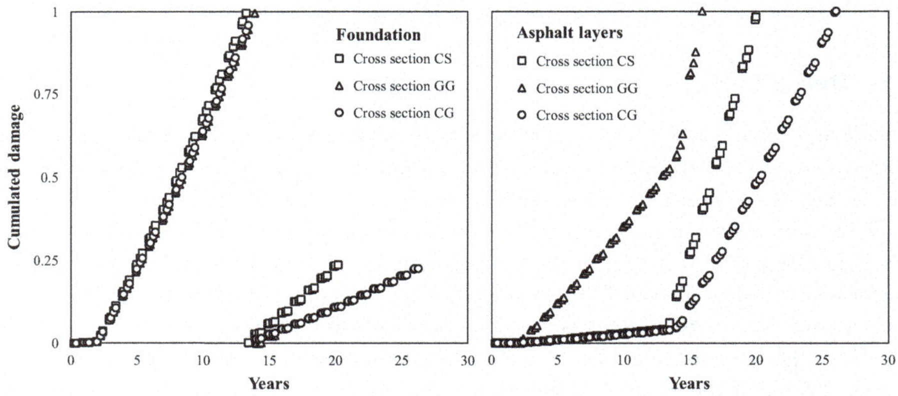

Fig. 1. Cumulated damage of considered pavement solutions

4 Conclusions

Based on the results obtained in the optimization activities illustrated in this paper, the Authors recommended the tunnel administration to adopt the innovative CG pavement cross section as the final construction solution. In fact, it was shown that such a choice leads to a non-negligible improvement of design life with respect to the CS cross section initially proposed in the preliminary design phase, thereby accommodating for unexpected variabilities occurring during construction and in service. Furthermore, it yields supplementary advantages which are related to the ease of construction (i.e. no need for subgrade compaction), to the better thermal conductivity (relevant due to the presence of a high-voltage buried line) and to the possibility of employing significant amounts of recycled materials. In such a context, it should be underlined that the laying of the SCM subgrade for a length of 1 km in the tunnel implies the use of 1,700 tons and 2,000 tons of RAP and aggregate sludge, respectively. This leads to a significant reduction of the consumption of virgin materials and of waste landfilling, thereby increasing the sustainability of construction operations.

The pavement cross section including two SCM layers (solution GG) proved to be inferior to other two solutions in terms of design life. Nevertheless, it should be considered for other future applications characterized by a design traffic corresponding to the calculated ESALs (equal to 11.6 million). Moreover, further improvements may be sought by employing high-performance asphalt mixtures which may allow designers to fully exploit the low damage rate exhibited by the SCM foundation. Finally, it should be emphasized that such a pavement solution may provide further advantages with respect to the use of recycled materials, with a further increase of the quantities of employed RAP and aggregate sludge of the order of 25%.

Although preliminary data suggest that the innovative CG and GG pavements are highly cost-effective, future investigations will include life-cycle cost analyses which will necessarily take into account environmental impacts.

References

ASTM D5334-14. Standard test method for determination of thermal conductivity of soil and soft rock by thermal needle probe procedure. ASTM International, USA

ASTM D 6103-00. Standard test method for flow consistency of controlled low-strength materials (CLSM). ASTM International, USA

Austroads (2017) Guide to Pavement Technology Part 2: Pavement Structural Design, AGPT02-17. Austroads Ltd., Sydney

Shell (1998) BISAR 3.0. Shell International Oil Products BV, The Netherlands

Bonnaure F, Gest G, Gravois A, Uge P (1977) A new method of predicting the stiffness of asphalt paving mixtures. In: Proceedings of the Association of Asphalt Paving Technologists, vol 49, pp 499–517

Choorackal E, Riviera PP, Dalmazzo D, Santagata E, Zichella L, Marini P (2019a) Performance-related characterization of fluidized thermal backfills containing recycled components. Waste Biomass Valoriz. https://doi.org/10.1007/s12649-019-00650-9

Choorackal E, Riviera PP, Santagata E (2019b) Mix design and mechanical characterization of self-compacting cement-stabilized mixtures for paving applications. Constr Build Mater (submitted)

EN 12390-3 (2003) Testing hardened concrete. Method of determination of compressive strength of concrete cubes. CEN, European Committee for Standardization, Brussels, Belgium

Ferne B (2006) Long-life pavements - a European study by ELLPAG. Int J Pavement Eng 7 (2):91–100

Merrill D, Van Dommelen A, Gaspar L (2006) A review of practical experience throughout Europe on deterioration in fully-flexible and semi-rigid long-life pavements. Int J Pavement Eng 7(2):101–109

Neher JH, McGrath MH (1957) The calculation of the temperature rise and load capability of cable systems. Trans Am Inst Electr Eng Part III 76(3):752–772

CNR (1995) Catalogo delle Pavimentazioni Stradali, CNR-BU 154. Consiglio Nazionale delle Ricerche, Rome

Riviera PP, Bertagnoli G, Choorackal E, Santagata E (2019) Controlled low-strength materials for pavement foundations in road tunnels: feasibility study and recommendations. Mater Struct 52(4). https://doi.org/10.1617/s11527-019-1367-4

SANRA (2014) South African Pavement Engineering Manual (SAPEM). South African National Roads Agency, Pretoria

Solanki P, Zaman M (2017) Design of semi-rigid type of flexible pavements. Int J Pavement Res Technol 10:99–111

WSDOT (2005) Everseries user's guide – pavement analysis computer software and case studies. Washington State Department of Transportation, Washington

Zheng J (2012) Design guide for semirigid pavements in China based on critical state of asphalt mixture. J Mater Civil Eng 25(7):899–906

Fast Falling Weight Accelerated Pavement Testing and Laboratory Analysis of Asphalt Pavements Reinforced with Geocomposites

Davide Ragni[1(✉)], Tony Montillo[2], Alessandro Marradi[2,3], and Francesco Canestrari[1]

[1] Department of Civil and Building Engineering and Architecture (DICEA), Università Politecnica delle Marche, Via Brecce Bianche, 60131 Ancona, Italy
d.ragni@pm.univpm.it
[2] Dynatest Italy Srl, Viale Togliatti 108, 50059 Vinci, Italy
[3] Department of Civil and Industrial Engineering, Università di Pisa, Largo Lucio Lazzarino 1, 56126 Pisa, Italy

Abstract. Many reinforcement systems have been designed to improve fatigue life and to prevent or mitigate reflective cracking and rutting in asphalt pavements. The goal of this research was to assess the effectiveness of asphalt pavement rehabilitation with geocomposites to limit fatigue cracking, reflective cracking and rutting by using Accelerated Pavement Testing (APT) and laboratory testing. APT was used to evaluate the pavement performance, simulating the effects of long-term vehicular traffic on the pavement structure in a relatively short period of time. Accelerated pavement tests were performed using the Fast Falling Weight Deflectometer (FastFWD) equipment in a trial section characterized by different types of interfaces (reinforced with geocomposites and unreinforced). In addition, two types of specimen were taken from the trial section to carry out laboratory testing: cores for Ancona Shear Test Research and Analysis (ASTRA) tests and beams for three point bending (3PB) tests. The APT and 3PB tests results demonstrated that the geocomposites investigated are an effective method to enhance asphalt pavement performance. Moreover, ASTRA tests showed that the application of geocomposite at the interface causes a debonding effect between asphalt layers that could be beneficial to promote stress-relieving in case of reflective cracking and thermal cracking.

Keywords: FastFWD APT · Reinforcement asphalt pavement · Geocomposites

1 Introduction

It is generally agreed that fatigue cracking, reflective cracking, thermal cracking and rutting are the major mechanisms of distress in asphalt pavements.

Maintenance and rehabilitation processes in the road networks are often performed by placing within asphalt layers reinforcement systems in order to prevent or delay the development of cracks. These systems can significantly increase the maintenance intervals of asphalt pavements resulting in a cost-effective and long-lasting pavement rehabilitation method. Several reinforcement systems from many manufacturers

M. Pasetto et al. (Eds.): ISAP APE 2019, LNCE 48, pp. 417–430, 2020.
https://doi.org/10.1007/978-3-030-29779-4_41

worldwide are available in the market. Among the others, geocomposites that are obtained by the combination of bituminous membranes and reinforcing materials represent a valid solution since such materials join the tensile properties of the reinforcements with the stress-relieving and the waterproofing effects of the bituminous membranes.

Different studies (Austin and Gilchrist 1996; Brown et al. 2001; Canestrari et al. 2015; Canestrari et al. 2018; Ferrotti et al. 2012; Nejad et al. 2016; Prieto et al. 2007; Saride and Kumar 2017; Shukla and Yin 2004; Sobhan and Tandon 2008) showed that geosynthetics can extend the pavement's fatigue life, improve resistance to reflective cracking, and increase rutting resistance. At the same time, the presence of a reinforcement inevitably causes a considerable reduction of the interlayer shear resistance at the interface (Brown et al. 2001; Ferrotti et al. 2011; Pasquini et al. 2013; Zamora-Barraza et al. 2010).

Recently, accelerated pavement testing (APT) has become an important tool to evaluate in-situ pavement performance. APT is carried out to reproduce the effects of long-term vehicular traffic and environmental loading on the pavement structure in a relatively short period of time, simulating the accumulation of damage throughout the service life (Saeed and Hall 2003).

The main goal of this research was to assess the effectiveness of asphalt pavement rehabilitation with geocomposites to limit fatigue cracking, reflective cracking and rutting. A full-scale trial section characterized by different types of interfaces (reinforced with geocomposites and unreinforced) were built and accelerated pavement tests were performed using Fast Falling Weight Deflectometer (FastFWD).

In addition, two sets of specimens were taken from the trial section to carry out laboratory testing: cores for Ancona Shear Test Research and Analysis (ASTRA) tests and beams for three point bending (3PB) tests. Laboratory results were compared with those derived from a previous experimental campaign carried out on specimens prepared in the laboratory by employing the same geocomposites.

2 Experimental Program

2.1 Trial Section

A full-scale trial section was constructed in July 2018 on an existing pavement inside an industrial area, including three test fields characterized by different interfaces as follows:

1. reinforced with geocomposite (coded as R1);
2. reinforced with geocomposite (coded as R2);
3. unreinforced with a tack-coat interface used as reference for comparison purposes (coded as UN).

Each test field (8 m long and 3 m wide) is in turn divided into 2 sections for studying the performance in terms of fatigue and reflective cracking, coded as FC (fatigue cracking) and RC (reflective cracking), respectively.

The existing pavement was composed of 90 mm of asphalt concrete laid on 300 mm thick granular subbase course. The new pavement required the milling of 60 mm asphalt concrete before the construction of the trial section. After completion of the milling activities, in each section RC, the milled surface was cut simulating a pre-existing cracks network, with a square 20 cm mesh, in order to analyze the initiation of reflective cracks (Fig. 1). Right after, the reinforced test fields were prepared by placing the geocomposites directly over the milled surface (R1 and R2), whereas a conventional bituminous emulsion was applied on the milled surface of the unreinforced section (UN). Afterwards, the wearing course layer (50 mm thick) was laid-down and compacted.

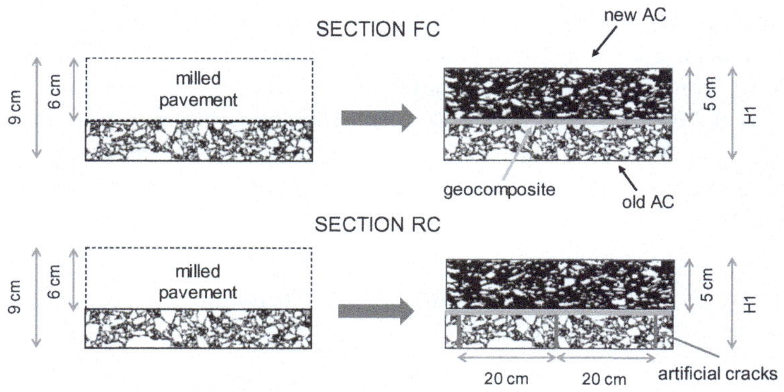

Fig. 1. Scheme of the experimental sections FC and RC

2.2 Materials

The asphalt concrete (AC) used for the wearing course was a typical dense graded asphalt mixture, with a maximum aggregate dimension size of 12.5 mm (AC 12.5) and bitumen content of 4.4% by aggregate mass.

Two geocomposites (R1 and R2), with a thickness of 2.5 mm, were used as reinforcement for this experimental study. Both geocomposites were manufactured with a Styrene-Butadiene-Styrene (SBS) polymer modified bitumen and the lower surface was characterized by an auto-adhesive film, whereas the upper surface was a polypropylene non-woven fabric. According to the product datasheet, the upper side was characterized by a melting temperature around 130–140 °C. The main difference between the two geocomposites was the type of geosynthetic within the geomembrane. R1 was reinforced with a stabilized continuous fiberglass fabric, which was characterized by a nominal tensile strength of 40 kN/m, whereas R2 was reinforced with a non-woven polyester fabric and multidirectional fiberglass, which was characterized by a nominal tensile strength of 35 kN/m. Moreover, R2 was characterized by a tensile elongation at rupture higher compared to R1 in both the longitudinal and transverse direction (i.e. 30% and 6%, respectively).

2.3 Testing Program and Procedures

The testing program was divided into two main parts. The first part was focused on the in-situ investigation of the full-scale trial section carried out by using APT tests. The second part was based on the laboratory investigation carried out by using ASTRA and 3PB tests on cores (diameter = 96 mm) and slabs (305 × 85 × 80 mm^3), respectively, taken from the three experimental test fields. It is worth noting that slabs were cut at the same locations where APT tests were performed. Moreover, laboratory results on field specimens were compared with the corresponding results of laboratory-compacted slabs prepared with the same geocomposites at the interface.

APT tests were performed by means of a FastFWD device developed by Dynatest in 2015 (Manosalvas-Paredes et al. 2017). In order to speed up the experimental procedures reaching a loading rate from 5 to 7.5 times faster than FWD device. The FastFWD was configured with a 300 mm diameter loading plate with nine geophones positioned at 0, 200, 300, 450, 600, 900, 1200, 1500 and 1800 mm from the center of the loading plate. Two APT test sessions were performed in each test field (both on section FC and section RC). The number of load applications and the load levels are shown in Table 1.

Table 1. Summary of APT tests program

Session	Section FC			Section RC		
	Drops	Fall height	Load	Drops	Fall height	Load
	[n°]	[–]	[kPa]	[–]	[–]	[kPa]
1*	13500	H2	750	6000	H2	500
	1000	H3	1050	500	H3	1050
2**	900	H1	500	600	H1	500
	2100	H4	1700	1400	H4	1700

*performed on 16–20 July 2018
**performed on 02–03 August 2018

The measured deflections from FastFWD tests were analyzed to estimate the equivalent half-space stiffness modulus Ee (using the Boussinesq equations according to Ullidtz 1998) as well as layer moduli (by means of Elmod 6 back-calculation software) and evaluate their evolution over testing time. Figure 2 shows the cross-sections of the test pavement used for the analysis. As regards back-calculated layer moduli, it worth noting that, E1 is not exactly the AC modulus, since the structural package, with thickness H1 (Fig. 1), was composed by two asphalt layers with different materials and an interface with or without geocomposite. Thus, the variations of E1 are representative of the overall response of the two AC layers allowing indirect evaluations on the behavior of the interface, in particular, in presence of geocomposites. The surface temperatures were measured by using an infrared thermometer, whereas the layer temperatures were measured through a thermometer placed in a drilled hole in the pavement.

In addition, the final permanent deformation of the pavement was measured at the end of test session 2 to determine the rutting performance of each test section (Fig. 3).

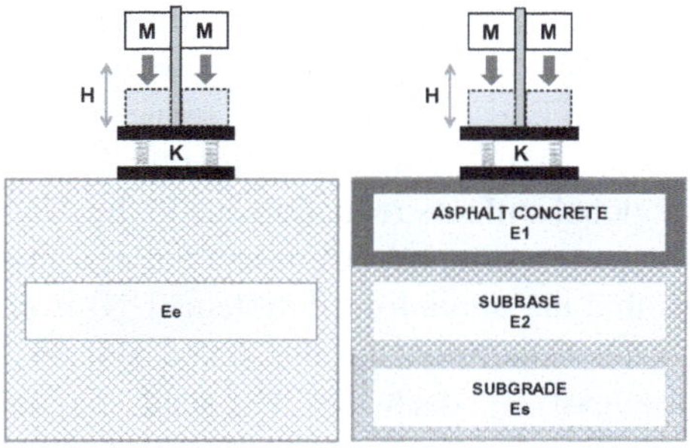

Fig. 2. Cross-sections of the test pavement used for the analysis

Fig. 3. Rutting at the end of APT tests

Table 2. Summary of laboratory tests program

Test field	Temperature	ASTRA test repetitions			3PB test repetitions
	[°C]	σ = 0.0 MPa	σ = 0.2 MPa	σ = 0.4 MPa	v = 50.8 mm/min
R1	20	3	3	3	3
	40	3	3	3	–
R2	20	3	3	3	3
	40	3	3	3	–
UN	20	3	3	3	3
	40	3	3	3	–

As regards the laboratory test program (Table 2), ASTRA tests were carried out at two temperatures (20, 40 °C) and three normal stress conditions (σ = 0.0, 0.2, 0.4 MPa) performing three repetitions for each test configuration. 3PB tests were carried out at 20 °C and at a constant rate of 50.8 mm/min performing three repetitions for each test configuration.

3 Results and Analysis

3.1 APT Tests

Figures 4 and 5 show the values of the equivalent stiffness modulus of the half-space Ee as a function of the number of drops and the test sessions, for the sections FC and RC, respectively. As shown in Fig. 4 (sections FC), reduced values of Ee are associated to the reinforced test fields (R1 and R2) highlighting how the application of the geocomposite leads to a global stiffness reduction due to the debonding phenomenon at the interface between the pre-existing pavement and the new layer in asphalt concrete. Results in Fig. 4 shows that the unreinforced system (UN) is characterized by a constant decrement rate of Ee. Contrarily, the reinforced systems (R1 and R2) show a decrement rate that progressively decreases and tends to stabilize asymptotically, highlighting the positive effect due to the presence of the geocomposite. Moreover, it can be observed that from session 1 to session 2 (i.e. increase in load), the difference between unreinforced and reinforced systems is reduced, due to the beneficial role given by the geocomposite which enhances its contribution starting from the initiation phase and subsequent propagation phase of the cracks. Analogous considerations can be drawn looking at the results of sections RC showed in Fig. 5.

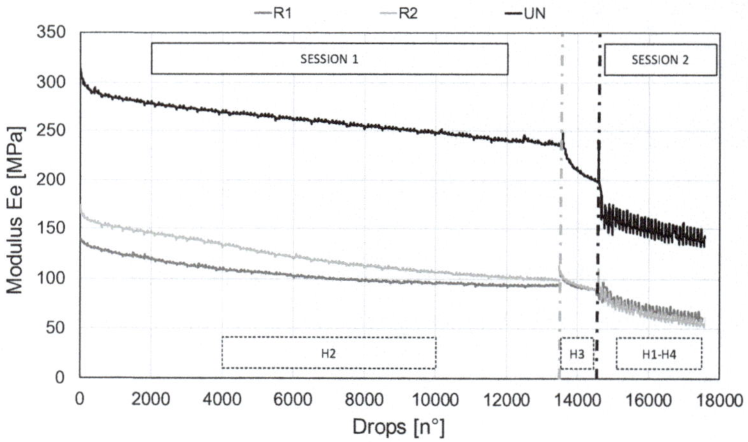

Fig. 4. Evolution of modulus Ee during APT tests on sections FC (test fields R1, R2, UN)

Comparing the sections FC and RC, it is possible to observe lower values of Ee for the sections RC compared to those calculated in the sections FC. This is due to the presence of the artificial cracks in the first layer of the pavement (see Fig. 1) that causes a structural weakening. However, it should be noted that the values of Ee are not sufficient for explaining the true contribution of the geocomposite within the layers in asphalt concrete since the pavement response (in terms of measured deflection) also depends on the values of E2 (subbase) and Es (subgrade).

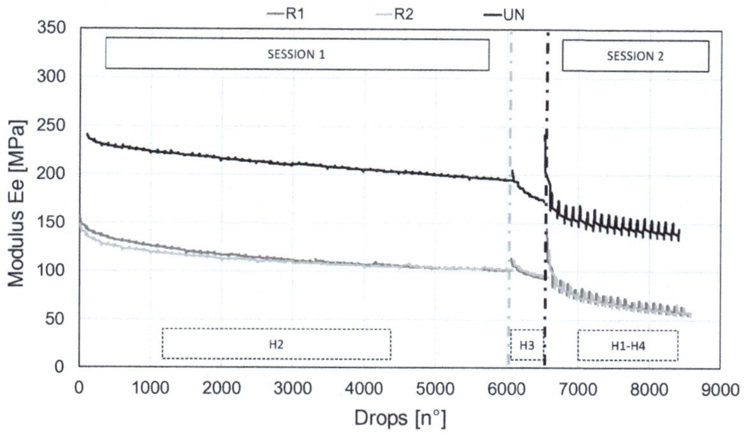

Fig. 5. Evolution of modulus Ee during APT tests on sections RC (test fields R1, R2, UN)

The data related to the modulus E1 were analyzed using a damage mechanics-based approach to better understand the behavior of the reinforced pavement allowing also a meaningful comparison between different test fields for both sections FC and RC. Damage can be described through a dimensionless scalar variable (D), whose value ranges from 0 (undamaged material) to 1 (complete failure). With this approach, all the values of the modulus E1 obtained during the FastFWD sessions were converted to the corresponding values related to a reference temperature (within the range recorded during the test) and normalized with respect to the initial value (undamaged modulus) at the same reference temperature. The abovementioned modulus ratio can be expressed as a function of the damage in order to detect the level of residual integrity of the material at a given point in the loading history (Eq. (1)).

$$(1 - D) = \frac{E_1}{E_{1,0}} \tag{1}$$

where D is the damage variable, E_1 is the AC modulus at a reference temperature and $E_{1,0}$ is the initial value of AC modulus at the same reference temperature.

Figures 6 and 7 show the parameter (1 – D) as a function of the number of drops at the test session 1 for the sections FC and RC, respectively, in which the dashed line indicates the conventional value of critical damage assumed equal to 0.5. As shown in Fig. 6, considering the load H2 (13500 cycles, 50 kN), it can be noticed that the unreinforced test field (UN) shows an unexpected trend in the final part in contrast with the rapid initial decrease of the parameter (1 – D), whereas R1 and R2 seems to be affected by a major damage. On the contrary, considering the load H3 (1000 cycles, 75 kN), UN appears to be affected negatively by the increased load level, showing a rapid decrease in the parameter (1 – D), whereas R1 and R2 show values of the parameter (1 – D) which tend to suffer less damage rate as the number of drops increases.

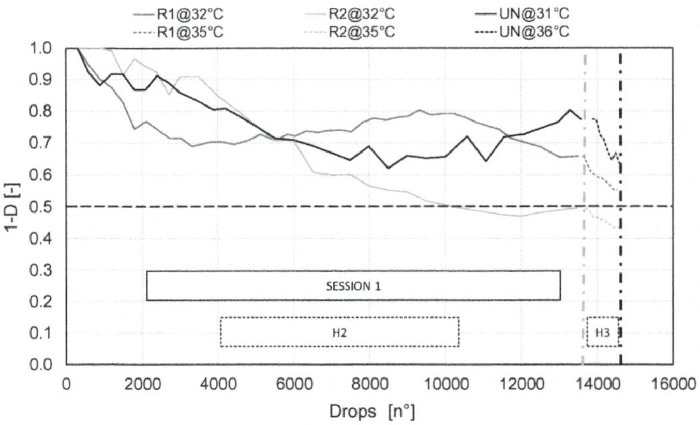

Fig. 6. Evolution of damage during the APT test (session 1) in the different test fields (sections FC)

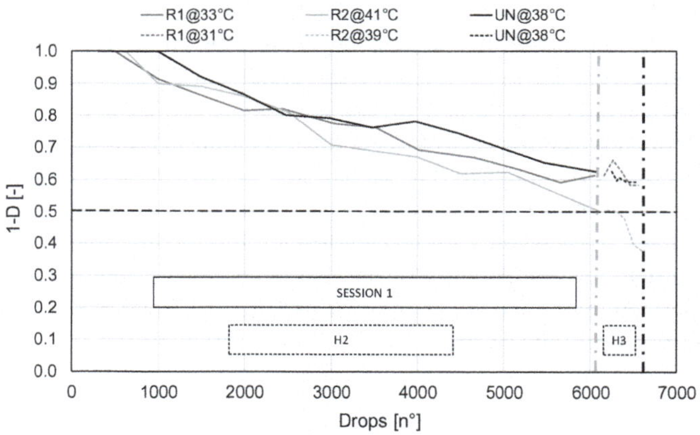

Fig. 7. Evolution of damage during the APT test (session 1) in the different test fields (sections RC)

With regard to sections RC, considering both load levels H2 (6000 cycles – 50 kN) and H3 (500 cycles – 75 kN), from Fig. 7 emerges a very similar behavior for all the test fields. The only exception is attributable to the behavior of R2 with the load H3 in which the decrease in the parameter (1 – D) is more pronounced.

It is important to highlight that no cracks were detected on the pavement surface at the end of FastFWD sessions for both sections FC and RC. These data suggest that geocomposites could offer a positive contribution only when a certain degree of damage in the pavement occurred. In fact, geocomposites are not able to provide increased stiffness to the pavement structure but, according to other studies (Austin and Gilchrist 1996; Shukla and Yin 2004; Pasquini et al. 2013), they could extend pavement life by retarding the propagation of fatigue and reflective cracking.

The FastFWD tests allowed also the evaluation of the behavior in terms of rutting resistance of the investigated test fields. Figure 8 shows the permanent deformation recorded beneath the load plate at the end of session 2, for all the interface configurations (i.e. test fields R1, R2, UN) and sections (FC and RC). In the sections FC (Fig. 8a), there is a maximum permanent deformation of about 20 mm, higher than the limit of 13 mm (half inch) usually associated with the pavement crisis due to the excessive accumulation of permanent deformations. Such a high value of permanent deformation can be mainly attributed to a plastic yielding of the unbounded layers, since asphalt pavement was composed of a total thickness of only 8 cm. Moreover, it is possible to observe that the reinforced systems show an increase in permanent deformation resistance respect of the unreinforced system, despite the presence of higher values of the modules E2 and Es of the unbound layers in the UN test field (Table 3).

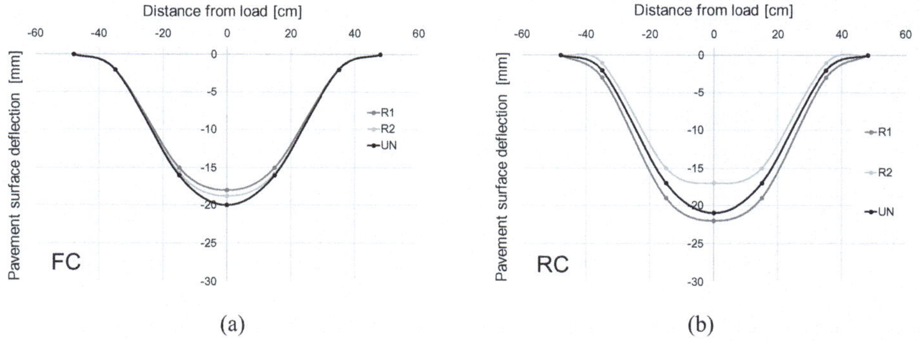

Fig. 8. Depth of permanent deformations in the sections FC (a) and RC (b) of the trial section

Table 3. Moduli of subbase (E2) and subgrade (Es) layers

Test fields	E2 [MPa]	Es [MPa]
R1	89	49
R2	81	57
UN	163	78

Permanent deformations higher than 13 mm can be still observed in Fig. 8b also for the sections RC. It is emphasized that these results were achieved for the RC sections by applying a number of load repetitions equal to about half of those selected for the sections FC (see Table 1), due to the weakening of the structural package induced by the realization of artificial cracks. In the case of the sections RC, the comparison of permanent deflection allows detecting diversified performance for the two geocomposites (about 10 mm). In particular, R2 provides better resistance against permanent deformations, nevertheless, the reinforcement was applied on subbase and subgrade layers with low stiffness moduli (Table 3).

As final comment, given the thin thickness of the AC layers, it can be stated that the APT tests were able to better investigate the performance behavior in terms of rutting resistance highlighting the positive effect induced by the application of the geocomposites. Such a result can be explained with the spreading of reduced and more even stress-strain distribution on the top of the subgrade in accordance with previous study (Manosalvas-Paredes et al. 2017).

3.2 Laboratory Tests

For each interface configuration, Fig. 9 shows the results of the ASTRA test in terms of the peak envelopes obtained applying three normal stresses (σ = 0.0, 0.2, 0.4 MPa) and two temperatures (20 and 40 °C). In accordance with several authors (Brown et al. 2001; Ferrotti et al. 2012; Pasquini et al. 2013; Zamora-Barraza et al. 2010), the results show how the presence of the reinforcement at the interface produces a debonding between the two asphalt concrete layers in contact. However, it is necessary to highlight that such difference is less marked at 40 °C (Fig. 9b). This indicates that, at higher temperatures, the maximum interlayer shear stress (τ_{peak}) is controlled mainly by the characteristics of the two AC layers in contact. However, the two geocomposites (R1 and R2) provide similar behavior at 20 and 40 °C.

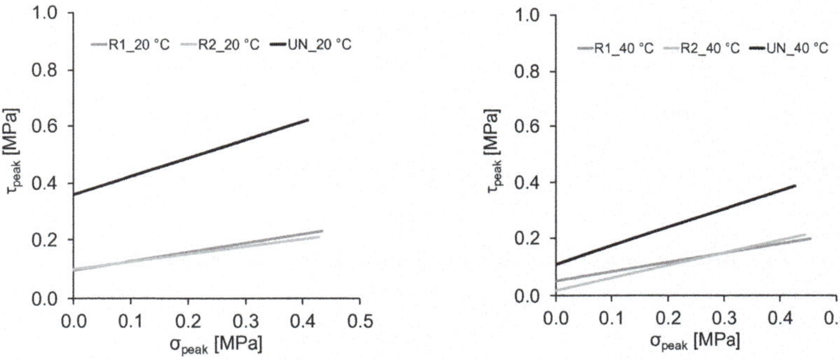

(a) (b)

Fig. 9. ASTRA test results: peak envelopes (IN-SITU) at 20 °C (a) and 40 °C (b)

ASTRA test results achieved in the present study were compared with those obtained in previous research (Fig. 10). The latter results were obtained performing ASTRA test, in the same testing conditions, on double-layered asphalt concrete specimens compacted in the laboratory using two different asphalt concrete (with polymer modified bitumen, coded BM, and plain bitumen, coded BU) and fabricated with the same geocomposites (R1, R2, UN). Looking at the results in Figs. 9 and 10, it can be observed the decrease of peak shear resistance for the in-situ specimens compared to the laboratory specimens, for all temperatures and interface conditions. This can be explained through non-optimal in-situ compaction and a non-perfect cohesion between the upper side of the geocomposites in polypropylene non-woven fabric and the AC due to a value of the melting temperature of the polypropylene (130–140 °C) too close to the compaction temperature. Although a certain grade of debonding is positive in presence of existing crack on the lower layer of an old milled pavement allowing the geocomposite to behave also as a stress-absorbing membrane interlayer (SAMI), further studies are currently in progress in order to evaluate possible improvement of the shear resistance of reinforced interfaces by coating the upper side of the geocomposite with a fine sand instead of polypropylene non-woven fabric.

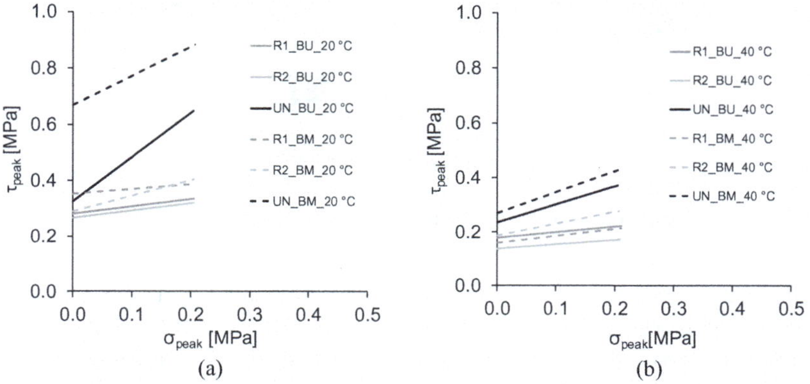

Fig. 10. ASTRA test results: peak envelopes (LAB) at 20 °C (a) and 40 °C (b)

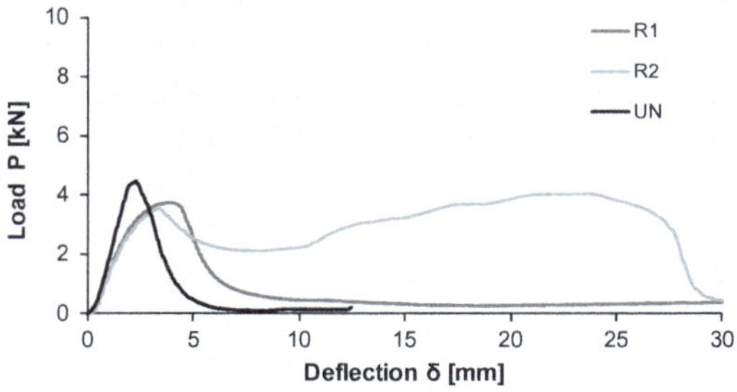

Fig. 11. Three point bending (3PB) test results

The flexural behavior of the double-layered specimens was investigated by carrying out 3PB tests whose results are shown in Fig. 11 in terms of load-deflection ($P - \delta$) curves. As can be observed, the unreinforced interface (UN) lead to slightly higher values of flexural strength point (P_{max}), occurring at lower deflection, compared to the reinforced interfaces (R1 and R2), this is ascribable to the de-bonding effect at the interface (Fig. 9). When UN interface reaches P_{max} value rapidly lose their resistance until complete failure (no residual flexural resistance), whereas R1 and, especially, R2 interfaces show a post-peak deformation phase. As expected, these finding suggests that the presence of the geocomposite at the interface delays the crack-propagation phase in the upper layer allowing a considerable increase in terms of flexural resistance.

A more direct comparison is shown in Fig. 12, where the failure mechanisms of the investigated 3PB test specimens are reported. It can be observed that the reinforced interfaces bring to horizontal crack propagation patterns (especially R2) coherently with the higher energy at failure registered during the 3PB tests (represented by area under the $P - \delta$ curve).

Fig. 12. Failure mechanisms of specimens in 3PB test

Also in this case (Fig. 13), 3PB results were compared with those obtained in previous research where the interface configurations (R1 and R2), prepared in laboratory by using two asphalt concrete (with polymer modified bitumen, coded as BM, and plain bitumen, coded as BU), were tested with 3PB equipment at 50.8 mm/min and 20 °C. From Fig. 13, it can be observed that in-situ specimens provide lower P_{max} with respect to laboratory specimens (both BM and BU). The lower interlayer shear resistance may have affected the reduction in the value of the P_{max} recorded on in-situ specimens for the geocomposites R1 and R2 (Fig. 9). In addition, since the in-situ specimens were taken from at the same locations where the FastFWD were performed, the same performance of an undamaged material cannot be expected due to the dissipated energy undergone by the in-situ specimens. This comment is supported also noting that both R1 and R2 geocomposites provide the same post-peak trend of the laboratory and in-situ compacted specimens.

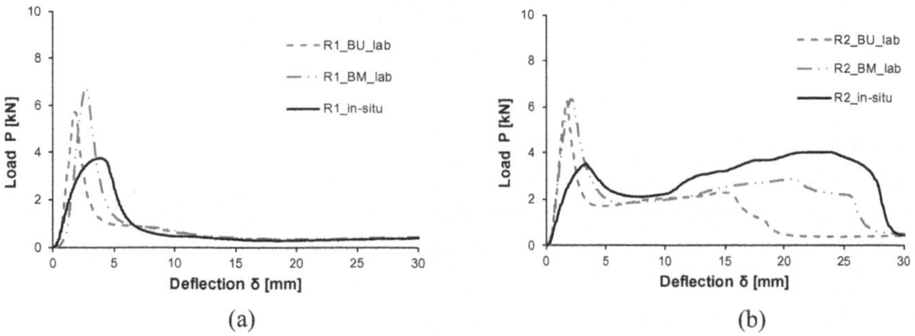

Fig. 13. 3PB test results comparison IN-SITU-LAB: R1 (a); R2 (b)

4 Conclusions

This paper summarizes the results of an experimental study focused on the evaluation of asphalt pavement rehabilitated with geocomposites. To this end, a full-scale test field was constructed with 3 different interfaces (R1, R2, UN), characterized by 2 sections (FC and RC). The experimental program involved APT tests carried out by means of FastFWD and laboratory tests (ASTRA and 3PB) performed on the specimens taken directly from each section. Laboratory results were compared with tests performed on laboratory-fabricated specimens in order to evaluate the reliability of laboratory tests for predicting the effectiveness of the geocomposites.

Based on the experimental results, the following main conclusions can be drawn:

- APT tests allowed to detect the positive effects due to the presence of the geocomposite, although the fatigue and reflection cracks have not spread to the surface due to the limited number of load repetitions. For both sections FC and RC, the results suggested that geocomposites did not contribute to increasing the overall modulus of pavement structure but provided a contribution to the delay of damage propagation.
- Given the thin thickness of the AC layers (less than 10 cm), the APT tests were able to better investigate the performance behavior in terms of rutting resistance highlighting the positive effect induced by the application of the geocomposites. Such a result can be explained with the spreading of reduced and more even stress-strain distribution on the top of the subgrade. This outcome is further supported taking into account that, despite the reduced stiffness of the subbase and subgrade layers, the test fields reinforced with the geocomposites (R1 and R2) exhibited high performances in terms of rutting resistance.
- In contrast to the specimens fabricated in the laboratory, the in-situ specimens 3PB tests showed an alteration of the P_{max} (reduced peak values) compatible with the damage induced by the FastFWD tests and with the reduction of interlayer shear resistance denoted during the ASTRA test. On the other hand, the behavior in the crack propagation phase for the in-situ specimens is perfectly comparable with the results of the specimens fabricated in the laboratory, providing excellent results especially for the geocomposite R2.

Acknowledgements. The activity presented in the paper was sponsored by Copernit S.p.A. (Italy) that gave both financial and technical support for the research project. The results and opinions presented are those of the authors.

References

Austin RA, Gilchrist AJT (1996) Enhanced performance of asphalt pavements using geocomposites. Geotext Geomembr 14(3–4):175–186

Brown SF, Thom NH, Sanders PJ (2001) A study of grid reinforced asphalt to combat reflection cracking. J Assoc Asphalt Paving Technol 70:543–569

Canestrari F, Belogi L, Ferrotti G, Graziani A (2015) Shear and flexural characterization of grid-reinforced asphalt pavements and relation with field distress evolution. Mater Struct 48 (4):959–975

Canestrari F, D'Andrea A, Ferrotti G, Graziani A, Partl MN, Petit C, Raab C, Sangiorgi C (2018) Advanced interface testing of grids in asphalt pavements. In: Partl M, Porot L, Di Benedetto H, Canestrari F, Marsac P, Tebaldi G (eds) Testing and characterization of sustainable innovative bituminous materials and systems RILEM state-of-the-art reports 24. Springer, Cham, pp 127–202

Ferrotti G, Canestrari F, Virgili A, Grilli A (2011) A strategic laboratory approach for the performance investigation of geogrids in flexible pavements. Constr Build Mater 25:2343–2348

Ferrotti G, Canestrari F, Pasquini E, Virgili A (2012) Experimental evaluation of the influence of surface coating on fiberglass geogrid performance in asphalt pavements. Geotext Geomembr 34:11–18

Manosalvas-Paredes M, Navarro Comes A, Francesconi M, Khosravifar S, Ullidtz P (2017) Fast falling weight deflectometer (FastFWD) for accelerated pavement testing (APT). In: Loizos et al (eds) 10th international conference on the bearing capacity of roads, railways and airfields, 28–30 June 2017. Taylor & Francis Group, Athens, pp 1267–1274

Nejad FM, Asadi S, Fallah S, Vadood M (2016) Statistical-experimental study of geosynthetics performance on reflection cracking phenomenon. Geotext Geomembr 44:178–187

Pasquini E, Bocci M, Ferrotti G, Canestrari F (2013) Laboratory characterisation and field validation of geogrid-reinforced asphalt pavements. Road Mater Pavement Des 14:17–35

Prieto JN, Gallego J, Perez I (2007) Application of the wheel reflective cracking test for assessing geosynthetics in anti-reflection pavement cracking systems. Geosynth Int 14(5):287–297

Saeed A, Hall JW (2003) NCHRP Report 512: Accelerated Pavement Testing: Data Guidelines. Transportation Research Board, National Research Council, Washington, D.C

Saride S, Kumar VV (2017) Influence of geosynthetic-interlayers on the performance of asphalt overlays on pre-cracked pavements. Geotext Geomembr 45:184–196

Shukla SK, Yin JH (2004) Functions and installation of paving geosynthetics. In: 3rd asian regional conference on geosynthetics, Seoul, 21–23 June 2004

Sobhan K, Tandon V (2008) Mitigating reflection cracking in asphalt overlay using geosynthetic reinforcements. Road Mater Pavement Des 9(3):367–387

Ullidtz P (1998) Modelling flexible pavement response and performance. Polyteknisk Forlag, Lyngby

Zamora-Barraza D, Calzada-Pérez MA, Castro-Fresno D, Vega-Zamanillo A (2010) New procedure for measuring adherence between a geosynthetic material and a bituminous mixture. Geotext Geomembr 28(5):483–489

Automatic Crack Detection Results Using a Novel Device for Survey and Analysis of Road Pavement Condition

Gaetano Bosurgi[1(✉)], Giuseppe Sollazzo[2], Nicola Bongiorno[1], and Orazio Pellegrino[1]

[1] Department of Engineering, University of Messina, Vill. S. Agata, C. da di Dio, 98166 Messina, Italy
gbosurgi@unime.it
[2] Department of Engineering, University of Palermo, Viale delle Scienze, Ed. 8, 90128 Palermo, Italy

Abstract. To accurately schedule maintenance operations, it is crucial to monitor the pavement state and, thus, evaluate the structural and functional indices during its entire service life. In particular, surface conditions (cracks, potholes, patches, rutting, etc.) must be properly checked – especially in terms of detection, identification, and classification of distresses – for avoiding reaching dangerous values for users (overpassing safety thresholds). However, monitoring and survey activities can be so costly (not only in economic terms, but further considering the execution time) to discourage their execution, with evident drawbacks for a proper maintenance management and the related intervention strategies. In view of the above, a novel device for real-time and continuous survey of road pavement surface distresses was developed in the Infrastructure Lab of the University of Messina. Using the potential of laser lighting systems and 3D cameras for optical triangulation, this device can acquire 3D images of the road surfaces, limiting the disturbance on vehicular traffic and speeding up condition survey operations. In this paper, the authors present some survey tests and the related outcomes of this novel device, evidencing the applications of specific algorithms for distress detection and quantification.

Keywords: Condition surveys · High-performance survey · 3D pavement data · Crack detection

1 Introduction

Pavement maintenance is a key aspect in the road asset management, for assuring high level of comfort for users and significant savings for road agencies. At this regard, continuous and accurate monitoring procedures to evaluate the state of the pavement and its defeats are essential. In general terms, a proper scheduling of the maintenance activities has to rely on reliable and frequent measurements of pavement structural and functional indices (Sollazzo et al. 2017). In particular, surface conditions (cracks, potholes, patches, rutting, etc.) must be properly checked – especially in terms of

M. Pasetto et al. (Eds.): ISAP APE 2019, LNCE 48, pp. 431–440, 2020.
https://doi.org/10.1007/978-3-030-29779-4_42

detection, identification, and classification of distresses – for avoiding reaching dangerous values for users (AASHTO 2008). Traditionally, pavement condition surveys rely on human visual inspection, performed by surveying operators walking by the side of the road and recording the observed distresses. However, human inspection is dangerous, slow, time-consuming, expensive, and also not objective, reliable, or robust (Zalama et al. 2014; Koch et al. 2015). Consequently, various systems and technologies have been developed aiming to perform frequent and economically efficient surveys with improved data quality and output results (Wang and Gong 2005). The first generation of digital image acquisition systems using customized lighting devices experienced inconsistent image quality due to the presence of shadow and sunlight (Wang 2011). Later, the introduction of laser illumination technology overpassed this problem, assuring consistent surface image data at up to 1-mm horizontal resolution in real time at any time of the day (Wang 2000). Nowadays, with progresses in hardware sensors and data storage, innovative survey systems permit to collect data at high speed and with consistent resolution. Moreover, recently, various researchers have further improved these high-efficiency survey devices, providing not only digital high-quality images of the pavement surface, but also high-resolution point clouds for reliable 3D surface reconstructing (Wang et al. 2011; Sollazzo et al. 2016). Customized 3D laser camera sensors and optics produce a complete point cloud with accurate height information on pavement surfaces. With the advances of the survey technology and system performances, the quality of the collected data has reached an adequate level (Wang 2011; Mathavan et al. 2015), while the detection and classification algorithms still need improvements and optimization.

In this paper, some specific results of a crack detection algorithm based on Matched Filtering (MF) are provided. Tests are performed on images acquired through a novel device for real-time and continuous survey of road pavement surface distresses, developed in the Infrastructure Lab of the University of Messina. Using a laser-based lightning system, it can acquire at traffic speed accurate 3D images of the pavement surface, with almost no disturbance on traffic. The collected data, after specific pre-processing procedures, are then used for automatic crack detection, using a MF-based algorithm. Despite the recent optimistic results provided by Deep Learning (DL) based techniques (Zhang et al. 2017, 2018) in crack and defeat identification and recognition, there is still a need for improving simpler but enough-accurate "low-cost" methodologies, able to detect cracks properly, limiting resource consumption in terms of economic and processing costs. The MF approach, manually tuned by expert operators for reliable adaptation to different scenarios and pavement textures, still represents an adequate solution for the crack recognition issue. Similarly, analogous simplified algorithms able to properly detect other kinds of distresses may represent a valuable alternative to the complex DL-based solutions.

2 The Novel Device for Pavement Condition Survey

As previously said, the presented experiments were carried out on a novel database of 3D images, acquired in the city of Messina (ITALY) through a custom equipment defined and developed in the Infrastructure Lab of the University of Messina. The device is installed on the mobile laboratory and represents a preliminary prototype.

The main elements of the equipment are:

- 3D cameras;
- Laser lightning system;
- Inertial Measurement Unit;
- GPS antenna and encoder.

Image acquisition is performed by a couple of 3D cameras, able to provide a 3D representation of the pavement surface, up to 3.70 m wide. According to the scheme provided in Fig. 1, using the optical triangulation principle (Ouyang and Xu 2013), they acquire perspective views of the pavement, further analysed and processed for determining the height of the different points of each single frame. In the actual configuration, the resolution of each camera is 2048 × 100 pixels. According to the position of the optical means, each pixel represents 1 mm of the pavement surface. Naturally, the length of each frame can be controlled, for specific adaptation to the vehicle speed, in compliance with the frame rate of the camera. These cameras provides both 3D and traditional 2D images of the pavement surface.

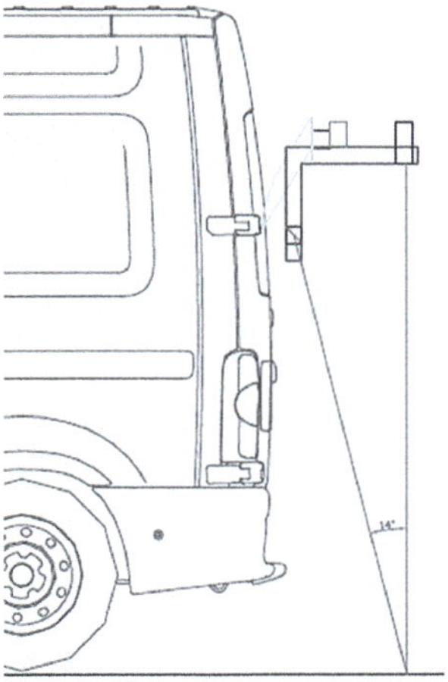

Fig. 1. Side view of the equipment with optical triangulation model (Color figure online)

The 3D reconstruction of the pavement exploits the laser lightning system. In detail, two different red lasers project two laser lines (about 2 mm wide) on the pavement surface, almost at the center of the acquired frame. The laser lightning assures a proper surface illumination. Moreover, according to the laser power and using specific filters for the cameras, acquisition quality and data accuracy are guaranteed in all the environmental conditions and the external lightning scenarios. In the actual configuration, 2 W lasers emit infrared light line (810 nm) for each camera, but different laser devices have been tested for improving response in the least favorable external conditions.

A double frequency GPS antenna and a high-precision encoder assure the accuracy of the positioning data of each acquired surface frame. Further, an Inertial Measurement Unit (IMU), including accelerometers, magnetometers, and gyroscope, provides continuous and precise data regarding the accelerations of the acquisition system. By processing this information, is then possible not only to improve the positioning estimate, but also to specifically correct the 3D representation by correcting the effects of the system relative movements from the pavement surface. This correction process (requiring not simple and specifically defined algorithms) is not relevant for crack detection, but becomes essential for other types of analyses (such as, for instance, roughness estimates).

3 Preliminary Crack Detection Results

In the development and optimization of the prototype, several configurations have been tested and numerous experiments have been performed in the city of Messina. For properly tuning all the different parameters affecting quality and accuracy of the final representation, equipment position, camera rates, frame size, laser power, and many other variables have been varied and analyzed, defining an actual stable configuration.

The frames acquired by the 3D cameras are specifically processed for providing a global reliable 3D representation of the pavement surface. For example, in Fig. 2 a raw image of a pavement surface section acquired on a rural road is shown. In practice, the model can be considered as a dense point cloud, in which each point represents 1 mm of the pavement surface. Furthermore, the same data can be also shown in a 2D representation (Fig. 3). However, in this case, the plane image is not a traditional digital reproduction of the pavement surface: in fact, each single pixel does not provide information regarding light intensity of the acquired picture, but the height of the related point. In this way, 3D data can be also processed by means of traditional computer vision techniques, but with a more consistent reliability of the visualized information (Sollazzo et al. 2016). In Fig. 4, as further example of 3D data representation, a different pavement surface section is provided, with a significant depression.

This kind of data, analyzed either in 3D formats or 2D, contains all the required information for properly performing condition surveys and analysis, in a computer-based approach. Identification and recognition of the different distresses (such as the following phases of classification and quantification) can be naturally performed in an automatic way, by means of innovative computer-vision and artificial-intelligence techniques.

At the moment, in its actual configuration the device defined in the previous paragraph can be used for:

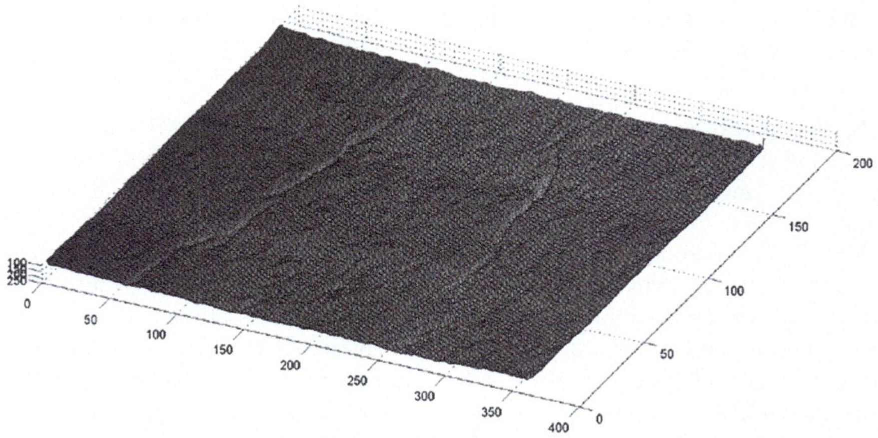

Fig. 2. 3D model of the pavement surface

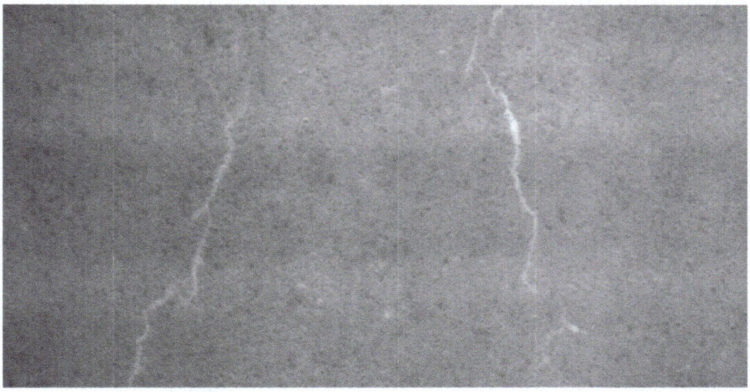

Fig. 3. 2D representation of the 3D data

Fig. 4. 3D representation of a pavement section with depression

- surface defeats detection, using both 2D and 3D information according to the specific distress features;
- rutting analysis, by processing the 3D data and, in particular, the available transversal sections;

- roughness evaluation, processing the several longitudinal profiles in the 3D point cloud and the IMU data;
- texture estimate.

In this context, avoiding details and discussions regarding the IMU-based correction of the acquired images, the focus is on crack detection. In particular, in the following the results of a MF-based algorithm for crack detection using 3D data are presented.

The MF methodology consists in designing a filter to match the main characteristics of an object, such as shape, orientation, and intensity features. In recent years, MF has been widely used for detection of fingerprints, retinal blood vessels and other objects in digital images. The methodology has also been proposed for crack detection in 2D pavement images (Zhang et al. 2013) with optimistic results. In Sollazzo et al. (2016), the methodology was applied for crack detection on 3D data as part of a hybrid approach including Tensor Voting and Minimum Spanning Tree techniques. In crack detection problems, a kernel representing a succession of Gaussian curves is generally applied to describe the intensity profile of a crack cross section (Fig. 5a and b). The filter is then rotated in different orientations for evidencing cracks in all the possible directions (Fig. 5c and d).

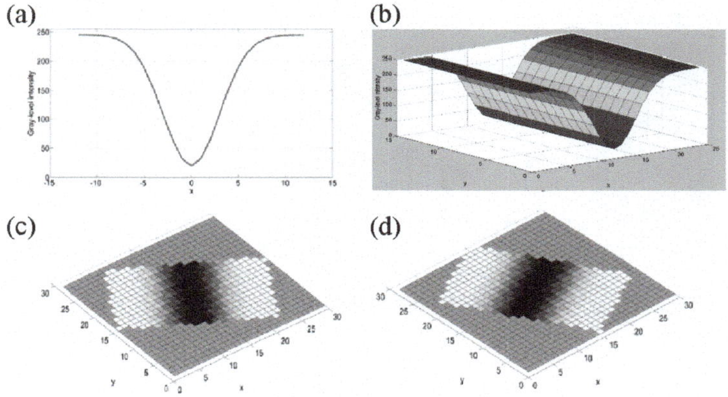

Fig. 5. Kernel of the MF: (a) Gray-level intensity profile model; (b) 3D view of the 2D kernel; (c, d) 2D kernels at different orientations

For this test, the raw MF results are further processed through a local or global thresholding algorithm for reducing noise and evidencing the final crack results. The different parameters adopted for the various tests have been initially tuned and fixed for all the collected images, as said acquired all along the same rural road in similar external lightning and texture conditions. The algorithm has been then properly coded in the Matlab © environment, designing a special software tool (Fig. 6) for parameter tuning, methodology application, and result visualization and analysis. As an example, in Figs. 7 and 8 two different tests are presented, comparing the original data (on top) with the crack results (on bottom). Images size is 5000 $\times$ 1920 pixels, or mm.

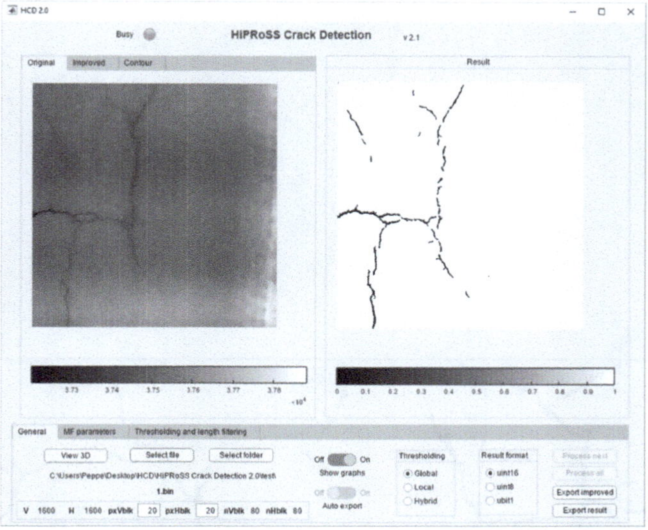

Fig. 6. The crack detection software

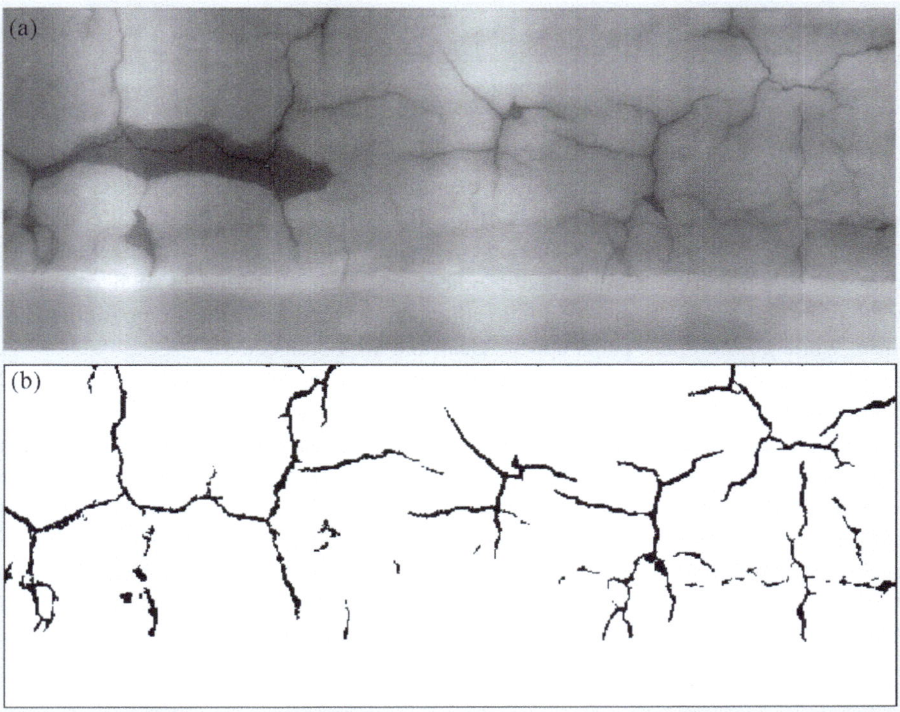

Fig. 7. Crack detection: test 1; (a) original image; results.

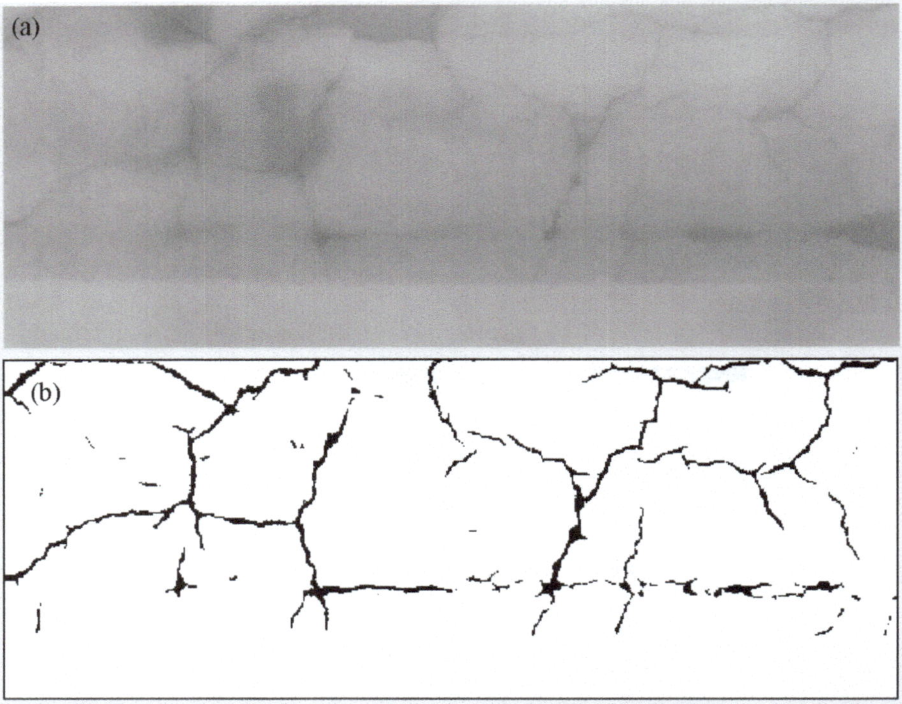

Fig. 8. Crack detection: test 2; (a) original image; results.

The results, even if only a reduced example of the method potentiality, are very optimistic and show the effectiveness of the algorithm in the experimental context. It seems that the MF method provides an acceptable approximation of the actual crack paths and produces coarse crack width indications. No segment, even small, is missed in the proposed results. Considering the adopted parameters, results are not affected by noise, while all the crack paths appear clear and well connected.

As previously said, despite the recent optimistic results provided by DL-based techniques in crack detection and analysis, there is still a need for improving simpler but enough-accurate "low-cost" methodologies. These alternative methodologies and procedures are able to detect cracks properly, with reduced acquisition and analysis costs. The MF approach, manually tuned by expert operators for reliable adaptation to different scenarios and pavement textures, still represents an adequate solution for the crack recognition issue. Moreover, coupling this approach with optimization tools for automatic parameter fine tuning can represent an affordable solution for adapting similar simple approaches to different conditions, especially in terms of lightning and pavement texture scenarios.

4 Conclusions

In this paper, the authors presented preliminary results of a MF-based crack detection algorithm on 3D images acquired by a novel survey devices developed in the Department of Engineering of University of Messina. The experiment outcomes show that the proposed cracking detection algorithm has a high accuracy and acceptable precision, appearing as a valid alternative to the more complex and costly DL methodologies. Although the methodology seems to require manual tuning for satisfactory adaptation to different scenarios and pavement textures, the detected crack paths are adequate for pavement condition estimate and maintenance planning purposes. It is found that the proposed procedure is robust in identifying the main crack paths in the road sections, providing well connected crack segments, with very low noise and false positive rate. The potentialities of the equipment and the algorithm represent only a preliminary attempt to deal with the problem. Further research could improve the reliability and adaptability of the procedure on one side, while, on the other, 3D data can be further processed by different algorithms for detecting other kinds of distresses.

References

AASHTO (2008) Mechanistic-Empirical Pavement Design Guide, Interim Edition: A Manual of Practice. AASHTO, Washington, D.C.

Koch C, Georgieva K, Kasireddy V, Akinci B, Fieguth P (2015) A review on computer vision based defect detection and condition assessment of concrete and asphalt civil infrastructure. Adv Eng Inform 29:196–210

Mathavan S, Kamal K, Rahman M (2015) A review of three-dimensional imaging technologies for pavement distress detection and measurements. IEEE Trans Intell Transp Syst 16 (5):2353–2362

Ouyang W, Xu B (2013) Pavement cracking measurements using 3D laser-scan images. Meas Sci Technol 24:105204

Sollazzo G, Fwa TF, Bosurgi G (2017) An ANN model to correlate roughness and structural performance in asphalt pavements. Constr Build Mater 134:684–693. https://doi.org/10.1016/j.conbuildmat.2016.12.186

Sollazzo G, Wang KCP, Bosurgi G, Li JQ (2016) Hybrid procedure for automated detection of cracking with 3D pavement data. J Comput Civil Eng 30(6):04016032. https://doi.org/10.1061/(ASCE)CP.1943-5487.0000597

Wang KCP, Li L, Luo W, Larkin A (2011) Potential measurement of pavement surface texture based on three-dimensional (3D) image data. In: Proceedings of the 91st transportation research board annual meeting, TRB, Washington, D.C.

Wang KCP (2000) Design and implementation of automated systems for pavement surface distress survey. J Infrastruct Syst 6(1):24–32

Wang KCP (2011) Elements of automated survey of pavements and a 3D methodology. J Mod Transp 19(1):51–57

Wang KCP, Gong W (2005) Real-time automated survey system of pavement cracking in parallel environment. J Infrastruct Syst 11(3):154–164

Zalama E, Gòmez-Garcià-Bermejo J, Medina R, Llamas J (2014) Road crack detection using visual features extracted by gabor filters. Comput-Aided Civil Infrastruct Eng 29:342–358

Zhang A, Li QJ, Wang KCP, Qiu S (2013) Matched filtering algorithm for pavement cracking detection. Transp Res Rec J Transp Res Board 2367:30–42. Transportation Research Board of the National Academies, Washington, D.C.

Zhang A, Wang KCP, Fei Y, Liu Y, Chen C, Yang G, Li JQ, Yang E, Qiu S (2018) Automated pixel-level pavement crack detection on 3D asphalt surfaces with a recurrent neural network. Comput-Aided Civil Infrastruct Eng 34(3):805–819

Zhang A, Wang KCP, Li B, Yang E, Dai X, Peng Y, Fei Y, Liu Y, Li JQ, Chen C (2017) Automated pixel-level pavement crack detection on 3D asphalt surfaces using a deep-learning network. Comput-Aided Civil Infrastruct Eng 32(10):805–819

Experimental Evaluation of Improving Effects of Thermal Environment of Water Retaining Pavement on Wheelchair Users

Masashige Aoki[1(✉)], Yasuhiro Shimazaki[2], and Kenji Karaki[1]

[1] Institute of Research and Development, Taisei Rotec Corporation, 1456 Kamiya, Kounosu City, Saitama 365-0027, Japan
masashige_aoki@taiseirotec.co.jp

[2] Department of Architecture and Civil Engineering, Toyohashi University of Technology, 1-1 Hibarigaoka, Tempaku-cho, Toyohashi, Aichi 441-8580, Japan

Abstract. The temperature on the asphalt pavement (AS) reaches nearly 60 °C in summer daytime, for example, and this may cause larger thermal loads on humans and heat-related disorders. In order to design a better environment by improving ground coverings, we investigated the basic characteristics of the water retentive block pavement (WR). WR is composed of porous blacks and exerts the function of suppressing the temperature-rise of the road surface by depriving the vaporization from the fully retained moisture in porous blocks. We previously developed a novel method for evaluating human thermal environment on pavements by observing typical climatic variables and additional physiological variables. Thus in this study, we firstly examined a subjective experiment on WR with standing-still posture under summer outdoor radiative condition for evaluating thermal environment based on the method. Next, we conducted another subject experiment on WR in a wheelchair. Our results showed that thermal loads on humans on WR tended to be lower than that on the conventional asphalt pavement. Since the height of person affects the amount of radiative heat receiving from ground surface, WR significantly improved thermal load on human in a wheelchair. Totally, we verified that WR was a possible solution for realizing the better thermal environment by improving ground coverings.

Keywords: Urban heat island (UHI) · Water retentive block pavement (WR) · Human thermal environment · Physiological variables · Universal design

1 Introduction

In recent years, the thermal environment has been harsh due to global warming and the influence of the urban heat-island phenomenon (UHI) in Japan. One cause of UHI is artificial ground surface coverings such as the asphalt pavement. Since the temperature on the asphalt pavement surface reaches nearly 60 °C in summertime, a person who stays on the conventional Asphalt Pavement (AS) receives a great amount of radiant heat

M. Pasetto et al. (Eds.): ISAP APE 2019, LNCE 48, pp. 441–449, 2020.
https://doi.org/10.1007/978-3-030-29779-4_43

from the road surface and the risk of the heat stroke increases. Radiant heat as referred here is mainly long-wave radiation that depends on road surface temperature conditions.

One paving technique for suppressing the "higher road surface temperature in summertime" is Water Retentive (WR) Block Pavement. WR pavement deprives the road surface of vaporization heat when water retained in the water retentive block is vaporized and subsequently suppresses a temperature rise in the road surface (Oguri et al. 2007). This is a technique utilizing energy exchange by latent heat transport similar to the effect rendered by soil or green spaces which are natural covering surfaces. So it is possible to reduce radiation from the environment and the air temperature by lowering the road surface temperature.

Attempts were therefore made to improve this harsh thermal environment of the road surface by temperature rise suppression pavement such as WR pavement. In previous studies, evaluations for the thermal environment were mainly based on environmental physical quantities (Kosaku et al. 2007), that is data from physical environmental factors such as road surface temperature, air temperature and humidity on the pavement. The WBGT (Wet Bulb Globe Temperature) value by black bulb thermometer are sometimes reported. However, just measuring the environmental physical quantities, does not allow directly evaluating how changes in these quantities affects people on the pavement.

Therefore, in addition to environmental physical quantities used for conventional evaluations, the authors have focused on a new quantitative evaluation of the thermal environment for people by way of a "human thermal load" (Shimazaki et al. 2011) obtained by integrating measured values from physiological responses (human physiological quantity) of persons on the pavement (Nitta et al. 2017; Aoki et al. 2017; Aoki et al. 2018). Up until now, the authors conducted experiments with subjects in standing-still posture on the test pavement of both WR and AS in an artificial climate chamber, which can stably recreate diurnal climatic conditions outdoors in the summertime, and also real outdoors in the summertime. These experiments confirmed that the human thermal load for subjects on WR pavement is significantly smaller than subjects on AS pavement. Since the human thermal load can express the degree of load of existing environment from human body physiological quantities which are a direct response from humans, WR is a human-friendly pavement.

Also, as described above, the radiation environment and the air temperature tend to worsen in the area close to the hotter road surface during daytime in summer. Compared to persons in a standing posture, wheelchair users who are usually closer to the road surface are more likely to be exposed to the harsher thermal environment. In addition, wheelchair users often have trouble controlling their body temperature. It is therefore meaningful from the viewpoint of universal design to improve the thermal environment. In this study, we therefore conducted the experiment for thermal environment and the human thermal load for subjects in standing-still postures on outdoor test pavement and subjects sitting on wheelchairs.

2 Materials and Methods

In this chapter, along with giving an overview of the water retentive block we will show the concept of human thermal load used as an index of the thermal environment for human beings, and results from previous experiment of subjects conducted on the test pavement in the indoor artificial climate chamber and the outdoor test pavement.

2.1 Material (Water Retentive Block)

A water retentive block is composed of waste materials (glass 10%, ceramic 90%) and thus contains fine voids all over which can hold water from rain (water retention capacity of 15% or more of the block volume). The general dimensions of the water retentive block are: length × width = 40 × 40, 30 × 30 or 10 × 20 (cm), thickness = 6 or 8 (cm). Therefore, it is possible for a block with a thickness of 6 cm to hold 12 L/m^2 or more, and for that of 8 cm to hold 9 L/m^2 of water or more. Since the degree of reflectance (albedo) is one of important factors for the formation of urban climate (Akbari et al. 2001), the values of albedo are also listed as a characteristic of pavements. The bending strength is confirmed to be more than 3.0 N/mm^2, for general interest.

2.2 Methods

2.2.1 Human Thermal Load

Finding the thermal condition of the human body is important for evaluating the thermal environment and preventing heat stroke. When the human body and surrounding environment are in a state of thermal equilibrium, the thermal condition of the human body can be expressed by the heat balance as

$$M + R_{net} = W + C + E \tag{1}$$

where, M is metabolic rate [W/m^2] (calculated from body surface area, O_2 intake, CO_2 production), W is mechanical workload [W/m^2] (external work done by the human body), R_{net} is net radiation [W/m^2] (amount of solar and infrared radiation received by the human body and transmitted from the human body), C is sensible heat loss [W/m^2] (heat loss due to difference between average skin temperature and air temperature, heat exchange by respiration), E is latent heat loss [W/m^2] (heat loss due to perspiration and exhalation, insensible perspiration), respectively.

When the heat is not dissipated sufficiently from the human body, people feel hot yet on the other hand they feel cold if body heat is excessively dissipated. If the thermal sensation is not a neutral level, a thermal load is applied to the human body. This load amount is referred to as the human thermal load F_{load}, and defined by the heat balance equation as

$$F_{load} = M - W + R_{net} - C - E \tag{2}$$

where F_{load} is human thermal load [W/m^2].

Each element of Eq. (2) includes consideration of environmental physical quantities such as temperature, humidity, amount of radiation, airflow, and human physiological quantities such as human body temperatures, perspiration, metabolic rate, and clothing insulation.

2.2.2 Physical Quantities and Physiological Quantities

(1) Physical quantities

Table 1 shows measurement items and measurement methods for environmental physical quantities. Every quantities were measured at 1-min interval.

Table 1. Measurement items and measurement methods of environmental physical quantities

Measured item	Procedure	Instrument	Measurement interval
Radiation (Long-, Short-wave radiation)	Six directions at 0.7 m, 1.0 m (North, south, east, west, upward, downward)	Radiometer; EKO MR-60	Every 1 min
Temperature and Humidity	Five points in vertical direction (at 0, 0.35, 0.5, 0.7, and 1.0 m)	Pt-100 resistance and capacitance hygrometer; T and D TR-73U, Shinyei Technology THP-728	
Wind velocity	3D wind profile at 1.5 m	Ultrasonic anemometer; Young CYG-81000	
Ground surface temperature	Five points (the center, and 0.5 m distant from the center in north, south, east and west direction)	Thermocouple J-type	

(2) Physiological quantities

Table 2 shows measurement items and measurement methods for human physiological quantities. As shown in Table 2, regarding human physiological quantities, the skin temperature at seven points, rectal temperature, sweat production at seven points, heart rate, and metabolic rate were measured for each subject for at 1-min interval. Subjects wore black and white tight-fitting clothes and shoes, whose thermophysical properties were measured beforehand. Subjects also wore the same standard caps and sunglasses to eliminate the effects from glare. In addition, only those persons who agreed by giving informed consent were allowed to participate in this experiment in order to comply with ethical regulations based on the Helsinki Declaration.

Table 2. Measurement items and measurement methods for human physiological quantities

Measured item	Procedure	Instrument	Measurement interval
Body temperature	Seven point at skin (Head, upper arm, hand, abdomen, thigh, leg, and foot), and rectum	Thermistor; Nikkiso Thermo N542	Every 1 min
Sweat evaporation	Seven point at skin (Head, upper arm, hand, abdomen, thigh, leg, and foot) Weight loss (human and clothes)	Hygrometer; SysCom Corp. SHTDL-3 Weight scale; A and D FG150KBM and Shimadzu ELB3000	
Heart rate	ECG signal from chest	Heat rate meter; Polar RS800CK	
Metabolism	O_2 intake and CO_2 exhaust	Thermocouple J-type	Every 20 s

2.3 Verification in a Standing-Still Posture

2.3.1 Experiment in Artificial Climate Chamber

(1) **Overview of experiment**

In order to quantitatively evaluate the effect from improving the thermal environment on WR, subject experiments were conducted in an artificial climate chamber during the summer day. Subjects were 56 healthy adult men (28 each for WR, AS) and experiments were carried out in each trial for 30 min. The WR was also watered down before the every trial for maintaining a uniform wet state. The environmental setting values and measured values from the artificial climate chamber are listed in Table 3 and an overview of WR for experiments and AS for comparison are shown in Table 4.

Table 3. Environment settings in artificial climate chamber

Measured item	Setting value	Measurement value		
		Average	Max	Min
Air temperature [°C]	35	36.2	40.0	29.8
Humidity [%]	50	43	57	32
Radiation [W/m^2]	900	895	921	865

Table 4. Overview of WR and AS

Pavement	AS	WR
Size [m^2]	2.5 (1.2 m × 2.1 m)	2.0 (1.0 m × 2.0 m)
Thickness [m]	0.04	0.09 (Block: 0.06, Laying sand: 0.03)
Appearance color	Black	White
Percentage of reflected solar radiation [%]	8.5	25.0
Maximum water retention capacity [kg/m^2]		8.76

(2) **Results**

Table 5 shows the human thermal load obtained by integrating the measurement results from the environmental physical quantities and the human physiological quantities. As shown in Table 5, the amount of human thermal load for subjects on WR pavement given as F_{load} was 224 W/m^2 and was 78 W/m^2 lower than on AS which was 302 W/m^2. These results show a remarkable effect in lowering the thermal load on the human body. Among the breakdowns, the largest influence was the amount of net radiation.

Table 5. Calculation results for human body thermal load

Physiological response	AS	WR	WR-AS
Metabolic rate: M [W/m^2]	89	83	−6
Net radiation: $Rnet$ [W/m^2]	243	173	−70
Sensible heat loss: C [W/m^2]	2	3	1
Latent heat loss: E [W/m^2]	28	29	1
Human thermal load: F_{load} [W/m^2]	302	224	−78

2.3.2 Experiment at Outdoor Test Pavement

(1) **Overview of experiment**

Outdoor experiments were conducted in the summertime in August. The pavement consisted of two types: WR and AS, and each area was 49 m^2 (7 m × 7 m). The test pavement outline is shown in Table 6. Subjects were 30 healthy adult men and experiments were carried out in each trial for 30 min. Environmental measurement results are shown in Table 7. According to the Japan Meteorological Agency, the average amount of rainfall for last 30 years is 161 mm in July and 87 mm in August around the measuring site of Okayama. Since the duration of effect of WR for lowering the surface temperature is at least 3 days without any rainfall or water supply in preliminary experiment, these amount of rainfall would be sufficient for WR pavement.

Table 6. Test pavement overview

Pavement	S	WR
Size[m^2]	49 (7 m × 7 m)	49 (7 m × 7 m)
Thickness[m]	0.04	0.09 (Block: 0.06, Laying sand: 0.03)
Appearance color	Black	White
Percentage of reflected solar radiation [%]	8.5	25.0
Maximum water retention capacity [kg/m^2]		8.76

Table 7. Environment measurement results

Measured item	Measurement value		
	Average	Max	Min
Air temperature [C]	33.2	36.1	28.4
Humidity [%]	52	82	33
Radiation [W/m^2]	583	1175	139

Table 8. Calculation results for human body thermal load

Physiological response	AS	WR	WR-AS
Metabolic rate: M [W/m^2]	87	87	0
Net radiation: $Rnet$ [W/m]	193	181	−12
Sensible heat loss: C [W/m^2]	15	18	3
Latent heat loss: E [W/m^2]	29	35	6
Human thermal load: F_{load} [W/m]	237	215	−22

(2) **Results**

The measurement results for human thermal load are shown in Table 8. The human thermal load for subjects on WR, or namely F_{load} was 215 W/m^2 and so was 22 W/m^2 lower than on AS which was 237 W/m^2, thus confirming the effect of lowering the human thermal load.

3 Validation in a Wheelchair User (Sitting Posture)

3.1 Outline of Experiment

A pavement overview is shown in Table 9. The subjects were a total of 60 people. Among them, 33 healthy adult men were selected, and the tests were conducted with 30 men in wheelchairs and 30 men in standing-still postures for comparison purposes. Experiments were carried out for 30 min for each trial. The experimental procedure is shown in Fig. 1. The height of the seat surface of the wheelchair was 50 cm.

Table 9. Test pavement outline

Pavement	AS	WR
Size[m^2]	49 (7 m × 7 m)	49 (7 m × 7 m)
Thickness[m]	0.04	0.09 (Block: 0.06, Laying sand: 0.03)
Appearance color	Black	Brown
Percentage of reflected solar radiation [%]	8.5	13.0
Maximum water retention capacity [kg/m^2]		8.76

Fig. 1. Outdoor experiment layout

3.2 Results

Table 10 shows the human thermal load obtained by integrating measurement results from environmental physical quantities with human physiological quantities. As shown in Table 10, the human thermal load for subjects in standing-still postures on WR or namely F_{load} was 211 W/m^2 and was 58 W/m^2 lower than on AS which was 269 W/m^2, thus confirming a remarkable effect in lowering the thermal load on the human body. The human thermal load for the wheelchair subject was 277 W/m^2 on AS and 230 W/m^2 on WR, and the same as with the standing posture was 47 W/m^2 lower on WR, thus also confirming a remarkable effect in lowering the thermal load on the human body. Next, comparing the amount of net radiation for the subjects in standing-still postures with wheelchair subjects, the net radiation for wheelchair subjects was larger for both AS and WR. This was because the wheelchair itself was more strongly affected by the road surface temperature as was originally expected. Since WR pavement used in the experiments have different albedos and experimental comparisons are not carried at the same time for Tables 8 and 10, each heat flux varies. Totally, the human thermal load for wheelchair subjects was larger than the subjects in a standing, and it was suggested that the effectiveness of WR pavement on lowering the road surface temperature could be even further enhanced.

Table 10. Calculation result for human body thermal load amount

Physiological response	Standing-still postures			Wheelchairs		
	AS	WR	WR-AS	AS	WR	WR-AS
Metabolic rate: M [W/m^2]	62	61	−1	57	59	2
Net radiation: $Rnet$ [W/m^2]	260	202	−58	267	222	−45
Sensible heat loss: C [W/m^2]	20	21	1	17	19	2
Latent heat loss: E [W/m^2]	32	31	−1	30	32	1
Human thermal load: F_{load} [W/m^2]	269	211	−58	277	230	−47

4 Summary

The effect from improving the thermal environment by way of WR is newly evaluated by the human thermal load obtained by integrating human physiological quantities obtained from subject's metabolic rate and net radiation amount and other factors. This approach made it possible to evaluate by a direct index taking human physiological response into consideration showing that WR is "human friendly pavement" since it significantly reduces the human body thermal load amount compared to AS.

Also, in the additional outdoor experiment was carried out for subjects in standing-still posture and sitting in wheelchairs. The wheelchair, which is closer to the road surface, was more strongly affected by the road surface temperature, the net radiation was larger, and as a result the human thermal load was larger. These results lead us to conclude that the effect of WR pavement in improving the thermal environment is more effective to wheelchair users.

References

Akbari H, Pomerantz M, Taha H (2001) Cool surfaces and shade trees to reduce energy use and improve air quality in urban areas. Sol Energy 70(3):295–310

Aoki M, Nakamura T, Shimazaki Y (2018) Experimental investigation of the effect of water retentive sidewalk block pavement on improving human thermal environment. In: Proceedings of 12th international conference on concrete block pavement, Seoul, South Korea, 16–19 October, pp 150–160

Aoki M, Shimazaki Y, Nitta J, Kamidouzono K (2017) Improving effects of water retaining pavement on human thermal environment. Verification based on human physiological responses. In: Proceedings of the fourth international symposium on asphalt pavements & environment, Tokyo, Japan, 20–21 November, pp 1–15

Kosaku Y, Kasugai T, Hashimoto I, Koga M, Takane K (2007) Effects on temperature, humidity, and wet-bulb globe thermometer in the case of watering water retaining pavement. 2007 Tokyo civil engineering center annual report, pp 153–164. (in Japanese)

Nitta J, Shimazaki Y, Aoki M, Kamidouzono K, Sakoi T, Satsumoto Y (2017) Experimental evaluation on cooling effects of water retaining pavement on human. In: Proceedings of 8th Japanese-German meeting on urban climatology, Osaka, Japan, 25–29 March, pp 217–222

Oguri N, Kariya T, Tamaki T (2007) The effect of improving on thermal environment of water retaining block pavement with water supply function. In: Japan road conference proceedings, Tokyo, Japan, 1–2 November, pp 12–44. (in Japanese)

Shimazaki Y, Yoshida A, Suzuki A, Kawabata T, Imai T, Kinoshita S (2001) Application of human thermal load into unsteady condition for improving outdoor thermal comfort. Build Environ 46(8):1716–1724

Airport Pavement Management Systems: An Open BIM Approach

Sara Guerra de Oliveira[1(✉)], Andrej Tibaut[2], and Gianluca Dell'Acqua[1]

[1] School of Polytechnic and Basic Sciences, Department of Civil, Building and Environmental Engineering, University of Naples Federico II, Naples, Italy
sara.guerradeoliveira@unina.it

[2] Faculty of Civil Engineering, Transportation Engineering and Architecture, University of Maribor, Maribor, Slovenia

Abstract. Building Information Modelling (BIM) offers the possibility to access and oversee information about the asset, throughout its lifecycle. One of the phases where BIM can provide key benefits is the Operation phase, e.g. facility management/maintenance, decommissioning and major re-programming, supported by the Information Management Process (IMP), which may include a Computer-Managed Maintenance System (CMMS). When discussing roads, highways or runways the maintenance phase is largely concentrated on the pavement, and a robust Pavement Management System (PMS) should be integral part of the IMP, focusing mainly on the evaluation of the pavement's present condition and prediction of its future condition. In the airport domain, the importance of an Airport Pavement Management System (APMS), falls both in a cost effectiveness and aviation safety viewpoint, providing consistent, objective and systematic procedures for determining priorities, schedules and resources' assignment. Introducing the case study of the Lamezia Terme International Airport, this paper presents possible improvements in the interoperability of maintenance data of the airport's runway with reference to the IFC Reference Processes, being that Maintenance Management is one of the defined projects within the Facilities Management (FM) domain inside BuildingSMART International.

Keywords: Airport Pavement Management System (APMS) · Building Information Modelling (BIM) · Industry Foundation Classes (IFC)

1 Introduction

In the Ministerial Decree 560 of 2017, Italy set the timeline for the mandatory but gradual introduction of the use of digital modelling methods and tools for the Construction Sector based on the asset's tender value. Starting from the first of January 2019, public works with a tender value equal or greater than 100 million euros will be obliged to apply innovative approach that includes Building Information Modelling (BIM), in 2025 all public works should fully integrate BIM.

Infrastructure projects should nowadays be supported by Building Information Modelling (BIM) processes and benefit from their numerous advantages, namely at the

M. Pasetto et al. (Eds.): ISAP APE 2019, LNCE 48, pp. 450–459, 2020.
https://doi.org/10.1007/978-3-030-29779-4_44

design (visualization, automatic low-level corrections when the design changes, consistency of the design intent verification), collaborative (project team is given at an early stage a better understanding of the project) and Operation and Management levels (Sacks et al. 2018).

Originally developed for the architectural field, the BIM application to transportation infrastructure projects still demands some research, particularly on fully interoperable specific standards, however the rewards of its application throughout the whole lifecycle of the asset(s) justifies the effort of the academic and industrial communities. Figure 1 presents a summary of the infrastructure domains and the phases of the assets' lifecycle where BIM can be applied.

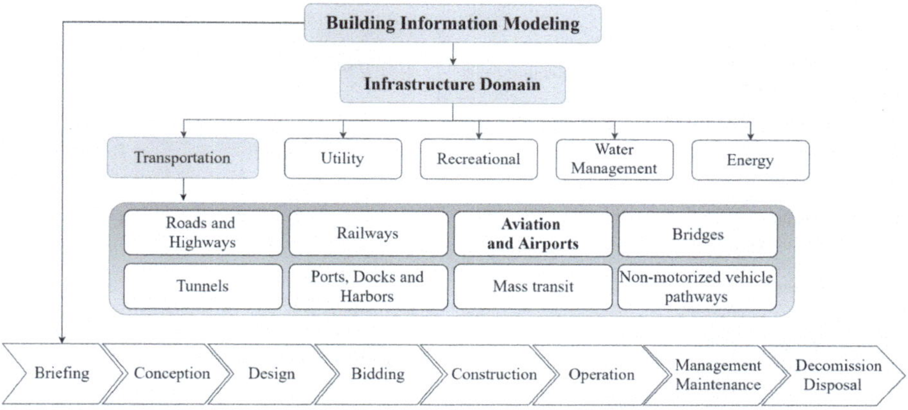

Fig. 1. BIM infrastructure domain and asset lifecycle phases.

BIM can be of crucial importance when applied in the Operation and facility management/maintenance phases. In the complex Airport domain, the importance of these stages cannot be overlooked, as it is both in terms of security and strategic importance to keep these infrastructures in optimal operating conditions.

Even in the case of airports where the BIM model has already been established concerning all the infrastructure (including the airside detailed pavement areas model), the integration of an eventual existing APMS is not absent of challenges. As highlighted in the Airport Room Roadmap Report (BuildingSMART 2015), a complete asset management plan still evidences problems that present challenges, namely in terms of inefficient and uncomplete handover information exchange, big data filtering, long-term planning needs and software application interoperability.

The importance of a well-established complete APMS as integrant part of the Airport BIM model is paramount, and this paper aims to provide guidelines as to better obtain its integration in an open BIM context.

BuildingSMART International (2019) establishes that Open BIM has as main goal the improvement of the process of exchanging non-proprietary BIM models and other data based on open standards and workflows.

2 BIM and the Information Management Process

An Information Management Process (IMP) is the method an organization uses to acquire/retrieve, organize and maintain information concerning an asset.

Figure 2 illustrates the integration of eventual already in-place enterprise systems, as an APMS, with the BIM process.

After validation procedures from the responsible team of the enterprise system, the information provided by such systems can go to the publish area of the Common Data Environment, as indicated in the Publicly Available Specifications, PAS 1192-2:2013 and PAS 1192-3:2014.

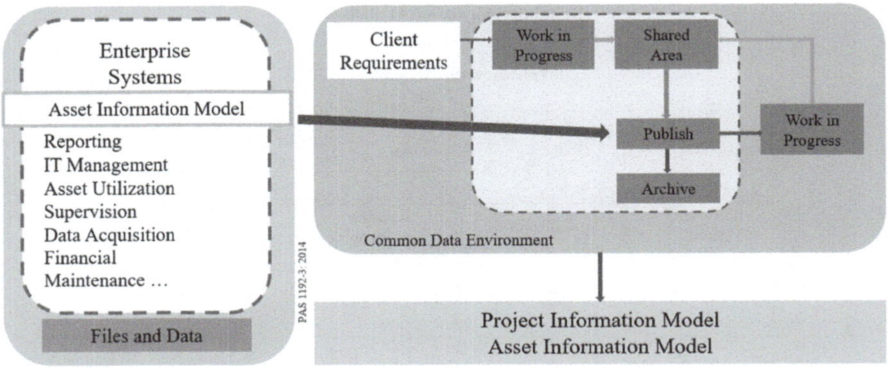

Fig. 2. Integration of an APMS system in the BIM process (Adapted from PAS 1192-2:2013 and PAS 1192-3:2014).

The process is therefore highly dependent of the quality and relevance of the contained information, the Computer-Managed Maintenance Systems (CMMS) provide a tool for all the data organization and can serve to reduce maintenance-related related costs and increase productivity of the management team.

The maintenance activities require an effective organization and an accurate, comprehensive and accessible database of relevant information. Recognizing that an airport air-side is submitted to rigorous and frequent inspections, and that any problem detected must be timely addressed, its clear that this type of infrastructure benefits significantly from well-structured models and robust management systems.

3 Airport Pavement Management Systems

Airports are divided in two main areas, landside and airside. The airfield pavements include the runway(s), taxiways, taxi lanes and aprons on the airside area.

Operation and Maintenance procedures of the pavement areas of the airport airside benefit and are binded to be sustained by the data contained in the APMS. Figure 3 illustrates that the pavement management can be applied at the network level, where the overall condition of the network and future state condition are evaluated and predicted, and the needs for intervention are prioritized for specific sections (either preventive maintenance, rehabilitation or reconstructive interventions), in sum, planning, budgeting and policy decisions. At the project level occurs the selection of project-specific pavement preservation treatments, (materials and procedures). The construction and monitoring of the treatment performance are also contained in this level (Hajek et al. 2011).

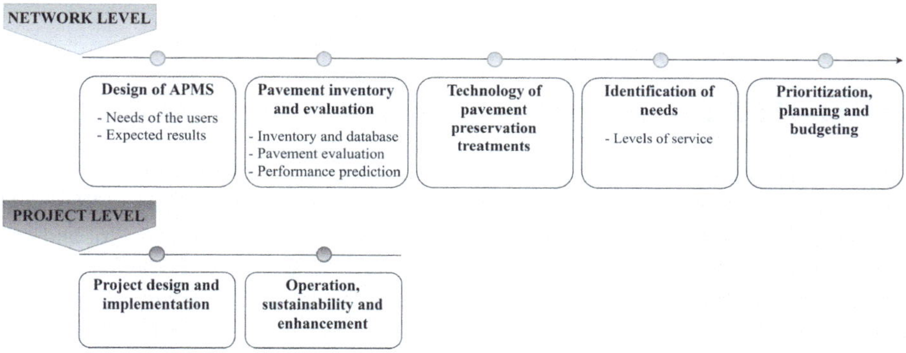

Fig. 3. Airport pavement management levels (Adapted from Hajek et al. 2011).

In Italy, the Italian Civil Aviation Authority, ENAC, carries out activities of regulation and control of the air sector, based on international (conventions, protocols, regulations and directives), national (laws and decrees) and internal regulations and circulars. Of particular interest to the present paper are the guidelines specific for the application of APMS (ENAC 2015). In this document the benefits of the implementation of these systems are listed; the development of a computerized database promoting the organization and storage of pavement related data, the regular and objective monitoring of the pavement condition, the establishment of deterioration rates and the planning and optimization of the treatments according to the available budget.

For a correct allocation and reference of all the information contained in the database, the pavement inventory divides the airport pavement network in branches (e.g. runways, taxiways), sections (with uniform pavement structure) and sample units as recommended in the ASTM standard D5340-12 (ASTM International 2018). A summary of the relevant data that should be contained on the database and connected to the division previously described is shown in Fig. 4.

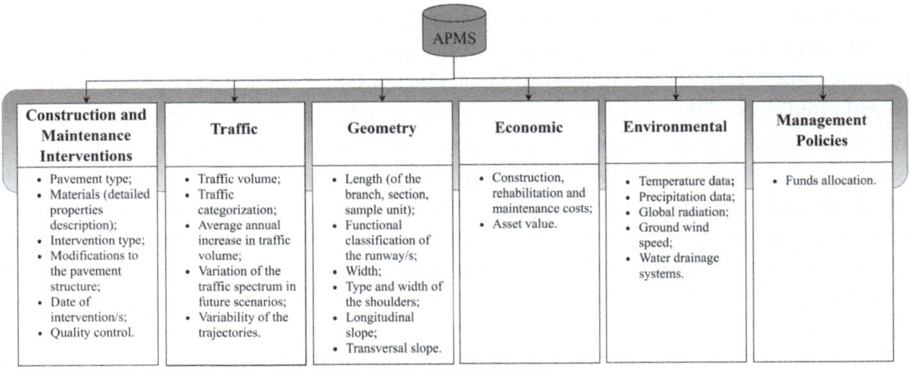

Fig. 4. APMS database information (Adapted from ENAC 2015).

4 Lamezia Terme International Airport

The Lamezia Terme International Airport is the main airport in the Calabria Region, Italy, with the identifier codes, International Air Transport Association: SUF and International Civil Aviation Organization: LICA. The airport opened in 1976, and has since then been subject to improvements, namely in 1982 and in 2007, the latter where a structural requalification of the air-side pavements took place. The airport airside is composed of one runway, one taxiway, four taxi lanes and aprons, distributed as can be seen on Fig. 5.

Fig. 5. Lamezia Terme International Airport.

The Airside pavement description and main characteristics are described as follows: The runway, is designated Rwy 10/28, according to its magnetic azimuth orientation. The runway has a flexible pavement type, with a loading-carrying capacity, pavement classification number PCN 58/F/B/W/T, where F: flexible pavement; B: subgrade strength (CBR between 8–13%); W: tire pressure supported unlimited and T: PCN estimated by technical evaluation (De Luca and Dell'Acqua 2018).

It presents a length of 3000 m, width of 45 m; shoulder width: 14 m, and variable longitudinal slope (ranging from 0.13% to 0.38%). The transversal slope presents values of 1.25% or 1.40% according to the section.

The taxiway, designated "Sierra", presents a flexible pavement type. The taxiway has a length of 1750 m, width of 30 m, shoulder width: 10 m, longitudinal slope: variable ranging from 0.09% to 0.49%, transversal slope: variable: 1.30%, 1.50%.

The airside contains four taxiway exits also with flexible pavement type structure and variable geometric characteristics, highly dependent of on the constraints imposed by the connection of the the runway to the taxiway (namely in terms of longitudinal and transversal slope). Generally, the sections of the exits present a width of 25 m and 10 m shoulder width.

5 Runway BIM Model

5.1 Geometry

The initial part of the Lamezia Terme airside case study, and main part of the present paper focuses on the construction of the runway 10/28 digital twin, i.e. a model that mirrors the asset (data-rich 3D model), and represents, reacts and can cause changes in the actual runway, when connected to the APMS. First and foremost, it is important to highlight that the runway model was produced with a tool dedicated to road design, from the Bentley Systems company, and for that motive, all the digital runway components had to be customized and detailed for the airport case, namely the transversal sections of the runway. The overall view of the airside model can be seen in Fig. 6, the model also includes the alignments for future incorporation of the taxiway and taxiway exits model.

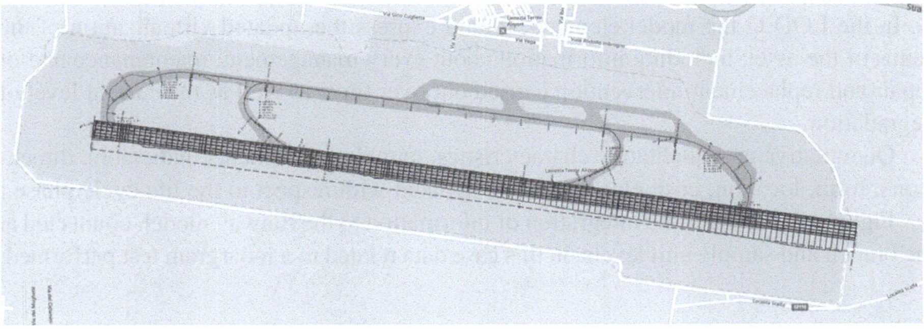

Fig. 6. Overall view of the airside model.

The corridor of the runway 10/28 was created based on a custom parametric cross-section that incorporates the designation of the pavement layers and specific identification point designations, as can be seen in Fig. 7. The main constraints imposed for the construction of the cross-section were of the horizontal, vertical and slope nature and the connection to the terrain was established defining an end-condition component for both the cut and the fill cases. It is important to mention that in the general model, at this stage, the connection to the terrain from the left side of the runway (West-East direction) was purposely excluded from the model, for future study of the intersections between runway and taxiway exits.

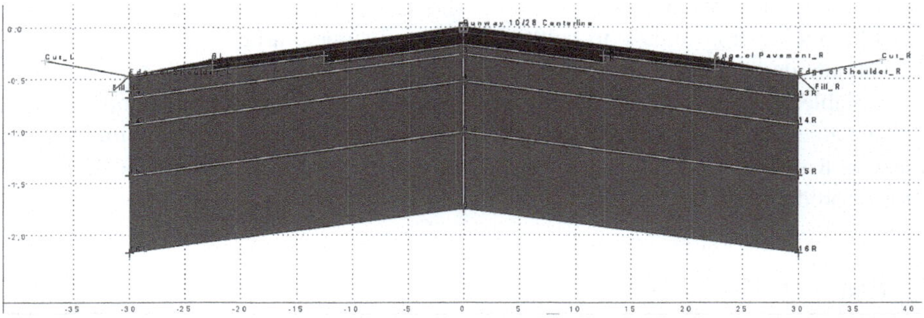

Fig. 7. Runway 10/28 custom cross-section.

5.2 Semantic Data

One of the main advantages of developing the BIM-model is the capacity of integrating non-graphical information. Even though the UNI 11337- 4 (U. Committee 2017), does not illustrate in detail the runway case application, the description of a Level of Development G, provides guidelines as to the information that a model built for operation and maintenance purposes should include. Although inspired by the USA scale (LOD 100, 200, …, 500), and integrating aspects of the UK LOD convention, the Italian scale is affected by national requirements not present in both systems. The national Italian LOD scale is represented by capital letters in alphabetic order from A to G. In the LOD G the model elements should express the updated virtualization of the status of the asset, including information about every management, maintenance and/or repair and replacement intervention carried out over time, as well as the current level of degradation.

Quantitative and qualitative characteristics, namely performance indicators, dimensions, form, location, cost, etc., should be updated with respect to the life cycle phase.

Figures 8 illustrates the integration of information to the runway model, connected at the branch and sample unit levels, in this case data related to a radargram test performed.

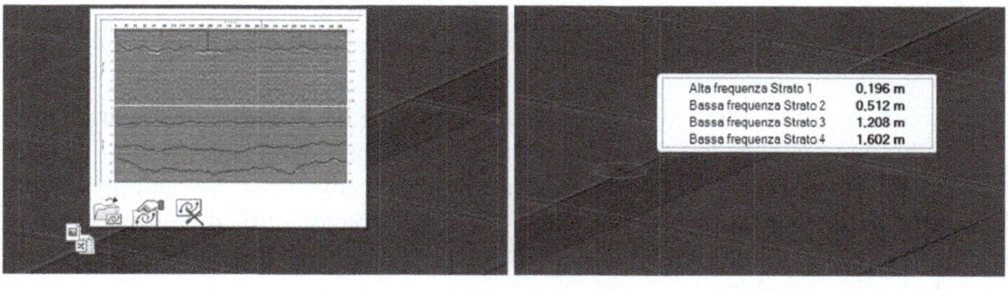

Fig. 8. Radargram field test data.

All data related to the pavement reported distresses should also be included, e.g. eventual alligator or fatigue cracking, bleeding, corrugation, depression, raveling, weathering. Indicators related to the pavement deterioration and characterization as the Pavement Condition Index (PCI) and the integrity of the top surface of the runway, are other important features that should be part of the records. The surface condition is evaluated by regular inspections, dependent on the environmental conditions, the wear and tear and on the accumulation of rubber deposits. The integrity of the runway can be evaluated through the study of its surface friction levels.

6 Interoperability Between a BIM Model and the APMS

In linear infrastructure projects, such as is the case of the runway, interoperability still poses some challenges. To this purpose, Common Data Standards, such as the Industry Foundation Classes (IFC), are currently in development, not yet including data schemas for infrastructure works (Tibaut et al. 2015).

However, some of the already published and validated IFC's can be applied to this specific case improving the connection between the APMS database and the model. This paper suggests the application of the ifcPropertySets mechanism to demonstrate how can the information can be connected in an open standardized way.

Although some properties can be defined directly within the schema of the IFC model, others can be added freely to the instance model. Being that, as referred, the schemas for linear infrastructures are not available yet, it is only normal that specific characteristics of these assets are not detailed in the most common properties of the defined IFC's. The already defined IfcCivilElement is a generalization of all elements within the civil engineering works and can be applied to occurrences of typical linear construction works as pavements.

So, using the already defined IfcCivilElement, connected to the wearing course of the runway, the possibility of integrating data to be used in the APMS becomes optimized. The already available Property Sets for the IfcCivilElement are: Pset_Condition, Pset_EnvironmentalImpactIndicators, Pset_EnvironmentalImpactValues, Pset_ManufacturerOccurrence, Pset_ManufacturerTypeInformation, Pset_PackingInstructions, Pset_ServiceLife and Pset_Warranty. The proposed amplification (Fig. 9), concerns data referent to the friction levels of the runway's surface, obtained through the Griptester, and the connection is made through a defined IfcPropertySet, named "WearingCourseConditionHistory". The left side of Fig. 9 shows the newly IfcPropertySet specific for the connection of the previous history records, such as the latest GripTester results, and the right side shows a standardized PropertySet, in the case "Pset_Condition", with examples of the correspondent information included.

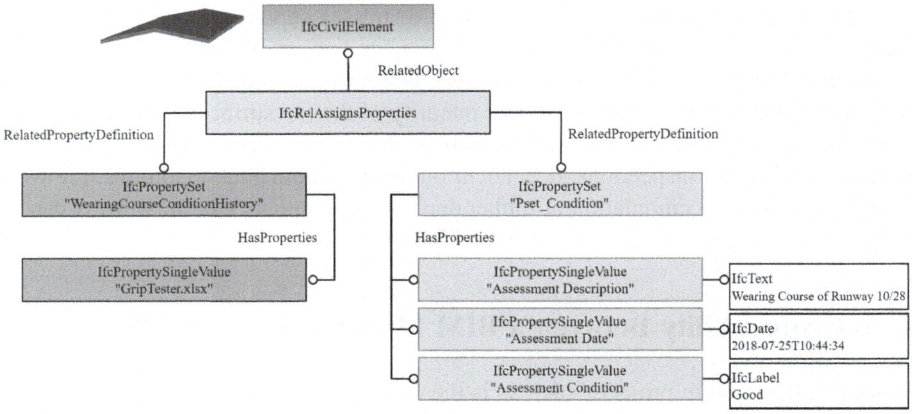

Fig. 9. Use of properties, integration of the IfcPropertySet "WearingCourseConditionHistory".

By using this mechanism when the IFC project file is shared between stakeholders the access to all this data is displayed, as well and integrated in the APMS. The connection to "GripTester.xlsx" assures that all previously friction field tests results are connected to the wearing course of the runway, organized by date, including information about the sample unit, distance from the runway centerline and corresponding grip-number obtained.

This mechanism can be further extended to include other data, serving the PropertySet "WearingCourseConditionHistory" as an example. The PropertySet allows the introduction of further properties.

7 Conclusions and Future Developments

The present paper presents a mechanism to further enhance the interoperability between the BIM model aggregated information and the APMS database. The detailed mechanism is based on the IFC use of properties and their connection to specific objects of the model, the example presented is the Runway 10/28 wearing course and the connection to friction related historic data. While the finished schemas are not available for the infrastructure field it is important to study options within the already published ones as they will certainly be applicable in the future. Given the importance of maintaining the airside pavement in optimal conditions, the connection of relevant information and its interoperability is crucial for an optimized management of the asset.

Future developments of the project presented in this paper will include the completion of the Airport's air-side pavement areas modelation, the further enrichment with non-geometric data, the study and implementation of semantic web technologies which can further improve the integration of relevant data to the model and the expansion of the IfcPropertySets mechanism for Airport pavement management purposes.

References

Sacks R, Eastman C, Lee G, Teicholz P (2018) BIM handbook: a guide to building information modeling for owners, designers, engineers, contractors, and facility managers. Wiley, Hoboken

BuildingSMART (2015) Airport Room Roadmap Report. https://www.buildingsmart.org/wp-content/uploads/2018/08/Airport-Roadmap-Report_FINAL.pdf

BuildingSMART International (2019) Technical Vision. https://www.buildingsmart.org/standards/technical-vision/

BSI (2013) PAS 1192-2:2013 Specification for information management for the capital/delivery phase of construction projects using building information modelling

BSI (2014) PAS 1192-3:2014: Specification for information management for the operational phase of assets using building information modelling

Hajek J, Hall J, Hein D (2011) Common airport pavement maintenance practices, ACRP synthesis 22. Transportation Research Board

ENAC, Ente Nazionale per l'Aviazione Civile (2015) Airport Pavement Management System Linee guida sulla implementazione del sistema di gestione della manutenzione delle pavimentazioni. N.° 3/2015-APT

ASTM D5340-12 (2018) Standard Test Method for Airport Pavement Condition Index Surveys, ASTM International, West Conshohocken, PA. www.astm.org. https://doi.org/10.1520/d5340-12r18

De Luca M, Dell'Acqua G (2018) Touchdown remaining lift on the wings and dynamic vertical force transmitted to the runway. Periodica Polytech Civil Eng 62(3):590–595

U. Committe, UNI 11337-4 (2017) Edilizia e opere di ingegneria civile - Gestione digitale dei processi informativi delle costruzioni - Parte 4: Evoluzione e sviluppo informativo di modelli, elaborati e oggetti (Construction and civil engineering works: Digital management of building information processes-Part 4: Evolution and information development of models, documentation and objects)

Tibaut A, Pečnik S, Korošec MR, Mihalič K, Zabreznik I (2015) BIM-based parametric modeling of roads and infrastructure objects. In: 32nd CIB W78 conference, Eindhoven, The Netherlands

Mixture Design for Recycled Porous Asphalt Pavement and Results of Follow-up Survey for Ten Years

Atsushi Kawakami[1(✉)], Iwao Sasaki[2], Hiroyuki Nitta[2], and Masayuki Yabu[1]

[1] Pavement Research Team, Public Works Research Institute (PWRI), 1-6 Minamihara, Tsukuba-Shi, Ibaraki 305-8516, Japan
kawakami@pwri.go.jp

[2] Innovative Materials and Resources Research Center, Public Works Research Institute (PWRI), 1-6 Minamihara, Tsukuba-Shi, Ibaraki 305-8516, Japan

Abstract. Porous asphalt (PA) pavement has been widely used mainly in national highways and expressways in Japan since the "Drainage Asphalt Pavement Technical Guidelines" was published in 1996. Also, more than 20 years have elapsed since PA pavement began to popularize, and the recycled aggregate due to reconstruction on PA pavement is expected to increase in the future. However, since polymer-modified bitumen with high adhesive strength is used for PA, and since aggregate gradation is largely different from dense graded asphalt (DG) mixture usually used, the recycling method of PA is considered necessary. In this study, the recycled PA pavement which was made of recycled PA mixture and DG mixture, were constructed in 2006 in National Highway No. 408 of Tsukuba City, Japan. And the results of follow-up survey on durability and serviceability for ten years of recycled PA pavement were reported.

Keywords: Porous asphalt pavement · Drainage asphalt pavement · Recycle · Mixture design · Follow-up survey

1 Introduction

Due to its ability to drain rainwater on road surfaces, porous asphalt pavement (PA pavement) is also referred to as drainage pavement in Japan. In addition, since PA pavement is capable of reducing tire-pavement noise, it has been widely used mainly for national highways and expressways after the Drainage Pavement Technical Guide (Japan Road Association 1996) was published. The construction area of national highways under the jurisdiction of the Ministry of Land, Infrastructure, Transport and Tourism (MLIT) reached 50 km^2 as of 2005 (Japan Road Association 2006), which has now reached 30% of the entire national highways. Also, it has been more than 20 years since PA pavement started to be widely used and so reclaimed asphalt pavement (RAP) containing PA mixture is also expected to increase through future replacement work for such PA pavement. However, since polymer-modified bitumen type-H that

M. Pasetto et al. (Eds.): ISAP APE 2019, LNCE 48, pp. 460–469, 2020.
https://doi.org/10.1007/978-3-030-29779-4_45

has high viscosity is used in PA mixture and the aggregate gradation of RAP from PA pavement is largely different from conventional mixtures, PA pavement using RAP produced from PA pavement requires a mix design method that is different from ordinary methods. However, at the moment, sufficient knowledge for the use of RAP for PA pavement has not been accumulated and a reclamation technique has yet to be established. Also, although study on long-term performance and durability of PA pavement using RAP on real roads has been conducted in various places (Sasaki et al. 2005) (Minegishi et al. 2008), they have not been adequately clarified. This paper reports on the results of a survey of pavement surface properties with respect to durability and performance of recycled PA pavement (Kawakami et al. 2013) that has been offered for use for ten years.

2 Study Method

2.1 Materials Used and Mix Design

First, the authors studied the mix design of recycled PA mixture using RAP produced from PA pavement. The RAP mix ratio in recycled PA mixture was set at 10% and 20%. The materials used in the test construction, the RAP particle size, and the composition of a mixture and mixture properties are shown in Tables 1 and 2. Note that PA pavement is normally constructed of a PA mixture on the surface course and a DG mixture on the binder course. Therefore, in order to reproduce RAP of PA pavement, the RAP used in this study was mixed two types of RAP. One was the milled material from porous asphalt mixture and the other one was milled material from the DG mixture at different construction site. And these were mixed in equal proportions.

Table 1. Pavement materials for test construction

Materials		Detail
Virgin material	Single-sized aggregate	Hard sandstone 13/5 mm
	Sand	River sand
	Filler	Lime stone
	Bitumen	Polymer modified bitumen type-H
	Rejuvenator	Liquid type
	Modifier	Thermo plastic elastomer (Plant-mixing type)
RAP	RAP from DG	13/0 mm
	RAP from PA	13/0 mm

Under the present circumstances, a mix design method for recycled PA mixture has not been established. Unlike a new PA mixture, it is important in the mix design of a recycled PA mixture that a rejuvenator and polymer-modified bitumen are added to restore the properties of aged bitumen contained in the RAP and that the aggregate grading is also adjusted. After studying some trials, the authors developed the mix design flow shown in Fig. 1 and followed it to produce a recycled PA mixture. The targeted performance values for the properties of the recycled PA mixture were set at

Table 2. Mixture design and characteristics of recycled PA mixture

Recycled PA mixture		Section A & B	Section C	Section D
Contents (%)	Single-sized aggregate	84	71	78
	Sand	10.5	5	7
	Filler	5.5	4	5
	Mixed RAP	-	20	10
	Designed bitumen contents	-	5.0	4.9
	Bitumen contents in RAP	-	0.88	0.44
	Rejuvenator	-	0.11	0
	Modifier	-	0.10	0.06
	New bitumen	4.8	3.9	4.4
	Optimum bitumen contents		4.8	4.9
Porosity (%)			20	19.1
Marshall stability (kN)		5.53	8.6	6.6
Cantabro loss late at 20 °C (%)		6.8	9.3	9.5
Cantabro loss late at −20 °C (%)		-	23.5	23.7
Permeability ($\times 10^{-2}$ cm/s)		20.3	15.4	22.1
Dynamic stability (Times/mm)		5,730	15,750	12,600

20% for porosity, 1×10^{-2} cm per sec or more for coefficient of water permeability, up to about 10% for Cantabro loss factor and 3,000 cycles per mm or more for dynamic stability. Also the optimum bitumen content was obtained by a binder run-off test. The Cantabro loss factor can be obtained from a Cantabro test using Los Angeles Tester, and the value was used to determine the amount of a rejuvenator. Dynamic stability is an indicator for resistance to permanent deformation and can be obtained from a wheel tracking tester. All of these tests were carried out in accordance with the Pavement Surveying and Testing Handbook, a testing standard in Japan.

2.2 Test Construction

Test construction for recycled PA pavement was carried out on the 400-meter-long in National highway Route 408 located in Tsukuba City. According to the Road Traffic Census Survey by the MLIT, as of 2015, the traffic volume of cars was approximately 8,000 vehicles per lane per day, of which 1,000 were large vehicles. And it is equivalent to N6 of the traffic classification in Japan. Figure 2 shows the types of an asphalt mixture and the pavement structure used for the test construction. In this Section D with a 10% mix ratio of the surface- and binder-course mixture RAP and Section C with a 20% mix ratio were set as recycled PA pavement construction sections. Also, as for comparison, a one-layer milling and overlaying (OL) construction section (Section A) and a two-layer OL construction section (Section B) were constructed as new PA pavement. Note that the date of test construction for Sections A and B was December 2006 and that of Sections C and D was March 2007, and by March 2017, these pavements have been in use for ten years.

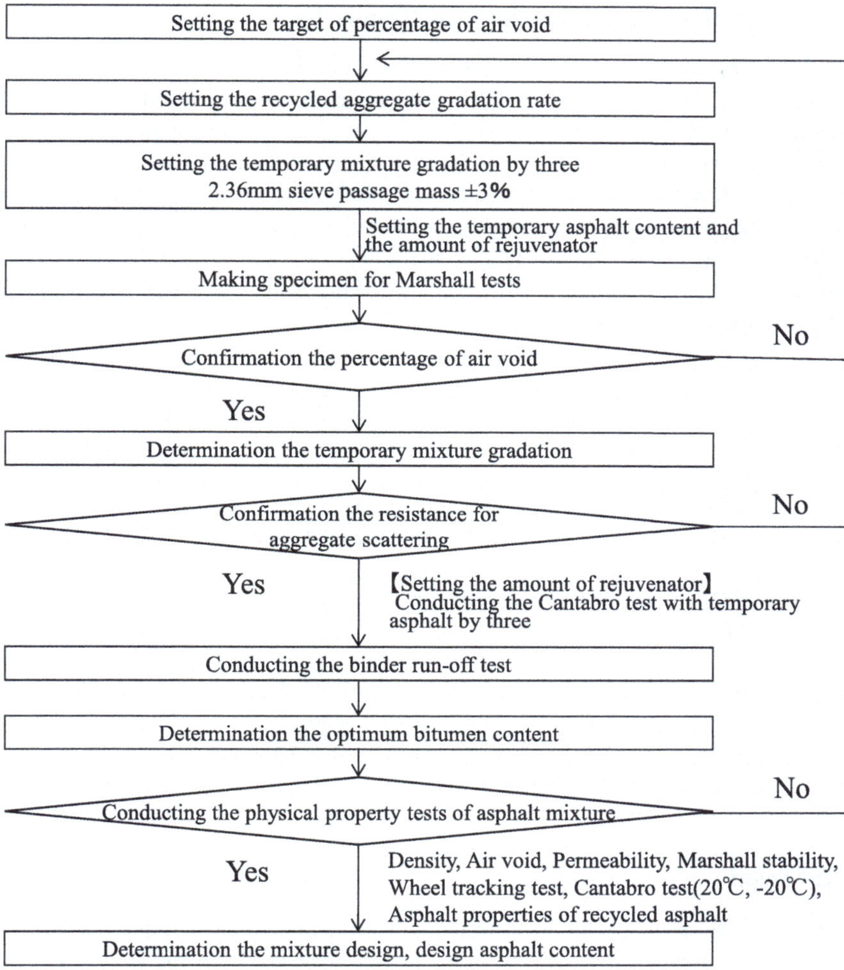

Fig. 1. Mix design flow for recycled PA mixture

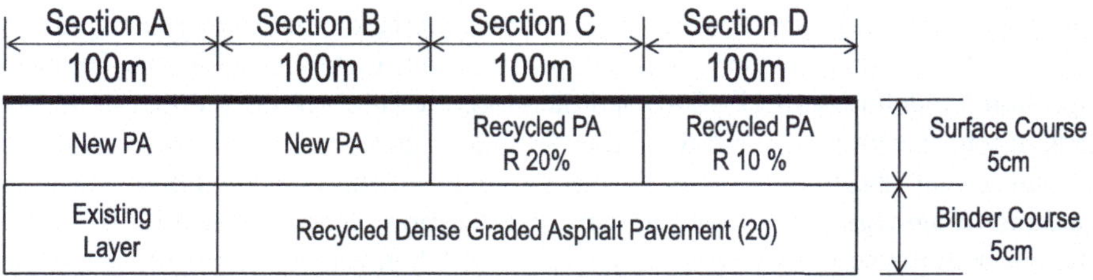

Fig. 2. Types of an asphalt mixture and the pavement structure used for the test construction

2.3 Survey Method for a Road that Has Been in Use

Table 3 shows the survey items after the road has been offered for use. With respect to the durability and performance of recycled PA pavement, road surface properties such as rut depth, evenness, surface texture, skid resistance, and tire-pavement noise (Fig. 3) were surveyed. The survey method was in accordance with the Pavement Surveying and Testing Handbook (Japan Road Association 2007). The measurement locations for each survey item are as shown in Table 3, and the survey results for each construction section were obtained by averaging measurement data for each Section. Note that crack ratio could not be measured due to difficulty in making initial judgment depending on the surface shape of PA pavement.

Table 3. Survey items and investigation methods

Survey items	Investigation methods		Location
Rut depth	S030	Method for measuring rut depth of pavement surface	25,50,75 m
Evenness	S028	Method for measuring roughness of pavement surface	OWP, IWP
Tire-pavement noise	S027-1T	Method for measuring tire road noise using normal tire	OWP
Texture	S022-2T	Method for measuring texture depth of pavement surface using contactless laser sensor	OWP
Skid resistance	S021-2	Method for measuring surface frictional properties using the british pendulum tester	25, 75 m OWP, IWP

2.4 Rut Depth

Figure 4 shows the results of the measurement of rut depth. As shown in the figure, the rut depth slightly increased in all the construction sections from the time when the pavement was newly installed. However, there were no remarkable differences in the rut depth between the new PA pavement sections (Sections A and B) and recycled PA pavement sections (Sections C and D) that have been in use for ten years. Also, the rut depth was 6 mm at the most, which was significantly below the setting example level of 20 to 40 mm for Category B (Management control standards) in the Pavement Inspection Guidelines (MLIT 2016) and therefore it can be considered that there is no problem with performance and durability for plastic deformation. However, although this was visual observation, cracks were found in all the construction sections and scattering of aggregate was found in some construction sections. Based on this result, although both the new PA pavement and recycled PA pavement have sufficient plastic deformation resistance, there is a concern about aging and stiffening of binders.

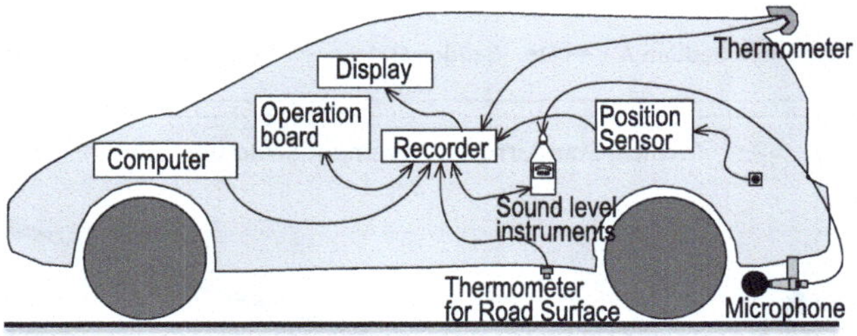

Fig. 3. Method of measuring tire-pavement noise

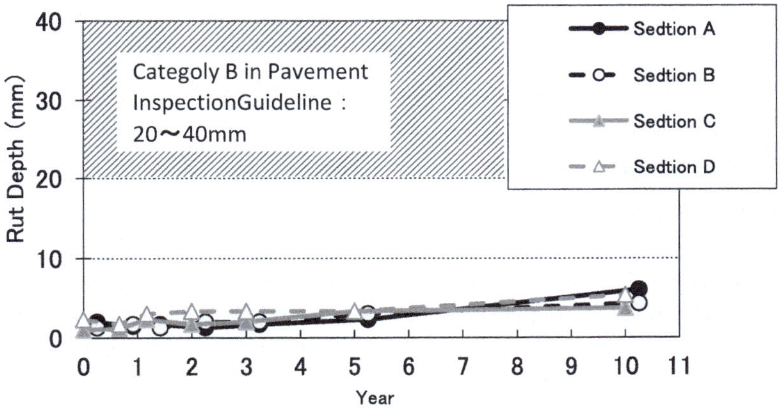

Fig. 4. Results of the measurement of rut depth.

2.5 Evenness

Figure 5 shows the results of the measurement of evenness. Evenness was most excellent in Section B (two-layer OL) of the new PA pavement. In addition, all the construction sections satisfied the evenness immediately following construction in accordance with the Technical Standard for Pavement Structure (MLIT 2001) of 2.4 mm or less. However, the evenness of Section A of the new PA pavement and Section C of the recycled PA pavement exceeded 2.0 mm. In regard to Section A, it is considered that part of the section facing a large vehicle garage and a rough road surface in some areas where vehicle access is increasing due to housing land development affected the result. As for Section C, however, the cause of the result could not be identified because there was no particular access by large vehicles, etc., and no noticeable differences with other sections were found from a survey of pavement bearing capacity by a falling weight deflectometer (FWD) survey.

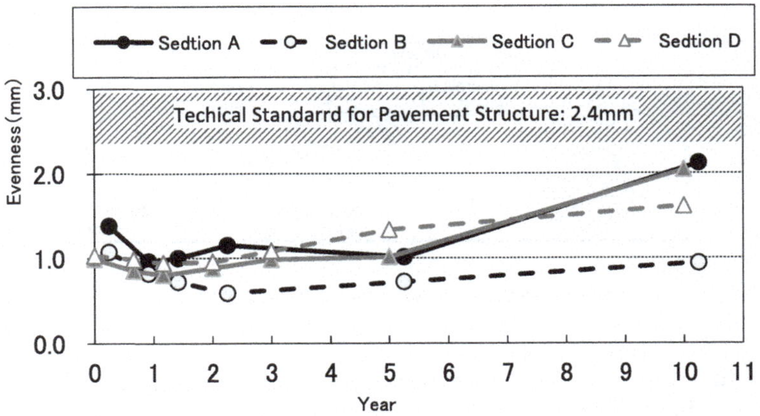

Fig. 5. Results of the measurement of evenness.

2.6 Tire-Pavement Noise

Figure 6 shows the results of the measurement of tire-pavement noise. As shown in the figure, the tire-pavement noise gradually increased in all the construction sections. After the pavement has been in use for five years, higher noise reduction effect was confirmed in the recycled PA pavement (Sections C and D) than the new PA pavement (Sections A and B). However, because the results were reversed after it has been in use for ten years, it could not be determined that there were noticeable differences between the new PA pavement and the recycled PA pavement, and so it is considered necessary to continue to confirm this matter in the future. Overall, collapse of voids was observed due to aging deterioration and it is therefore presumed that this has caused the tire-pavement noise to increase. However, since the tire-pavement noise on the DG pavement was nearly 98 dB, there is a greater noise reduction effect than the DG pavement.

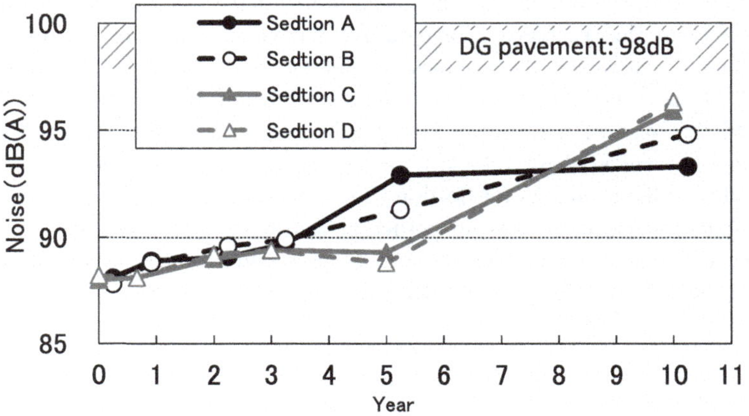

Fig. 6. Results of the measurement of tire-pavement noise.

2.7 Texture Depth

Figure 7 shows the results of the measurement of texture depth (MTM: Mini Texture Meter). Although texture depth was on a declining trend until after the pavement has been in use for three years, it increased thereafter. This is thought to be due to the surface texture slightly being smoothed by traffic load in three years and the road surface became rougher after five years of use as a result of scattered aggregate due to the effect of aged and stiffened binders. Although the results indicate texture depth of the recycled PA pavement (Sections C and D) was slightly lower than the new PA pavement (Sections A and B), a difference in the road surface conditions between the recycled PA construction section and the new PA construction section could not be visually observed on site.

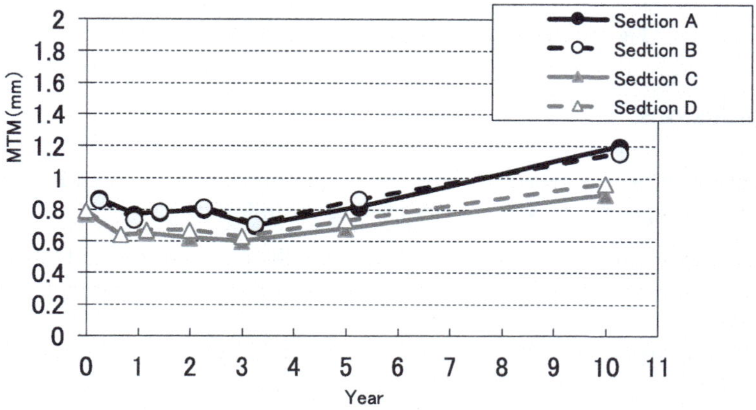

Fig. 7. Results of the measurement of texture depth.

2.8 Skid Resistance

Figure 8 shows the results of the measurement of skid resistance. No significant differences in the skid resistance value (BPN: British Pendulum Number) were observed between the new PA pavement construction section and the recycled PA pavement construction section. Since the value is trending downward in some sections, it is considered necessary to continue to survey this item.

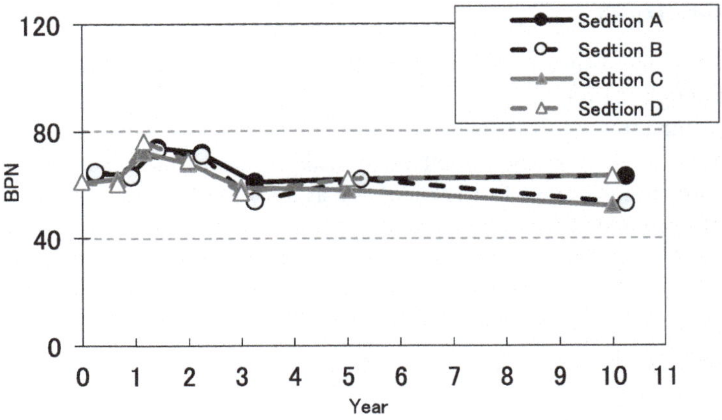

Fig. 8. Results of the measurement of skid resistance.

3 Conclusions

The durability and performance of recycled PA pavement that has been in use for ten years found through this survey are summarized as follows:

(1) As for the mix design of recycled PA mixture, the Cantabro test and bitumen run-off test can be used to determine the quantity of rejuvenator and optimum binder contents.

(2) There was no problem in performance and durability because the rut depth was only 6 mm at the most, though it slightly increased from the time of new installation and there was no significant difference between the new PA pavement section and the recycled PA pavement section. Although this was a visual observation, cracks were found in all the construction sections and scattering of aggregate was found in some construction sections. For these reasons, although the pavements have sufficient plastic deformation resistance, there is a concern about aging and stiffening of binders.

(3) Since the tire-pavement noise gradually increased in all the construction sections, it can be considered that the performance property has reduced over time as it was in use. However, it is considered that the noise reduction effect is greater than DG pavement even after ten years of use.

(4) With regard to evenness, texture depth and skid resistance, no noticeable differences were confirmed between the new PA pavement and the recycled PA pavement construction sections.

(5) There were no noticeable differences in the comparison of RAP mix ratio (Section C: 20%, Section D: 10%) in the recycled PA pavement or in the comparison of milling thickness (Section A: one-layer OL; Section B: two-layer OL) in the new PA pavement.

Acknowledgements. This test construction and follow-up survey were conducted in cooperation with the Tsuchiura Public Works Office, Department of Public Works of Ibaraki Prefecture to whom I am grateful for their assistance.

References

Japan Road Association (1996) Drainage pavement technical guide. Maruzen Publishing Co., Ltd.

Japan Road Association (2006) Interim report of test pavement on national road toward establishment of recycling technology for drainage pavement

Japan Road Association (2007) Handbook of pavement investigation and examination method. Maruzen Publishing Co., Ltd.

Kawakami A, Kubo K, Sasaki I, Kano T (2013) Study on the recycling method for drainage asphalt pavement and evaluation for its durability. In: Advanced materials research, ICPT 2013, vol 723, pp 664–669

Minegishi J, Ueno S, Kobayashi K (2008) Study on pavement recycling by test construction using recycled materials from low noise pavement. Annual Report of Tokyo Metropolitan Civil Engineering Center, pp 63–68

Ministry of Land, Infrastructure, Transport and Tourism (2001) Technical standards for pavement structure

Ministry of Land, Infrastructure, Transport and Tourism (2016) Procedure of Pavement Inspection

Sasaki I, Nitta H, Kubo K (2005) Changes in pavement surface properties of test construction in national highway using recycled aggregates of drainage pavement. In: Proceedings of Japan road congress. 12P64

Performance Related Quality Assurance in Pavement Construction

Salvatore Damiano Cafiso(✉), Brunella Capace, and Alessandro Di Graziano

Department of Civil Engineering and Architecture (DICAR), University of Catania, Via S. Sofia 64, 95125 Catania, Italy
dcafiso@unict.it

Abstract. In pavement construction and management, there is the need to identify measures in order to guarantee high levels of performance over time.

More specifically, QA/QC performance-related specifications (PRS) describe the desired levels of construction quality characteristics correlate with engineering properties that predict as-built pavement performance in order to guarantee the expected performances over the design period. They thus provide the basis for rational acceptance/pay adjustment decisions to be correctly related to actual changes of future maintenance costs.

Considering asphalt pavements, numerical simulations were carried out assuming changes in material's moduli with the aim to develop PRS for QC in pavement construction. Results allowed to estimate effects of construction structural deficits in the perspective of future pavement performance. Basing on Life Cycle Cost Analysis, changes in maintenance costs were estimated.

Keywords: Life cycle cost analysis · QA/QC · Performance related specifications

1 Introduction

Nowadays in an international context where sustainable social-economic-environmental development is the cornerstone for the growth of transport infrastructures, the quality requirement is essential to guarantee high levels of road performance. A comprehensive understanding of issues pertaining to the quality of a project is needed in order to achieve high quality that not only gives acceptable return value to society but also satisfies the needs of all the stakeholders of infrastructure projects (Warsame 2013).

The "Quality Control" (QC) of an infrastructure is determined by the compliance with the technical requirements included into "Quality Assurance" (QA) system. Quality Assurance refers to all those planned and systematic actions necessary to provide confidence that a road infrastructure will perform as by design.

Quality Control is comprehensive of those QA actions and considerations necessary to assess and adjust production and construction processes so as to control the level of quality being produced in the end product (Banerjee et al. 2012).

In the last ten years there has been interest in the paving industry in defining the quality assurance in terms of performance, where the contractor responsible for

M. Pasetto et al. (Eds.): ISAP APE 2019, LNCE 48, pp. 470–479, 2020.
https://doi.org/10.1007/978-3-030-29779-4_46

production and placement would then be paid on the basis of the difference in service life between the as-designed and as-built pavement. In this case, the specifications state the AQC (Acceptance Quality Characteristics) as a measure of pavement performance (NCHRP 2009).

In this way, the construction companies have at their disposal a series of tools through which to improve the production process and reduce the risk of non-compliance that could result in penalties or the rejection of the road authorities (Mensching et al. 2013).

In this research study the quality control during pavement construction are treated after the placement of pavement layers with the aim to develop performance related quality controls for the pavement structure.

The AASHTO empirical design method has been used to address the performance of the pavement in terms of residual life and needs for future overlay treatments. Even if AASHTO empirical method presents same limitations when compared to the Mechanistic approach, in the framework of the materials and pavement typologies considered in the study it has proved to be an immediate tool both for the assessment of the structural performance of the as built-pavement and of maintenance interventions with the overlay design procedure.

The paper starts with the definition of the reference database of pavement structures and material properties, the AASHTO empirical design is then applied to calculate the changes in the residual life and overlay treatments due to changes in the properties of construction material, economic costs are then estimated to develop Penalty/Bonus criteria.

2 Database Selection

The performance analysis of road pavements in Quality Control requires a thorough study in terms of relationships between measures of load bearing capacity and residual life and also reliability of the method proposed in the contract specifications.

In order to make an analysis on a wide spectrum of pavement structures, a database of pavements type was created, looking to a range of solutions applicable to roads with different traffic volumes and subgrade conditions. To this aim, the Italian Catalog of Road Pavements (1995) is a useful reference because it includes a wide range of pavement design solutions reporting the traffic volume and composition and the assumed values of reliability and PSI (Present Serviceability Index) according to the AASHTO Guide for Design of Pavement Structure (1993).

Table 1 shows the flexible road pavements used in this study, where, for each of them, it's reported: the ID; the layer coefficient factors a_1 (wear + binder), a_2 (base), a_3 (subbase); the thickness of layers h_1, h_2, h_3 and the asphalt concrete resilient moduli M_1, M_2 of the first two layers and the resilient modulus M_3 of the granular subbase layer; the subgrade module M_S. Basing on the previous pavement structural characteristics, the Structural Number (SN) and ESALs were calculated and reported in Table 1, as well. Table 1 data represents the design characteristics that will be used as reference to analyze the compliance of as built-pavement with the design project.

Table 1. Database pavement types

Pav Id	a_1	a_2	a_3	h_1 mm	h_2 mm	h3 mm	M_1 MPa	M_2 MPa	M_3 MPa	M_S MPa	SN cm	SN inch	$ESAL_0$
P1	0.42	0.33	0.12	110	170	350	2766	2505	110	30	14.11	5.55	2.40E+07
P2	0.42	0.33	0.12	110	220	350	2766	2505	110	30	16.12	6.35	6.55E+07
P3	0.42	0.33	0.12	130	160	150	2766	2505	110	90	12.41	4.89	6.23E+07
P4	0.42	0.33	0.12	110	170	150	2766	2505	110	90	12.06	4.75	1.01E+08
P5	0.42	0.33	0.12	90	150	350	2766	2505	110	30	12.61	4.96	1.07E+07
P6	0.42	0.33	0.12	90	120	150	2766	2505	110	90	9.41	3.70	2.02E+07
P7	0.42	0.33	0.12	130	200	350	2766	2505	110	30	15.94	6.28	6.02E+07
P8	0.42	0.33	0.12	110	180	150	2766	2505	110	90	12.39	4.88	1.22E+08
P9	0.42	0.33	0.12	110	250	150	2766	2505	110	90	14.72	5.80	4.21E+08
P10	0.42	0.33	0.12	100	80	150	2766	2505	110	90	8.50	3.34	1.43E+07

Moreover, in order to simulate a variability in as-built pavement characteristics, the layer moduli M_1 (surface + binder), M_2 (base), M_3 (subbase), M_s (subgrade) were modified in different percentages of −20%, −10%, 0%, +15%, obtaining 13 different solutions for each single type of road analyzed for a total of 130 pavement composing the case study.

3 Structural Capacity and Performance

Road pavements are designed to withstand during the design period to traffic loads (ESALs) and external environmental conditions, and for this reason they must have a structural capacity that depends on the layers' thickness and the structural characteristics of the materials.

In the framework of the AASHTO empirical design, during the design period, a progressive reduction of SN, from the initial value SN_0, is produced by the traffic loads till the final value of Present Serviceability Index (PSI_f) at the design number of $ESAL_0$ (Fig. 1). It is expected that an as-built reduced structural number SN_i, due to construction nonconformities, produces an early decay of the pavement performance which can be associated to a SN curve that reaches early the final value of PSI_f at $ESAL_i$ with a residual capacity SN_{eff}. To restore the design conditions (i.e. Design period and $ESAL_0$) an unplanned anticipated maintenance (which generally translates into an overlay treatment) is needed to provide the structural capacity (ΔSN_{eff}) in order to make possible the retrieval of the design $ESAL_0$ (ΔESAL).

Applying the overlay design procedure reported in AASHTO (1993) SN_{eff}, ΔSN_{eff} and ΔESAL can be calculated starting from the initial values SN_0 and SN_i, as described in the following flow chart (Fig. 2).

Assuming that the reference pavements are characterized by a design SN_0 in the range 3.0 ÷6.0 inches, a final PSI_f in the range 2.0 ÷3.0 and a percentage of as-built nonconformities (ΔSN_i) from 0% to 20%, Fig. 3 shows the reduction in the residual life (RL) that can be expressed as percentage of the design ESALs: $\Delta RL = \Delta ESAL/ESAL_0$.

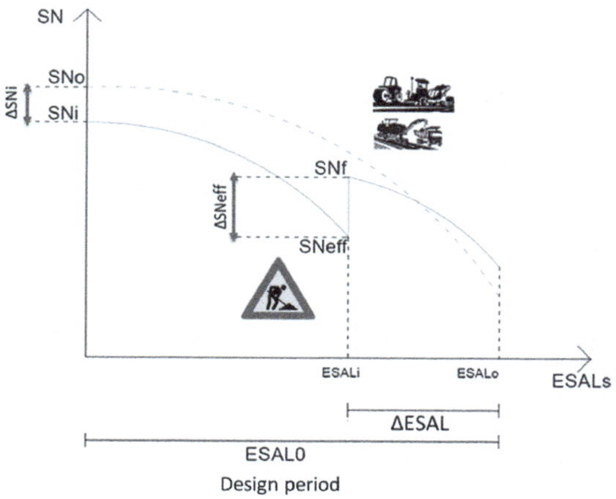

Fig. 1. Overlay intervention to recover construction defects. Relationship between SN and Esals.

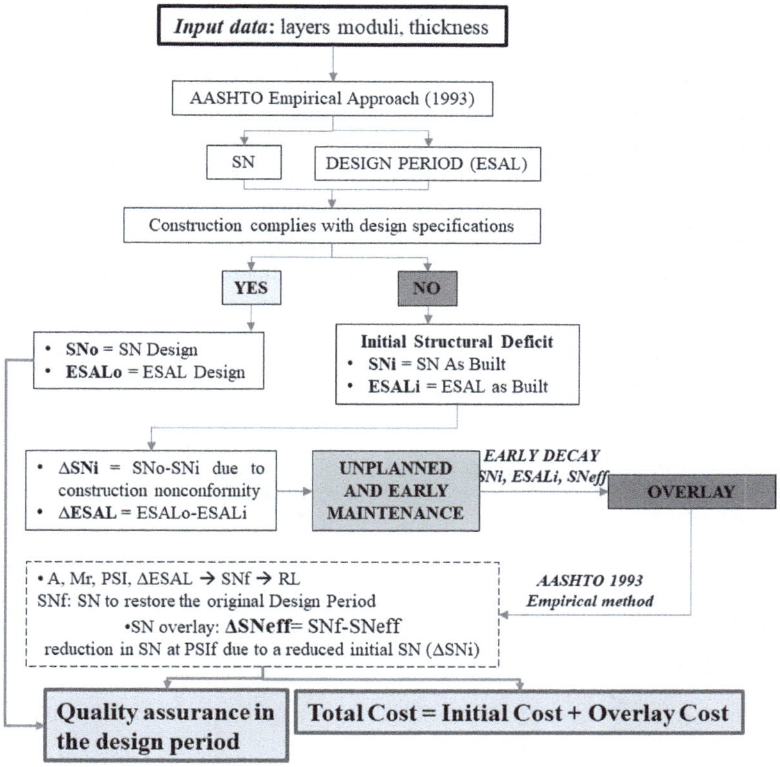

Fig. 2. Procedure flow chart.

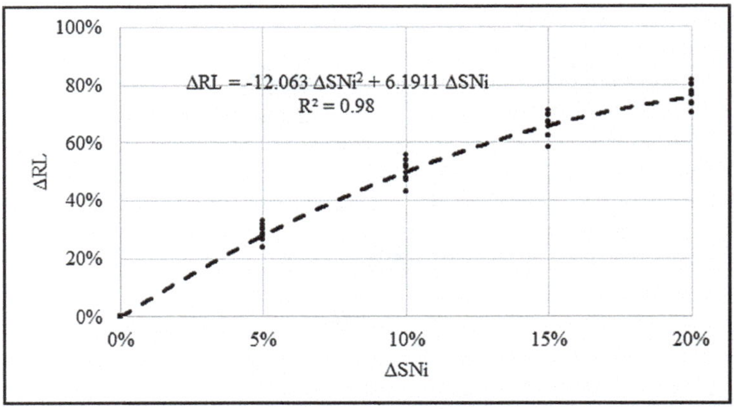

Fig. 3. ΔSNi vs. ΔRL [%]

As expected, ΔRL is well correlated to ΔSN_i and a ratio of 4 was estimated between the two factors that means a 4% reduction in Residual Life (RL) for each 1% deficit in SN_0 design value. The result is referred to the flexible pavements used as reference in the present paper. In general, if a similar pavement type with SNo and PSI in the same range of the analysis are considered, the correlation can be considered valid.

Less intuitive is the correlation between ΔSN_i and ΔSN_{eff} because of the limitation to predict the decay in the structural capacity. The AASHTO overlay design procedure make possible to calculate the expected SN_{eff} at the PSI_f and the structural capacity of the overlay (ΔSN_{eff}) to restore the original design period. A good linear correlation was found between ΔSN_i and ΔSN_{eff} with a R-squared of 0.90 (Fig. 4) and a statistical significance at the 95.0% level of confidence in the coefficient estimate (Table 2).

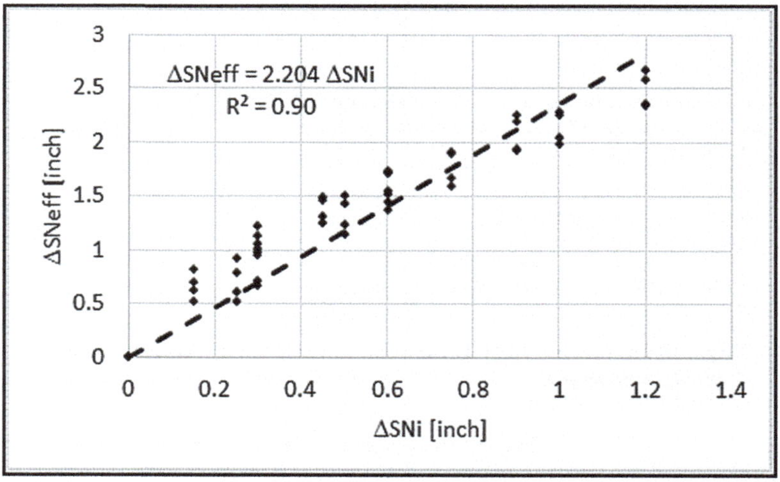

Fig. 4. ΔSNi vs. ΔSNeff

Table 2. Linear regression - coefficient estimate

	Least squares	Standard	T	
Parameter	*Estimate*	*Error*	*Statistic*	*P-Value*
Slope	2.20362	0.0530668	41.5254	0.0000

A part of the statistical linear correlation, that will be used in the following analysis, the result is meaningful because ΔSN_{eff} is more than the double of the initial deficit in structural capacity ΔSN_i. Therefore, to evaluate the economic losses the measure of initial deficit must be related not only to the reduction of the Residual Life but also to the increase in the future maintenance costs to restore performance and design period that are more related to ΔSN_{eff}.

4 Impact on Residual Life and Maintenance

To better understand the effects of a deficit in initial structural capacity on residual life and maintenance treatment, numerical simulations have been carried out on sample of pavement structures extracted from the overall database. Specifically, three pavements have been selected as representative of the whole database in Table 1 because characterized by values of SN equals to the maximum, average and minimum: 6.35, 4.96, 3.34 inches, respectively (Fig. 5).

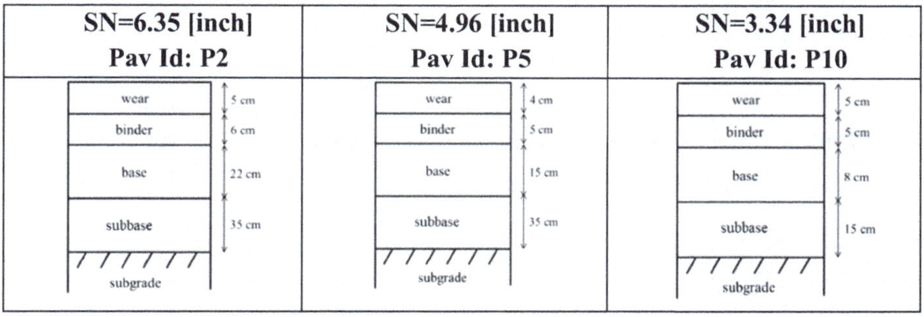

Fig. 5. Road pavements of the database with the max, average and minimum value of SN

As maintenance treatment made at the year when the PSI_f is reached (before the year 15th), two different applications of overlay were considered to provide the ΔSNeff in order to restore the pavement performance within the original design period assumed equal to 15 years:

– *Case study 1*: new overlay with Asphalt Concrete (AC) over the existing road pavement.

- *Case study 2*: Milling of the existing pavement to maintain constant the level of the pavement surface and overlay composed by 4 cm of wear course realized with virgin AC in order to restore surface characteristics (Cafiso and Taormina 2007) and the base layer composed by Reclaimed Asphalt Pavement (RAP) at 40%.

Even if the case study 1 is not a practiced solution, it was useful to consider for providing the minimum values estimation for construction costs. For each design pavement in Fig. 5, four different applications of overlay were considered basing on initial deficit of SN for the as-built pavement

$$\Delta SNi = SN_0 - SN_i$$

Using the methodological approach showed in Fig. 1, for each pavement was calculated the year in which to carry out the maintenance treatment in advance of a fixed 15-years design period and the overlay thickness (cm) of each layer to be placed in, to restore the design conditions in terms of design period and PSI_f.

Results of the overlay design as reported in Table 3 for the case study 1 and Table 4 for case study 2.

Table 3. Net overlay interventions calculated on the 3 road pavements

		Overlay 3 cm			Overlay 6 cm			Overlay 8 cm			Overlay 9 cm		
Pav Id	SN_0	SN_i	ΔSN_{eff}	Year	SN_i	ΔSN_{eff}	Year	SN_i	ΔSN_{eff}	Year	SN_i	ΔSN_{eff}	Year
P2	6.35	6.14	0.48	11	5.9	0.97	8	5.75	1.32	7	5.67	1.50	6
P5	4.96	4.75	0.48	10	4.5	0.97	7	4.36	1.32	6			
P10	3.34	3.13	0.48	9	2.9	0.97	5	2.74	1.32	4			

Table 4. Milling and overlay interventions calculated on the 3 road pavements

		Overlay 4 cm (4 + 0)			Overlay 8 cm (4 + 4)			Overlay 10 cm (4 + 6)			Overlay 13 cm (4 + 9)		
Pav Id	SN_0	SN_i	ΔSN_{eff}	Year	SN_i	ΔSN_{eff}	Year	SN_i	ΔSN_{eff}	Year	SN_i	ΔSN_{eff}	Year
P2	6.35	6.23	0.27	13	6.1	0.54	11	6.05	0.67	10	5.96	0.87	9
P5	4.96	4.84	0.27	12	4.7	0.54	10	4.66	0.67	9			
P10	3.34	3.22	0.27	11	3.1	0.54	9	3.04	0.67	8			

5 Economic Costs

Life Cycle Cost Analysis of the pavement construction and maintenance was used to compute economic costs taking into consideration standard unit costs and net present value of maintenance works at the time of construction.

Figure 6 shows the increase in economic costs related to a reduction of the SNi, where:

$$\Delta COST = (\text{Total cost} - \text{Initial cost})/\text{Initial cost}$$

with

- Total cost includes the construction and the unplanned maintenance cost;
- Initial cost represents the cost of design pavement construction.

The increase in the total agency costs was estimated in the range 1.78 ÷ 4.55 times of the deficit in structural capacity of the as built pavement for unplanned maintenance works due to the overlay construction with reference respectively to the net overlay (case 1) and milling and overlay treatments (case 2). The cost increment (ΔCOST) takes into account the unplanned maintenance costs to restore the original design life of the pavement.

Because, previous studies (Praticò et al. 2016) highlighted research needs in the application of the AASHTO empirical approach for the estimation of penalties and

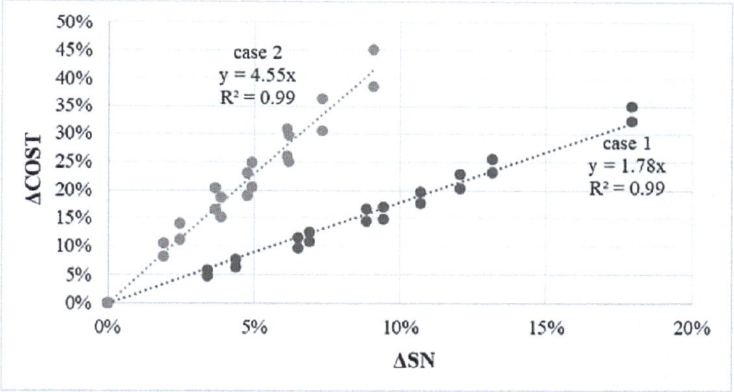

Fig. 6. Comparison of economic cost for the two case studies: (1) net overlay vs. (2) milling and overlay (AC + RAP).

bonuses, results were compared with the NCHRP Performance-Related Specification. NCHRP (2011) applied the Mechanistic Empirical Pavement Design Guide (MEPDG) for the evaluation of penalties and bonuses.

The Predicted Life Difference (PLD) is used as the basis for establishing the Pay Factor Penalty/Bonus. A part of the cut off values (8% maximum bonus and 20% reject quality level), as shown in the Fig. 7, results in the present study are comparable to those of NCHRP for penalties with maintenance strategy (2) and for bonuses with case (1). Moreover, both regression models between the residual life RL and the penalty/bonus factors (Δcost) showed a very good value of R-square equal to 99%.

It is meaningful to note that the maintenance strategy plays an important role in the estimation of costs necessary to restore the as-designed performance and therefore in the application of penalty or bonus to nonconformities (Buddhavarapu et al. 2014).

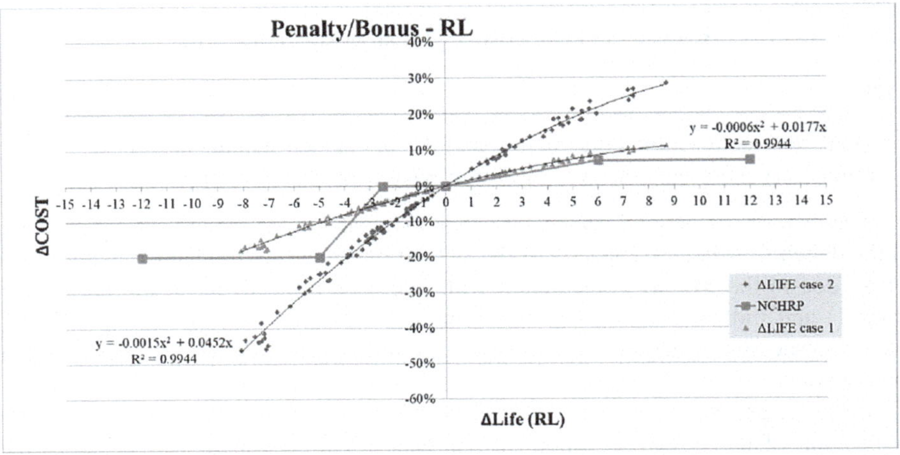

Fig. 7. Comparison between the models developed and the NCHRP performance related approach.

6 Conclusions

QA/QC performance-related specifications describe the desired levels of construction quality characteristics correlate with engineering properties that predict as-built pavement performance to be guaranteed over the design period.

In this study, numerical simulations and Life Cycle Cost analysis were carried out in order to estimate the maintenance costs associated with defects in as-built pavements. The AASHTO Empirical Design guide approach resulted an effective tool because able to evaluate the changes in the residual life due to initial reduction in the structural capacity and the needs for overlay treatments to restore the design period and performance, as well. The use of the Empirical approach should be considered not up to date when compared to the new MEPDG. Anyway, if contract requirements have to be defined, some considerations are in favor of the empirical one:

- traffic composition and climatic factors, to be considered in MEPDG, are not unique and each project should be specifically evaluated;
- distress models in MEPDG need calibrations.

Analogously, AASHTO pure empirical has limitation with respect to:

- empirical nature and functional form of the model;
- heavy truck traffic levels over 50 million ESAls are not considered;
- new asphalt mixes such as SMA, modified, RAP are not directly incorporated;
- climate conditions not appropriate for serious frost or other local problems.

Despite these limitations, the simplicity and consistency of the empirical approach represents an added value for the understanding and transparency issues in the framework of construction contracts. Moreover, a comparative analysis of results limits the absolute bias in the model estimations. Taking into account those considerations in the field of application, AASHTO empirical results were consistent with MEPDG as showed in Fig. 7 by the comparison with NCHRP penalty/bonus criteria.

Results highlighted as not only the reduction in Residual Life is the performance criterion to be considered, but also the increase in maintenance costs have to be explicitly evaluated because correlated with different factors to the design requirements.

References

AASHTO (1993) Guide for Design of Pavement Structures 1993

Banerjee A, Smit A, Prozzi JA (2012) Influence of operational tolerances on HMA performance. Constr Build Mater 27(1):15–23

Buddhavarapu P, Smit A, Prozzi J, Fan W, Gurmu Z (2014) Revised Pay Adjustment Factors for HMA and Concrete Pavements, FHWA/TX-14/0-6675-1, Austin

Cafiso S, Taormina S (2007) Texture analysis of aggregates for wearing courses in asphalt pavements. Int J Pavement Eng 8(1):45–54

CNR (1995) Catalogo delle Pavimentazioni Stradali (Italian Pavement Catalog)

Mensching DJ, Myers McCarthy L, Mehta Y, Byrne M (2013) Modeling flexible pavement overlay performance for use with quality-related specifications. Constr Build Mater 48:1072–1080

NCHRP (2009) Report 626: NDT Technology for Quality Assurance of HMA Pavement Construction. Transportation Research Board, Washington D.C.

NCHRP (2011) Report 704: A Performance-Related Specification for Hot-Mixed Asphalt. Transportation Research Board, Washington D.C.

Praticò FG, Noto S, Astolfi A (2016) Issues and perspectives in the application of different pavement design methods to life cycle cost analysis. MAIREPAV 2016

Warsame A (2013) Framework for Quality Improvement of Infrastructure Projects. J Civil Eng Archit 7(12) (Serial No. 73):1529–1539

The BIM (Building Information Modeling)-Based Approach for Road Pavement Maintenance

Gaetano Bosurgi[1(✉)], Clara Celauro[2], Orazio Pellegrino[1], Nicola Rustica[1], and Sollazzo Giuseppe[2]

[1] Department of Engineering, University of Messina, Vill. S. Agata, C. da di Dio, 98166 Messina, Italy
gbosurgi@unime.it

[2] Department of Engineering, University of Palermo, Viale delle Scienze, Edificio n. 8, 90128 Palermo, Italy

Abstract. The recent developments of road pavement survey technologies guarantee acquisition of detailed data regarding the main functional and structural indicators, at a very high frequency and levels of accuracy and precision unconceivable few years ago. The relevant information, regarding surface distresses too, is thus available in economical, high-speed and high-resolution forms, with limited disturbance on traffic and road agencies issues. However, the availability of such a mass of data poses interesting questions concerning the most efficient processing and analysis methods, for optimizing maintenance operation planning. Unfortunately, the PMS adoption is not very common and widespread, as they are significantly complicated and sophisticated, with critical consequences in their actual and intense practical use. Then, there is a need of simplifying the pavement maintenance process, by defining smart tools actually supporting the operators in the analysis and decisional phases. BIM (Building Information Modeling) procedures can absolutely represent a strategical solution. Indeed, BIM models, already widely adopted in the structural fields, have been recently introduced in the infrastructural design phases, simplifying the entire life-cycle of the construction, with representation clarity, mistake reduction, and money savings. However, although the BIM advantages have been significantly exploited in the design phases, pavement maintenance has not been directly involved in the BIM processes. Then, this paper focuses on a methodological analysis of the problem, aiming to propose a novel BIM-based approach for managing pavement maintenance. The goal is to define proper "smart objects", based on specific relational databases, that may represent a more suitable instrument for handling pavement condition and quality information, in a user-friendly software environment. The potentialities of the modern survey technologies may be integrated and optimized in a BIM process, involving in parallel both design and maintenance.

Keywords: BIM · I-BIM · Pavement maintenance · Pavement condition

M. Pasetto et al. (Eds.): ISAP APE 2019, LNCE 48, pp. 480–490, 2020.
https://doi.org/10.1007/978-3-030-29779-4_47

1 Introduction

In order to allow transport infrastructures to effectively perform their task as driving force of the economic growth, road pavements have to assure adequate structural and functional characteristics for their entire service life. To achieve this goal, it is essential, from one side, to properly design pavements and, from the other, to adequately deal with the complex maintenance issues, guaranteeing performance and safety reliability (Celauro et al. 2017). Then, it is important to plan a reliable and continuous monitoring activity for evaluating and anticipating the evolution of pavement condition and state during its service life.

Considering the recent significant development achieved in the estimation of pavement structural and functional indices by the modern survey and monitoring systems (Sollazzo et al. 2016; Zhang et al. 2017), the focus has currently moved, affecting mostly management and analysis procedures of such big and detailed data acquired on the field.

At this regard, although the modern pavement survey systems assure high-speed and high-quality acquisition of big-data concerning the pavement state in continuous, the analytical and technical methodologies to fully exploit similar data are not equally evolved. Indeed, managing survey acquisition and results in user-friendly software, to simplify querying and elaboration and favour their use in flexible and reliable analyses and simulations, is not currently practical.

Consequently, a revision of data management procedures is fundamental for simultaneously achieving two different goals:

- simplifying management, handling, and visualization of the survey results (not only in terms of simplified and global indices, but also representing virtually and in real-time 2D and 3D models of surfaces and distresses);
- proposing a decision support model, derived from traditional PMS, easier to use and optimize.

Then, the modern BIM (Building Information Modeling) technologies – that have represented reliable and convenient solutions in many civil engineering fields (Bryde et al. 2013) for decades – seem perfectly fitting to this issue. However, although BIM solutions are nowadays ready for guaranteeing the required production standards in the building architecture, structural, and MEP (mechanical, electrical, and plumbing) fields, their transposal in the infrastructural area (I-BIM – Infrastructure-Building Information Modeling) cannot actually be considered satisfactory yet (Dell'Acqua 2016). Despite this, the continuous development of I-BIM models and procedures has started providing the first responses in terms of effectiveness and technical-economical productivity, especially if focused on the design phases only. On the contrary, issues related to infrastructure service and operation (management and maintenance, with particular reference to pavements) are relegated to the limits of this innovation process, even if the related potential benefits may be immediate and very remarkable for both road agencies management needs and users' effective quality levels.

The paper aim focuses on the definition of novel protocols of analysis, elaboration, and processing of survey data, to be adopted into appropriate I-BIM environments, in order to strengthen and optimize the traditional planning procedures of road pavement management and maintenance. In the following sections, indications concerning the definition of appropriate "smart objects", stored and managed in specific relational databases and characterized by several pavement features derived from surface surveys, are provided and discussed. In this way, a powerful and flexible tool would be realized, for querying and processing information, defining a rapid, specific, and optimized methodology which supports road agencies in the delicate decisional phases related to road maintenance and management.

2 General Notes on the BIM-Based Approach for Infrastructures

Infrastructure engineering is traditionally considered as a complex topic requiring multidisciplinary skills, owing to issue and analysis etherogeneity. Further, for defining technical solutions in compliance with all the imposed boundaries, it generally determines a remarkable time and resource consumption. This complexity mainly lies in variety and contemporaneity of the involved aspects, introducing economic, environmental, geometrical, social, etc. constraints and limits, which naturally influence and condition the technical choices in every design scenario.

These aspects evidence how traditional approaches for defining the "best" technical alternatives in the infrastructure field are characterized by unacceptable uncertainty levels, considering both time and resource consumption and exposure to error risk. At this regard, in recent years, many efforts have been spent in developing novel tools and algorithms that may actually speed up and improve (even in terms of reliability) the organization of civil engineering activities. From one side, many attempts focused on innovative decision support algorithms, able to autonomously analyse and compare numerous technical alternatives, for optimizing the involved variables and identifying the "best solution" that minimizes diseconomies, costs (not only in construction, but also in management) and the most relevant eventual critical issues (Bosurgi and Trifirò 2005; Bosurgi et al. 2013; Moreira et al. 2017).

In parallel, similar targets have been pursued by another fundamental innovation that has modified the civil engineering field from roots, i.e. the so-called "BIM revolution". The BIM technologies, recently introduced in the civil engineering field thanks to both hardware and software computer developments, represent a novel way of thinking design, construction activity planning, and maintenance of the construction for its entire life cycle (Bryde et al. 2013). Furthermore, they guarantee high levels of precision, detail quality, productivity, and efficiency in information managing and sharing, producing a remarkable improvement in terms of representation clarity, reduction of error risk, and greater process global savings (Lee et al. 2003; FHWA 2013). The main principles and advantages of BIM are provided in Fig. 1.

From an operative perspective, BIM basically relies on a shared software environment, in which all the involved subjects may operate and play in real-time and in parallel, with evident benefits in sharing technical choices and, in general, information. This approach assures positive effects in terms of error and inaccuracy reduction and team efficiency increase (Li et al. 2014).

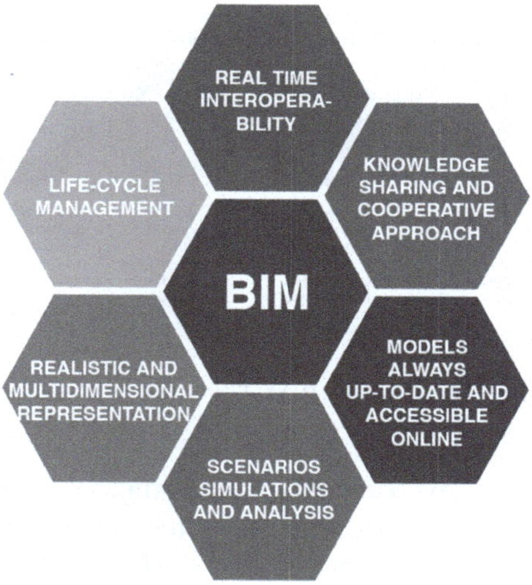

Fig. 1. BIM principles

The BIM core is represented by the introduction of specific "smart objects", which provide a real, clear, and interactive 3D representation of all the single elements of a construction (Chen et al. 2016). They also contain several features that fully define all the characteristics and functional aspects, even their mutual interaction and with the external context. Relying on the considerable storage, processing, and analysis capability of very large amount of data available in reliable relational databases, the most advanced BIM instruments represent a very helpful support tool, useful to simplify cost and time evaluations. In addition, they support the operators in organization and operative planning during all phases of the construction life cycle, including management and maintenance issues too.

According to this, nowadays the number of representative dimensions of a construction can be increased from the traditional 3D for spatial representation, up to *n* dimensions (Lee et al. 2003; Ding et al. 2012; FHWA 2013), including and involving in the model further information related to other technically relevant aspects (time, cost, energy, maintenance, safety, etc.), as shown in Fig. 2.

In building architectural, structural, and MEP fields, the adoption of BIM technologies has been extremely rapid and widespread, also because of the definition of a common interchange standard represented by the file format IFC (Industry Foundation Classes, developed by building SMART - Liebich et al. 2010). Although this technology is still evolving and the involved process improvements are not fully exhaustive, the operators have been gradually leaving the traditional CAD-based approach towards BIM methodologies, especially for the advantage of reducing inaccuracies and design errors (and, thus, legal disputes), but also of drastically decreasing delays due to modifications and corrections.

In the infrastructural field, instead, generally, similar results have not been achieved yet. It should be noticed that, especially in road infrastructures, the horizontal

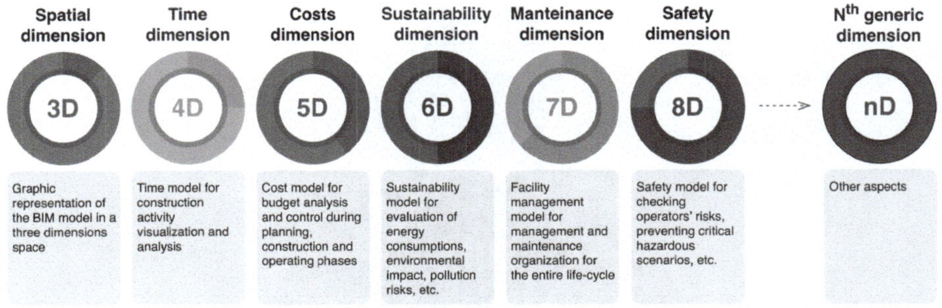

Fig. 2. BIM dimensions

dimensions of the models, involving very wide portions of territories, imply serious technical and operative differences respect to the traditional "vertical" BIM.

This is also due to the complications in a direct adaptation of the IFC standard (Lee and Kim 2011; Dell'Acqua 2018), because of the implicit differences in the object typologies representing the models, despite some attempts in their adjustment or in the definition of novel format specifically conceived for infrastructural modelling (for instance LandXML - INFRAMODEL 2017).

Despite these conceptual and practical delays, the attention on the topic is remarkable and the development is so rapid and incessant that I-BIM solutions can nowadays be applied to actually realize the conceptual model of a road in a specific 3D realistic territorial context. In a "user-friendly" environment, the operator can fully design a highway alignment by introducing and combining simple smart objects typical of a road section (Tibaut et al. 2015), with opportunities to handle in real-time their actual interactions with the external territorial conditions and with the other elements of the road design.

Limited to the design phase, then, I-BIM models are almost ready to offer the same advantages previously described for traditional BIM concerning an interactive, integrated, and flexible operation management, mainly because of their approach properly structured to favour collaboration among different specialized operators. In fact, they can operate in parallel on different aspects of the infrastructure (geometry, structural constructions, walls, bridges, drainage, hydraulic elements, etc.) in a georeferenced and shared 3D model. Moreover, the I-BIM approach may represent a very useful way to handle and solve alignment optimization problem, by means of automatic and smart algorithms (Bongiorno et al. 2019).

However, the aspects of the I-BIM technology related to the phases of the construction life cycle following the design step appear in marked delay in the formulation of the operative procedures and, thus, in their computer-based development. These phases regard control in execution phase, service life management, and maintenance activity planning.

3 Ideas and Preliminary Tests for Road Maintenance in an I-BIM Environment

From previous considerations, it appears that, if on one hand I-BIM procedures applied on design and execution phases reached an acceptable development level, similar results have not regarded, to this point, issues related to road maintenance and its management (Chong et al. 2016).

However, for the same operative philosophy based on which they were conceived, these tools show huge potentialities for a quick interface with traditional instruments adopted in the complex process going from control and survey phases to the identification of the optimal maintenance strategies. In summary, referring to a PMS, the authors would propose a hybrid procedure useful to optimize its architecture (including its models and databases) in a BIM environment. This would simplify its interoperability, dynamism, sharing, and especially handling and analysis of the same data in a quick and error-safe way. The introduction of the BIM concept in a PMS can rely on two main aspects, strongly correlated:

1. the possibility to acquire and manage, graphically and in an interactive way, the results of pavement condition surveys, by means of the commonly adopted indices of geometrical, structural, and functional quality;
2. the possibility to present graphically and in a way easy to be modified in real-time, the selected rehabilitation interventions - according to the type and gravity of the distresses detected in each section.

At this regard, exploiting 3D representation potentialities on which BIM procedures relies – assuring extreme dynamism in handling data and objects typically involved in a road infrastructure (geometry, pavement, structures, signage, drainage, etc.) -, further data and features derived by survey and monitoring activities could be integrated, aiming to optimize maintenance strategies and properly design the rehabilitation activities. In particular, the modern pavement 3D survey techniques, for automatic and georeferenced identification, classification, and quantification – also in terms of severity - of surface distresses (cracks, irregularities, holes, etc.), seem to well integrate with BIM procedures. Indeed, the survey object, suddenly visualized and located in the realistic representation of the road segment, may become interactive, providing and exploiting all the included information. Similar details and data represent the input of a decisional process aiming to optimize the final maintenance activities.

Then, through the definition of specific smart objects, properly linked with the reference infrastructure, the operator is able to actually visualize the effective state of the pavement, in a realistic and accurate manner, by using real 3D images or, better, their elaborations evidencing distresses. For example, Fig. 3 shows a 3D image of a pavement section acquired and processed by a novel device for high-performance pavement surface surveys developed in the Infrastructure Lab of University of Messina.

The operator can handle as needed the graphical and interactive representation of the various structural and functional indices derived from the survey data in a BIM environment, with different detail and precision levels. Indeed, it would be possible to operate locally or averaging on specific areas (as large as preferred) of the road

Fig. 3. 3D model of a pavement section with cracks, acquired and processed by a novel device for high-performance pavement surface surveys developed in the Infrastructure Lab of University of Messina

pavement or of the entire alignment. These graphical and interactive modes of processing and handling outcomes of the survey activities performed on field determine in this process characteristics of operative rapidity, high simplicity, and accessibility that cannot be assured by traditional approaches.

Although this procedure has to be properly studied and defined in all its various phases, in this first approach to the problem, simply methodological, a preliminary application has been designed and analysed, through a commercial software. In this experimental test, the authors considered only a very synthetic representation of pavement sections, characterized by different performance indices derived from a specific survey. In particular, relying on the results of a detailed survey campaign performed along an Italian highway (A20 Messina-Palermo), a single trait of the infrastructure has been reported in a BIM environment. On this trait, through specific and customized proper operations, further information related to the survey activity has been added to the base "lane" component. In Fig. 4, a graphical scheme of the pavement condition along the selected trait is provided. In this figure, each single coloured segment represents different pavement quality indicators, directly determined through the analysis of data acquired during surveys.

As easily perceived, beyond the high clarity of the involved information, the quick and simple tasks required to calculate and provide all the quantities and related metrical values, linked to the represented quality indices, represent one of the main benefit produced by implementing survey data in a BIM environment.

Equally, if not more, relevant are the advantages of this approach in handling maintenance alternatives derived from the analysis of such distress and quality indices, easily identified and suddenly located in the same BIM model. In detail, as for quality indices, further specific smart objects – with characteristic attributes related to the intervention typologies (intervention name and category, involved materials and resources, unit costs, manpower and machines required, duration, etc.) -, able to

Fig. 4. Graphical representation using Autodesk® Infraworks®, in their exact position, of specific survey results regarding structural and functional indices of the pavement along a trait of the A20 highway "Messina-Palermo"

graphically represent the various maintenance and rehabilitation activities and, at the same time, quantify in real-time costs and other significant variables. For example, Fig. 5 provides the extension of some intervention strategies suggested for different types and levels of distress evaluated through the survey activities. In this way, management and optimization of a complex maintenance planning procedure would be simplified and easy to control, with a remarkable operative flexibility.

As further support to the proposed hybrid methodology, once the alternative typologies of maintenance intervention are integrated in a BIM environment, their use in the construction phases can maximize the several operative and analytical possibilities assured by this approach. For example, it would be interesting to study, in a relatively short time, the planning of the maintenance working-zone (location, extension, time scheduling, etc.) in several different alternative configurations. Moreover, advanced traffic simulation tools and algorithms, already available, can process the proposed alternatives to check maintenance effects on the considered section and the entire infrastructure network (in terms of lags, queues, accident increase, and other relevant social or economic aspects).

Consequently, the choice of the most satisfactory maintenance solution will be based not on simplistic evaluations related to the activity costs, but could take into account the strategical availability of more information and specific simulations. This approach, even in compliance with the traditional PMS procedures, exploits their efficiency and productivity, maximizing the practical flexibility of the BIM

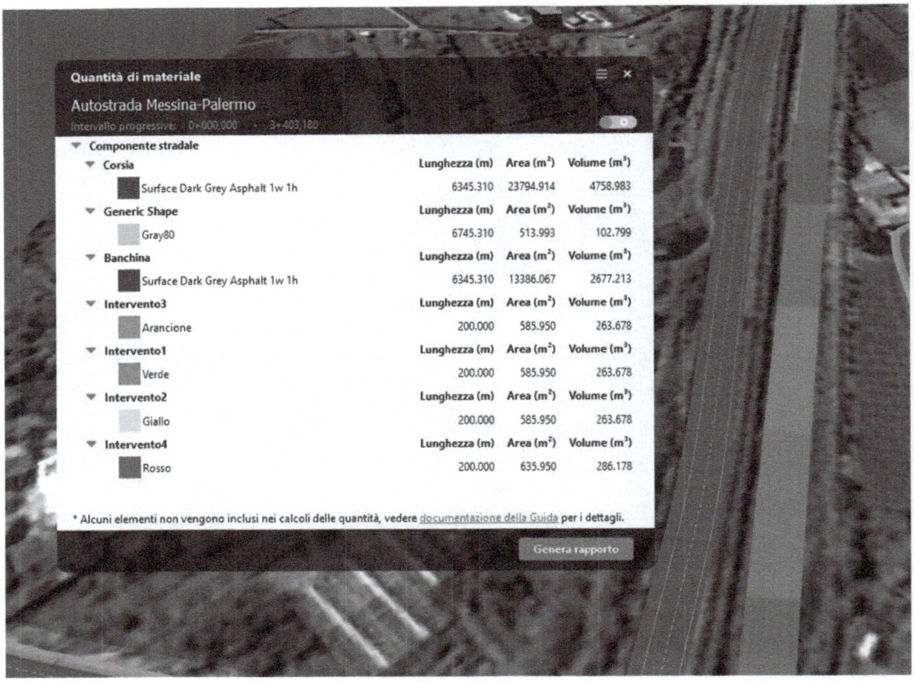

Fig. 5. Representative figure reporting reference data and dimensions of the specific smart objects related to different maintenance solutions in an I-BIM environment.

methodology and the comparisons of the effects of several maintenance and rehabilitation activities.

These applications, absolutely preliminary, evidence how BIM procedures potentially represent an actual response to the simplification needs of the road administrations, for which traditional PMS procedures appear very complicated, intricate, and absolutely not user-friendly. At the same time, they can represent a stimulus and incentive to deal with the maintenance management issue at large scale, with the required rapidity and operative flexibility, making finally possible a generalized and effective adoption of similar procedures, methodologies and instruments.

4 Conclusions

In this paper, the authors described and analysed the possibility to extend the advantages of I-BIM procedures and methodologies to road maintenance management, even considering elaboration and handling of detailed survey data related to pavement condition. In particular, in view of the potentialities of both modern survey systems and I-BIM procedures, the actual applicative advantages produced by this solution are evidenced, aiming to higher levels of efficiency for road agencies and of quality for the final infrastructure users. This issue, analysed from a methodological point of view, seems to be actually applicable in an I-BIM environment, since BIM principles and philosophy are fully coherent and useful for road maintenance goals. Furthermore, this solution could overpass some practical operative and management problems typical of traditional PMS, favouring, instead, organization and analysis of data in a more flexible and functional environment and through highly collaborative and user-friendly platforms. At this regard, as a preliminary experimental approach, specific smart objects representing strategical aspects for road maintenance have been defined and introduced in an I-BIM model. In details, these objects include specific attributes useful to identify and quantify the different maintenance needs. In conclusion, this preliminary experiment evidences how this solution may effectively represent a simplified and efficient decisional support system, for maximizing quality and benefits of all the available data.

Acknowledgements. This paper is independent of Autodesk, Inc., and is not authorized by, endorsed by, sponsored by, affiliated with, or otherwise approved by Autodesk, Inc. Autodesk, Infraworks are registered trademarks or trademarks of Autodesk, Inc., and/or its subsidiaries and/or affiliates in the USA and/or other countries.

References

Bongiorno N, Bosurgi G, Carbone F, Pellegrino O, Sollazzo G (2019) Potentialities of a highway alignment optimization method in an I-BIM environment. Periodica Polytech Civil Eng https://doi.org/10.3311/ppci.12220

Bosurgi G, Trifirò F (2005) A model based on artificial neural networks and genetic algorithms for pavement maintenance management. Int J Pavement Eng 6(3):201–209

Bosurgi G, Pellegrino O, Sollazzo G (2013) A PSO highway alignment optimization algorithm considering environmental constraints. Adv Transp Stud Int J 31:63–80

Bryde D, Broquetas M, Volm J (2013) The project benefits of building information modelling (BIM). Int J Project Manag 31:971–980

Celauro C, Corriere F, Guerrieri M, Lo Casto B, Rizzo A (2017) Environmental analysis of different construction techniques and maintenance activities for a typical local road. J Clean Prod 142(4):3482–3489

Chen S, Lok K, Jeng T (2016) Smart BIM object for design intelligence. In living systems and micro-utopias: towards continuous designing. In: Proceedings of the 21st international conference of the association for computer-aided architectural design research in Asia CAADRIA, Hong Kong

Chong HY, Lopez R, Wang J, Wang X, Zhao Z (2016) Comparative analysis on the adoption and use of BIM in road infrastructure projects. J Manag Eng 32:05016021-1

Dell'Acqua G (2018) Bim per infrastrutture – il Building Information Modeling per le grandi opere lineari. EPC Editore

Dell'Acqua G (2016) I-BIM Infrastructure-Building Information Modeling: Stato dell'Arte. Le Strade, vol 1521

Ding L, Zhou Y, Luo H, Wu X (2012) Using nD technology to develop an integrated construction management system for city rail transit construction. Autom Constr 21:64–73

Schneider C (2013) 3D, 4D, and 5D engineered models for construction. Techbrief FHWA-HIF-13-048. Executive Summary, March 2013

INFRAMODEL (2017) Finnish Inframodel application documentation for LandXML v1.2 – Inframodel Version 4: 2017 - schema version 4.0.3. https://buildingsmart.fi/infra/inframodel/

Lee SH, Kim BG (2011) IFC extension for road structures and digital modeling. Proceedia Eng 14:1037–1042

Lee A, Marshall-Ponting A, Aouad G, Wu S, Koh I, Fu C, Cooper R, Betts M, Kagioglou M, Fischer M (2003) Developing a vision of nD-enabled construction. Construct IT Report

Li J, Hou L, Wang X, Wang J, Guo J, Zhang S, Jiao Y (2014) A project-based quantification of BIM benefits. Int J Adv Robotic Syst 11(8):123

Liebich T, Adachi Y, Forester J, Hyvarinen J, Richter S, Chipman T, Weise M, Wix J (2010) Industry Foundation Classes IFC2xEdition4 Release Candidate 1. BuildingSMART

Moreira AV, Fwa TF, Oliveira JRM, Costa L (2017) Coordination of user and agency costs using two-level approach for pavement management optimization. Transp Res Rec J Transp Res Board 2639:110–118

Sollazzo G, Wang KCP, Bosurgi G, Li JQ (2016) Hybrid procedure for automated detection of cracking with 3D pavement data. J Comput Civil Eng 30(6):04016032. https://doi.org/10.1061/(ASCE)CP.1943-5487.0000597

Tibaut A, Pecnik S, Rozenicnik Korosec M, Mihalic K, Zabreznik I (2015) BIM-based parametric modeling of roads and infrastructure. In: Proceedings of the 32nd CIB W78 conference 2015, Eindhoven, The Netherlands

Zhang A, Wang KCP, Li B, Yang E, Dai X, Peng Y, Fei Y, Liu Y, Li JQ, Chen C (2017) Automated pixel-level pavement crack detection on 3D asphalt surfaces using a deep-learning network. Comput-Aided Civil Infrastruct Eng 32(10):805–819

A New Design Methodology for Improving Porous Concrete Properties to Achieve Multifunctional and Sustainable Pavements

Eduardo Javier Elizondo-Martinez[1(✉)], Valerio Carlos Andrés-Valeri[2], Jorge Rodriguez-Hernandez[1], and Daniel Castro-Fresno[1]

[1] GITECO Research Group, Universidad de Cantabria, 39005 Santander, Spain
eduardo-javier.elizondo@alumnos.unican.es
[2] Institute of Civil Works, Faculty of Engineering Sciences, Universidad Austral de Chile (UACh), Campus Miraflores, 2060 Valdivia, Chile

Abstract. Porous concrete (PC) pavements have gained a lot of attention in recent years because of the multiple advantages they offer to mitigate the climate change effect while giving comfort and safety to citizens. However, they are still not very often used because of the low strength they obtain due to the high air void amount they have for the infiltration of rainwater. The following research introduces a modified methodology of design, named PCD (Porous Concrete Design), based on ACI 522R-10 and ACI 211.3R-02 standards, with the purpose of increasing strength of PC surface layers. The study demonstrated an increase of about 30% in strength, with a decrease in infiltration capacity of about 40%. Nevertheless, permeability results are good enough to handle stormwater events.

Keywords: Porous concrete · Permeable pavements · Green infrastructure · SUDS · Sustainability

1 Introduction

Construction has been one of the main issues of the environmental problems the world population is facing nowadays. Since construction is very important, useful, necessary, and is always in constant execution for the development of countries, it is very difficult to replace the conventional methods for new ones that are more environmentally friendly (Rodríguez-Hernández et al. 2013). Among the different constructions in cities, pavements are one of the main works that are a big environmental issue: decrease of the groundwater levels due to a blockage of the natural water cyclic course, rise of cities temperatures, water and air pollution, among other problems. As well as causing some impact in citizens safety: water ponding in streets, potholes, noise pollution, among other problems (American Concrete Institute ACI Committee 522 2010; Eriskin et al. 2017). Therefore, Porous Pavements have become a powerful solution to mitigate some of these problems, allowing to infiltrate rainwater into the ground, refilling groundwater deposits and decreasing its pollution.

Porous Concrete (PC) Pavements are considered one of the best solutions to mitigate environmental problems overall (Lian and Zhuge 2010). They are a special type

M. Pasetto et al. (Eds.): ISAP APE 2019, LNCE 48, pp. 491–499, 2020.
https://doi.org/10.1007/978-3-030-29779-4_48

of pavement design due to the air voids in their structure, in a range of 15–30%, to allow rainwater infiltration (Khankhaje et al. 2017; Giustozzi 2016). This can be achieved mainly by avoiding the use of sand in the mixture design. Besides, PC pavements can become multifunctional with the use of certain additives. As they present a light colour, PC pavements can reflect, rather than absorb, the sun radiation, decreasing cities temperatures, preventing the Heat Island Effect (Li et al. 2014). Also, other additives can help to mitigate the air pollution caused by the smog of cars through a process of photocatalysis (similar to photosynthesis in plants) (Hasan et al. 2017). However, PC pavements, as well as Porous Pavements in general, are still not widely used because they present limited mechanical resistance to traffic loads, restricting their use to sidewalks, parking lots and minor roads.

Many investigations have pointed out at the lack of sand as the main reason for their low mechanical capacity (Agar-Ozbek et al. 2013). Therefore, different kinds of additives are used in the mixture to improve strength (Giustozzi 2016; Bonicelli et al. 2016; Li et al. 2017). Other researchers have tried to strengthen up the coarse aggregate with stronger polymers (Chen et al. 2013) such as cellulose fibers (Shen et al. 2013; Lee et al. 2011; Rehder et al. 2014), even replacing cement with other additions such as silica fume or fly ash (Chen et al. 2013; Zhong and Wille 2016). Very good and interesting results have been obtained, but they are considered very difficult to replicate in other regions of the world because of the prevailing materials in every country. With this, a modified methodology, derived from ACI 522R-10 (American Concrete Institute ACI Committee 522 2010) and ACI 211.3R-02 (ACI 211 Committee 2002) norms, named PCD (Porous Concrete Design) is proposed. The aim of this new methodology is to produce a more workable mixture with a better adhesion between particles, increasing the mechanical capacity while maintaining a good enough permeability. The methodology developed is based on fixing the target void ratio in compacted porous concrete mixtures, and dosing the raw materials based on the cement mortar strength and the voids in compacted mineral aggregates.

2 Materials and Methods

2.1 Materials

For the preparation of the PC samples, Portland Cement CEM 1 52.5R UltraVal was used as a cementitious material. This is considered a high resistance cement in a short period of time, and according to EN 1907-6, it has a specific weight of 3.144 gr/cm^3. A water-cement ratio (w/c) of 0.40 was employed for the comparison between the PCD and the ACI methodologies.

Standard EN 1097-3:1998 was used to determine the voids in mineral aggregate (VMA). The test was made twice (compacted and uncompacted) in order to establish a parameter of the air void content. The parameters ranged from 34.64% to 48.16%. For the design a VMA of 44.3% was chosen.

Porphyric material was employed as sand in a sand-cement ratio (s/c) of 0.00 (mixtures with “0” in Table 1), 0.50 (“5”), and 1.00 (“1”) to compare the influence of sand in the mixture properties. This material was also used as coarse aggregate, in a size of 8-12 mm for the methodologies comparison. The U.S. FHWA 0.45 power chart gradation curve theory was employed.

Table 1. Mixture dosages for comparing ACI 522R-10 and PCD methodologies

Mixture	w/c	s/c	Cement (kg/m^3)	Water (kg/m^3)	Sand (kg/m^3)
PCD-0	0.40	0.00	341.47	76.39	0.00
PCD-0.5	0.40	0.50	275.53	63.68	28.61
PCD-1	0.40	1.00	230.93	55.09	47.95
ACI-0	0.40	0.00	209.97	83.99	0.00
ACI-0.5	0.40	0.50	195.97	78.39	95.14
ACI-1	0.40	1.00	181.97	72.79	190.28

2.2 PCD Methodology

A new methodology, named PCD (Porous Concrete Design), based on the normative ACI 522R-10 and ACI 211.3R-02, was employed. Consisting in breaking the relation between coarse aggregate and sand design, introducing sand in the cement paste design, making a mortar. Therefore, the amount of sand modifies cement and water quantities. Coarse aggregate depends on the particle density and its porosity. This methodology enables to keep the s/c and w/c ratios defined.

Both ACI 211.3R-02 and ACI 522R-10 norms showed almost the same dosage designs, with very small difference (negligible), despite designing with different order in each methodology. With this, it was decided to elaborate mixtures with the ACI 522R-10 normative mainly because it is a more recent normative (ACI 211.3R-02 was released in 2002, and ACI 522R-10 in 2010). Figure 1 demonstrates the difference in the design steps for both ACI 522R-10 and PCD normative.

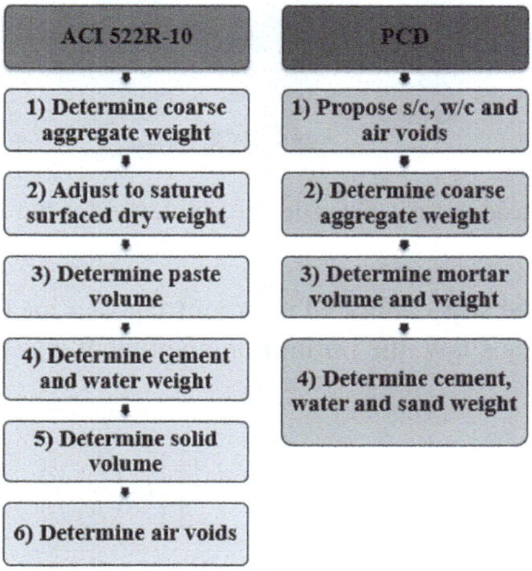

Fig. 1. Comparative steps between the ACI and PCD methodologies

2.3 Production

For the comparison of methodologies phase, three different mixtures were elaborated for each methodology, varying the s/c ratio (0, 0.5, and 1). An air void content of 20% was considered for all the mixtures in both methodologies. It is important to clarify that the ACI 522R-10 norm states that in porous mixtures with sand, cement paste amount must be reduced by 2% for every 10% of sand of the total aggregate amount for well-compacted concrete; and by 1% for every 10% of sand for slightly-compacted concrete. Since a mechanical press with a controlled load of 200 N/mm was employed for compaction, it was considered as a well-compacted concrete. In Table 1, the mixture dosages employed are shown. Specimens were cylindrical in shape, with a diameter of 10.20 cm and a height of 6.5 cm compacted in Marshall Molds with a mechanical press and introduced in curing for 28 days.

2.4 Tests

2.4.1 Permeability Test

Permeability capacity of the ACI 522R-10 and PCD samples was evaluated with a falling head permeameter that was adapted to cylindrical specimens, as seen in Fig. 2. A PVC tube of 4" diameter with rubber inside was used as a mold. The sample is placed inside the mold and is adjusted with metal clamps. To measure the permeability capacity, a calibrated methacrylate tube is placed on top of the mold where water is introduced and permeability is measured with a chronometer.

Fig. 2. Falling head permeameter device employed for the permeability test

The methacrylate tube was calibrated in order to have a measurement fall of 20 cm. Finally, employing Darcy's law, the infiltration coefficient was calculated, according to Eq. (1):

$$K = \left[\frac{(A_{sample})(h_{sample})}{(A_{tube})(t)}\right]\left[ln\left(\frac{h_1}{h_2}\right)\right] \tag{1}$$

Where K is the infiltration coefficient (cm/s) A_{sample} is the area of contact of the sample, in cm^2, h_{sample} is the height of the sample, in cm, A_{tube} is the area of the tube's gap, in cm^2, t is the time it takes the water to go from the higher point, h_1 to h_2, in seconds. For the calculation of the real porosity (P), once the mixture is elaborated, Eq. (2) is employed:

$$P = \frac{V_{Tot} - \left[W_{Dry*}\left(\frac{\%_{CA}}{\rho_{CA}} + \frac{\%_S}{\rho_S} + \frac{\%_C}{\rho_C}\right)\right] - W_{Dry} * \frac{\%_W}{\rho_W}}{V_{Tot}} * 100 \qquad (2)$$

Where W_{DRY} corresponds to the mixtures' weights under dry conditions, in grams. $\%_{CA}$, $\%_S$, $\%_C$ and $\%_W$ represent the percentage of the total mixture of the coarse aggregate, sand, cement and water respectively. P_{CA}, P_S, P_C, and P_W, represent the density of the mixture's components mentioned above, in gr/cm^3.

2.4.2 Indirect Tensile Test

The Indirect Tensile (IT) test was employed in order to represent the behavior of the pavement when vehicles transit on it. Norms EN 13286-42 (AENOR 2003), EN 12390-6 (AENOR 2010) and EN 12390-1 (AENOR 2014) explain the IT test and show the equipment and equations for the test. This test consists of applying load along the cross section of the sample, causing a perpendicular deformation in it, leading to failure.

2.4.3 Compression Strength Test

Some authors have demonstrated that IT strength represents between 12-15% of compressive strength (CS) (Bonicelli et al. 2015). Employing some equations correlating IT strength and CS, theoretical CS can be calculated (Behnood et al. 2015). However, specimens were later cut into cubical shapes of 1:1 proportions (6.50 × 6.50 × 6.50 cm) with a mechanical saw to fulfill the EN 12390-3 (AENOR 2009) norm's requirements and evaluate the compressive strength values, despite not fulfilling the minimum dimensions of norm. Nevertheless, this enabled a comparison to be made between the mixtures and methodologies according to the purposes of this investigation. The same device used for the IT test was employed for this test.

3 Results and Discussion

3.1 Permeability Results

As seen in Fig. 3, when comparing ACI 522R-10 and PCD methodologies, two main characteristics can be noticed. First, ACI 522R-10 mixtures obtained a higher permeability than PCD mixtures, where results where around 40% higher. Second, a contrary behavior between the two methodologies is noticed, as in ACI 522R-10 samples, the mixture without sand obtained the higher permeability, while in PCD samples, the mixture with an s/c of 1 obtained the best permeability. This can be explained because of the differences in the design methodology, where the mortars design in PCD mixtures does not affect the coarse aggregate amount and the air voids established. On the contrary, in ACI 522R-10, the use of sand affects the amount of coarse aggregate, decreasing the air voids. In addition, cement paste and sand are not related in the design. Although permeability rates in PCD mixtures were lower, with a minimum permeability of 0.30 cm/s, they are considered good enough to handle stormwater events according to porous pavement materials standards, where a minimum permeability rate of 0.12 cm/s (100 m/day) is suggested by some authors (Alvarez et al. 2011).

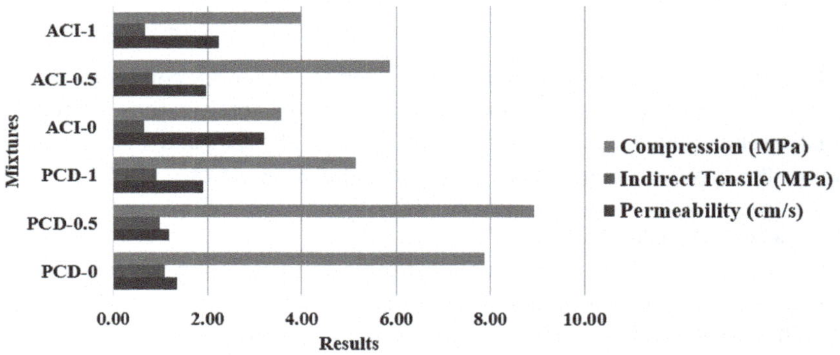

Fig. 3. ACI 522R-10 vs PCD Methodology results

3.2 Indirect Tensile Test

Unlike infiltration, mechanical capacity was higher in PCD mixtures, with around 30% higher indirect tensile resistance than ACI 522R-10 mixtures, as seen in Fig. 3. It is noticed that even the strongest sample of the ACI 522R-10 norm was weaker than the worst sample of the PCD normative, with around 9% of difference. In addition, the mixture without sand in the PCD normative obtained the best result, with 1.14 MPa of strength. This means that when removing sand from the dosage, with this methodology of design, the cement paste becomes more adhesive and there is a greater adhesion between the coarse aggregate particles.

3.3 Compression Strength Test

In terms of compression strength, PCD mixtures were around 28% stronger than ACI 522R-10 samples, where a s/c ratio of 0.50 turned out to be the one with the highest value in PCD, while in ACI 522R-10, a s/c ratio of 1. This means that for both methodologies, the addition of sand helps to increase the samples strength. Besides, in terms of PCD methodology, with more sand, less cement is employed, so the mixture tends to be less adhesive; therefore, the aggregate particles are not sufficiently well united. In ACI 522R-10 mixtures, strength decreases gradually as sand amount is decreased, because the same amount of cement is used despite the sand ratio employed.

3.4 Multicriteria Decision Making Analysis

In order to find out what methodology works better, the analytical hierarchy process (AHP) multicriteria decision-making method was employed, in order to merge the properties analyzed previously separated. Table 2 demonstrates the values of importance given to the variables studied (permeability, indirect tensile, and compression). Indirect tensile and compression strength were considered equal among each other, with a value of importance of 50-50, but considered a little higher than permeability, with a value of 0.40–0.60. This happened because permeability results were considered good enough to handle storm water events in both methodologies.

Table 2. Variables values of importance

Mixture	Indirect tensile	Permeability	Compression
Indirect tensile		0.60-0.40	0.50-0.50
Permeability	0.40-0.60		0.40-0.60
Compression	0.50-0.50	0.60-0.40	

Table 3 shows the weights obtained from the analysis, where the last column is the average weight. It can be seen in the Table 3 that the PCD methodology obtained higher values and can be more helpful when trying to fulfill all the properties. Among the samples, mixture PCD-0.5 obtained the best result, as its compression strength resistance was quite higher than the rest of the mixtures. This means that a s/c ratio of 0.5 gives the best results, where a lack and an excess of sand can weaken the mixture. Mixture ACI-0 obtained a good result in the analysis, but this was thanks to the high permeability capacity it obtained, as its mechanical results where the lowest ones. PCD-1 was not that bad in general, but its low result in the analysis is caused because of the excess of sand in the mixture, where the mortar became less adhesive as the cement amount was lower.

Table 3. ACI 522R-10 and PCD mixtures analysis results

Mixture	Permeability	Indirect tensile	Compression	Average weight
PCD-0	0.0376	0.28114	0.23513	0.18462
PCD-0.5	0.02503	0.1687	0.40287	0.19887
PCD-1	0.10131	0.125	0.06195	0.09609
ACI-0	0.47463	0.04988	0.0271	0.18387
ACI-0.5	0.11923	0.09351	0.09508	0.10261
ACI-1	0.19972	0.05826	0.03389	0.09729

4 Conclusions

This paper introduces a modified methodology of design for PC mixtures, named PCD, used as pavement surface layer with the main objective of improving its mechanical characteristics maintaining a good hydraulic capacity. This is fundamental for the future implementation of additives that can help increase safety and other environmental aspects of the mixture. The following conclusions can be drawn:

The PCD methodology tends to increase the mechanical capacity of the mixtures, around 30% more than ACI 522R-10 methodology, maintaining good permeability values.

The PCD methodology helps the mixture to obtain a good permeability capacity when sand is added, improving the mechanical capacity. However, an excess in the sand amount can lead to a decrease in the mixtures strength.

In terms of Indirect Tensile resistance, the PCD mixture without sand obtained the highest value, decreasing its capacity in around 13% when sand was added in an s/c ratio of 0.50. Removing sand from the design, increases the cement amount, the mortar becomes more adhesive, and the mixtures strength increases, resisting heavier loads.

Future research is being made in this line of investigation in order to continue with the improvement of the mechanical properties of mixtures with the PCD methodology, as well as introducing safety properties and new positive environmental characteristics, by the addition of different admixtures in different dosages.

Acknowledgements. This study was funded by the Spanish Ministry of Economy and Competitiveness and the European Union (ERDF) through the project SUPRIS-SUReS (Ref. BIA2015-65240-C2-1-R). The authors would like to thank Grupo Cementos Portland Valderrivas for providing the cement material used in the investigation.

References

ACI 211 Committee (2002) Guide for Selecting Proportions for No-Slump Concrete Reported by ACI Committee 211. American Concrete Institute 02:1–26

AENOR (2003) UNE-EN 13286-42: Mixtures with Aggregate and Hydraulic Binder. Part 42: Test Method for Determining the Indirect Tensile Strength of Mixtures with Aggregate and Hydraulic Binder

AENOR (2009) UNE-EN 12390-3: Hardened Concrete Testing. Part 3: Specimens Compression Strength

AENOR (2010) UNE-EN 12390-6: Hardened Concrete Testing. Part 6: Indirect Tensile Strength of Specimens

AENOR (2014) UNE-EN 12390-1. Hardened Concrete Testing. Part 1: Shape, Dimensions and Other Characteristics of the Specimens and Molds

Agar-Ozbek AS, Weerheijm J, Schlangen E, Van Breugel K (2013) Investigating porous concrete with improved strength: testing at different scales. Constr Build Mater 41:480–490. https://doi.org/10.1016/j.conbuildmat.2012.12.040

Alvarez AE, Martin AE, Estakhri C (2011) A review of mix design and evaluation research for permeable friction course mixtures. Constr Build Mater 25(3):1159–1166. https://doi.org/10.1016/j.conbuildmat.2010.09.038

American Concrete Institute ACI Committee 522 (2010) Report on Pervious Concrete

Behnood A, Verian KP, Gharehveran MM (2015) Evaluation of the splitting tensile strength in plain and steel fiber-reinforced concrete based on the compressive strength. Constr Build Mater 98:519–529. https://doi.org/10.1016/j.conbuildmat.2015.08.124

Bonicelli A, Arguelles GM, Pumarejo LGF (2016) Improving pervious concrete pavements for achieving more sustainable urban roads. Procedia Eng 161:1568–1573. https://doi.org/10.1016/j.proeng.2016.08.628

Bonicelli A, Giustozzi F, Crispino M (2015) Experimental study on the effects of fine sand addition on differentially compacted pervious concrete. Constr Build Mater 91:102–110. https://doi.org/10.1016/j.conbuildmat.2015.05.012

Chen Y, Wang K, Wang X, Zhou W (2013) Strength, fracture and fatigue of pervious concrete. Constr Build Mater 42:97–104. https://doi.org/10.1016/j.conbuildmat.2013.01.006

Eriskin E, Karahancer S, Terzi S, Saltan M (2017) Examination of the effect of superhydrophobic coated pavement under wet conditions. Procedia Eng 187:532–537. https://doi.org/10.1016/j.proeng.2017.04.411

Giustozzi F (2016) Polymer-modified pervious concrete for durable and sustainable transportation infrastructures. Constr Build Mater 111:502–512. https://doi.org/10.1016/j.conbuildmat.2016.02.136

Hasan MR, Zain MFM, Hamid R, Kaish ABMA, Nahar S (2017) A comprehensive study on sustainable photocatalytic pervious concrete for storm water pollution mitigation: a review. Mater Today Proc 4(9):9773–9776. https://doi.org/10.1016/j.matpr.2017.06.265

Khankhaje E, Salim MR, Mirza J, Salmiati MW, Hussin RK, Rafieizonooz M (2017) Properties of quiet pervious concrete containing oil palm kernel shell and cockleshell. Appl Acoust 122:113–120. https://doi.org/10.1016/j.apacoust.2017.02.014

Lee M, Huang Y, Chang T, Pao C (2011) Experimental study of pervious concrete pavement, no 218, pp 93–99. https://doi.org/10.1061/47629(408)12

Li H, Harvey J, Ge Z (2014) Experimental investigation on evaporation rate for enhancing evaporative cooling effect of permeable pavement materials. Constr Build Mater 65:367–375. https://doi.org/10.1016/j.conbuildmat.2014.05.004

Li J, Zhang Y, Liu G, Peng X (2017) Preparation and performance evaluation of an innovative pervious concrete pavement. Constr Build Mater 138:479–485. https://doi.org/10.1016/j.conbuildmat.2017.01.137

Lian C, Zhuge Y (2010) Optimum mix design of enhanced permeable concrete - an experimental investigation. Constr Build Mater 24(12):2664–2671. https://doi.org/10.1016/j.conbuildmat.2010.04.057

Rehder B, Banh K, Neithalath N (2014) Fracture behavior of pervious concretes: the effects of pore structure and fibers. Eng Fract Mech 118:1–16. https://doi.org/10.1016/j.engfracmech.2014.01.015

Rodríguez-Hernández J, Fernández-Barrera A, Andrés-Valeri V, Vega-Zamanillo A, Castro-Fresno D (2013) Relationship between urban runoff pollutant and catchment characteristics. J Irrig Drainage Eng 139:833–840

Shen W, Shan L, Zhang T, Ma H, Cai Z, Shi H (2013) Investigation on polymer-rubber aggregate modified porous concrete. Constr Build Mater 38:667–674. https://doi.org/10.1016/j.conbuildmat.2012.09.006

Zhong R, Wille K (2016) Compression response of normal and high strength pervious concrete. Constr Build Mater 109:177–187. https://doi.org/10.1016/j.conbuildmat.2016.01.051

Correction to: Properties of Asphalt Binders with Increasing SBS Polymer Modification

Mike Aurilio, Peter Mikhailenko, Hassan Baaj,
and Lily D. Poulikakos

Correction to:
Chapter "Properties of Asphalt Binders with Increasing SBS Polymer Modification" in: M. Pasetto et al. (Eds.):
***Proceedings of the 5th International Symposium on Asphalt Pavements & Environment (APE)*, LNCE 48,**
https://doi.org/10.1007/978-3-030-29779-4_6

In the original version of the book, the author name "Mike Aurillio" has been changed to "Mike Aurilio" in the Frontmatter, Backmatter and Chapter 6. The correction chapter and book have been updated with the changes.

The updated version of this chapter can be found at
https://doi.org/10.1007/978-3-030-29779-4_6

M. Pasetto et al. (Eds.): ISAP APE 2019, LNCE 48, p. C1, 2020.
https://doi.org/10.1007/978-3-030-29779-4_49

Author Index

M. Pasetto et al. (Eds.): ISAP APE 2019, LNCE 48, pp. 501–503, 2020.
https://doi.org/10.1007/978-3-030-29779-4

The manufacturer's authorised representative in the EU is Springer
Nature Customer Service Centre GmbH, Europaplatz 3, 69115 Heidelberg,
Germany. If you have any concerns regarding our products, please
contact ProductSafety@springernature.com

Printed and bound by CPI Group (UK) Ltd, Croydon, CR0 4YY
05/10/2026
02228131-0011